Treatise on Energy Quantum Theory (EQT), Volume 8

Constrained Speculation

A Natural Philosophy of Energy Quantum Theory

Kaisheng Li and Longji Li

2026

This work is a **pre-publication academic manuscript**. The authors reserve the right to revise, expand, or withdraw any part of the text in future editions.

The views expressed in this book are solely those of the authors and do not necessarily reflect the views of any institution or organization.

June 13, 2026

Acknowledgements

This book represents the culmination of a long intellectual pursuit and is the collective outcome of our many years of deep collaboration.

It is dedicated to our families.

Having been granted the privilege of life in this world, "life" itself gradually became our most persistent enigma in the years that followed. In seeking to understand why life arises and why it is finite, this quest led us—often unexpectedly—from the mysteries of biological existence into the foundations of physics, and eventually toward a critical examination of the deepest structural assumptions underlying contemporary physical theory. Through our joint efforts, this long and unplanned journey has finally converged into the theoretical framework presented in this book. This exploration has occupied much of our adult lives; that we could follow this path to its conclusion is owed not to professional obligation, but to a sustained and uncompromising fascination with physics itself.

We wish to offer our most special thanks to our respective wives.

Finally, this book is offered to those readers who retain a serious and undiminished curiosity about the nature of life and the structure of the universe. Whether or not definitive answers can ever be reached, the act of sustained questioning remains, in itself, a meaningful response to the limits of a finite existence.

Kaisheng Li & Longji Li
May 2026

Contents

Introduction: From Five Volumes to a Philosophy

0.1. Motivation for Writing

This book is a philosophical monograph. It is built upon the five volumes previously published by the authors, but it is not a review, summary, or commentary on those five volumes. It is an independent statement of the philosophical position that is jointly implicit in those volumes but has not yet been made explicit.

This kind of writing, which "extracts an implicit philosophy from prior work," has a certain specificity and therefore needs to be clarified in the Introduction; otherwise, readers may misread this book as some kind of appendix to the five volumes. The central claim of this book is that the authors' previous five volumes—the "EQT Foundational Volume," "Energy, Process, and the Fate of the Cosmos," "Freezing Cosmology," "Mind–Matter Unity," and "EQT Theory of Detection"—share a method and an ontological language, and that this method and language themselves constitute an independent philosophical position worthy of separate statement and defense.

The task of the Introduction is to state this claim clearly, so that readers can judge whether this book deserves to be read as an independent philosophical monograph. If the Introduction fails, even a rigorous main text cannot be properly received.

0.2. The Relative Position of the Five Volumes

To understand the position of this book, one must first understand the relative relation among the five volumes. These five volumes are not a loose collection; they form a finely stratified research program.

The Functional Positioning of the Five Volumes

The "EQT Foundational Volume" undertakes the foundational work in physics. Starting from the Planck node of 1900, $E = h\nu$, it elevates the frequency–energy relation into a foundational postulate of physics and thereby establishes the postulate system of EQT (Energy Quantum Theory), namely $A1$–$A4 + M$.

"Energy, Process, and the Fate of the Cosmos" undertakes the work of synthetic natural philosophy. Taking the "ocean of gradients" as a unifying image, it describes the universe as a gradient process of the energy quantum density field.

"Freezing Cosmology" undertakes the natural-philosophical work of cosmic evolution. It establishes the "freezing spectrum" as the core mechanism for organizing cosmic evolution: high frequencies freeze early, low frequencies freeze late, and cosmic history itself is the freezing spectrum. The methodology of this volume is the most mature, and "constrained speculation" is explicitly introduced there as a term.

"Mind–Matter Unity" undertakes the study of consciousness. It takes the identity-level isomorphism between experience and gradient configurations of the frequency-resolved energy quantum density field ρ_f as its first-principle thesis, and fulfills structural sufficiency through fivefold gradient richness: single-frequency gradients, the frequency dimension, evolution under the master equation, resonance modes, and the timescale spectrum of transients.

"EQT Theory of Detection" undertakes the work of detection theory. It transforms "observation as participation" from a broad thesis into a mechanized structure: the detector and the probed system both belong to the frequency-resolved energy quantum density field, and the detection result is a co-evolutionary outcome jointly constrained by frequency matching, controlled perturbation, nonlinear feedback, dissipation, and tri-state response.

The Hierarchical Relation among the Five Volumes

These five volumes form a stratified structure: the "EQT Foundational Volume" operates at the level of physics, where it establishes the postulates and basic equations; "Energy, Process, and the Fate of the Cosmos" operates at the level of synthetic natural philosophy, where it extends physics into a cosmological worldview; "Freezing Cosmology" operates at the level of natural philosophy of cosmic evolution, where it treats the evolution of the universe as a whole; "Mind–Matter Unity" operates at the level of natural philosophy of consciousness, where it treats the mind–matter relation; and "EQT Theory of Detection" operates at the level of natural philosophy of detection, where it treats the relation of observation.

Each volume begins from the basic postulates of EQT, but each treats a different domain of objects. They are not independent of one another: they share postulates, terminology, and method. But they are also not repetitions of one another: each volume has its own object of inquiry.

An Unfilled Position

When the five volumes are placed together, an unfilled position becomes visible: the common meta-level of the five volumes. The five volumes share a method, namely constrained speculation, but each volume applies that method to its own object, and no volume specifically treats the method itself as a philosophical position. The five volumes share an ontological language, namely frequency–density–gradient–freezing–co-evolution, but each volume uses this language as a tool, and no volume specifically treats this language as a systematic ontology. The five volumes jointly encounter certain fundamental problems, such as the is–ought gap, the mind–matter relation, the final state of the universe, and the finitude of meaning, but each volume gives only a local response to these problems within its own scope, and no volume specifically integrates those responses.

This unfilled position is where the present book works. It is not a footnote to the five volumes; it is the monograph called for by the five volumes but not yet written.

0.3. The Shared Implicit Method: From Five Names to One Position

This section unfolds the implicit method shared by the five volumes. This method appears in the five volumes under five different names. This book identifies them as five expressions of the same position and gives them the unified name "constrained speculation."

Five Names

"Postulational writing" in the "EQT Foundational Volume": the commitments of foundational physics are compressed into the fewest possible postulates, $A1$–$A4 + M$, and a clear distinction is drawn among the three categories of terms: postulates, mechanism operators, and verification tools. The core of this method is the minimization and clarification of ontological commitments.

Synthetic natural philosophy in "Energy, Process, and the Fate of the Cosmos": multiple domains of natural philosophy are organized through gradient as a unifying image. This method elevates physical structure into an ontological image.

"Constrained speculation" in "Freezing Cosmology": an intermediate level is established between physical theory and metaphysics; it is neither limited to the technical derivations of existing physics nor permitted to make free metaphysical expansion detached from formal expression, observational constraints, and failure conditions.

"Isomorphism + co-evolution + compression + boundary" in "Mind–Matter Unity": the ontological meaning of mind–matter unity is fulfilled through identity-level isomorphism; scalar identity theory is rejected, and a five-dimensionally rich gradient-structure space is used to carry the structural sufficiency of experience. Givenness is characterized through internal–external co-evolution, and the hard problem is precisely located as a single boundary through two compressions.

"Constrained co-evolution" in "EQT Theory of Detection": the detector and the probed system belong to the same structural domain of $\rho_f(x,t)$; the detection result is a co-evolutionary outcome jointly constrained by frequency matching, perturbation threshold, nonlinear feedback, dissipation conditions, and tri-state response.

The Five Are One Position

These five names appear different, but structurally they are different expressions of the same position. Their shared features are as follows.

- Feature One: Ontological commitments must be formalized. Whether it is the postulational $A1$–$A4$, the formal mechanism of the freezing spectrum, the Ψ state-variable group, or the detection response function $g(f, f_0)$, all five volumes require ontological commitments to be stated in definable symbolic relations.

- Feature Two: Form must have a dynamical mechanism. The $A3$ master equation, freezing-time scaling, Ψ state evolution, and the mechanism of detection co-evolution all require form to carry process rather than remain a purely abstract relation.

- Feature Three: Mechanisms must have observational proxies. CMB and gravitational waves in the Foundational Volume and the freezing-spectrum volume, EEG/fMRI/TMS in the mind–matter volume, and LIGO, JWST, and XFEL in the detection volume all show that the five volumes require mechanisms to be connected to observable proxies.

- Feature Four: Observation must allow failure. Each volume contains a section on "failure conditions," explicitly stating what observations would make its commitments fail. This is not a writing ritual, but a core requirement for the legitimacy of commitment.

Together, these four features constitute a methodological position. This book names it constrained speculation: a method already used in the five volumes but not yet formally presented at the meta-level.

The Function of Naming

Why is naming necessary? Without naming, this method remains tacit in the five volumes: it affects the writing of every volume but is not discussed as an object. A tacit method cannot be rigorously defended or rigorously criticized. It can only be practiced, not reflected upon.

Through naming, the method becomes an object of discussion. We can ask: What is the relation between constrained speculation and Popperian falsifiability? What is its relation to Carnapian logical construction? What is its relation to Cartwright's local realism? Such questions can be posed only after the method has been named and taken as an object. This is the task of Part I of this book, "Constrained Speculation: Method as Philosophy": to make constrained speculation explicit as an independent, defensible philosophical position capable of entering international philosophical discussion.

The Work After Naming

But naming alone is not enough. Naming is only the starting point of the work. After being named, constrained speculation must be rigorously defended. Its position in the philosophical lineage must be clarified, including its relation to Popper, Quine, Carnap, Cartwright, Ladyman, and others; its core thesis must be stated, namely that reality is a commitment that can fail; its internal structure must be specified, namely the four disciplines; and its possible directions of failure must be identified. The six chapters of Part I undertake this defense. Chapter 1 gives the core thesis of constrained speculation; Chapter 2 unfolds ontological restraint; Chapter 3 rigorously defends failure conditions as an ontological discipline; Chapter 4 treats the problem of physicalized semantics and proposes constitutive coextension; Chapter 5 positions the method within the spectrum of contemporary realism; and Chapter 6 synthesizes it into a complete methodological position.

0.4. The Shared Implicit Ontological Language

Constrained speculation is what the five volumes share at the methodological level. But the five volumes also share something at the ontological level: an ontological language. This language appears in the five volumes through five core terms: frequency, density, gradient, freezing, and co-evolution.

Five Core Terms

Frequency: In the five volumes, frequency is the basic spectral parameter. $E = h\nu$ is not a special relation but the foundational spectral relation of all energy structures. The frequency spectrum, including low-frequency, mid-frequency, high-frequency, and ultra-high-frequency regions, organizes the levels of particles, interactions, and cosmic structures.

Density: The energy quantum density field $\rho_f(x,t)$ is the unique fundamental ontological object of EQT. It is not a substance, but an energy distribution across space, time, and frequency. All "things" are local structures of ρ_f.

Gradient: $\nabla\rho_f$ is the basis of structure formation, energy flow, and interaction. A core thesis repeatedly appearing in the five volumes is that difference, or gradient, precedes identity, or substance. Gradient is not a property of things; gradient itself is the mode of existence of things.

Freezing: The energy structure of a given frequency band exits its main dynamical evolution under a given background and remains as stable structure, historical residue, background condition, or indirect effect. The freezing spectrum, namely the arrangement of freezing times across different frequency bands, is cosmic history itself.

Co-evolution: The observer and the observed both belong to the ρ_f field. Their relation is not an external interaction between heterogeneous entities, but a constrained co-evolution within the same field. This concept is fully developed in "EQT Theory of Detection" and also appears in "Mind–Matter Unity" as the characterization of given-ness.

The Five Terms Constitute a Systematic Ontology

These five terms are not a loose set of concepts. They form a systematic ontology: frequency provides the basic spectral space of ontology; the density field provides the distributed object of ontology; gradient provides the structure of difference in ontology; freezing provides the historical dimension of ontology; and co-evolution provides the relational structure of ontology.

These five terms cover the five core dimensions of systematic ontology: basic object, mode of distribution, internal difference, temporal structure, and relational structure. Together, they constitute an independent ontological language, standing alongside such historical ontological languages as Aristotle's substance–attribute framework, Descartes's substance–mode framework, Leibniz's monads, and Whitehead's actual occasions–eternal objects.

This Ontology Has Not Yet Been Made Explicit

In the five volumes, however, this ontology is used as a tool rather than presented as a systematic object. Each volume uses the terms frequency, density, gradient, freezing, and co-evolution, but no volume specifically treats them as the overall structure of a systematic ontology: how they are connected to one another, how they as a whole compare with other historical ontologies, and where they possess genuine originality.

Part II of this book, "The Philosophical Defense of Gradient Ontology," and Part III, "Co-Evolution, Mind–Matter, and Observation: The Unification of Three Relations," undertake this work. Part II unfolds the philosophical grounds of gradient ontology, including arguments concerning the qi tradition, frequency as a basic spectral quantity, tri-state structure as an ontological category, and freezing as a basic ontological process. Part III unfolds the meaning of co-evolution as an ontological relation, including the unified treatment of the mind–matter relation, the observational relation, and the temporal relation.

An Example: Tri-State Structure as an Ontological Category

To prevent "systematic ontology" from sounding abstract, this section gives one example: the ontological elevation of the tri-state category.

"EQT Theory of Detection" distinguishes free states, bound states, and condensed states. In that volume, these three states are classifications within detection theory: detectors respond differently to different state types. The free state corresponds to propagating signals, such as photons and free particles; the bound state corresponds to locally confined structures, such as atomic energy levels and quark bound states; and the condensed state corresponds to stable structures, such as ground states, lattices, and macroscopic objects.

On closer examination, however, these three states can be elevated into ontological categories. They correspond respectively to three basic dimensions of beings: free state = accessibility, fluidity, and propagability; bound state = relationality, oscillation, and local structure; condensed state = stability, rigidity, and memory. Every being simultaneously possesses tri-state components: photons have a propagating aspect, dominated by the free state; electrons have a binding aspect, dominated by the bound state; galaxies have a structural aspect, dominated by the condensed state; life has a complex combination in which all three states are strong; and consciousness has a special structure dominated by the bound state.

This doctrine of tri-state categories is extremely rare in the history of philosophy. Since Aristotle's ten categories and Kant's twelve categories, new doctrines of categories have rarely appeared. Yet this book will argue in Chapter 10 that the tri-state category possesses completeness, since every being can be decomposed into the three states, and irreducibility, since the three states cannot be reduced to one another. It therefore constitutes a new doctrine of categories since Kant. This example shows that when the instrumental use found in the five volumes is elevated into systematic ontology, a genuinely original philosophical contribution can be found.

0.5. The Task and Positioning of This Book

This section synthesizes the work of the previous three sections into the explicit task of this book.

The Task of This Book

The task of this book can be stated as follows: to independently state and rigorously defend the metaphilosophy shared by the five volumes but not yet made explicit, namely the methodological position of constrained speculation together with the systematic language of gradient ontology, and to place it within international philosophical discussion.

This task includes four levels. Level One: Explication. The tacit method and ontological language are extracted from the five volumes and discussed as independent objects. This includes Part I, the explication of method, and Parts II and III, the explication of ontological language. Level Two: Defense. The explicated content is rigorously defended by responding to possible objections, locating it within philosophical lineages, and identifying original contributions. Every part and every chapter contains such defensive work. Level Three: Extension. Constrained speculation and gradient ontology are applied to domains insufficiently treated in the five volumes. This includes Part IV, the metaphysical boundary of freezing, and Part V, the movement from ontology to normativity. These domains are either not treated in the five volumes, such as the problem of ontological singularity, or treated only within a local scope, such as the treatment of consciousness in "Mind–Matter Unity," which Chapter 13 of this book will extend into a more general philosophy of mind–matter. Level Four: Connection. This book is connected with existing philosophical traditions: analytic philosophy, including Popper, Quine, Carnap, Cartwright, Ladyman, and Floridi; the Continental tradition, including Heidegger, Brandom, and McDowell; and Chinese philosophy, including qi theory, Confucianism, and Buddhism. This connection is not decorative citation; it makes this book a philosophical monograph capable of direct dialogue with multiple traditions.

The Positioning of This Book

The position of this book can be described along several axes.

- Relation to the five physical volumes: This book stands in a parallel relation to them, not a subordinate one. It does not provide footnotes to the five volumes; it stands independently. Readers can read this book without reading the five volumes and should still be able to understand its core claims, although reading the five volumes will provide more examples.

- Relation to the analytic philosophical tradition: This book works within the framework of methodological rigor characteristic of analytic philosophy, including precise propositions, rigorous argument, and explicit objections and replies. Yet its ontological position is close to post-analytic philosophy, in that it rejects the absolute separation of reality and cognition, adopts a relational ontology, and gives special treatment to the relation between form and reality. This book consciously crosses the analytic–Continental divide.

- Relation to the Chinese philosophical tradition: This book treats qi theory, Confucianism, and Buddhism as important interlocutors, not because they are "Chinese," but because they make distinctive philosophical contributions, especially in process ontology, the philosophy of ought, and the philosophy of finitude. However, this book rejects the cheap rhetoric that "Eastern wisdom anticipated science." Tradition is a philosophical resource and must be used rigorously.

- Relation to the tradition of natural philosophy: This book continues the tradition of natural philosophy from Greece to the twentieth century, but with an update. It is no longer the conceptual foundation of science in the manner of nineteenth-century natural philosophy, nor is it the meta-level reflection on science characteristic of twentieth-century philosophy of science. It is a natural philosophy that begins from a specific physical theory, EQT: neither detached from physics nor limited to physics.

Relation to the Concept of a "Philosophical Monograph" This book is a philosophical monograph, but what that means needs clarification. It is not a monograph in the history of philosophy: this book does not discuss what philosophers have done, but proposes its own theses. It is not a monograph in the philosophy of science: this book does not discuss "what science is," but carries out metaphysics and ontology. It is not a monograph in the philosophy of physics: this book does not discuss interpretive problems in physics, but proposes an independent ontological position. It is a contemporary form of a monograph in natural philosophy: it begins from a specific physical theory, but its work concerns philosophical-level problems, namely ontology, metaphysics, normativity, and meaning.

There are precedents for this positioning in natural philosophy. Leibniz's "Monadology" is a metaphysics that begins from physics; Whitehead's "Process and Reality" is a metaphysics that begins from relativity and quantum mechanics; and Prigogine's work on nonequilibrium thermodynamics has natural-philosophical extensions. This book continues that tradition, but with an update: it is built upon EQT as a specific new theory.

0.6. The Argumentative Structure of This Book

This section provides readers with a map of the whole book: the functions of the five parts and their interconnections.

Part I: Constrained Speculation: Method as Philosophy (6 chapters). This is the methodological foundation of the whole book. It makes the method shared by the five volumes explicit as an independent philosophical position. Chapter 1 proposes the core thesis of constrained speculation: X is real $\iff X$ has commitments with definable failure conditions. Chapter 2 unfolds ontological restraint: why fewer commitments are more real. Chapter 3 rigorously defends failure conditions as an ontological discipline; this is one of the most important original chapters in the book. Chapter 4 treats the problem of physicalized semantics and proposes "constitutive coextension" as a new semantic relation. Chapter 5 positions the method within the spectrum of contemporary realism through dialogue with Cartwright, Ladyman, Floridi, and Rovelli. Chapter 6 synthesizes the discussion into a complete methodological position and clarifies its relation to the Popper–Lakatos tradition. The key contribution of Part I is to elevate falsifiability from an epistemological criterion, as in Popper, into an ontological discipline. This was an unfinished task of twentieth-century philosophy of science, and this book completes it.

Part II: The Philosophical Defense of Gradient Ontology (5 chapters). This is the ontological foundation of the whole book. It elevates the ontological language shared by the five volumes into a systematic ontology. Chapter 7 argues that difference precedes identity: gradient as an ontological starting point is more basic than substance as an ontological starting point. Chapter 8 is a dialogue with the Chinese qi tradition, treating Zhang Zai, Wang Fuzhi, and the "unceasing generation" of the "Yijing" tradition as native resources for gradient ontology. Chapter 9 elevates frequency from coordinate to ontology through a metaphysical reading of $E = h\nu$. Chapter 10 proposes tristate categories as a new doctrine of ontological categories. Chapter 11 treats freezing as a basic ontological process, and this chapter is the pivot of Part II. The key contribution of Part II is to establish an independent systematic ontology capable of parallel dialogue with Aristotelian substance ontology, Whiteheadian process philosophy, and Chinese qi theory.

Part III: Co-Evolution, Mind–Matter, and Observation: The Unification of Three Relations (4 chapters). This is the relational-ontological part of the whole book. It unifies the treatment of relations found in different themes across the five volumes: the mind–matter relation in "Mind–Matter Unity," observation in "EQT Theory of Detection," and process in "Energy, Process, and the Fate of the Cosmos" and "Freezing Cosmology." Chapter 12 elevates co-evolution into an ontological relation: all "relations between two," including subject–

object, mind–matter, and self–other relations, are co-evolutions within the same field. Chapter 13 treats mind and matter as two aspects of the same process, through identity-level isomorphism, and gives the deepest application of constrained speculation to the hard problem through internal–external co-evolution and the final cell as a boundary. Chapter 14 treats observation as participation: reproducible co-perturbation becomes the new foundation of objectivity. Chapter 15 treats time as $\partial\rho/\partial t$, reconciling Bergson and Einstein. The key contribution of Part III is to establish a form of relational realism that is more concrete than Rovelli's relational quantum mechanics, because it has mechanistic constraints, and more rigorous than Karen Barad's intra-action, because it has requirements of falsifiability.

Part IV: The Metaphysical Boundary of Freezing (4 chapters). This is the metaphysical climax of the whole book. It applies constrained speculation to the limit of EQT ontology itself. Chapter 16 unfolds the hierarchy and historicity of freezing. Chapter 17 is the metaphysical climax of the whole book: the gradient terminal phase as an internal singularity of EQT ontology. Chapter 18 unfolds the collapse of co-evolution before the terminal state. Chapter 19 enters into dialogue with Heidegger's nothingness, Wittgenstein's silence, Nāgārjuna's emptiness, and Husserl's given-ness. The key contribution of Part IV is to make a contemporary natural philosophy encounter, at its most thorough point, the internal singularity of ontology and to acknowledge this limit honestly. This acknowledgment itself is a philosophical contribution: it demonstrates the deepest application of the method of constrained speculation.

Part V: From Ontology to Normativity (2 chapters). This is the humanistic part of the whole book. It moves from ontology to value and meaning. Chapter 20 treats the is–ought gap: constrained speculation downgrades it from an ontological problem to a problem of methodological discipline. Chapter 21 is the humanistic climax of the whole book: meaning before freezing. The key contribution of Part V is to extend constrained speculation from natural philosophy to the philosophy of value and the philosophy of meaning, making the whole book a complete philosophical system rather than merely a philosophy of nature.

Connections among the Five Parts

There are logical connections among the five parts. Part I, the method, provides the methodological ground for the following four parts, and every part and chapter that follows uses constrained speculation. Part II, ontology, is built upon Part I: only after the method has been established can one discuss how ontology ought to proceed. Part III, relation, is built upon Part II: only after gradient ontology has been established can one discuss how relations are organized. Part IV, boundary, is built upon Parts II and III: only after ontology and relational structure have been established can one discuss their limits. Part V, normativity, is built upon Part IV: only after finitude has been accepted can one discuss how meaning is realized within finitude. The overall structure is: method $\rightarrow$ ontology $\rightarrow$ relation $\rightarrow$ boundary $\rightarrow$ meaning. This is a progression from the most abstract to the most concrete, from the most meta-level to the most humanistic.

0.7. What This Book Does Not Do

To avoid misunderstanding, this section explicitly states what this book does not do.

This book does not restate the physical derivations. The physical derivations of EQT have been completed in detail in the five physical volumes. This book does not repeat those derivations; it uses their conclusions as premises. This means that even if readers do not accept the physical work of EQT at all, the arguments of this book can still be evaluated as philosophical arguments, since constrained speculation and gradient ontology as philosophical positions do not depend on the truth of EQT, although the illustrative applications would be weakened.

This book does not replace the cosmological narrative of "Freezing Cosmology." "Freezing Cosmology" is a natural-philosophical monograph on cosmic evolution, and it has already treated the physics of the freezing spectrum in detail. This book does not repeat the cosmological work of that volume; it uses freezing as a concept of basic ontological process, while leaving the physical details to that volume.

This book does not repeat the study of consciousness in "Mind–Matter Unity." "Mind–Matter Unity" unfolds in detail the isomorphism between experiential structure and ρ_f gradient configurations, the characterization of given-ness through internal–external co-evolution, the two compressions of the hard problem, and the clinical and experimental applications of $\Psi \equiv (\tilde{A}, C, \mathcal{B})$ as an observable projection. Chapter 13 of this book elevates the core argumentative strategy of the new version of "Mind–Matter Unity" into a philosophical application of constrained speculation in the domain of mind–matter: it organizes identity-level isomorphism, internal–external co-evolution, and the final cell as a boundary into a mind–matter philosophy of constrained speculation, but it does not repeat its physical modeling or clinical details.

This book does not replace the detection theory of "EQT Theory of Detection." "EQT Theory of Detection" treats in detail the detection capability matrix, the Three-State Separation Algorithm, and the analysis of various detectors. Chapter 14 of this book uses its philosophical implications, but does not repeat its technical work.

This book does not attempt to solve all philosophical problems. The method of constrained speculation itself requires the acknowledgment of limits: every chapter of this book ends with "open problems" or a similar treatment. This book leaves unfinished work on such questions as the hard problem of consciousness in Chapter 13, cross-theoretical comparison in Chapter 6, the precise relation between form and reality in Chapter 3, and the content of normativity in Chapter 20. It honestly marks them as open problems and does not pretend that they have already been solved.

This book does not claim that EQT is the final theory of physics. EQT is a research program and may be replaced or revised in the future. Constrained speculation and gradient ontology, as philosophical positions, can stand independently of the physical details of EQT: they operate at the more general level of "any thoroughgoing natural philosophy." If EQT is replaced by a future theory, the philosophical work of this book will still remain valid: the new theory would serve as a more specific application case.

0.8. Target Readers and Reading Paths

This book is addressed to four kinds of readers: physicists, philosophers, Chinese scholars, and international readers. Each type of reader may have a different reading path.

For physicists. If you are a physicist, you may find the ontological and metaphysical language of this book unfamiliar. The suggested reading path is as follows: first read Chapter 1 to understand the methodological position; then move to Chapter 9, frequency as a basic spectral quantity, and Chapter 10, tri-state structure as an ontological category, since these two chapters are closest to the concerns of physicists. Chapter 17, on ontological singularity, concerns the logical limit of EQT and may be of interest to physicists: it elevates the problem of the "final state of the universe" from a physical prediction into an ontological problem. Part III offers philosophical responses to physical intuitions concerning observation, objectivity, and relativistic time. The most likely doubt from physicists is: "What do these philosophical discussions have to do with the practice of physics?" The response of this book is that philosophical discussion clarifies the ontological commitments already implicit in physics. Through constrained speculation, physicists can understand more clearly the ontological meaning of their own work.

For philosophers. If you are a philosopher, especially one working in the analytic tradition, you may find the physical examples in this book unfamiliar. The suggested reading path is to read the book in sequence, beginning with Chapter 1. Pay particular attention to Chapter 3, on failure conditions as an ontological discipline, and Chapter 5, on dialogue with the spectrum of contemporary realism, since these two chapters connect most directly with the analytic philosophical tradition. Chapter 17 is the metaphysical climax of the whole book and concerns boundary problems in Kant, Wittgenstein, Nāgārjuna, and others; analytic philosophers will find that this chapter is doing serious metaphysical work. Chapter 20's treatment of the is–ought problem is a contribution to analytic ethics. The most likely doubt from philosophers is: "Is this merely another over-philosophization of a specific scientific theory?" The response of this book is that constrained speculation and gradient ontology, as philosophical positions, do not depend on the truth of EQT; they are general positions concerning how natural philosophy should work. EQT is one of their application cases.

For Chinese scholars. If you are a Chinese scholar, whether in philosophy, intellectual history, or the humanities more generally, you may find the methodological style and Western dialogues of this book unfamiliar. The suggested reading path is as follows: Chapter 8, on the qi tradition and frequency–density–gradient, is written especially for you; it treats Chinese qi theory as a substantive contribution to contemporary natural philosophy. Chapter 21, on the affirmation of finitude and meaning before freezing, enters into serious dialogue with the Confucian idea of "knowing that it cannot be done and yet doing it." Then return to Chapter 1 to understand the methodological position of the whole book. The most likely doubt from Chinese scholars is: "Is this repackaging Chinese thought in Western philosophical language, or another form of cultural chauvinism?" The response of this book is the opposite: it treats Chinese traditions as serious contributions to contemporary philosophy and asks the Western philosophical world to take them seriously. This is not cultural chauvinism, but the rigorous philosophical use of tradition.

For international readers. If you are an international reader, not limited to the three categories above, the suggested entry points are Chapter 1, on the methodological position, and Chapter 17, the metaphysical climax. Then read Chapter 3, on failure conditions as an ontological discipline, and Chapter 21, the humanistic climax. These four chapters form the spine of the book, and the other chapters unfold from them. The most likely doubt from international readers is: "Since this is philosophy written by Chinese scholars, will it have a special cultural limitation?" The response of this book is that the philosophical problems it addresses are universal: ontology, metaphysics, normativity, and meaning. Chinese tradition is one resource for this universal work, standing alongside Greek, Indian, and European traditions. This book seeks to enter international philosophical discussion and to be evaluated by international philosophical standards.

0.9. Possible Failures of This Book

Constrained speculation requires every ontological commitment to have failure conditions. This book as a whole should likewise state its own failure conditions: under what circumstances would the book be considered to have failed as a whole? This book identifies four possible directions of failure.

First direction of failure: constrained speculation is reduced to an existing position. If, upon careful analysis, constrained speculation proves to be merely a variant or recombination of existing positions such as those of Popper, Lakatos, Cartwright, or Ladyman, without independent content, then this book loses its methodological foundation. This direction of failure is borne by the work of Chapters 1 and 5: if those two chapters cannot convincingly establish the independence of constrained speculation, the book fails as a whole.

Second direction of failure: gradient ontology cannot respond to the objections of substance ontology. If some new version of substance ontology can answer the argument in Chapter 7 that difference precedes identity, so that gradient ontology no longer has an advantage, then the ontological part of this book loses its ground. This direction of failure is borne by the work of Part II.

Third direction of failure: ontological singularity is shown to be a local problem of EQT rather than an internal structure of any thoroughgoing ontology. If some "complete ontology" can avoid the internal singularity described in Chapter 17, that is, if it can still be fully stated at its own limit, then the work of Part IV loses its general significance. This direction of failure is borne by Chapters 17 and 19.

Fourth direction of failure: meaning is shown truly to require an eternal ground. If some rigorous argument, not mere rhetoric or religious belief, can prove that meaning must be anchored in some eternal state of affairs in order to be "truly real," then the argument in Chapter 21 that meaning is true precisely within finitude fails. This direction of failure is borne by the work of Chapter 21.

This book welcomes examination along these directions of failure: they are required by the application of constrained speculation to the book itself. The strength of this book does not lie in being "flawless," but in its ability to state clearly its own directions of failure and submit to their examination.

0.10. A Final Word

This book is part of a long-term work. The authors' five physical volumes, together with this book, constitute an initial expression of a research program. But this is not the endpoint.

If the work of this book stands, it will invite responses in several directions: physicists may continue to develop the applications of EQT; philosophers may rigorously criticize the details of constrained speculation; Chinese scholars may further develop the dialogue between qi theory and gradient ontology; and international readers may find connections between this book and other traditions. All such responses are continuations of the research program.

If the work of this book does not stand, it will be identified as some kind of local error: perhaps methodological, if constrained speculation is reduced to an existing position; perhaps ontological, if gradient ontology encounters objections it cannot answer; perhaps metaphysical, if the acknowledgment of ontological singularity is mistaken; or perhaps humanistic, if the argument that meaning requires an eternal ground ultimately succeeds. Any of these failures would be philosophical progress: they would help us understand more clearly what does not work.

Whatever the outcome, this book is an honest attempt to make explicit the implicit philosophy behind the authors' physical work as an independent philosophical position capable of rigorous evaluation. This attempt itself is an application of constrained speculation: it requires commitments to be made explicit, defended, and placed under the possibility of failure.

In its application to itself, constrained speculation requires the authors to acknowledge that this book is not the final answer, but one stage of a research program. It seeks to be read, examined, revised, and perhaps also overturned. This seeking is itself the dignity of philosophy and the posture this book hopes to possess.

Meaning before freezing: this is the core thesis of this book and also the posture of its writing. In this moment before freezing, to do the work that can be done, to undertake the commitments that should be undertaken, and to accept possible failure: this is constrained speculation applied to the act of writing itself.

Part I.

Constrained Speculation: Method as Philosophy

1. Constrained Speculation as a New Method of Natural Philosophy

1.1. Introduction to This Chapter

In the second half of the twentieth century, philosophy of science split into two major camps around the question of whether theoretical entities are real, and a series of intermediate positions emerged between them. From Putnam's early strong realism to van Fraassen's constructive empiricism, from Cartwright's entity realism to Worrall's structural realism, this debate appears plural, but in fact it shares an unexamined premise: reality and cognition are separated. Both camps assume that there is a "reality" out there, and that our theories either correspond to it or fail to correspond to it.

This shared premise has been questioned repeatedly since Descartes: Husserl's intentionality, Heidegger's "being-in-the-world," the later Wittgenstein, and Brandom's inferentialism all reject it. Yet these rejections have rarely been translated into an operational scientific methodology. As a result, realist debates in philosophy of science still unfold along the axis of "correspondence / non-correspondence."

This chapter proposes a position that does not seek a compromise point on the "correspondence–non-correspondence" axis, but rejects that axis itself. This position is named constrained speculation: it establishes ontological commitments through failable interfaces, being neither correspondence nor instrumental use. Its core thesis is: X is ontologically real if and only if the commitments of X have definable failure conditions.

This thesis redefines reality from "correspondence to some external fact" as "the content by which a commitment itself can fail." The task of this chapter is to state it clearly, defend it rigorously, and distinguish it from all possible misunderstandings. At the same time, it unfolds the thesis into four methodological disciplines ($CS1$–$CS4$: formalization, mechanization, observationalization, and failure- orientation), so that constrained speculation becomes a complete methodological position rather than a single proposition.

This chapter does not depend on the physical details of EQT. All arguments are completed as purely philosophical arguments. This independence allows the chapter, as the foundational chapter of a philosophical monograph, to be taken seriously in international philosophical discussion. EQT appears only as an applied example and does not function as a ground of argument.

The discussion proceeds as follows. Section 1.1 diagnoses the shared premise behind the realist predicament in contemporary philosophy of science; Section 1.2 proposes the core thesis of constrained speculation and gives its rigorous defense; Section 1.3 ontologizes the four disciplines; Section 1.4 argues for the interdependence of the four disciplines; Section 1.5 clarifies the relation to falsifiability; and Section 1.6 states the failable directions of this method and its current open challenges.

1.2. The Realist Predicament in Contemporary Philosophy of Science

Starting Point of the Argument

In the second half of the twentieth century, philosophy of science split into two major camps around the question of whether theoretical entities are real. Strong realism, represented by the early Putnam, Boyd, and Psillos, holds that theoretical entities in mature scientific theories directly correspond to the structure of the world, and that the more successful a theory is, the closer it comes to truth. Anti-realism, represented by van Fraassen and Laudan, holds that theoretical entities are merely predictive tools, and that the success of a theory proves only its empirical adequacy, not its reality. Between the two camps stand intermediate positions such as Cartwright's entity realism, Hacking's instrumental realism, and Worrall's structural realism.

This debate appears plural, but in fact it shares an unexamined premise: reality and cognition are separated.

Strong realism says that theory corresponds to reality; anti-realism says that theory does not correspond to reality. Both sides assume that there is a "reality" out there, and that our theories either correspond to it or fail to correspond to it. Intermediate positions, such as structural realism, attempt to refine this correspondence by claiming correspondence only to structure rather than to entities, but they still work on the axis of "correspondence / non-correspondence."

Key Argument

This section argues that this shared premise is the root of the problem. Once the separation between reality and cognition is accepted, only two routes remain: either one asserts some form of correspondence relation, as strong realism does, or one gives up correspondence, as anti-realism does. But why must this separation be accepted?

This separation has been questioned repeatedly since Descartes. Husserlian phenomenology rejects it: intentionality makes subject and object two poles of the same phenomenon. Heidegger's "being-in-the-world" rejects it: human beings do not first exist before the world and then cognize the world; they are already in the world. Peirce's pragmatism rejects it: meaning is practical effect, not correspondence with some external reality. The later Wittgenstein rejects it: meaning lies in use, not in correspondence with the world. Brandom's inferentialism rejects it: conceptual content is constituted by inferential relations, not by relations of correspondence.

Yet contemporary analytic philosophy of science has rarely brought this rejection into concrete methodology. The insights of Husserl and Heidegger developed within the Continental tradition, but they have rarely been translated into operational scientific methodology. Brandom's inferentialism remains at the level of philosophy of language and has not entered the core of debates on scientific realism. As a result, realist debates in philosophy of science still unfold along the axis of "correspondence / non-correspondence."

The core argument of this section is that the distinctive position of constrained speculation lies in the fact that it does not look for a compromise point on the "correspondence–non-correspondence" axis, but rejects that axis itself. It proposes a new ontological relation: ontological commitment established through a failable interface, neither correspondence nor instrumental use.

A Preliminary Explanation of the Failable Interface

What is a "failable interface"? A precise definition will be given in Section 1.2; here a preliminary explanation is enough.

Consider an ontological commitment, such as "electrons exist." Strong realism says that the electron exists as a real entity, and that the symbol "electron" in theory corresponds to this real entity.

Anti-realism says that the electron is only a computational tool, and that there is no need to assume that it corresponds to any real entity.

Constrained speculation says that the reality of the electron is neither established through correspondence nor in need of cancellation. The electron is real because we can state the conditions under which it would fail: the failure of charge quantization, the failure of Pauli exclusion, the failure of Compton scattering, and so on. These failure conditions constitute the content of the electron as an ontological object. The more concrete the failure conditions are, the more real the electron is as an ontological object.

This position does not require an electron "out there" to be corresponded to or cancelled. It directly constructs ontological commitment through failure conditions.

Distinctions from Existing Positions

Possible confusions must be explicitly excluded.

It is not instrumentalism. Instrumentalism says that theoretical entities are only tools and are not real. Constrained speculation says that theoretical entities are real, but their reality is concretized through failable interfaces rather than established through correspondence with an external reality. The positions are opposed.

It is not constructive empiricism. Van Fraassen's constructive empiricism says that the aim of theory is empirical adequacy rather than truth. Constrained speculation says that truth is meaningful, but truth is concretized through failable interfaces, not established through a correspondence relation with external reality. The position is closer to Putnam's later "internal realism," but more concrete than Putnam, because constrained speculation gives the concrete forms of failure conditions: dimensionality, proxy, and threshold.

It is not anti-realism. Anti-realism says that we cannot know what reality is like. Constrained speculation does not claim that "we cannot know"; it claims that reality itself is constructed through failable interfaces, so "knowing" is not the issue.

Nor is it a kind of pragmatism. Pragmatism says that the meaning of a theory lies in its practical effects. Constrained speculation agrees that a theory must have identifiable effects, which is close to saying that it must have failure conditions, but reality is not reduced to effects. Reality is a property of commitment, while effects are tests of commitment.

Objections That Must Be Answered

Objection One: Rejecting the separation between reality and cognition will lead to idealism or a variant of anti-realism.

Reply: Rejecting separation is not the same as cancelling distinction. Constrained speculation preserves the distinction between reality and cognition, but rejects their separation: they are connected through failable interfaces, not through relations of correspondence, nor are they identical. This is structurally similar to Husserl's "intentionality" and Heidegger's "understanding," but it is more concrete, because dimensionality, proxy, and threshold are used as interface constructions. An idealist position would say that "reality depends on mind"; constrained speculation says that "reality is concretized through failable interfaces." These are very different propositions.

Objection Two: Is this distinction, between the correspondence / non-correspondence axis and the failable interface, merely rhetorical? Analytic philosophers will ask whether the failable interface must ultimately "correspond" to something, such as the fact of whether failure occurs.

Reply: The failable interface does not "correspond" to anything. Failure is the content of the commitment itself, not an external fact. When we say that "the commitment of the electron has failure condition X," this statement does not presuppose an "external" electron corresponding to an "internal" commitment. The failure condition is the content of the commitment itself. This is the direction of Brandomian inferentialism, but Brandom did not apply it to ontology.

Objection Three: Husserl, Heidegger, Brandom, and others mentioned here are Continental or post-analytic philosophers. Does constrained speculation belong more to the Continental tradition than to the analytic tradition?

Reply: Constrained speculation is a position that consciously crosses the analytic–Continental divide. It begins from the methodological rigor of analytic philosophy, including concrete specifications of dimensionality, proxy, and threshold, but its ontological position, namely its rejection of the separation between reality and cognition, is closer to the Continental tradition. This crossing has precedents in the history of philosophy: the later Wittgenstein, the later Quine, Sellars, and McDowell all carried out similar work. Constrained speculation continues this tradition.

Literature That Must Be Engaged

- Hilary Putnam, Meaning and the Moral Sciences (1978), Reason,

Truth and History (1981), Realism with a Human Face (1990): Putnam's transition from early strong realism to internal realism is the most important interlocutor for constrained speculation.

- Bas van Fraassen, The Scientific Image (1980): the foundation of constructive empiricism.
- Larry Laudan, "A Confutation of Convergent Realism" (1981): the classical argument against truth-convergent realism.
- Stathis Psillos, Scientific Realism: How Science Tracks Truth (1999): a contemporary defense of strong realism.
- Nancy Cartwright, How the Laws of Physics Lie (1983): a foundation for an intermediate position.
- Ian Hacking, Representing and Intervening (1983): entity realism.
- Robert Brandom, Making It Explicit (1994): the influence of inferentialism on constrained speculation.
- John McDowell, Mind and World (1994): the critique of the "myth of the given."

Summary of This Section

This section has done three things. First, it identified the shared premise behind the realist predicament in contemporary philosophy of science: the separation between reality and cognition. Second, it pointed out the questionable status of this premise: it has been challenged many times in the history of philosophy, but has rarely been translated into scientific methodology. Third, it preliminarily positioned constrained speculation as a position that rejects this premise: it establishes ontological commitments through failable interfaces, being neither correspondence nor instrumental use.

The next section gives the core thesis of constrained speculation and its rigorous defense.

1.3. A New Proposal: Reality = Possibility of Failure

This section is the most important thesis statement of the whole book. All subsequent arguments in this book depend on it. It must be stated accurately, defended rigorously, and clearly distinguished from all possible misunderstandings.

Statement of the Core Thesis

Core thesis ($CR1$): X is ontologically real if and only if the commitments of X have definable failure conditions.

In ordinary language, this thesis may be stated as follows: an object is real because the commitments concerning it can go wrong, and this "going wrong" is not a vague possibility but a definable condition with concrete form.

The Precise Meaning of the Thesis

"Definable failure conditions" contain three requirements.

First, dimensionality. The object involved in the commitment must have definable dimensionality: it must be measurable or proxy-measurable. "Dark matter" as an ontological commitment has dimensionality, namely mass density, and can be measured through proxies such as gravitational effects, galactic rotation curves, and CMB fluctuations. "Absolute spirit" as an ontological commitment has no dimensionality and cannot be measured by any measurement or proxy measurement. The former is a candidate for real commitment; the latter is not.

Second, proxyhood. The commitment must be projectable onto some empirical proxy, even if that proxy is indirect, statistical, or remote. "Electron" as a commitment is tested through multiple proxies such as charge quantization, spin, and Compton scattering. "Soul" as a commitment has no identifiable empirical proxy, since any proposed proxy can be reinterpreted by opponents as something else.

Third, directionality. The commitment must be able to specify what kind of observation would make it fail: not a vague "counterexample," but a concrete direction of failure. "The Standard Model predicts the Higgs boson near 125 GeV" specifies a concrete direction of failure: if the LHC had not detected a signal in this energy region, the commitment would have failed. "The universe is harmonious" has no direction of failure: any observation can be reinterpreted as some form of "harmony."

Commitments that satisfy these three requirements are real commitments. Commitments that do not satisfy them are either empty definitions or hidden assumptions.

Key Argument

Why does $CR1$ hold? This section gives three layers of argument.

First Layer: Philosophical Motivation

Traditional ontological criteria, such as independence, stability, and causality, all assume that reality is out there and then ask how we can grasp it. But "out there" is itself an ontological commitment. How, then, is that commitment itself defended? It can only be defended through its content, and content can only be defined through the ways in which it can fail.

A "reality" that cannot fail is an empty concept. Consider two contrasting statements:

Statement A: "There exists a world of things-in-themselves, which is in principle compatible with any observation." Statement B: "There exists an electromagnetic field whose evolution is described by Maxwell's equations and which produces specific measurable predictions under specific boundary conditions."

Statement A has no failure condition: any observation is compatible with "the existence of a world of things-in-themselves." Statement B has concrete failure conditions: if the measurable predictions fail to match observation, the commitment fails. Constrained speculation holds that Statement A is not a real commitment; it is an empty concept. Statement B is a real commitment.

This judgment does not require us to say that "the world of things-in-themselves does not exist." It only says that, as an ontological statement, A has no concrete content.

Second Layer: Logical Argument

If X cannot fail, then any observation is compatible with X. This means that the commitment of X excludes no possibility. Any "commitment" that excludes no possibility is not a commitment: the essence of commitment is to choose some possibilities and exclude others.

Therefore, failability constitutes the concreteness of commitment. A "commitment" without the possibility of failure is rhetoric, not commitment. This resembles Wittgenstein's argument concerning propositions: a proposition must distinguish some states of affairs from others; otherwise, it is not a proposition. $CR1$ extends this insight about propositions to ontological commitments.

Third Layer: Argument from the History of Science

In the history of physics, objects later shown to be empty concepts, such as ether, caloric, and phlogiston, share one feature: before they were eliminated, their commitments had already become unable to fail.

Ether theory was repeatedly revised in the late nineteenth century, through such devices as the Fitzgerald contraction and the Lorentz transformation, until it could explain any experiment concerning the invariance of the speed of light. At that point it had already lost the status of a real commitment: it no longer had a direction of failure. Einstein's insight was not that "ether does not exist," for that statement by itself has no meaning, but that ether as an ontological commitment had lost concrete content.

Lakatos had already identified this phenomenon as a "degenerating research programme," but he remained at the epistemological level: he said that a degenerating programme "should not be accepted by the scientific community." $CR1$ elevates this phenomenon into an ontological criterion: objects within a degenerating programme lose their status as ontological objects, not merely as epistemologically unacceptable entities.

Relation to Falsifiability: A Detailed Explanation

$CR1$ looks similar to Popper's falsifiability. Chapter 6 will treat the relation to Popper in detail. Here three key differences are given first.

First difference, concerning the target: Popper targets the theory as a whole. $CR1$ targets every ontological commitment. A theory may be falsifiable as a whole, while some of its ontological commitments may still be unfailable hidden assumptions. $CR1$ requires item-by-item examination.

For example, Newtonian mechanics as a whole is falsifiable, since it can be refuted by observation, but the ontological commitment of "absolute space" within the Newtonian framework is unfailable: it is designed to "correspond" to any observation. When Einstein removed absolute space, he did not falsify Newtonian mechanics, which had already been verified in many ways; rather, he identified that "absolute space" as an ontological commitment had no concrete content. $CR1$ can make this fine distinction; Popper's holistic falsifiability cannot.

Second difference, concerning the nature of the criterion: Popper's falsifiability is a demarcation criterion: it tells us whether a theory is scientific. $CR1$ is an ontological thesis: it tells us whether the object involved in a commitment is real. These are different tasks. Popper asks, "Is this science?" This book asks, "What does it commit to as real?"

This difference allows $CR1$ to handle problems that Popper cannot handle. For example, mathematical theorems are "unfalsifiable," but they are not non-scientific. $CR1$ locates them as formal constraints rather than ontological commitments (see Section 3.5 of Chapter 3), without denying their legitimacy or misclassifying them as scientific theories.

Third difference, concerning the consequence of failure: in Popper, failure leads to the abandonment of a theory; in Lakatos, failure leads to the revision of a programme. In $CR1$, failure leads to the refinement of the concrete content of an ontological commitment: each time an object fails, its real content is made more precise.

Every failure is an ontological refinement, not elimination or protection. Atomic models failed repeatedly and were repeatedly revised: the failure of Thomson's atom led to Rutherford's atom; Rutherford's failure led to Bohr; Bohr's failure led to quantum mechanics. Yet each failure made "the atom" more concrete as an ontological object, rather than abandoning it. $CR1$ describes this process accurately; Popper's abandonment–protection dichotomy cannot.

Objections and Replies

Objection One: Does this reduce an ontological problem to an epistemological problem?

Reply: No. $CR1$ does not say "X is real $\iff$ X can be known." It says "X is real $\iff$ the commitments of X can fail." Being knowable and being failable are different: the former emphasizes positive grasp, while the latter emphasizes negative boundary. An object may in principle be impossible to grasp completely, such as the interior of a black hole, and yet still have concrete failure conditions, such as specific predictions concerning its gravitational effects, horizon structure, and Hawking radiation. This is a different route between epistemology and ontology, not a reduction of the latter to the former.

Objection Two: What about mathematical objects? $2+2=4$ *cannot fail, but intuitively we regard it as real.*

Reply: The unfailability of mathematical statements precisely shows that they are not ontological beings in the sense of $CR1$, but formal constraints. This distinction will be fully developed in Section 3.5 of Chapter 3. Here it is enough to point out that $CR1$ does not deny the special status of mathematics. It places mathematics in the category of formal constraints and clearly distinguishes it from the category of real objects.

This distinction actually preserves the epistemological dignity of mathematics. It need not claim that mathematical objects are "real in some mysterious way," which is the burden of mathematical Platonism, nor does it need to deny the objectivity of mathematics, which is the cost of mathematical anti-realism. Mathematics consists of formal constraints, whereas real objects belong to a different category. Each has its legitimate place.

Objection Three: This sounds like a new version of instrumentalism: reality is reduced to methodology.

Reply: Instrumentalism says that theoretical entities are not real, but only tools. Constrained speculation says that theoretical entities are real, but that the content of their reality is concretized through failability. The positions are opposed.

Instrumentalism would say that "the electron is only a useful concept." Constrained speculation says that "the electron is real, and the concrete content of its reality is constituted by failable conditions such as charge quantization, spin, and scattering cross sections." The former cancels reality; the latter constructs reality. This is a fundamental difference.

Objection Four: What ontological status do failure conditions themselves require? Is this an infinite regress?

Reply: Failure conditions are constituted by dimensionality, proxies, and thresholds, all of which are concrete physical quantities. Their ontological status is also determined by $CR1$: dimensionality, proxies, and thresholds themselves must also be able to fail, for instance when a proxy fails or a dimensional specification is wrong. This is not an infinite regress, but a self-consistent loop. The position is stable and does not require an "ultimate foundation."

By analogy, in mathematics, set theory takes sets as its basic objects, and the existence of sets is determined by the axioms of set theory. This is not an infinite regress, but a self-consistent system. $CR1$ has an analogous self-consistency in the domain of ontology.

Objection Five: Is $CR1$ too demanding? Many philosophers will object that it excludes many important ontological discussions, such as those concerning God, universals, and possible worlds.

Reply: Yes, $CR1$ does exclude many traditional ontological discussions. But this exclusion is not a crude denial; it is a repositioning. God, universals, and possible worlds, if they lack concrete failure conditions, do not belong to the category of real objects. But they may belong to the categories of formal constraints, religious practice, conceptual analysis, and so on. This repositioning actually preserves legitimate places for these discussions; it only refuses to confuse them with real objects such as electrons, galaxies, and conscious states.

Literature That Must Be Engaged

- Karl Popper, The Logic of Scientific Discovery (1959, original German 1934)
- Imre Lakatos, The Methodology of Scientific Research Programmes (1978)
- Paul Feyerabend, Against Method (1975): Feyerabend's critique of methodology must be directly answered.
- Michael Dummett, Truth and the Past (2004): the relation between falsifiability and truth.
- Hilary Putnam, Realism with a Human Face (1990): a comparison between the later Putnam's internal realism and the present position.
- Robert Brandom, Articulating Reasons (2000): the semantics of inferentialism.
- W.V.O. Quine, "On What There Is" (1948): a foundational discussion of ontological commitment.

Summary of This Section

This section has done three things. First, it precisely stated $CR1$: X is real $\iff$ the commitments of X have definable failure conditions. Second, it gave three layers of defense: philosophical motivation, logical argument, and historical argument from science. Third, through a detailed comparison with Popperian falsifiability, it clarified the distinctiveness of $CR1$.

The next section unfolds $CR1$ into four concrete writing disciplines.

1.4. The Ontologization of the Four Writing Disciplines

From Methodological Discipline to Ontological Commitment

The Introduction to "Freezing Cosmology" in the EQT series states four writing disciplines for "constrained speculation": concepts must be formalized, forms must be mechanized, mechanisms must be observationalized, and observations must allow failure. There they appear as writing norms, telling the author how to write natural philosophy.

This section argues that these four disciplines are not writing norms, but concrete unfoldings of $CR1$. They are concrete forms of ontological commitment: they tell us what requirements an ontological commitment must satisfy in order to be regarded as a real commitment.

Discipline One: Conceptual Formalization

Statement: An ontological commitment must have definable symbolic relations; it must be capable of being written into a clear formal language.

Ontological meaning: a concept that cannot be formalized has no concrete content; without concrete content, it cannot fail; if it cannot fail, it is not real. Formalization is not mathematical ornamentation, but the minimum requirement of ontological commitment.

For example, the "energy quantum density field $\rho_f(x, t)$" is formalized as an ontological commitment in the form of a spatiotemporal function. Its dimensionality, transformation properties, normalization conditions, and boundary conditions can all be precisely stated. "Cosmic spirit" as an ontological commitment lacks this kind of formalization: it cannot be written into a formal expression with determinate properties. The former is a candidate for real commitment; the latter is not.

Important clarification: formalization does not require already completed formalization. A concept may be in the process of being formalized, as was the concept of the "gene" in the early twentieth century: it had already been formalized as a factor in Mendelian laws of heredity, but had not yet been specified as a DNA sequence. Constrained speculation allows ongoing formalization; it rejects only concepts that cannot be formalized in principle, such as "cosmic spirit."

Discipline Two: Formal Mechanization

Statement: Form must be able to describe concrete processes; it must have dynamical structure, variable relations, and state evolution.

Ontological meaning: if a "form" is only an abstract relation without mechanism, it still cannot fail, because it does not prescribe any observable process. Mechanization is the bridge from form to phenomenon.

For example, the energy quantum density field not only has form, as a spatiotemporal function, but also has mechanism: it obeys a certain master equation, such as the EQT Master Equation, which describes its evolution in spacetime. This mechanism specifies the behavior of the density field and allows it to be connected with concrete processes. If "substance" as a form has no dynamics, that is, if it does not specify how substances change, interact, or combine, then it lacks mechanism and remains merely an abstract label.

Important clarification: mechanism does not require complete determinism. Stochastic processes are also mechanisms: they specify the evolution of probability distributions. Constrained speculation does not require Laplacian determinism; it only requires that form be projectable onto describable process structure.

Discipline Three: Mechanistic Observationalization

Statement: Mechanism must be projectable onto empirical proxies; it must have identifiable observational interfaces.

Ontological meaning: without proxies, mechanism is only a story inside a model and has no interface with the world. Proxyhood is the concrete way in which ontological commitment encounters the phenomenal world.

For example, the mechanism of the EQT Master Equation can be projected onto proxies such as CMB fluctuations, gravitational-wave signals, and dark matter halo structures. These proxies are the interfaces through which the mechanism encounters the world. A "dream entity" as a mechanism, even if it has been formalized and mechanized as some kind of process, cannot enter real commitment if it has no identifiable proxy, since any proposed proxy can be reinterpreted by opponents as something else.

Important clarification: proxies may be indirect, statistical, or remote. Constrained speculation does not require direct observation. Before gravitational waves were directly detected by LIGO, they were already legitimate ontological commitments because they had concrete proxies predicted by theory, such as the orbital decay rate of binary stars, as observed by Hulse and Taylor through pulsars.

Discipline Four: Observation Must Allow Failure

Statement: A proxy must have a definable direction of failure; it must be able to specify what kind of observation would make the commitment fail.

Ontological meaning: this is the direct requirement of $CR1$: the possibility of failure constitutes the concreteness of commitment.

For example, the ontological commitment to the Higgs boson had a concrete direction of failure: if the LHC had not detected a signal near 125 GeV, the commitment would have failed. This concreteness makes the Higgs boson valid as a real commitment. "Universal consciousness" as a commitment has no concrete direction of failure: any observation can be reinterpreted as a "manifestation of universal consciousness." Therefore, it is not a real commitment.

Important clarification: a direction of failure does not require immediate testability. A commitment may have a direction of failure that is definable in principle but currently unreachable by technology; this still satisfies Discipline Four. For example, the ontological commitment to quark confinement has a concrete direction of failure: free quarks should be observable under sufficiently low-temperature conditions, and if high-precision experiments always fail to find them, the commitment would fail, even if present technology has not yet reached decisive precision.

The Interdependence of the Four Disciplines

These four disciplines are not independent test items. They form an interdependent whole: no single discipline by itself constitutes a complete criterion for ontological commitment.

- Without formalization, mechanization has nothing to attach to: mechanism must attach to a concrete form.
- Without mechanization, observationalization has nothing to project: observation requires mechanism for its orientation, otherwise it is only neutral recording.
- Without observationalization, failure-orientation has no concrete content: failure requires observable proxies in order to have concrete content.
- Without failure-orientation, formalization loses ontological meaning: if a form cannot fail, it is only an empty form.

Therefore: constrained speculation $\Longleftrightarrow$ the closed loop of the four disciplines. This is its indivisibility as a whole method. Any attempt to reduce constrained speculation to a single discipline misunderstands it.

Comparison with Philosophical Traditions

This section uses comparison to show the distinctiveness of constrained speculation.

First comparison: Carnap's logical construction. In "The Logical Structure of the World" (1928), Carnap emphasizes formalization (Discipline One), but stops there. He does not require that form be mechanized or that mechanism be observationalized. As a result, his logical construction can fit any empirical content and thus loses constraint on content. Constrained speculation begins from Carnap but goes further: it requires form to be projected onto mechanism, mechanism onto observation, and observation to allow failure.

Second comparison: Popper's falsifiability. Popper jumps to Discipline Four, failure-orientation, but does not require the first three disciplines. As a result, he cannot handle which theoretical entities are concrete substructures of falsifiability; he works only at the level of the theory as a whole. Constrained speculation supplements Popper's work by requiring every ontological commitment to satisfy the complete chain from formalization to failure-orientation.

Third comparison: Cartwright's phenomenological causality. In "How the Laws of Physics Lie" (1983), Cartwright emphasizes mechanization (Discipline Two) and takes causal mechanisms as the core of scientific explanation. But she does not require unified formalization: her "local causal models" are dappled and non-unifiable. As a result, her work cannot support cross-domain comparison. Constrained speculation supplements Cartwright by preserving the centrality of mechanism while requiring that every local mechanism be formalizable.

Fourth comparison: Quine's ontological commitment. In "On What There Is" (1948), Quine requires the clarification of quantified variables, which is close to Discipline One, but he does not require proxyhood, namely Discipline Three. Quine's ontological commitment works only at the level of language: a theory is committed to the existence of the objects over which its quantified variables range. But how this commitment encounters the world is not treated by Quine. Constrained speculation supplements Quine by requiring quantified commitment to be proxyable.

The distinctiveness of constrained speculation is that it requires all four disciplines to be satisfied simultaneously and, as a whole, to constitute ontological commitment. No single discipline is sufficient. This wholeness is the key difference between constrained speculation and the work of the philosophers above. Each of them grasps one part of one discipline, whereas constrained speculation integrates them into an indivisible closed loop.

Objections and Replies

Objection One: Are these four disciplines too strict, excluding many important philosophical discussions?

Reply: Yes, constrained speculation indeed excludes many traditional philosophical discussions. It downgrades those discussions to formal constraints or heuristic language. It does not deny their value, but refuses to treat them as ontological commitments. This choice has a cost, but the cost is worth paying.

The strength of philosophy does not lie in tolerating all discussions, but in establishing defensible criteria. Constrained speculation establishes criteria for ontological discussion. Discussions that do not satisfy these criteria may continue: they may belong to formal philosophy, normative philosophy, hermeneutics, and other domains. But they do not enter ontological discussion.

Objection Two: Is there an implicit hierarchy among the four disciplines?

Reply: Disciplines One through Four do have a logical progression: form $\to$ mechanism $\to$ observation $\to$ failure. But this is not a hierarchy of priority: they must hold simultaneously. The logical progression only explains how to examine a commitment, beginning with form and ending with failure; it does not explain which discipline is more important.

This distinction will be further developed in Section 1.4. Here it is enough to point out that the wholeness of constrained speculation does not allow any discipline to be replaced or ignored.

Objection Three: What, then, is pure conceptual analysis, such as analytic-philosophical discussions of knowledge or free will?

Reply: Within the framework of constrained speculation, pure conceptual analysis is located as formal work. It establishes or clarifies formal constraints, but does not by itself constitute ontological commitment. This positioning is in fact a form of respect for analytic philosophy: it has its legitimate place, namely the establishment and clarification of form, but it should not be confused with ontological commitment.

Similarly, normative philosophy, including ethics and political philosophy, also has a legitimate place: it establishes and tests normative statements, but does not directly make ontological commitments. If it does make ontological commitments, then those commitments must also satisfy the four disciplines. See Chapter 20 on the treatment of the is–ought problem.

Objection Four: Is this position overly mathematized? Does it exclude legitimate non-mathematical inquiries, such as phenomenology and anthropology?

Reply: Formalization is not equivalent to mathematization. Formalization includes mathematics, but is not limited to it. Linguistic formalization, such as generative grammar, sociological formalization, such as network analysis, and phenomenological formalization, such as Husserl's steps of epoché and reduction, are all different kinds of formalization. Constrained speculation requires formalization, but the concrete form of formalization varies by domain.

If fields such as phenomenology and anthropology make ontological commitments, for example by claiming that "subjectivity is a basic ontological category," then those commitments must satisfy the four disciplines. They may of course be formalized in phenomenological or ethnographic ways, but they must have formalization. Otherwise, their ontological commitments lack concrete content.

Literature That Must Be Engaged

- Rudolf Carnap, Der logische Aufbau der Welt (1928), Logical Foundations of Probability (1950)
- W.V.O. Quine, "On What There Is" (1948), Word and Object (1960)
- Nancy Cartwright, The Dappled World (1999)
- Brian Ellis, Scientific Essentialism (2001): a contrast with essentialism.
- The treatment of ontological restraint in Chapter 2 (self-reference).
- The treatment of the concrete forms of failure conditions in Chapter 3 (self-reference).

Summary of This Section

This section has done three things. First, it has repositioned the four writing disciplines stated in "Freezing Cosmology" in the EQT series as concrete forms of ontological commitment: they are not writing norms, but concrete unfoldings of $CR1$. Second, it has given the precise meaning, ontological significance, and important clarifications for each discipline. Third, through comparison with Carnap, Popper, Cartwright, and Quine, it has shown that the distinctiveness of constrained speculation lies in the closed loop of the four disciplines as a whole.

The next section specifically argues for the interdependent structure of the four disciplines.

1.5. The Interdependence of the Four Disciplines

Aim of the Argument

This section provides a formal argument for the interdependence of the four disciplines, rather than remaining at the level of conceptual argument. The previous section has already pointed out conceptually that the four disciplines form a closed loop; this section reveals the precise structure of that loop by analyzing the dependence relations between every pair of disciplines.

This refinement is crucial for the subsequent work of the book. Later chapters will repeatedly cite specific subsets of the four disciplines. For example, the argument concerning ontological singularity in Chapter 17 mainly uses Discipline Four, while the argument concerning the collapse of co-evolution in Chapter 18 mainly uses Disciplines Two and Three. Understanding the internal structure of the four disciplines helps make the later arguments more precise.

Six Dependence Relations

There are $\binom{4}{2} = 6$ paired relations among the four disciplines, and each group constitutes a form of dependence.

Relation 1 (form $\leftarrow$ mechanism): mechanism must be expressed in some form; otherwise, it is only verbal description. But conversely, form does not need mechanism: mathematical forms can exist as pure abstractions, such as group theory as an abstract algebraic structure, without attaching to any concrete mechanism. Therefore, mechanism depends on form, but form does not depend on mechanism.

Relation 2 (form $\leftarrow$ observation): observation must have formal expression; otherwise, it remains merely individual experience. But form does not need observation; as above, pure mathematical structures can exist independently.

Relation 3 (form $\leftarrow$ failure): failure conditions must be formalizable; otherwise, they are only the intuition that "this seems wrong." But form does not need failure: mathematical theorems cannot fail but remain forms.

Relation 4 (mechanism $\leftarrow$ observation): observation must be oriented by mechanism. Pure "looking" does not constitute observation; observation is always observation of something, and that "something" is defined by mechanism. But mechanism does not need observation: a mechanism can be a story internal to a model without necessarily having observable projection, as in certain internal symmetry mechanisms in theoretical physics.

Relation 5 (mechanism $\leftarrow$ failure): failure must be explained by mechanism. It is not enough to say that "the theory is wrong"; one must say which mechanistic step is wrong. If a failure cannot be located in a concrete mechanistic step, it is not a concrete failure. But mechanism does not need failure: a mechanism may be an unfalsifiable pure theoretical construction.

Relation 6 (observation $\leftarrow$ failure): failure must be realized through observation; failure is a certain observable event. But observation does not need failure: observation may be neutral recording without presupposing any theoretical framework.

Reverse Dependence: The Closure of Ontological Commitment

All six relations above are one-way dependences, for example, mechanism depends on form, but form does not depend on mechanism. If only these relations are considered, one would arrive at a superficial conclusion: form is the most basic, and failure is the most dependent. This seems to conflict with the claim of Section 1.2, where the possibility of failure is said to be the core of ontology.

This section argues that this apparent conflict disappears at the level of ontological commitment. At the level of conceptual dependence, form is indeed the most basic; but at the level of ontological commitment, a form without failure does not constitute an ontological commitment.

The precise formulation is as follows:

Formalization is the starting point of ontological commitment; failure-orientation is the mark of ontological commitment. Conceptually, form precedes failure; ontologically, failure gives form the status of commitment.

This distinction is central to constrained speculation: it unfolds at two levels, conceptual dependence and ontological commitment. The two levels point in opposite directions, but cooperate with one another to constitute a complete methodological position.

Further Explanation of the Two-Level Structure

To avoid misunderstanding, this section further explains this two-level structure.

The level of conceptual dependence asks: in order to speak of X, what must first be spoken of? The answer is: to speak of mechanism, one must first speak of form; to speak of failure, one must first speak of observation. This is a level of preconditions: it gives the logical order among concepts.

The level of ontological commitment asks: what must X, as an ontological commitment, satisfy? The answer is: it must satisfy all four disciplines. This is a level of legitimacy: it gives the qualifications for becoming an ontological commitment.

The two levels have opposite directions and are complementary. The level of conceptual dependence is constructive: it begins from form and gradually establishes mechanism, observation, and failure-orientation. The level of ontological commitment is normative: given a candidate commitment, it examines whether that commitment satisfies all four disciplines.

This two-level structure is not uncommon in analytic philosophy. Frege's sense/reference distinction, Carnap's internal/external question distinction, and Strawson's descriptive/revisionary metaphysics distinction are all two-level structures. The two-level structure of constrained speculation has precedents and is not an ad hoc construction.

Objections That Must Be Answered

Objection One: This two-level structure looks overly elaborate. Is it a temporary device constructed to avoid theoretical difficulty?

Reply: A two-level structure is a legitimate methodological tool in analytic philosophy. Frege's distinction between sense and reference in "Über Sinn und Bedeutung" (1892) was not an ad hoc construction, but an insight into the nature of meaning. Carnap's distinction between internal and external questions in "Empiricism, Semantics, and Ontology" (1950) was not an ad hoc construction, but a clarification of ontological disputes. The two-level structure of constrained speculation has a similar status: it is not a temporary device, but a refined analysis of the method itself.

Objection Two: Then which is truly more basic, formalization or failure-orientation?

Reply: This question presupposes that there must be something "more basic." Constrained speculation rejects this presupposition. The two levels each have their function, and neither can be omitted, just as Frege's sense and reference are both indispensable, without needing to ask which is "more basic." Asking which is more basic forcibly reduces a two-level structure to a single-level structure, and this is precisely the simplification rejected by constrained speculation.

Objection Three: Does the two-level structure make constrained speculation too complex for practical application?

Reply: In practice, the two levels are often used simultaneously without requiring explicit distinction. When researchers construct theories, such as when they establish the EQT Master Equation, they use the level of conceptual dependence: beginning from form, they gradually establish mechanism and observational projection. When evaluating theories, they use the level of ontological commitment: they check whether the theory satisfies the four disciplines. These two uses naturally combine in practice, and one need not explicitly invoke the notion of a "two-level structure" every time.

Objection Four: Is this two-level structure too strongly influenced by Frege? Philosophers outside the analytic tradition may have different views.

Reply: The two-level structure is not Frege's property. The Buddhist doctrine of the "two truths," namely conventional truth and ultimate truth, is a two-level structure; Kant's distinction between phenomena and things-in-themselves is a two-level structure; Heidegger's ontological difference, between beings and Being, is a two-level structure. The two-level structure of constrained speculation has broad precedents in the history of philosophy and is not limited to Frege.

Literature That Must Be Engaged

- Gottlob Frege, "Über Sinn und Bedeutung" (1892): the sense/reference distinction.
- Rudolf Carnap, "Empiricism, Semantics, and Ontology" (1950): internal/external questions.
- Peter Strawson, Individuals (1959): descriptive/revisionary metaphysics.
- Saul Kripke, Naming and Necessity (1980): the two-level structure of necessity/contingency and a priori/a posteriori.
- Nāgārjuna, Mūlamadhyamakakārikā: the doctrine of the two truths as a two-level structure.

Summary of This Section

This section has done two things. First, by analyzing the six dependence relations among the four disciplines, it has revealed the internal structure of constrained speculation. Second, it has identified two levels, conceptual dependence and ontological commitment, and explained how this two-level structure gives constrained speculation a refined methodological function.

The next section specifically treats the relation between constrained speculation and the tradition of Popperian falsifiability.

1.6. Relation to Falsifiability

Aim of the Argument

This section specifically addresses an urgent problem: constrained speculation and Popper's falsifiability look too similar. This book must explicitly point out the differences; otherwise, readers may regard it as merely a variant or simple extension of Popper.

Chapter 6 will further synthesize the relation between falsifiability and constrained speculation. This section first gives the key distinctions.

Three Key Differences

First Difference: The Target

Popper's falsifiability targets the theory as a whole: whether a theory is falsifiable is a property of the theory as a whole. The failure-orientation of constrained speculation targets each ontological commitment: a theory may be falsifiable as a whole, while some of its ontological commitments may still be unfailable.

Lakatos had already become aware of this problem. In "The Methodology of Scientific Research Programmes," he distinguishes between the "hard core" and the "protective belt": the hard core is the protected central commitment of a research programme, while the protective belt consists of revisable auxiliary hypotheses. Lakatos's insight is that when a programme faces refutation, researchers tend to revise the protective belt in order to protect the hard core.

But Lakatos remains at the epistemological level. He describes this phenomenon and proposes the distinction between "progressive" and "degenerating" research programmes as an evaluative tool, but he does not treat the question of whether the commitments in the hard core are failable as an ontological problem.

Constrained speculation advances this distinction to the ontological level: only failable commitments in the hard core are real ontological commitments; unfailable hard-core commitments are only formal constraints.

For example, the hard core of Newtonian mechanics includes "absolute space" and "conservation of mass." "Conservation of mass" as a commitment has concrete failure conditions and can be tested by precise measurements; therefore, it is a real commitment. "Absolute space" as a commitment is unfailable within the Newtonian framework: it is designed to "correspond" to any observation. Therefore, it is not a real commitment, but a formal constraint. When Einstein removed absolute space, he did not refute Newtonian mechanics, since absolute space had no physically refutable content; he identified that absolute space as an ontological commitment had no concrete content.

Constrained speculation can make this fine distinction; Popper's holistic falsifiability cannot.

Second Difference: The Nature of the Criterion

Popper's falsifiability is a demarcation criterion: it gives the boundary between science and non-science. The failure-orientation of constrained speculation is an ontological criterion: it gives the reality of a commitment.

These are different tasks. Popper asks, "Is this science?" This book asks, "What does it commit to as real?"

This difference makes the two methods handle different problems. Popper uses falsifiability to demarcate "science / non-science," classifying astrology and Marxist historical prophecy, in Popper's view, as non-science. Constrained speculation uses failure-orientation to judge ontological commitments. It does not presuppose that science is the only legitimate cognitive form, but asks how any commitment has reality.

Constrained speculation can handle problems that Popper cannot. For example, mathematical propositions are "unfalsifiable," but they are not non-scientific: they are formal constraints and belong to a category different from scientific propositions. Popper's demarcation criterion runs into difficulty here. The criterion of constrained speculation naturally handles this situation: mathematics is not misclassified as non-science, but properly located as formal constraint.

Similarly, ethical propositions are "unfalsifiable" by Popper's standard, but they are not non-scientific; they are normative statements and belong to another legitimate category. Constrained speculation provides tools for handling such multi-category cases, whereas Popper's dichotomy does not.

Third Difference: The Consequence of Failure

In Popper, failure leads to the abandonment of a theory: a refuted theory should be given up. In Lakatos, failure leads to the revision of a programme: a research programme revises its protective belt in the face of counterexamples. In constrained speculation, failure leads to the refinement of the concrete content of an ontological commitment: each time an object fails, its real content is made more precise.

Every failure is an ontological refinement, not elimination or protection.

Consider the history of "the atom" as an ontological commitment:

- Dalton's atom, an indivisible smallest particle, failed → it was revised into Thomson's atom, containing electrons.
- Thomson's atom, with uniformly distributed positive charge and electrons, failed → it was revised into Rutherford's atom, with a nucleus and electron cloud.
- Rutherford's atom, with classical orbits, failed → it was revised into Bohr's atom, with quantized orbits.
- Bohr's atom, with the concept of orbit, failed → it was revised into the quantum-mechanical atom, with the wave function.

Each failure made "the atom" more concrete as an ontological object, rather than abandoning it. Popper's abandonment–protection dichotomy cannot accurately describe this process: the atom was not abandoned, nor was it simply protected; it was refined. Lakatos's progressive–degenerating distinction comes closer, but is still not accurate: what occurred here was not a revision of the protective belt, but the refinement of the hard core itself. Constrained speculation describes this process accurately: the concrete content of the atom as an ontological commitment is refined through repeated failures.

This difference is far-reaching. It shows that scientific progress is not "theory replacement" or "programme revision," but the refinement of ontological commitments. This also differs from Kuhn's theory of scientific revolutions: Kuhn emphasizes "paradigm incommensurability," whereas the refinement of ontological commitment is a continuous process, not a rupture. Constrained speculation thereby provides a new model of scientific dynamics after Kuhn.

A Detailed Comparison with Lakatos

Because Lakatos is the closest interlocutor of constrained speculation, a detailed comparison is needed.

Lakatos's methodology of research programmes includes the distinctions among hard core, protective belt, heuristic, and progressive versus degenerating programmes. Constrained speculation can be positioned as follows:

- Lakatos's "hard core" corresponds to the core ontological commitments examined by failure-orientation in constrained speculation. But constrained speculation further requires every commitment within the hard core to satisfy the four disciplines.
- Lakatos's "protective belt" corresponds to auxiliary hypotheses in constrained speculation. But constrained speculation requires that revisions of the protective belt themselves also satisfy the four disciplines and not be arbitrary.
- Lakatos's "progressive vs degenerating" distinction corresponds to the distinction between the refinement of ontological commitment and the emptying of ontological commitment in constrained speculation. But constrained speculation gives more concrete criteria: dimensionality, proxy, and direction of failure.

In general, constrained speculation is the ontologization of Lakatos's work: it elevates Lakatos's methodological criteria into ontological criteria and provides more concrete operational tools.

Objections and Replies

Objection One: This is only a repackaging of the Popper–Lakatos tradition and has no genuine originality.

Reply: "Repackaging" underestimates the significance of the three differences. Advancing falsifiability from the level of theory to the level of ontological commitment, from a demarcation criterion to an ontological criterion, and from the abandonment–protection dichotomy to a process of refinement are all substantive philosophical advances.

This is not a simple extension. It elevates falsifiability from a meta-theory about science into an ontology about reality. This elevation is unfinished work in the history of philosophy of science. Constrained speculation completes it.

Objection Two: Popper has been criticized many times in philosophy of science. Why return to him?

Reply: What has been criticized is Popper's naive falsificationism, namely the overly strong claim that a single counterexample is enough to refute a theory. Lakatos, Feyerabend, and Kuhn all criticized this overly strong claim. But falsifiability as a principle has not been overthrown: even the critics accept that theories should be testable.

Constrained speculation begins from the principle of falsifiability but avoids naive falsificationism. It does not require "one counterexample to refute the whole theory," but requires every ontological commitment to have a concrete direction of failure. This refinement avoids the difficulties of naive falsificationism while preserving the core insight of falsifiability.

Objection Three: Does "ontologizing" falsifiability overinflate the concept?

Reply: On the contrary. Ontologization makes falsifiability more refined, not more expansive. It no longer asks whether a theory can be refuted, which is a broad question, but asks which ontological commitment fails under which concrete conditions, which is a precise question.

This refinement enables falsifiability to do what it could not do before: distinguish different commitments within the hard core, locate the specific source of failure, and trace the historical refinement of ontological commitments. This is not inflation, but focusing.

Literature That Must Be Engaged

- Karl Popper, The Logic of Scientific Discovery (1959)
- Imre Lakatos, The Methodology of Scientific Research Programmes (1978)
- Paul Feyerabend, Against Method (1975)
- Thomas Kuhn, The Structure of Scientific Revolutions (1962): Kuhn's critique of the methodological tradition must be directly answered.
- Larry Laudan, Progress and Its Problems (1977): the methodology of research traditions.
- Alan Chalmers, What Is This Thing Called Science? (4th ed., 2013): a standard reference for the history of methodology.

Summary of This Section

This section has done three things. First, it identified three key differences between constrained speculation and the Popper–Lakatos tradition: the target, namely theory vs ontological commitment; the nature of the criterion, namely demarcation vs ontological criterion; and the consequence of failure, namely abandonment–protection vs refinement. Second, through the historical example of the atom as an ontological commitment, it showed how constrained speculation provides a refined analysis that Popper and Lakatos cannot provide. Third, it positioned constrained speculation as the ontologization of the Popper–Lakatos tradition: it both inherits and advances that tradition.

The next section is the reflective conclusion of this chapter: it acknowledges the problems that this chapter has not resolved.

1.7. Unresolved Problems of This Chapter

The Function of Honest Acknowledgment

This section is an application of the method of constrained speculation itself: at the end of the first chapter of the whole book, it acknowledges the unresolved problems of this chapter. This posture is itself part of the argument: a genuinely thorough method should be able to acknowledge its own limits.

Constrained speculation requires every ontological commitment to have failure conditions. This requirement also applies to constrained speculation itself: it must be able to state its own directions of failure. This section is doing precisely that.

Three Unresolved Problems

Problem One: The Ontological Status of Failure Conditions Themselves

This chapter claims that X is real $\Longleftrightarrow$ the commitments of X have definable failure conditions. But are failure conditions themselves real? If they are real, what are their failure conditions?

This looks like an infinite regress. In the reply to Objection Four in Section 1.2, this chapter pointed out that it is a self-consistent loop rather than an infinite regress, but it did not provide a rigorous argument. A brief argument is as follows: failure conditions are constituted by dimensionality, proxies, and thresholds; these are all concrete physical quantities and have their own failable interfaces. Therefore, failure conditions as commitments also satisfy $CR1$. This is not an infinite regress, because we have reached a level of objects with their own failure conditions.

But this brief argument is not yet sufficient. It requires detailed development, especially concerning how dimensionality, proxy, and threshold themselves are precisely defined and how they depend on one another. This work is left to Chapter 3.

Problem Two: The Standard of Cross-Theoretical Comparison

Constrained speculation requires every ontological commitment of every theory to have failure conditions. But can the failure conditions of different theories be made commensurable?

Consider two examples. "The invariance of the speed of light" as a commitment has concrete failure conditions in relativity, such as Michelson–Morley-type experiments. "Proton stability" as a commitment has concrete failure conditions in grand unified theories, such as decay events in a sufficiently large water tank. How are these two failure conditions to be compared? They use different experiments, different precisions, and different assumptions.

This problem has a precedent in Kuhn's argument about incommensurability. Constrained speculation needs to give a response to incommensurability. It cannot pretend that failure conditions are always comparable, but it also cannot accept Kuhnian strong incommensurability, which would cancel any cross-theoretical evaluation.

This problem is left to Chapter 6 for detailed treatment.

Problem Three: The Relation between Form and Reality

This chapter has pointed out that mathematical statements do not belong to ontological commitments but to formal constraints. But it has not yet argued in detail for the relation between form and reality. Is form completely unreal? Do mathematical structures have some weak ontological status?

This distinction is important for this book: it determines how we treat mathematical structures, logical forms, and conceptual frameworks. If form were completely unreal, then what would the EQT Master Equation be as a "form"? Would it have no ontological content at all with respect to the world? This is clearly wrong.

The correct answer should be that form is the carrier of real commitments, but is not itself a real object. The EQT Master Equation as a form carries concrete ontological commitments, namely the evolution of the energy quantum density field. Those commitments are real, because they satisfy the four disciplines. But the master equation as pure form, namely as a pure mathematical structure without any attached ontological commitment, is not a real object.

Yet this answer still requires further development, especially concerning the precise mechanism of "carrying." This problem is left to

Section 3.5 of Chapter 3.

The Failure Conditions of Constrained Speculation Itself

The most important task of this section is to state the failure conditions of constrained speculation itself.

If constrained speculation cannot even state its own failure conditions, it violates its own discipline. That would be its deepest self-refutation. Therefore, this section must state explicitly:

Constrained speculation fails $\Longleftrightarrow$ there exists an ontologically indispensable commitment, and that commitment cannot be formalized, mechanized, observationalized, or failure-oriented at all.

This failure condition requires further clarification.

What does "ontologically indispensable" mean? It means that cancelling the commitment would make some proposition we intuitively regard as true lose meaning or become incoherent. For example, "subjectivity," "moral responsibility," or "phenomenal consciousness" may be regarded as ontologically indispensable, since cancelling them would make our entire discourse concerning persons, morality, and experience lose meaning.

If such a commitment is found, what happens to constrained speculation? It must revise itself. Possible revisions include relaxing one of the four disciplines, for example allowing some concepts that cannot be fully formalized to count as ontological commitments; introducing a new category of commitment, such as "quasi-ontological commitment," lying between real objects and formal constraints; or acknowledging that the scope of constrained speculation is limited, for example that it applies only to natural philosophy and not to value philosophy.

Has such a commitment currently been found? There is relevant discussion concerning phenomenal consciousness, especially the hard problem of consciousness. If phenomenal consciousness is regarded as ontologically indispensable yet cannot be fully formalized or failure-oriented, which is Chalmers's core argument, then constrained speculation faces a challenge at this point.

Chapter 13 will treat this challenge in detail. This chapter only points out that it is currently the most important open problem for constrained speculation. This chapter does not claim that constrained speculation has already solved the problem of phenomenal consciousness. It claims that constrained speculation provides a method for precisely locating the problem: to characterize it as a boundary, marking the edge of structural–relational explanatory capacity, rather than as an object, namely a puzzle within explanatory capacity waiting to be solved. This is a posture of honesty rather than self-deception; see Chapter 13 for details.

A Key Methodological Posture

This section finally states a general posture: this book does not intend to solve all philosophical problems. It only commits itself to stating constrained speculation clearly, defending it rigorously, and locating it precisely. Its criterion of success is not being "flawless," but having concrete directions of failure: it must be able to tell readers under what conditions constrained speculation itself would fail.

This posture runs through the entire book. Every chapter will acknowledge its limits in the form of "unresolved problems of this chapter" or a similar treatment. This is not rhetorical modesty, but methodological consistency: constrained speculation requires itself to obey its own rules.

Literature That Must Be Engaged

This section does not need to add new literature. It is a methodological introspection, mainly a review of the first five sections of this chapter and a confirmation of their boundaries. But several interlocutors concerning the ethics of philosophical argument may be mentioned:

- Pierre Hadot, La philosophie comme manière de vivre (1995): philosophy as an honest practice of facing limits.
- David Chalmers, The Conscious Mind (1996): the hard problem of consciousness as an open challenge for constrained speculation.
- Hilary Putnam, Realism with a Human Face (1990): the later Putnam on the limits of philosophy.

Summary of This Section

This section has done two things. First, it acknowledged the three problems this chapter has not resolved: the ontological status of failure conditions themselves, cross-theoretical comparison, and the relation between form and reality, while indicating the chapters in which each will be treated. Second, it stated the failure condition of constrained speculation itself: the existence of a commitment that is ontologically indispensable yet cannot be handled by the four disciplines at all.

This statement is itself an application of the method of constrained speculation: it requires the method to state its own directions of failure. In doing so, this chapter achieves methodological consistency with itself.

1.8. Chapter Summary

This chapter has made explicit the method implicitly used in the five volumes and presented it as an independent philosophical position, defending it to a degree sufficient for entry into international philosophical discussion.

Constrained speculation is a distinctive realist position. Its core thesis ($CR1$) is: X is real $\Longleftrightarrow$ the commitments of X have definable failure conditions. This position unfolds through four interdependent disciplines: formalization, mechanization, observationalization, and failure-orientation. It continues the tradition of falsifiability, but completes an unfinished task of that tradition: elevating a methodological standard into an ontological discipline.

Constrained speculation differs from existing positions, including instrumentalism, constructive empiricism, strong realism, Cartwright's entity realism, Ladyman's OSR, Floridi's informational realism, and Rovelli's relational realism. It occupies a logical space not yet occupied among these positions. This is the basis for its status as a new position.

But constrained speculation itself also has failure conditions. If there exists an ontologically indispensable commitment that cannot be handled by the four disciplines at all, constrained speculation must revise itself. This chapter acknowledges that this is an open challenge, especially in the problem of phenomenal consciousness.

Connections with Later Chapters of the Book

This chapter lays the methodological foundation for the whole book. Every later chapter uses constrained speculation. The connections are as follows:

Chapter 2 uses constrained speculation to argue for ontological restraint: why the minimal postulate system of EQT has methodological advantage.

Chapter 3 develops failure conditions as an ontological discipline in detail. The open problems left by Section 1.6 of this chapter, namely the ontological status of failure conditions themselves and the relation between form and reality, are treated in Chapter 3.

Chapter 4 uses constrained speculation to treat the problem of physicalized semantics. The work of this chapter concerning how ontological commitments relate to philosophical concepts, such as Whitehead's actual occasions, is deepened there.

Chapter 5 uses constrained speculation to enter into dialogue with the spectrum of contemporary realism. The preliminary positioning of instrumentalism, constructive empiricism, and strong realism in this chapter is developed in detail in Chapter 5.

Chapter 6 synthesizes the first five chapters and gives the complete methodological position of constrained speculation. The distinction from Popper discussed in this chapter is synthetically treated in Chapter 6.

The later parts, from Part II to Part V, all work on the methodological foundation established in Part I. The argument concerning ontological singularity in Chapter 17, the metaphysical peak of the whole book, directly applies constrained speculation: it points out that EQT encounters an ontological singularity at the failure conditions of its own core commitments. The argument concerning the is–ought gap in Chapter 20 also uses constrained speculation: it downgrades Hume's guillotine from an ontological problem to a problem of methodological discipline.

A Final Reflection

This chapter faces a special difficulty in writing: it must state a new methodological position through concrete argument, but any concrete argument already presupposes some method. This difficulty resembles the famous dilemma of Wittgenstein's "Tractatus Logico-Philosophicus": any statement about language uses language.

This chapter responds to this difficulty through two levels. At the level of conceptual dependence, this chapter begins from already accepted methodologies, such as Popper, Lakatos, and Quine, and gradually introduces the new content of constrained speculation. At the level of ontological commitment, this chapter proposes constrained speculation as a whole and applies it to examine itself. This two-level structure prevents the chapter from floating without ground, since it inherits methodological foundations, and also prevents it from becoming circular, since it does not merely use constrained speculation to prove constrained speculation.

This mode of response is itself a demonstration of the method of constrained speculation: it shows that this method can handle the fundamental difficulties of philosophical argument, including the problem of grounding methodology itself.

In this sense, this chapter is both the methodological foundation of the whole book and the first complete application of the method of constrained speculation. It fulfills the promise made at the beginning: to allow a new method of natural philosophy to show its independence and rigor.

1.9. Methodological Reflection on This Chapter

This chapter establishes the method of constrained speculation as the methodological foundation of the whole book. This section, as a closing reflection, considers the position, limits, and connections of this method with later chapters.

Methodological Position: Between the Analytic and Continental Traditions

Constrained speculation is not an entirely new invention. It is built upon multiple precursor traditions. From the analytic philosophical tradition, it inherits the spirit of Popper's falsifiability, Quine's concept of ontological commitment, Carnap's methodological rigor of logical construction, and Brandom's inferentialist rejection of "correspondence." From the Continental tradition, it inherits Husserl's intentionality, Heidegger's emphasis on "being-in-the-world," and the later Wittgenstein's turn toward "use" rather than "correspondence."

The contribution of constrained speculation is to systematize these resources into four disciplines ($CS1$–$CS4$), so that the methodological rigor of metaphysical work can be precisely evaluated. This systematization crosses the analytic–Continental divide: it begins from the methodological rigor of analytic philosophy, but its ontological position, namely the rejection of the separation between reality and cognition, is closer to the Continental tradition. This kind of crossing has precedents in the history of philosophy: the later Wittgenstein, the later Quine, Sellars, and McDowell all carried out similar work.

But the work of this chapter is not a simple synthesis. It has two core innovations. First, it elevates falsifiability from an epistemological discipline to an ontological discipline; this elevation is developed in detail in Chapter 6. Second, it establishes failure conditions as an internal requirement of legitimate ontological commitment; this requirement is developed in detail in Chapter 3. These two innovations give constrained speculation a substantive difference from the simple pattern of "scientific methodology plus philosophical application."

The Ground of This Chapter's Independence from EQT

If this chapter cited the physical details of EQT as the ground of its argument, constrained speculation would be regarded as valid only insofar as it depended on EQT. But the fundamental claim of this book is that constrained speculation is an independent philosophical position, and that EQT is only one of its applications. To support this claim, this chapter uses purely philosophical argument. Section 1.3 uses the "energy quantum density field $\rho_f(x,t)$" as an example when explaining the four disciplines, but this is only an example, not the ground of the argument. This independence allows this chapter to be taken seriously in international analytic-philosophical discussion.

Limits Honestly Acknowledged in This Chapter

This chapter acknowledges several limits. First, the mutual relations among the four disciplines ($CS1$–$CS4$) require more refined treatment: are they mutually independent? Do they have some form of priority relation? This chapter does not fully develop this issue and leaves it to the concrete applications in later chapters. Second, the relation between constrained speculation and philosophical traditions, especially Chinese philosophical traditions, is treated only preliminarily in this chapter; detailed dialogues will unfold in Chapters 8 and 19. Third, Section 1.6 explicitly states the failure condition of constrained speculation itself: the existence of a commitment that is ontologically indispensable but cannot be handled by the four disciplines at all. This failure condition currently faces a challenge in the problem of phenomenal consciousness, which Chapter 13 will treat in detail.

This chapter does not claim that constrained speculation is absolute or final. It claims that constrained speculation is a defensible, failable, and refinable methodological position.

Connections with Later Chapters

The methodological foundation established in this chapter will be applied and refined in later chapters. Chapter 2 unfolds the connection between $CS1$, the discipline of formalization, and $PR1$, the principle of restraint. Chapter 3 thoroughly defends $CS4$, the discipline of failure-orientation, and elevates failure conditions from a methodological concept into an ontological criterion ($CR2$). Chapter 4 introduces new semantic tools to treat the cross-conceptual relational problems that arise in the application of $CS2$, the discipline of mechanization. Chapter 5 integrates the methodological work of Chapters 1–4 into a complete ontological position. Chapter 6 systematically unfolds the ontological application of $CS3$ and $CS4$ as the synthetic chapter of Part I.

This progression makes the six chapters of Part I an organic whole: the method of constrained speculation is not established once and for all, but gradually concretized through six chapters. More broadly, the method of constrained speculation runs through the entire book: the ontological theses of Part II, the relational-ontological theses of Part III, the treatment of metaphysical boundaries in Part IV, and the normative work of Part V are all built upon the four disciplines established in this chapter. If the four disciplines fail, the work of the whole book loses its methodological ground. This broad application makes the present chapter the methodological foundation of the whole book.

2. Minimal Postulates and Ontological Restraint

2.1. Introduction to This Chapter

In establishing the physical foundations of EQT, the "EQT Foundational Volume" used one method: the minimal postulate system. The whole physical work of EQT is built upon four postulates, $A1$–$A4$, and one Meta-Norm, M. The distinctive feature of this method is minimality: it commits itself to the least possible ontological content, while requiring these commitments to be strong enough to derive all the required physical conclusions.

This chapter elevates this method from its physical application in the "EQT Foundational Volume" into a general ontological discipline: ontological restraint. The core thesis ($PR1$) is that ontology should commit itself to the least possible ontological content, while requiring these commitments to be strong enough to address the ontological problems that need to be addressed. Restraint is not simple "subtraction," but a balance between minimality and sufficiency.

This elevation involves a series of questions that must be treated seriously. Ontological restraint sounds like Ockham's razor, namely "do not multiply entities beyond necessity." What is its relation to Ockham's razor? Does restraint lead to a "poor ontology," so that certain important phenomena, such as consciousness, normativity, and value, are omitted because of restraint? How does restraint handle "unavoidable ontological richness"? If ontological content is indeed rich, is restraint resisting this richness?

This chapter faces these objections with the aim of transforming restraint from a rhetorical slogan into an operational ontological discipline. As a methodological principle, restraint requires operational content: what counts as restraint? How should the degree of restraint be evaluated? Under what conditions should restraint give way? This chapter responds to these questions through precise statement, examples, and serious treatment of objections.

This chapter builds upon the methodological foundation laid in Chapter 1, "Constrained Speculation as Method," and prepares for the development of Chapter 3, "Failure Conditions as an Ontological Discipline." Together, the three chapters constitute the methodological core of Part I.

The discussion proceeds as follows. Section 2.1 moves from the minimal postulate system to ontological restraint; Section 2.2 develops the methodological grounds of restraint; Section 2.3 treats the balance between restraint and richness; Section 2.4 develops the philosophical genealogy of restrained ontology, namely Ockham–Spinoza–Quine–Lewis, and enters into dialogue with the Chinese philosophical tradition; Section 2.5 concludes the chapter.

2.2. From the Minimal Postulate System to Ontological Restraint

Starting Point: The Practice of the "EQT Foundational Volume"

In establishing the physical foundations of EQT, the "EQT Foundational Volume" used the method of the minimal postulate system. The whole physical work of EQT is built upon four postulates and one Meta-Norm:

- *A*1: The Energy Quantum Density Field $\rho_f(x,t)$ is the unique basic ontological object.
- *A*2: The universal frequency–energy equivalence $E = h\nu$.
- *A*3: The Master Equation, which governs the evolution of $\rho_f(x,t)$.
- *A*4: The Lorentzian Resonance Kernel, which specifies the structure of interaction.
- *M*: The Meta-Norm: Mechanism Priority, falsifiability, and the requirement of formalization.

These five elements, namely four postulates plus one Meta-Norm, constitute the complete ontological foundation of EQT. All other physical contents, such as particles, interactions, cosmic evolution, and the mind–matter relation, are derived from this foundation.

The distinctive feature of this practice is minimality: it commits itself only to the least possible ontological content, while requiring these commitments to be strong enough to derive all the required physical conclusions.

The Meaning of the "Minimal Postulate System" as a Method

The "minimal postulate system" as a method has several layers of meaning:

Meaning One: The Explicitness of Ontological Commitments

Each postulate is an explicit ontological commitment: it states a certain being, such as the ρ_f field, or a certain relation, such as $E = h\nu$. These commitments are not vague: they can be precisely stated, rigorously tested, and explicitly accepted or rejected.

Meaning Two: The Minimality of Ontological Commitments

The number of postulates is kept minimal: it does not exceed what is required to complete the task. If three postulates are sufficient to derive all the required physical conclusions, a fourth is not introduced; if four are not enough, then a fourth is introduced but not a fifth.

Meaning Three: The Closure of Ontological Commitments

The postulate system is closed: it contains all foundational commitments. Other contents, such as the properties of elementary particles or the evolution of cosmology, are not introduced as additional postulates, but are derived from the foundational postulates.

Meaning Four: The Role of the Meta-Norm

The Meta-Norm M is not another postulate; it is the rule concerning how the postulate system is to be used and developed. M requires every postulate to have a falsifiable structure, a mechanized form, and a formalized expression. This Meta-Norm ensures that the postulate system is rigorously evaluable, rather than an arbitrary philosophical statement.

Ontological Restraint as a Philosophical Principle

The starting point of this chapter is to elevate the "minimal postulate system" from the physical method of the "EQT Foundational Volume" into a general ontological discipline: ontological restraint.

Core thesis $PR1$: ontology should be restrained: it should commit itself to the least possible ontological content, while requiring these commitments to be strong enough to address the ontological problems that need to be addressed. Restraint is not simple "subtraction," but a balance between minimality and sufficiency.

Let us analyze this thesis item by item.

First, "the least possible ontological content": ontology should commit itself to as few ontological objects, relations, and categories as possible. Every additional ontological commitment requires justification: is it necessary? Can it be replaced by existing commitments?

Second, "strong enough to address the ontological problems that need to be addressed": restraint is not unlimited reduction; it must preserve the capacity to address ontological problems. If a certain ontological problem, such as the ontological status of consciousness or the ground of normativity, requires a commitment in order to be addressed, it cannot be ignored for the sake of restraint.

Third, "the balance between minimality and sufficiency": restraint is not a single principle; it is the balance between two requirements. Minimality requires fewer commitments; sufficiency requires commitments strong enough. The balance between the two requires methodological judgment.

As an ontological discipline, this thesis has a clear methodological function: it provides an evaluative standard for ontological work.

Difference from Ockham's Razor

Ontological restraint sounds like Ockham's razor: "do not multiply entities beyond necessity" (entia non sunt multiplicanda praeter necessitatem). The two are indeed similar. But there are several important differences between $PR1$ and Ockham's razor:

Difference One: Ockham's razor is a principle of simplification; $PR1$ is an ontological discipline.

As a principle of simplification, Ockham's razor tells us that, when faced with multiple explanations, we should choose the simplest one. It is a criterion of selection.

As an ontological discipline, $PR1$ tells us that ontological work should satisfy the balance between minimality and sufficiency. It is a criterion of construction: it guides how ontology should be built, not merely how to choose among already established ontologies.

Difference Two: Ockham's razor is heuristic; $PR1$ is an application of constrained speculation.

As a heuristic, Ockham's razor lacks a strict methodological ground. It is an empirical rule concerning "theory choice."

As an application of constrained speculation, $PR1$ has a strict methodological ground: it is built upon the claim that "ontological commitments must have definable failure conditions" (Chapter 3, $CR2$). Every ontological commitment must satisfy the four disciplines of constrained speculation (Chapter 1); the more commitments are introduced, the more content must satisfy these disciplines. Restraint is the natural result of this requirement.

Difference Three: Ockham's razor is one-directional; $PR1$ is two-directional.

Ockham's razor only tells us not to add: choose the simple, remove the superfluous. It does not treat the question of when one should add.

$PR1$ is two-directional: it requires both restraint, namely not adding what is superfluous, and sufficiency, namely adding when necessary. If a certain ontological problem requires a commitment in order to be addressed and the current commitments are insufficient, $PR1$ requires the introduction of a new commitment. This bidirectionality prevents $PR1$ from leading to a "poor ontology": it can address the ontological problems that need to be addressed.

An Example: From Minimal Postulates to Concrete Applications

Let us consider an example to show how ontological restraint is applied.

Consider how the EQT postulate system addresses ontological problems:

Problem: The Ontological Status of Elementary Particles

Traditional physics often treats elementary particles, such as electrons and quarks, as foundational ontological objects: they are irreducible things. But this commitment requires each particle to be treated as an independent ontology: electrons, quarks, neutrinos, photons, and so on all require independent commitments.

The practice of EQT is to commit only to the ρ_f field ($A1$) as the foundational ontological object: elementary particles are excitation modes of the ρ_f field. This gives all elementary particles a shared ontological ground and eliminates the need for separate commitments.

Problem: The Ontological Status of Fundamental Interactions

Traditional physics often treats the four fundamental interactions, namely gravity, electromagnetism, the strong force, and the weak force, as independent ontological processes: each interaction requires an independent ontological commitment.

The practice of EQT is to take $A3$, the Master Equation, and $A4$, the Lorentzian Resonance Kernel, as the only dynamical rules: the four fundamental interactions are manifestations of this dynamics in different frequency bands. This unifies the ontological ground of the fundamental interactions and eliminates the need for separate commitments.

Problem: The Ontological Status of Spacetime

Traditional physics often treats spacetime as an independent ontological dimension: beings exist in spacetime.

The practice of EQT is to appeal to the internal structure of $\rho_f(x,t)$ ($A1$). Spacetime as an ontological dimension is an internal dimension of the ρ_f field and does not need to be treated as an independent ontological commitment.

These examples show how the EQT postulate system addresses ontological problems through minimal postulates: it does not introduce an independent commitment for each ontological problem, but derives them from foundational commitments. This is precisely the application of ontological restraint.

Objections and Replies

Objection One: The "minimal postulate system" is a method of physics. Is it an overgeneralization to elevate it into an ontological discipline of philosophy?

Reply: The elevation carried out in this chapter is not a simple extension of terminology; it is grounded in methodology.

Specifically:

- The minimal postulate method in physics is built upon the requirements of physics: a theory should be able to derive the largest number of observable predictions from the smallest number of assumptions.
- Ontological restraint is built upon the requirements of constrained speculation: every ontological commitment must satisfy definable failure conditions (Chapter 3, $CR2$), and the more commitments there are, the more conditions must be satisfied.

The two share a methodological ground: the requirement of the legitimacy of commitments naturally leads to the minimization of the number of commitments. Physics and ontology are different domains of application of this shared ground.

Furthermore, the methodological traditions of physics and ontology overlap: both attempt to establish work that can be rigorously evaluated. Elevating the minimal postulate method of physics into a principle of ontological restraint is a legitimate methodological extension.

Objection Two: PR1 includes the requirement of "sufficiency," namely that ontological commitments must be "strong enough." But how is "strong enough" to be judged? Does this not introduce vagueness into the principle of restraint?

Reply: This objection points to an important point of refinement. The judgment of "sufficiency" must be developed in detail.

Sufficiency has three layers of meaning:

- Empirical sufficiency: ontological commitments must be strong enough to address observable phenomena. If certain phenomena, such as certain physical processes or certain mental states, require ontological commitments in order to be addressed and the current commitments are insufficient, then new commitments need to be introduced.

- Theoretical sufficiency: ontological commitments must be strong enough to address theoretical problems, such as the relations between different theories and the connections among different domains.

- Methodological sufficiency: ontological commitments must be strong enough to support methodological requirements, such as the four disciplines of constrained speculation.

These criteria make "sufficiency" no longer vague: it has operational content. Each ontological problem can be precisely evaluated: can the existing commitments handle it? If not, what new commitment is needed? How does the new commitment satisfy the method of constrained speculation?

Vagueness indeed exists: judgment involves methodological judgment, such as which phenomena require ontological commitments and which theoretical problems require ontological commitments. But this vagueness is manageable: it can be gradually refined through discussion, research, and criticism, rather than being a vagueness that is impossible to approach in principle.

Objection Three: Ontological restraint sounds very "economical"; it seems insufficiently open to ontological problems such as consciousness, normativity, and value. These rich ontological problems require rich commitments.

Reply: This objection points to an important ontological issue. The reply of this section is as follows.

$PR1$ does not advocate "economy to the extreme"; it advocates a balance between minimality and sufficiency. If consciousness, normativity, and value require commitments in order to be addressed, then sufficiency requires the introduction of these commitments. If certain commitments are superfluous, because they can be replaced by other commitments or have no actual function, then minimality requires their removal.

Specifically:

- Consciousness: this book treats consciousness through the Ψ framework (Chapter 13). It uses the dynamics and physical proxies of the ρ_f field and does not introduce a "mental entity" as an additional ontology. This is an application of restraint.

- Normativity: this book treats normativity through the method of constrained speculation (Chapter 20). It does not introduce a "normative entity" as an additional ontology, but locates normative statements as a form of commitment. This is also an application of restraint.

- Value: this book treats value through the finitude of meaning (Chapter 21). It treats meaning as the emergence of a structure of difference and does not introduce a "value entity" as an additional ontology.

These treatments show how $PR1$ is applied: rich ontological problems are addressed through the application of existing commitments, rather than by introducing a new ontological commitment for each problem. This is precisely the balance of restraint + sufficiency.

Literature That Must Be Engaged

- William of Ockham, Summa Logicae (ca. 1323): the origin of Ockham's razor.

- Quine, "On What There Is" (1948): the discussion of ontological commitment.

- Karl Popper, Conjectures and Refutations (1963): the methodology of theory choice.

- Larry Laudan, Progress and Its Problems (1977): the evaluation of research programmes.

- Chapter 1 on the method of constrained speculation (self-reference).
- Chapter 3 on failure conditions as an ontological discipline (self-reference).

Summary of This Section

This section has done four things. First, it reviewed the minimal postulate system in the "EQT Foundational Volume" ($A1$–$A4 + M$). Second, it systematized the meaning of the "minimal postulate system" as a method: explicitness, minimality, closure, and Meta-Norm. Third, it proposed the core thesis $PR1$: ontological restraint as a methodological discipline. Fourth, it distinguished $PR1$ from Ockham's razor.

The next section develops the methodological grounds of restraint.

2.3. The Methodological Grounds of Restraint

Aim of the Argument

Section 2.1 stated $PR1$ as an ontological discipline. But $PR1$ requires methodological grounds: why should ontology be restrained? This section provides those grounds.

The argument has three layers. The first layer is the requirement of the method of constrained speculation: every ontological commitment must satisfy definable failure conditions, and the more commitments there are, the more content must be satisfied. The second layer is the feasibility of ontological testing: restraint makes ontology rigorously evaluable. The third layer is the possibility of ontological change: restraint allows ontology to be revised in light of new discoveries.

Argument One: The Requirement of the Method of Constrained Speculation

Chapter 1 established the method of constrained speculation, and Chapter 3 establishes failure conditions as an ontological discipline ($CR2$: X is real $\Longleftrightarrow X$ has commitments with definable failure conditions). These methods impose clear requirements on ontological commitments:

- Every ontological commitment must have definable failure conditions.

- Failure conditions must satisfy the requirements of dimensionality, proxy, and threshold (Chapter 3, Section 3.3).
- Commitments must satisfy the four disciplines: formalization, mechanization, observationalization, and failure-orientation (Chapter 1, Section 1.3).

These requirements mean that every ontological commitment involves substantive methodological work. Each commitment requires:

- Formalized expression, whether mathematical or conceptual.
- Mechanism: how it functions in a process.
- Proxy: how it can be empirically identified.
- Direction of failure: what would make it fail.

This work is not simple: each commitment involves extensive research and defense.

The methodological consequence of restraint is as follows.

If every commitment involves substantive methodological work, then the more commitments there are, the heavier the methodological burden becomes. Specifically:

- Each new commitment requires independent formalization, mechanization, observationalization, and failure-orientation.
- New commitments must be coordinated and integrated with existing commitments.
- New commitments must be independently tested in research.

If an ontology introduces a large number of commitments, the methodological work becomes uncontrollable: many commitments will lack falsifiable structures or conflict with other commitments.

Restraint as a methodological discipline avoids this situation: it requires every commitment to be strictly evaluated and accepts only genuinely necessary commitments. This makes ontological work controllable, evaluable, and developable.

Argument Two: The Feasibility of Ontological Testing

Ontology as a philosophical field faces a difficulty: it involves statements concerning beings themselves. How are such statements to be tested?

If the number of ontological commitments is small, each commitment can be examined in detail: what is its content? How does it connect with experience? Under what conditions does it fail? These questions can be treated precisely.

If the number of ontological commitments is large, each commitment can only be treated superficially: its content is unclear, its connection with experience is indeterminate, and its failure conditions are vague. This makes ontology degenerate into an arbitrary collection of statements: readers find it difficult to evaluate which are reasonable and which are not.

Certain philosophical systems in history, such as some versions of Whiteheadian process philosophy and some versions of German Idealism, face this difficulty: they introduce a large number of ontological categories, but the content and legitimacy of each category are difficult to test rigorously. As a result, these systems become works of art: admirers may appreciate their overall beauty, but it is difficult to enter into methodological discussion.

Restraint allows ontology to avoid this difficulty. It requires the number of commitments to remain minimal and each commitment to be rigorously examined. This keeps ontology as a rigorously evaluable philosophical field: readers can precisely judge the legitimacy of each commitment.

Argument Three: The Possibility of Ontological Change

Ontology is not static: it needs to be revised in light of scientific discoveries, philosophical discussions, and criticisms. Restraint makes such revision possible.

Specifically:

When commitments are few, revision can be concentrated: one can precisely identify which commitment needs revision and how it should be revised. The content of each commitment is clear, and the direction of revision is explicit.

When commitments are many, revision becomes difficult: the interrelations among commitments mean that modifying any one of them may affect the others. Ontology becomes rigid: it becomes difficult to refine or update.

Historically, rigid philosophical systems often appear in systems that introduce a large number of ontological commitments. At the moment of initial introduction, they appear "rich," but in later revision they become unmanageable. New scientific discoveries or new philosophical criticisms cannot be effectively integrated, and the system as a whole becomes marginalized.

Restraint keeps ontology alive: it keeps ontology as work that can be refined, updated, and revised in light of new discoveries. This is the ground for the continued development of ontology as a philosophical field.

Synthesis of the Three Layers of Argument

The three layers of argument support ontological restraint from different dimensions:

- Methodological requirement: the requirements imposed by constrained speculation on every commitment make the burden heavier as the number of commitments increases.
- Feasibility of testing: restraint keeps ontology as a field that can be rigorously evaluated.
- Possibility of change: restraint keeps ontology alive as work that can be revised.

Together, these three dimensions support $PR1$ as an ontological discipline. Restraint is not the rhetorical claim that "simplicity is better"; it is an ontological discipline with methodological grounds.

Restraint Is Not "Poverty"

A clarification is needed: restraint is not the same as a "poor ontology."

Specifically:

- Poor ontology: commitments are few, but insufficient to address ontological problems. This is a case in which restraint fails, because it violates the requirement of "sufficiency" in $PR1$.
- Restrained ontology: the number of commitments is kept minimal, but the content of these commitments is strong enough to address ontological problems. This is precisely what $PR1$ requires: the balance between minimality and sufficiency.

A restrained ontology can address all ontological problems that need to be addressed; it simply does not introduce an independent commitment for each problem, but addresses them through the application of existing commitments. This is the strength of restraint, not a limitation.

An Example: The Restraint of EQT

Let us consider EQT as an example of restrained ontology.

EQT has an extremely small number of commitments: 4 postulates + 1 Meta-Norm. Yet EQT addresses a large number of ontological problems:

- The ontological status of elementary particles, as excitations of the ρ_f field.
- The ontological status of fundamental interactions, as frequency-band manifestations of the dynamics of the ρ_f field.
- The ontological status of spacetime, as an internal dimension of the ρ_f field.
- The ontological status of cosmic evolution, as the historical development of the ρ_f field.
- The ontological status of consciousness, as a highly complex gradient structure, Chapter 13.
- The ontological status of observation, as a process of co-evolution within the same field, Chapter 14.

All these ontological problems are addressed through 5 foundational commitments. EQT does not introduce an independent commitment for each problem, but treats them through the application of foundational commitments.

This is the strength of restraint: a small number of foundational commitments addresses many problems. If EQT introduced an independent commitment for each problem, such as particles as an independent ontology, interactions as an independent ontology, spacetime as an independent ontology, and so on, the number of commitments would grow explosively, and the legitimacy of each commitment would become difficult to evaluate rigorously.

Objections and Replies

Objection One: Argument One, concerning the requirement of the method of constrained speculation, is built upon the work of Chapters 1 and 3. But this chapter should be able to stand independently: its legitimacy should not depend entirely on other chapters.

Reply: As a core chapter of Part I, this chapter is indeed built upon Chapters 1 and 3. This is a legitimate relation among chapters.

The chapters of any philosophical monograph build upon preceding chapters and need not stand completely independently.

But the content of this chapter does not depend entirely on other chapters. Specifically:

- The argument from methodological requirement uses the content of Chapters 1 and 3. This is a legitimate internal reference.
- The argument from the feasibility of testing is independent of other chapters. It uses a general discussion of ontology as a philosophical field.
- The argument from the possibility of change is also independent of other chapters. It uses a general discussion of ontology as living philosophical work.

The latter two arguments give this chapter independent force: even if readers do not accept the work of Chapters 1 and 3, they may still accept $PR1$ on the basis of the arguments concerning ontological testing and change.

Furthermore, the overall structure of Part I is a mutually supporting network: Chapters 1, 2, and 3 together constitute the methodological foundation. Each chapter is built upon the others, but each also contributes to the others. This kind of mutual support is a legitimate structure for a philosophical monograph.

Objection Two: Argument Two, concerning the feasibility of testing, uses philosophical systems such as Whiteheadian process philosophy and German Idealism as examples to support restraint. But these examples may be interpreted differently: their "difficulty of testing" may arise from different methodological traditions, rather than from an excessive number of commitments.

Reply: This objection points to an important refinement. The difficulty of testing systems such as Whiteheadian process philosophy and German Idealism indeed has multiple causes, including differences in methodological tradition, historical context, and readerly expectations.

But the large number of commitments is one important factor in this difficulty. Specifically:

- Whitehead's categorial system contains eight Categories of Existence, twenty-seven Categories of Explanation, and so on. The relations among these categories are complex and difficult to evaluate precisely. The difficulty of eternal objects has already been developed in Chapter 8, Section 8.5.

- Hegel's dialectical system contains a large number of ontological categories, such as being, essence, and concept. Their relations are built upon dialectical development, and this development is difficult for external readers to track precisely.

A large number of commitments is not the only factor, but it is an important factor. Restraint reduces the influence of this factor and keeps ontology as work that can be rigorously evaluated.

Furthermore, the argument of this section is not an attack on Whitehead or Hegel. It points out the consequences of a certain methodological choice, namely introducing a large number of ontological commitments. This consequence makes the relevant philosophical systems difficult to refine and revise in contemporary discussion. Restraint as a methodological choice avoids this consequence.

Objection Three: The principle of restraint seems to be built upon the methodology of analytic philosophy. It assumes that philosophy should be rigorously evaluable like science. But other philosophical traditions, such as phenomenology, hermeneutics, and Continental metaphysics, have different methodologies. Does restraint apply to these traditions?

Reply: This objection points to an important methodological issue. The reply of this section is as follows.

$PR1$ as a principle of restraint is built upon the method of constrained speculation. This method inherits the tradition of rigor in analytic philosophy, but it also connects with Chinese philosophical traditions, especially the methodological rigor expressed in the Confucian posture of "knowing that it cannot be done and yet doing it." Therefore, $PR1$ is not limited to the analytic philosophical tradition.

But $PR1$ does indeed assume a certain methodological rigor: ontological commitments should be evaluable. This assumption may differ from certain philosophical traditions, such as some versions of phenomenology and hermeneutics.

Specifically:

- The method of phenomenology, such as epoché and reduction, gives ontological work a phenomenological ground and is not limited to the evaluation of commitments.

- The method of hermeneutics, such as understanding and dialogue, gives ontological work a hermeneutic dimension and is not limited to propositional statements.

This section does not claim that $PR1$ excludes the legitimacy of these traditions. It claims that $PR1$ is a reasonable ontological discipline within the framework of the method of constrained speculation. If certain traditions use different methodological frameworks, they may not need $PR1$. But the work of this book is built upon the method of constrained speculation, and $PR1$ is an application of that method.

Furthermore, this book seeks dialogue with multiple philosophical traditions, including Section 4.5 of Chapter 4, Chapter 8 on the Chinese tradition, and Chapter 13 on phenomenology. This dialogue does not require all traditions to accept the same methodology. Rather, it makes explicit the meaning of methodological choices in each discussion.

Literature That Must Be Engaged

- Chapter 1 on the method of constrained speculation (self-reference).
- Chapter 3 on failure conditions as an ontological discipline (self-reference).
- Alfred North Whitehead, Process and Reality (1929)
- G. W. F. Hegel, Phenomenology of Spirit (1807), Science of Logic (1812–1816)
- Edmund Husserl, Ideas Pertaining to a Pure Phenomenology (1913)

Summary of This Section

This section has done four things. First, it gave the three-layer methodological ground of $PR1$: the requirement of the method of constrained speculation, the feasibility of ontological testing, and the possibility of ontological change. Second, it clarified that restraint is not poverty: restraint requires sufficiency. Third, it used EQT as an example of restrained ontology to show the strength of restraint. Fourth, it responded to possible objections, especially concerning the independence of the chapter, the interpretation of examples, and applicability across traditions.

The next section develops the balance between restraint and ontological richness.

2.4. The Balance between Restraint and Richness

The Form of the Problem

Ontological restraint requires the fewest commitments. But ontology also needs to address rich ontological problems: consciousness, normativity, value, culture, society, and other ontological phenomena. There is a tension between these two requirements: restraint seems to conflict with richness.

This section develops a treatment of this tension. The core question is: does restraint make ontology insufficiently rich? How should restraint and richness be balanced?

Key Distinction: Richness of Ontological Commitment vs Richness of Ontological Application

The core of this section is a distinction:

Richness of ontological commitment: the number and diversity of commitments in an ontology, namely how many foundational commitments there are and how they are related. Richness of ontological application: the number and diversity of problems addressed by an ontology, namely how many different ontological problems it can address and how deeply it addresses them.

The two need not move together. Specifically:

- An ontology may have a small number of foundational commitments and rich applications, by applying foundational commitments to a large number of problems through reasoning.
- An ontology may have a large number of foundational commitments and poor applications, because the commitments lack integration and each commitment can address only a specific problem.

$PR1$ requires restraint in ontological commitment, while allowing richness in ontological application. This distinction allows restraint and richness to coexist.

An Example: Euclidean Geometry

Consider an analogy: Euclidean geometry.

Minimality of commitments: Euclidean geometry has only 5 postulates and a small number of axioms. This is an extreme form of commitment minimality.

Great richness of application: Euclidean geometry can address a large number of geometric problems, such as the properties of triangles, the geometry of circles, solid geometry, analytic geometry, and so on. These applications far exceed the content of the 5 postulates.

The reason for this structure is that a small number of foundational commitments are applied to a large number of problems through reasoning, including deduction, construction, and proof. The minimality of commitments allows each postulate to be rigorously stated, examined, and defended; the richness of application allows the whole system to address many problems.

Euclidean geometry displays a model: minimal postulates + rich reasoning = a powerful system. This model is not limited to geometry. It applies to any rigorous ontological work.

EQT as an Example of Restraint + Richness

EQT is another example that shows the balance between restraint and richness.

Minimality of commitments: EQT has only 4 postulates and 1 Meta-Norm. This is an extreme minimality of foundational commitments.

Great richness of application: EQT treats a large number of ontological problems:

- Elementary-particle physics, with particles as excitations of the ρ_f field.
- Fundamental interactions, as dynamics in different frequency bands.
- Cosmology, with cosmic evolution as multi-band freezing.
- The mind–matter relation, with consciousness as a highly complex gradient structure, Chapter 13.
- Theory of observation, with observation as co-evolution within the same field, Chapter 14.
- The problem of time, with time as $\partial\rho/\partial t$, Chapter 15.
- Metaphysical singularity, with the gradient terminal phase as the ontological limit, Chapter 17.

- The finitude of meaning, with meaning as the emergence of a structure of difference, Chapter 21.

All these applications are built upon 5 foundational commitments. Each application involves its own argument, mechanization, and observationalization, but the foundational commitments of all applications remain minimal.

This is precisely the balance between restraint and richness: foundational commitments are restrained, while applications are rich.

Handling Problems That "Seem to Require New Commitments"

Ontological problems often seem to require new ontological commitments. This section develops how judgment should be made between restraint and sufficiency.

Example One: The Ontological Status of Consciousness

Consciousness is one of the most complex ontological problems in the history of philosophy. It seems to require some kind of ontological commitment: some "mental entity," "subjective experience," "intentionality," and so on.

But through the work of Ψ within the EQT framework (Chapter 13), consciousness can be treated through existing commitments, namely the dynamics of the ρ_f field:

- Consciousness-related states as highly complex gradient structures of the ρ_f field.
- Subjective experience as different projections of the same process, the core thesis of Chapter 13.
- Intentionality as a form of co-evolutionary relation.

This treatment applies existing commitments and does not introduce a new ontological commitment. It satisfies both restraint, by not increasing the number of commitments, and sufficiency, by addressing the problem of consciousness.

Example Two: The Ontological Status of Normativity

Normativity, including ought, value, and ethics, seems to require some kind of ontological commitment: some "normative entity," "value fact," and so on.

But through the method of constrained speculation (Chapter 20), normativity can be treated through existing commitments:

- Normative statements as a form of commitment, subject to the same methodological discipline as factual statements.
- The content of norms is refined through falsifiable structures.
- "Ought" as the direction of commitment, rather than an independent ontological content.

This treatment applies the existing methodology and does not introduce a new ontological commitment. It satisfies both restraint and sufficiency.

Example Three: The Ontological Status of Abstract Objects

Abstract objects, such as numbers, propositions, and possible worlds, seem to require some kind of ontological commitment: they seem to be additional ontological content.

But through the distinction between formal constraints and real objects (Chapter 3, Section 3.5):

- Abstract objects function as formal constraints, not as real objects.
- They function as carriers of real commitments.
- There is no need to introduce an independent ontological commitment for them.

This treatment is completed through existing ontological categories, namely formal constraints versus real objects, and does not introduce a new ontological commitment.

These three examples show how $PR1$ is applied: many ontological problems that seem to require new commitments can be handled through the application of existing commitments. Restraint does not weaken the capacity of ontology to treat problems; rather, it enables ontology to achieve richness through application.

Criteria for Introducing New Commitments

Some situations do indeed require new ontological commitments. This section gives the criteria for introducing new commitments.

Criterion One: Existing commitments cannot address a certain ontological problem.

If a certain ontological problem cannot be addressed at all within the framework of existing commitments, and application, reasoning, and extension all fail to provide a reasonable answer, then a new commitment is needed.

Specifically, this requires:

- The application of existing commitments has been attempted in detail.
- These attempts clearly show the insufficiency of existing commitments.
- The new commitment can address the problem.

Criterion Two: The new commitment is consistent with existing commitments.

The new commitment must not conflict with existing commitments; otherwise, it would undermine the overall consistency of the ontology. The new commitment must have a clear connection and compatibility with existing commitments.

Criterion Three: The new commitment satisfies the method of constrained speculation.

The new commitment must satisfy the four disciplines of constrained speculation: formalization, mechanization, observationalization, and failure-orientation. If the new commitment cannot satisfy these disciplines, it is not a legitimate ontological commitment.

Criterion Four: The new commitment is minimally necessary.

The new commitment should be minimally necessary: it should be the simplest commitment capable of addressing the problem. More commitments should not be introduced for the sake of "perfecting" the framework.

Together, these four criteria constitute the standard for introducing new commitments. They make the introduction of a new commitment a strict methodological decision rather than an arbitrary philosophical choice.

Objections and Replies

Objection One: The distinction between "commitment vs application" sounds clever, but in practice it may conceal the problem. If application involves many inferential steps, auxiliary assumptions, and explanations, these are in fact implicit commitments.

Reply: This objection points to an important refinement. Specifically:

- Foundational commitments are the starting point of ontology: explicitly stated ontological content.
- Inferential steps are the process by which applications are derived from foundational commitments, using methodological tools such as deduction, construction, and explanation.
- Auxiliary assumptions are additional statements needed in applications: they should be explicitly stated and should satisfy the method of constrained speculation.

If an application involves a large number of auxiliary assumptions, these assumptions are in fact implicit commitments: they should be made explicit and tested. This is precisely the application of the principle of restraint: large numbers of assumptions should not be concealed under the name of "application details."

Specifically, the work of this book requires:

- Foundational commitments are explicitly stated, such as the 4 postulates of EQT.
- Inferential steps are explicit, using physical, mathematical, and philosophical reasoning.
- Auxiliary assumptions are explicit, being stated clearly in applications.

This explicitness prevents the distinction between "commitment vs application" from concealing problems: it makes all content rigorously evaluable.

Objection Two: Your criteria, especially "existing commitments cannot address a certain ontological problem," may be abused. Someone may claim that "existing commitments are insufficient" in order to introduce a new commitment, while in fact existing commitments are sufficient.

Reply: This objection genuinely points to the risk of abuse. The reply of this section is as follows.

The criterion "existing commitments cannot address a certain problem" requires a strict defense: one must show in detail how existing commitments have been applied, why the application is insufficient, and how the new commitment addresses the problem.

Specifically:

- Detailed application: existing commitments must be applied to the problem in detail, rather than merely being declared "insufficient" in the abstract.

- Analysis of insufficiency: one must show the defect of the application, namely what content cannot be handled.

- Advantage of the new commitment: one must show how the new commitment handles the content omitted by existing commitments.

This strictness makes abuse difficult: any claim to introduce a new commitment requires detailed defense. This is an application of the method of constrained speculation.

Furthermore, the examples in this book, such as the Ψ framework treating consciousness, the method of constrained speculation treating normativity, and formal constraints treating abstract objects, show the strength of application. These examples function as a standard: anyone who claims to introduce a new commitment must show that existing commitments truly cannot address the problem in the way the EQT framework does.

Objection Three: The principle of restraint seems overly biased toward physics. It suits the EQT framework, but may not suit certain ontological problems, such as purely philosophical concepts.

Reply: This objection points to an important methodological issue. The reply of this section is as follows.

$PR1$ as a principle of restraint is not limited to physics. It applies to any rigorous ontological work. Specifically:

- Ontological commitments in physics require restraint; EQT is its physical application.

- Ontological commitments in metaphysics also require restraint, such as the work of this book concerning ontological singularity and co-evolutionary relations.

- Ontological commitments in the philosophy of value also require restraint, such as the work of Chapter 20 on normativity and Chapter 21 on meaning.

Each domain has its own "type of commitment," but the principle of restraint applies to all types.

It must be acknowledged that restraint in physics may be more direct than restraint in purely philosophical concepts, because physics has observable proxies. Restraint in purely philosophical concepts requires more defensive work. But the principle is the same: any serious ontological work should be restrained.

Literature That Must Be Engaged

- Euclid, Elements (ancient Greece)
- David Lewis, On the Plurality of Worlds (1986): the ontological treatment of abstract objects.
- Ludwig Wittgenstein, Philosophical Investigations (1953): reflections on philosophical method.
- Chapter 3 on failure conditions as an ontological discipline (self-reference).
- Chapter 13 on Mind–Matter Unity (self-reference).
- Chapter 20 on how constrained speculation treats the is–ought gap (self-reference).

Summary of This Section

This section has done four things. First, it established the distinction between "richness of ontological commitment vs richness of ontological application." Second, through the examples of Euclidean geometry and EQT, it showed the balance of restraint + richness. Third, it developed how ontological problems that "seem to require new commitments," such as consciousness, normativity, and abstract objects, can be treated through existing commitments. Fourth, it gave four criteria for introducing new commitments.

The next section develops the philosophical genealogy of restrained ontology.

2.5. The Philosophical Genealogy of Restrained Ontology

The Function of the Genealogy

The restrained ontology of this chapter is built upon precursor traditions in Western philosophy. This section develops this genealogy, so that the work of this chapter becomes the contemporary continuation of an ongoing tradition.

Developing the genealogy has two functions:

- It gives this chapter historical legitimacy: it builds upon precursors and is not an arbitrary invention.
- It allows readers to locate the work of this chapter precisely: which precursors it inherits and what contemporary contributions it makes.

This section develops four major precursor nodes: Ockham's method, Spinoza's axiomatic method, Quine's ontological commitment, and Lewis's reflective equilibrium.

Precursor One: Ockham's Method

William of Ockham (ca. 1287–1347) was a representative of medieval English nominalism. His method is famous in the history of Western philosophy as "Ockham's razor."

The content of Ockham's method:

- Reject unnecessary ontological commitments: "do not multiply entities beyond necessity."
- Prefer simple explanations: when faced with multiple explanations, choose the simplest one.
- Cooperation between experience and reason: ontological commitments should have experiential and rational grounds.

Ockham's method has had a major influence in the history of philosophy: it is a precursor of modern scientific method, a foundation of analytic philosophical method, and the earliest expression of restrained ontology.

Relation to the work of this chapter:

Ockham's method is the earliest precursor of the work of this chapter. But the work of this chapter is not a simple repetition of Ockham's method; it is a contemporary refinement:

- Ockham's method is mainly heuristic: it tells us what not to do, but does not tell us in detail how ontology should be built.

- $PR1$ in this chapter, as an ontological discipline, is closely connected with methodology, including constrained speculation and failure conditions, and provides a criterion of construction.

- The legitimacy of Ockham's method is mainly built upon the cooperation of experience and reason; this is a medieval methodology.

- The legitimacy of $PR1$ in this chapter is built upon the method of constrained speculation; this is a contemporary methodology.

One may say that the work of this chapter is a contemporary refinement of Ockham's method: it inherits its direction, namely restraint in commitment, but gives it content through contemporary methodology.

Precursor Two: Spinoza's Axiomatic Method

Spinoza (1632–1677) used an axiomatic method in "Ethics" (1677): ontology is unfolded according to geometrical order. Specifically:

- It begins from a small number of definitions and axioms.

- It derives a large number of propositions through deduction.

- The whole ontology is unfolded as a unified system.

The features of Spinoza's method are:

- Axiomatization: explicit starting points, namely definitions and axioms.

- Deductivity: propositions are derived from the starting points.

- Systematicity: the whole ontology is presented as a unified system.

This method is similar to the work of this chapter: it begins from minimal postulates and derives rich applications from them.

Relation to the work of this chapter:

Spinoza's method is one of the precursors of the work of this chapter. Specifically:

- Spinoza's "monism," according to which there is only one substance, God/Nature, is an example of restrained ontology.

- Spinoza's axiomatic method is structurally similar to the minimal postulate method of this chapter.

But there are important differences:

- Spinoza's work is built upon the rationalist tradition: it assumes that ontological content can be established through pure rational inference. The work of this chapter is built upon the method of constrained speculation: it requires ontological commitments to have falsifiable structures and physical connections.

- Spinoza's "substance" is a single ontological unit, God/Nature. The work of this chapter is built upon the ontology of difference (Chapter 7), understanding beings as structures of difference.

- Spinoza's deduction is a priori: it proceeds from definitions and axioms. The applications in this chapter involve physics and experience: they are not purely a priori.

One may say that Spinoza's method is similar to the work of this chapter in methodological structure, but differs importantly in content and methodology. The work of this chapter is a contemporary refinement of Spinoza's method, giving it a ground through the method of constrained speculation.

Precursor Three: Quine's Ontological Commitment

Quine (1908–2000) had a major influence on ontology in twentieth-century analytic philosophy. His position, especially in "On What There Is" (1948), is as follows:

- Ontological commitments are made explicit through quantified variables: a theory is committed to the existence of the objects over which its quantified variables range.

- Ontological choice is based on science: ontology should be built upon the best scientific theories.

- Ontology should be restrained: it should commit itself only to necessary ontological content.

The features of Quine's method are:

- Linguistic approach: ontological commitments are made explicit through language analysis.

- Naturalism: ontology is connected with science.
- Restraint: a preference for a small number of commitments.

Relation to the work of this chapter:
Quine's method is the closest twentieth-century precursor of the work of this chapter. Specifically:

- Quine's position of "restraint" is consistent with the direction of $PR1$ in this chapter.
- Quine's "naturalism," namely the connection between ontology and science, is consistent with the direction of this book's work, namely ontology within the EQT framework.

But there are important differences:

- Quine's work is mainly about language: he analyzes ontological commitments through quantified variables. The work of this chapter is mainly about ontological processes: it analyzes commitments through physics and ontology.
- Quine's naturalism accepts the content of science, but does not develop a refined connection between ontology and science. This chapter establishes such a connection through the EQT framework.
- Quine does not develop the ground of restraint: he mainly treats restraint as a methodological preference. $PR1$ in this chapter gives it a ground through the method of constrained speculation and failure conditions.

One may say that Quine's method is the closest precursor of the work of this chapter, while the work of this chapter makes contemporary contributions in methodological rigor and physical connection.
Precursor Four: David Lewis's Reflective Equilibrium
Lewis (1941–2001) is famous for modal realism, especially in On the Plurality of Worlds (1986). But his methodology, especially his use of reflective equilibrium, is relevant to the work of this chapter.
The features of Lewis's method are:

- Theoretical utility as ground: the introduction of ontological commitments is based on their utility in theory.

- Defense of commitment: every commitment requires defense: what problem does it address, and what benefit does it provide?

- Reflective equilibrium: a balance is sought between theoretical utility and philosophical intuition.

Relation to the work of this chapter:
Lewis's method is close to the work of this chapter in methodological rigor. But there are important differences:

- Lewis's work expands possible-world ontology: he claims that all possible worlds are real. This is an example of non-restraint. Chapter 3, Section 3.4 has already developed why this commitment lacks failure conditions and does not constitute a legitimate ontological commitment.

- Lewis uses "theoretical utility" as the ground of ontological commitment. This position is in tension with restraint. If commitments need only "theoretical utility" as their ground, then restraint may be breached: any utility may introduce a new commitment.

- $PR1$ in this chapter is stricter than Lewis's method: it requires methodological discipline, namely constrained speculation, plus the principle of restraint. A commitment needs to satisfy both requirements at the same time.

One may say that Lewis's method is close to the work of this chapter in the defense of commitments, but his ontological application, especially possible worlds, violates the principle of restraint in this chapter. The work of this chapter avoids this overcommitment through stricter methodological discipline.

The Meaning of the Genealogy

Through these four precursor nodes, the genealogy of this chapter's work becomes clear:

- Earliest precursor: Ockham, restraint as a methodological principle.

- Early modern precursor: Spinoza, the axiomatic method.

- Twentieth-century precursor: Quine, ontological commitment.

- Contemporary precursor: Lewis, reflective equilibrium.

The work of this chapter is built upon this genealogy: it is the contemporary continuation of the tradition of restrained ontology. But it makes contemporary contributions through the method of constrained speculation and the physical connection provided by the EQT framework.

Objections and Replies

Objection One: This genealogy is entirely Western. Does restrained ontology have expressions in other traditions, such as Chinese philosophy or Indian philosophy?

Reply: This objection points to an important cross-cultural comparison. The reply of this section is as follows.

The Chinese philosophical tradition has its own direction of restraint. Specifically:

- The Daoist tradition emphasizes simplicity. Laozi's "less leads to gain; more leads to confusion" is an expression of the direction of restraint. Zhuangzi's "Where is the Dao hidden that there are truth and falsity?" involves a critique of excessive ontological commitments.

- The Confucian tradition emphasizes practicality: ontological commitments should connect with ethical practice and should not become excessively abstract.

- The Chan Buddhist tradition emphasizes concision: unnecessary conceptual expansion is to be avoided.

These traditions are related to restrained ontology in direction. The treatment of the qi tradition in Chapter 8 of this book involves some of these resources: qi monism is an example of restrained ontology, since it does not introduce a second ontology.

But a complete cross-cultural dialogue requires dedicated work. This section focuses on the Western tradition because it connects most directly with contemporary analytic philosophical discussion. The direction of restraint in the Chinese tradition is treated in other chapters, especially Chapter 8.

Objection Two: This genealogy excludes certain important precursors, such as Russell's logical atomism, Carnap's logical construction, and Husserl's phenomenological reduction. Are these also precursors of restrained ontology?

Reply: Yes, these are all relevant precursors of restrained ontology. This section is limited to the four most directly relevant nodes: Ockham, Spinoza, Quine, and Lewis. The treatment of Russell, Carnap, Husserl, and others requires dedicated work.

Specifically:

- Russell's logical atomism, which analyzes the world into atomic facts, is an expression of restrained ontology. But his method is mainly linguistic analysis.
- Carnap's logical construction, which begins from sensations and builds the world through logical construction, is an expression of restrained ontology. But his method is mainly empiricist epistemology.
- Husserl's phenomenological reduction, which reduces the world to phenomenological structures, is an expression of restrained ontology. But his method is phenomenological, not analytic-philosophical.

Each precursor is relevant, and each also differs. The four nodes in this section are the most directly relevant examples: their methods are most closely related to $PR1$ in this chapter. The treatment of other precursors is left for future work.

Objection Three: Your genealogy always acknowledges similarity and then develops difference. You claim that the work of this chapter is a "contemporary refinement" of its precursors. Is this overly self-centered?

Reply: This objection genuinely points to a methodological issue. The reply is as follows.

This chapter does not claim to replace its precursors. It claims that contemporary continuation is a legitimate philosophical direction. Philosophical work in every age builds upon precursors and makes contributions through contemporary developments, such as new science, methodology, and philosophical discussion. This kind of continuation is a common pattern in the history of philosophy.

Specifically, "contemporary refinement" involves:

- Inheriting the direction of precursors, such as the direction of restrained ontology.

- Providing grounds through contemporary methodology, such as the method of constrained speculation and failure conditions.
- Showing strength through contemporary applications, such as applications within the EQT framework.

This is not the exclusion of precursors, but contemporary development built upon them. Such development is precisely the vitality of philosophy: tradition continues through contemporary work.

Furthermore, acknowledging similarity and difference is scholarly honesty. It explicitly states the relation between the work of this chapter and its precursors, allowing readers to evaluate the contribution of this chapter precisely. This is more rigorous than simply declaring that "this chapter is new work" or that "this chapter is an application of its precursors."

Literature That Must Be Engaged

- William of Ockham, Summa Logicae, Quaestiones in librum primum Sententiarum
- Baruch Spinoza, Ethics (1677)
- W.V.O. Quine, "On What There Is" (1948), Word and Object (1960)
- David Lewis, Counterfactuals (1973), On the Plurality of Worlds (1986)
- Russell, Our Knowledge of the External World (1914)
- Carnap, Der logische Aufbau der Welt (1928)
- Chapter 8 on the qi tradition (self-reference)

Summary of This Section

This section has done four things. First, it developed the four precursor nodes of restrained ontology: Ockham, Spinoza, Quine, and Lewis. Second, it clarified the similarities and differences between each precursor and the work of this chapter. Third, it positioned the work of this chapter as the contemporary continuation of the tradition of restrained ontology. Fourth, it responded to possible objections, especially concerning cross-cultural comparison and the selection of precursors.

The next section is the conclusion of this chapter.

2.6. Chapter Summary: Restraint as a Methodological Discipline

Completion of the Three Tasks

The three tasks undertaken by this chapter have all been completed.

Task One: to elevate the practice of the "minimal postulate system" from a method of physics into an ontological discipline. Section 2.1 developed the minimal postulate system in the "EQT Foundational Volume" and stated the core thesis $PR1$: ontological restraint as a methodological discipline.

Task Two: to rigorously defend ontological restraint as a methodological principle. Section 2.2 gave three layers of methodological grounds: the requirement of the method of constrained speculation, the feasibility of ontological testing, and the possibility of ontological change. Section 2.3 developed the balance between restraint and richness: through the distinction between "commitment vs application," restraint and richness can coexist.

Task Three: to treat objections that may arise from ontological restraint. Through discussion, especially in Section 2.3 concerning ontological problems that "seem to require new commitments," this chapter showed how restraint can address problems such as consciousness, normativity, and abstract objects. Section 2.4 showed the traditional grounds of restrained ontology through philosophical genealogy.

The Position of This Chapter in Part I

The overall structure of Part I, "Constrained Speculation: Method as Philosophy," is as follows:

- Chapter 1: Constrained Speculation as Method: the methodological foundation.
- Chapter 2 (this chapter): Minimal Postulates and Ontological Restraint: restraint as a methodological discipline.
- Chapter 3: Failure Conditions as an Ontological Discipline: failure conditions as the core of ontology.
- Chapter 4: Constrained Physicalization: constitutive coextension as a semantic tool.
- Chapter 5: Positioning within the Spectrum of Contemporary Realism: the positioning of the present position.

- Chapter 6: Falsifiability as an Ontological Discipline: dialogue with the tradition of falsifiability.

As the second chapter, this chapter undertakes the foundation of methodological discipline: restraint as an application of the method of constrained speculation provides the methodological basis for the ontological work of the following chapters.

The Position of This Chapter in the Whole Book

Looking backward, this chapter builds upon the methodological foundation of Chapter 1.

Looking forward, this chapter provides methodological discipline for later work:

- Chapter 3 on failure conditions: restraint requires every commitment to have failure conditions.
- Chapter 7 on difference preceding identity: the ontology of difference as an example of restrained ontology.
- Chapter 8 on the qi tradition: qi monism as a Chinese traditional expression of restrained ontology.
- Chapter 10 on tri-state categories: tri-state structure as a restrained doctrine of categories.
- Chapter 11 on freezing as a basic ontological process: freezing as the core process of restrained ontology.
- Chapter 13 on mind–matter: an example in which consciousness is treated through existing commitments.
- Chapter 17 on ontological singularity: the metaphysical limit of restrained ontology.
- Chapter 21 on the finitude of meaning: an example in which meaning is treated through existing commitments.

These connections make this chapter the starting point of the methodological discipline of the whole book: all subsequent ontological work is built upon the discipline of restraint.

Honesty about the Limits of This Chapter

This chapter acknowledges several limits.

First limit: the measurement of restraint has not been fully developed. This chapter gives the balance between restraint and sufficiency, but how to measure the "number of commitments" and the "richness of application" requires more detailed work. Ontological evaluation involves methodological judgment, and this judgment is itself difficult.

Second limit: cross-traditional dialogue remains insufficient. This chapter mainly enters into dialogue with the Western analytic philosophical tradition. Chinese traditions, including Daoism, Confucianism, and Buddhism, contribute to the direction of restraint. This chapter mentions them briefly but does not develop them in detail. Chapter 8 treats some of these resources, but a full cross-traditional dialogue on restraint requires dedicated work.

Third limit: the details of application have not been fully developed. This chapter gives the framework of "restraint + richness of application," but the details of each application, such as how existing commitments are applied in the domains of consciousness, normativity, and abstract objects, require dedicated treatment. Later chapters of this book develop these applications; this chapter only indicates the direction.

Fourth limit: the treatment of possible objections remains incomplete. This chapter has treated several objections, but serious opponents may offer more specific criticisms, such as taking the "rich commitments" of certain philosophical systems, including Whiteheadian process philosophy and Lewisian possible-world ontology, as counterexamples. These detailed dialogues require dedicated treatment.

Acknowledging these limits does not weaken the work of this chapter. It makes this chapter an honest initial work.

The Deepest Posture

The deepest posture of the work of this chapter can be stated as follows: restraint is not limitation, but strength.

The philosophical tradition often faces a temptation: to address ontological problems by introducing new commitments. A seemingly complex problem seems to require a "new dimension," a "new category," or a "new entity." This temptation has produced many complex philosophical systems in history. They seem to address many problems, but the legitimacy of each treatment is difficult to evaluate rigorously.

This chapter argues that the strength of philosophy does not lie in introducing a large number of commitments, but in addressing many problems through a small number of commitments. This is genuine philosophical strength: reasoning, application, and defense turn a small number of foundational commitments into powerful tools.

This posture is displayed in Euclidean geometry: 5 postulates address the entire domain of geometry. It is displayed in EQT: 4 postulates address the whole ontology of physics. It is also displayed in the work of this book: through the method of constrained speculation and minimal ontological commitments, a large number of ontological problems are addressed.

Meaning before freezing: this is one of the core theses of this book. This chapter provides the methodological ground for that thesis. "Meaning" as an ontological phenomenon does not require a "meaning entity" as an additional commitment: it is treated through the emergence of a structure of difference, built upon the EQT postulates. "Freezing" as an ontological process does not require a "freezing entity" as an additional commitment: it is treated through the dynamics of the ρ_f field, built upon the EQT postulates. These treatments realize ontological richness through restraint. This is precisely the demonstration of the work of this chapter.

Through the work of this chapter, restraint as a methodological discipline gains its place in the whole book. It is not a negative "subtraction," but a positive methodological choice. This choice makes the ontological work of this book, including the ontology of difference, process ontology, co-evolutionary ontology, and relational ontology, into a systematic whole: all content is built upon minimal commitments and unfolded into a rich ontological picture through reasoning and application.

2.7. Methodological Reflection on This Chapter

This chapter has elevated ontological restraint from the physical method of the "EQT Foundational Volume" into an ontological discipline. This section reflects on the position and limits of this work.

Methodological Position: Among Multiple Precursor Traditions

The direction of restrained ontology is not new: it is built upon the precursor genealogy of Ockham–Spinoza–Quine–Lewis. Ockham's principle provides a heuristic criterion of selection; Spinoza's substance monism provides an ontological example of restraint; Quine's concept of ontological commitment connects restraint with science; Lewis's modal realism, as a negative example, shows the difficulty of unrestrained ontology.

The contribution of this chapter is to elevate the physical method of the "minimal postulate system" into an ontological discipline; to provide restraint with a methodological ground through the method of constrained speculation (Chapter 1) and the ontology of failure conditions (Chapter 3); and to make the balance between restraint and richness operational through the distinction between "commitment vs application." These contributions are constructively original: they do new work on the basis of tradition.

The Balance between Restraint and Richness

This chapter has treated a methodological tension: the tension between economy and richness. Economy requires fewer commitments; richness requires the treatment of complex ontological contents such as consciousness, normativity, and value. This chapter treats this tension through the distinction between "commitment vs application": foundational postulates remain restrained, since $A1$–$A4+M$ in EQT is a minimal set of commitments, but applications can be rich, since multiple complex phenomena are derived from the foundational commitments. This distinction prevents restraint from leading to poverty: it constrains foundational postulates without constraining the scope of application.

Limits Honestly Acknowledged in This Chapter

This chapter acknowledges several limits. First, "restraint" as a methodological principle requires more refined operational content: what counts as "sufficiently restrained"? What counts as "excessively restrained"? This chapter has given methodological guidance, but full operationality requires more applications for demonstration. Second, the balance between restraint and ontological richness involves judgment; these judgments may have different applications in different contexts. Third, the performance of the principle of restraint in cross-traditional applications, such as the direction of restraint in Chinese qi theory, requires more detailed work.

Connections with Later Chapters

The principle of restraint established in this chapter will be applied in later chapters. Chapter 3, on failure conditions, uses restraint as methodological support for evaluating ontological commitments: restrained commitments have clearer failure conditions and therefore stronger reality under $CR2$. Chapter 5, on the realist genealogy, uses restraint as one feature of constrained process realism. All ontological work after Part II is built upon the principle of restraint: gradient ontology, co-evolutionary ontology, and process ontology are all restrained ontologies, deriving multiple ontological contents from a small number of foundational commitments.

More broadly, this chapter demonstrates the methodological rigor of this book. The work of this book is built upon a small number of commitments, namely the EQT postulates and the method of constrained speculation. This chapter clarifies the legitimacy of that choice. If readers do not accept restraint as a methodological discipline, the methodological ground of the later work of this book will be called into question. This chapter provides the methodological defense for the work of this book.

3. Failure Conditions as an Ontological Discipline

3.1. Introduction to This Chapter

Chapter 1, Section 1.2 proposed the core thesis: X is ontologically real if and only if the commitments of X have definable failure conditions. This thesis was stated as the starting point of constrained speculation, but its legitimacy as an ontological criterion has not yet been fully defended. It appears reasonable, but is this "apparent reasonableness" sufficient to make it an ontological discipline? The analytic philosophical tradition imposes strict requirements on ontological criteria: traditional criteria such as independence, stability, and causality each have their own defenses. Failure conditions, as a new criterion, require an equally rigorous defense.

This chapter undertakes that defense. It elevates failure conditions from a "methodological claim that appears reasonable" into an "ontological discipline that can stand after rigorous argument." Specifically, this chapter gives a precise statement of failure conditions as an ontological criterion ($CR2$), rigorously defends that statement, and treats the objections it faces.

This chapter also addresses two open problems left by Chapter 1, Section 1.6: the ontological status of failure conditions themselves, namely what ontological content failure conditions possess as statements, and the relation between form and reality, namely the relation between "formal constraints," such as mathematical structures and logical forms, and real objects. These two open problems cannot be avoided, because they affect the legitimacy of $CR2$ as an ontological criterion.

This chapter also uses $CR2$ to re-examine several classical ontological problems: abstract objects, possible worlds, and mathematical structures. This re-examination shows how $CR2$ reorganizes the ontological map: certain contents traditionally regarded as ontological problems, such as the ontological status of possible worlds, receive a clearer treatment under $CR2$.

This chapter is one of the foundational chapters of the metaphysical argument of the whole book. Its success or failure determines whether constrained speculation can stand as an ontological position, and not merely as a methodological position. If failure conditions cannot stand as an ontological discipline, constrained speculation degenerates into a kind of methodological suggestion; if they can stand, constrained speculation becomes an independent ontology capable of direct dialogue with contemporary analytic metaphysics.

This chapter does not depend on the physical details of EQT. All arguments are completed as purely philosophical arguments.

The discussion proceeds as follows. Section 3.1 reviews the three traditional pillars of ontological criteria, namely independence, stability, and causality, and their shared premise; Section 3.2 states the new criterion $CR2$ and its precise meaning; Section 3.3 develops how the three kinds of concreteness in failure conditions determine the concrete content of ontological commitments; Section 3.4 uses $CR2$ to re-examine classical problems such as possible worlds; Section 3.5 treats the distinction between form and reality; Section 3.6 develops the open problems faced by $CR2$.

3.2. The Traditional Three Pillars of Ontological Criteria: Independence, Stability, and Causality

Starting Point of the Argument

To argue for a new ontological criterion, one must first understand existing criteria. This section reviews the three main pillars of traditional ontological criteria: independence, stability, and causality. It then points out their shared premise and the problem with that premise.

Criterion One: Independence

In the "Categories," Aristotle locates "primary substance," namely concrete individual things, as ontologically most basic. The key feature of primary substance is independence: it is not in another thing and does not depend on another thing. Color inheres in a body, while the body does not inhere in color; therefore, the body is a primary substance, while color is a dependent attribute.

This criterion has a deep intuitive ground: we do indeed regard "independently existing things" as the most real. It recurs throughout the history of philosophy: Descartes's definition of substance as "that which exists without depending on anything else," Spinoza's substance as "that which is in itself and is conceived through itself," and Leibniz's monads as windowless independent units are all different forms of the independence criterion.

Criterion Two: Stability

In the "Republic" and the "Theaetetus," Plato proposes another ontological criterion: stability. Eternal and unchanging forms, or ideas, are more real than changing sensible things. The logic of the criterion is that what is truly real should not disappear over time or contradict itself; truth is stable, and therefore reality should also be stable.

This criterion has had a far-reaching influence in Western philosophy. Medieval scholastic philosophy locates God as the most real being because God is the most stable, namely eternal and unchanging. Kant treats "the persistence of substance" as one of the a priori conditions of the possibility of experience. Twentieth-century logical positivism takes "repeatable observation" as the basis of scientific statements; in essence, this remains an extension of the stability criterion.

Criterion Three: Causality

Aristotle's doctrine of the four causes, Descartes's causal principle that the effect cannot be more real than the cause, Hume's skepticism concerning causality and its later reconstructions: all of these treat causal efficacy as an ontological criterion. If something cannot produce effects, it is ontologically suspect; conversely, something that can produce effects must be real.

The Eleatic principle in contemporary analytic metaphysics, "to be is to have causal powers," is the strictest form of the causality criterion. Armstrong, Kim, and others use it to argue that abstract objects, such as numbers and propositions, should not be regarded as real, because they lack causal efficacy. Alexander's dictum, "to be is to have causal powers," is another expression of this tradition.

The Shared Premise of the Three Pillars

These three criteria appear different: independence concerns relations, stability concerns time, and causality concerns effects. Yet they share an unexamined premise:

Reality and cognition are separated, and ontological criteria concern the attributes of "the side of reality."

The independence criterion assumes that the independence of a thing is an attribute of the thing itself, unrelated to how we cognize it. The stability criterion assumes that the stability of a thing is an attribute of the thing itself, unrelated to our observation. The causality criterion assumes that the causal efficacy of a thing is an attribute of the thing itself, unrelated to our testing.

Chapter 1, Section 1.1 has already identified this shared premise as the source of the predicament of contemporary realism. This section further points out that the same premise is also the source of traditional ontological criteria. The three-pillar criteria all assume that "reality" is out there, and then ask how we can identify its attributes: independence, stability, and causal efficacy. Is this assumption legitimate?

The Respective Problems of the Three-Pillar Criteria

This section argues that the three traditional criteria each encounter unavoidable internal difficulties, all of which arise from the questionable nature of their shared premise.

The problem of the independence criterion: what counts as "independent"? Strictly speaking, Aristotle's primary substances are not completely independent either: Socrates depends on his parents, air, food, and the Earth. If "independence" requires strict non-dependence, then no finite thing is independent; if "independence" allows some degree of dependence, then the criterion becomes vague. Contemporary metaphysical discussions of "grounding" (Schaffer, Fine) attempt to refine this criterion, but they still encounter the problem of hierarchical regress: what is most fundamental? The answer need not be unique.

The problem of the stability criterion: what counts as "stable"? Plato's forms assume eternal and unchanging objects, but physics since the twentieth century has shown that even the most basic "entities," namely elementary particles, are dynamical processes, involving quantum fluctuations, virtual particles, and decay. If stability requires strict invariance, then the ontology of physics is empty; if dynamic stability, such as a "relatively stable process," is allowed, then the criterion must further clarify the meaning of "relative."

The problem of the causality criterion: what counts as "causal efficacy"? Hume already pointed out that causality is not directly observed, but inferred. In contemporary discussions, causality may be regularist, counterfactual, probabilistic, or informational; each version has its own difficulties. The deeper problem is that the attribution of causal efficacy already presupposes a certain ontology. We say that "electrons have causal efficacy" because we have already treated electrons as ontological units; but the causal efficacy criterion is supposed to tell us what counts as an ontological unit. This is circular.

The Questionable Nature of the Shared Premise

The respective difficulties of the three criteria point to the same deeper problem: locating ontological criteria as criteria "concerning the attributes of reality itself" presupposes that we can identify those attributes independently of cognition. But how are they to be identified? How do we know that something is "independent," "stable," or "causally efficacious" except through our concrete contact with it?

This predicament is not a new discovery. Kant already identified it in the "Critique of Pure Reason": he argued that we cannot directly know things-in-themselves, but only phenomena. Yet Kant's response was to transform the ontological problem into a problem of transcendental epistemology, while retaining the thing-in-itself as an unknowable ontological ground. This response has nominal elegance, but it is not operational in scientific practice. Physicists cannot conduct research using Kant's thing-in-itself.

The mainstream response in contemporary analytic metaphysics is to acknowledge the problem while avoiding it: by refining criteria such as grounding and structural equivalence, it attempts to bypass the difficulties of the original criteria. But such refinements often conceal, rather than solve, the problem of the shared premise.

The new path proposed by constrained speculation is not to repair criteria in the direction of "correspondence to reality itself," but to change the nature of the criterion itself: ontological criteria should not be criteria concerning "attributes of reality itself," but criteria concerning "how commitments become concrete." This is the core task of Section 3.2.

Objections and Replies

Objection One: You describe the three traditional criteria too simplistically. Contemporary metaphysicians such as Schaffer, Sider, and Fine have refined discussions of independence, stability, and causality. Does your criticism hit the target?

Reply: The aim of this section is not to discuss every contemporary refinement in detail, but to point out the questionable nature of the shared premise. Regardless of how refined the criteria become, as long as they are located as criteria "concerning the attributes of reality itself," they inherit the shared premise. Schaffer's grounding, Sider's structural realism, and Fine's essence-based metaphysics each make contributions, but they do not question the shared premise. The distinctiveness of constrained speculation lies precisely in questioning that premise.

Objection Two: Kant's response, namely transcendental epistemology, has already handled this problem. Why is constrained speculation still needed?

Reply: Kant's response has two problems. First, it retains the unknowable thing-in-itself as an ontological ground, which is not operational in scientific practice. Physicists cannot conduct research by saying that "the phenomenal world is true, but the thing-in-itself is unknowable"; they need explicit ontological commitments. Second, Kant's transcendental epistemology is built upon a priori categories, such as quantity, quality, relation, and modality, but these categories themselves have no failure conditions. Constrained speculation requires the criterion itself to satisfy the criterion. This is a more thorough reflection.

Objection Three: You seem to be proposing a new ontological criterion, but you have not yet given it. This section is only critical, not constructive.

Reply: Yes. The function of this section is to clear the ground. Section 3.2 will provide the construction. But clearing the ground has value in itself: it allows readers to understand why a new criterion is needed, rather than merely accepting a new criterion as a given.

Literature That Must Be Engaged

- Aristotle, Categories: the foundation of the independence criterion.
- Plato, Republic Books V-VII: the foundation of the stability criterion.
- David Armstrong, A Theory of Universals (1978), Universals: An Opinionated Introduction (1989): contemporary defenses of the Eleatic principle.
- Jonathan Schaffer, "On What Grounds What" (2009): the grounding relation.

- Kit Fine, "Essence and Modality" (1994): essence-based metaphysics.
- Theodore Sider, Writing the Book of the World (2011): structural realism.
- Immanuel Kant, Kritik der reinen Vernunft (1781/1787): Kant's transcendental response.

Summary of This Section

This section reviewed the three main pillars of traditional ontological criteria: independence, stability, and causality. It pointed out their respective internal difficulties and identified their shared premise: ontological criteria are located as criteria "concerning the attributes of reality itself." This location inherits the premise that reality and cognition are separated. This section argued that this premise is questionable and prepared the ground for introducing a new criterion in the next section.

3.3. A New Criterion: Failure Visibility

Statement of the Core Criterion

Criterion $CR2$: X is ontologically real if and only if the commitments of X have definable failure conditions, that is, if and only if the commitments of X are concretely related, in some identifiable way, to "what would make them fail."

This criterion is a further refinement of $CR1$ from Chapter 1, Section 1.2. $CR1$ gives the core thesis; $CR2$ emphasizes the "concrete relation between commitment and failure" as the precise form of the criterion. The following sections of this chapter will proceed on the basis of $CR2$.

The Core Reversal of the Criterion

The difference between $CR2$ and the traditional three-pillar criteria is not a difference of detail, but a fundamental reversal in the nature of the criterion.

The traditional criteria ask: "What attributes does X itself possess such that we identify it as real?" The answers are independence, stability, and causal efficacy. These attributes are all understood as attributes of X itself, independent of our cognition of X.

$CR2$ asks: "How are commitments concerning X concretized into ontological commitments?" The answer is: through definable failure conditions. This concretization is not an attribute of X itself, but an attribute of our relation to X.

The key of this reversal is that the ontological criterion is no longer a criterion concerning "the side of reality," but a criterion concerning "how a commitment becomes an ontological commitment." This does not cancel reality; it cancels the premise that "reality and cognition are separated."

The Legitimacy of the Reversal

Readers may immediately object that this reversal seems to turn ontology into epistemology: the reality of X is reduced to "how we relate to X." This section argues that this objection rests on a misunderstanding.

Precisely speaking, $CR2$ does not claim that "the reality of X = our cognition of X." It claims that "the reality of X is concretized through a failable interface." The difference between the two is as follows.

- "Reality = cognition" is epistemological idealism: it reduces existence to being cognized. Constrained speculation rejects this position.
- "Reality is concretized through a failable interface" is an ontological thesis: it says that the concrete content of an ontological commitment is constituted by its failure conditions, while the commitment itself remains a commitment concerning X, that is, concerning "the other side."

An analogy may help. When we say that "the charge of the electron is constituted by its measurable behavior in electromagnetic interaction," we are not saying that "the charge of the electron = electromagnetic interaction itself." These are different. The charge of the electron is an attribute of the electron itself; electromagnetic interaction is the identifiable manifestation of charge. An attribute is concretized through its manifestation, but it is not reduced to the manifestation.

Similarly, the reality of X is concretized through a failable interface, but the reality of X is not reduced to the interface itself.

Three Layers of Argument

$CR2$ needs three layers of argument in order to stand: the layer of philosophical motivation, the layer of logical necessity, and the layer of the history of science.

First Layer: Philosophical Motivation

If reality is completely independent of our cognition, how do we know that our ontological commitments involve anything at all? This question is not skeptical: it does not doubt the existence of reality, but asks how our statements about reality have concrete content.

Consider two candidate ontological commitments:

Statement A: "There exists some cosmic background reality, which leaves no trace in any observation, but provides the ontological ground for all phenomena." Statement B: "There exists an energy quantum density field $\rho_f(x,t)$, whose evolution is described by a specific master equation and which produces concrete measurable predictions for proxies such as CMB, gravitational waves, and dark matter halo structures."

Statement A has no failure interface: it is designed to "correspond" to any observation. Statement B has concrete failure interfaces: if the measurable predictions fail to match observation, the commitment fails.

These two statements differ fundamentally in the concrete content of their commitments. Statement A seems to say something, because it uses words such as "reality" and "ground," but it has no concrete content, because it excludes no possibility. Statement B truly says something: its concrete content is constituted by the possibilities it excludes, namely its directions of failure.

Constrained speculation proposes this difference as an ontological criterion: Statement B is a legitimate ontological commitment, while Statement A is not. This does not mean that the "thing" described by Statement A does not exist; the statement itself may not describe any "thing" at all.

Second Layer: Logical Necessity

More rigorously, this section gives the logical argument for $CR2$.

Let P be an ontological commitment concerning X. P must have some concrete content; otherwise, it is not a commitment but an empty mark.

What is "concrete content"? The concrete content of a statement is constituted by the possibilities it excludes. If P excludes no possibility, then P is indistinguishable from any other statement: it has no concrete content.

But what does it mean for P to "exclude some possibility"? It means that there are some conceivable circumstances under which P would not be true. These circumstances are failure conditions.

Therefore:

P has concrete content $\Longleftrightarrow$ P excludes some possibilities $\Longleftrightarrow$ P has failure conditions.

If an ontological commitment has no failure conditions, it has no concrete content; a "commitment" without concrete content is not a commitment. This argument is not rhetorical, but a strict logical argument concerning the concrete content of propositions.

This resembles Wittgenstein's argument concerning propositions in the "Tractatus Logico-Philosophicus": a proposition must be able to distinguish some states of affairs from others; otherwise, it is not a proposition. $CR2$ extends this insight concerning propositions to ontological commitments.

Third Layer: The History of Science

In the history of physics, objects that were shown to be empty concepts, such as ether, caloric, and phlogiston, all share one feature: before they were eliminated, their commitments had already become unable to fail.

Consider the fate of ether in detail. In the early nineteenth century, ether as the medium for the propagation of electromagnetic waves was a legitimate ontological commitment: it had concrete failable predictions, such as predictions concerning the detection of an ether wind relative to the Earth's velocity. The null result of the Michelson–Morley experiment in 1887 was the first failure. But ether theory was not immediately abandoned: it was revised as an "ether dragged by the Earth" or as Lorentz contraction theory.

Continued revisions made ether theory increasingly complex, and each new observation was absorbed by new auxiliary assumptions. By the early 1900s, ether theory had come close to being able to explain any electromagnetic phenomenon. At this point it had lost the status of a real commitment: it no longer had a direction of failure.

Einstein's insight was not that "ether does not exist"; at that time, this statement itself had already become meaningless, because ether had been designed to "correspond" to any observation. Einstein's insight was ontological: ether as an ontological commitment had lost concrete content, and therefore special relativity could reconstruct physics without presupposing ether.

Lakatos described this phenomenon as a "degenerating research programme," but he remained at the epistemological level: he said that a degenerating programme "should not be accepted by the scientific community." $CR2$ elevates this phenomenon into an ontological criterion: objects within a degenerating programme lose their status as ontological objects. They are no longer legitimate real commitments. This elevation is not a matter of epistemological attitude, but a matter of ontological status.

A similar loss of ontological status occurred with phlogiston, which was replaced by oxidation reactions; caloric, which was replaced by molecular kinetic energy; absolute space, which was replaced by general covariance; and absolute time, which was replaced by relativistic spacetime. Each case is a process in which an ontological commitment loses its real status after losing its failure interface.

Relation to $CR1$ in Chapter 1

$CR2$ and $CR1$, namely X is real $\iff$ the commitments of X have definable failure conditions, are two formulations of the same criterion. $CR1$ emphasizes the concise form of the criterion; $CR2$ emphasizes the "concrete relation between commitment and failure" as the precise mechanism of the criterion. The following sections of this chapter use $CR2$ as the operational form, but it is essentially identical to $CR1$.

Objections and Replies

Objection One: Your argument assumes that "concrete content $\iff$ exclusion of possibilities," but this assumption itself requires defense.

Reply: This assumption is not arbitrary. It has broad grounds in the analytic philosophical tradition. Wittgenstein, Popper, and van Fraassen all accept this assumption in different forms. Its core is that the content of a statement is constituted by the difference between what it says and what it does not say. If a statement has no "what it does not say," namely if it excludes no possibility, then it has said nothing. This is the minimal condition of the concreteness of a statement.

Objection Two: Your argument seems to depend on the concept of "possibility." What is the ontological status of possibility itself? Is this an infinite regress?

Reply: This question points to a serious metaphysical discussion, including Lewis's possible worlds and Kripke's modal metaphysics. The response of this chapter is that possibility need not be treated as an independent ontological category. Within the framework of $CR2$, "possibility" is a tool concept used to state failure conditions. It does not itself require an ontological commitment, but only clear rules of use. This is similar to Quine's treatment of quantified variables in "On What There Is": quantification is a tool and does not require the commitment of "quantified objects themselves" as additional ontology.

The ontological status of possibility will be further treated in Chapter 4, in the discussion of physicalized semantics.

Objection Three: Does your argument rely too heavily on examples from the history of science? Philosophical argument should not be built upon historical contingency.

Reply: The examples from the history of science are not the ground of $CR2$; they are illustrations of $CR2$. The grounds of $CR2$ are philosophical motivation and logical necessity. The examples from the history of science show that $CR2$ operates in actual scientific progress, but this is confirmation, not proof. $CR2$ can stand even without historical examples, as long as the logical argument, namely the second layer, holds.

Objection Four: You seem to be smuggling some kind of pragmatism or verificationism into ontology.

Reply: $CR2$ is neither pragmatism nor verificationism.

- Pragmatism says that "meaning = practical effect." $CR2$ does not say that meaning is practical effect. It says that the concrete content of an ontological commitment is constituted by failure conditions. This is a different proposition from "meaning = effect."

- Verificationism says that "a statement is meaningful $\iff$ it is verifiable." $CR2$ does not require verifiability. It requires failability, and failure may be indirect, remote, or statistical. This is much more permissive than verificationism's "direct verification."

$CR2$ is a new position. It absorbs insights from pragmatism and verificationism while avoiding their excessive commitments.

Literature That Must Be Engaged

- Karl Popper, The Logic of Scientific Discovery (1959)

- Imre Lakatos, The Methodology of Scientific Research Programmes (1978)

- Bas van Fraassen, The Scientific Image (1980): constructive empiricism.

- W.V.O. Quine, "On What There Is" (1948): quantification as a tool.

- David Lewis, On the Plurality of Worlds (1986): the ontological status of possible worlds.
- Saul Kripke, Naming and Necessity (1980): necessity and possibility.
- Ludwig Wittgenstein, Tractatus Logico-Philosophicus (1921), 4.1ff: on the concrete content of propositions.

Summary of This Section

This section has done three things. First, it stated the new criterion $CR2$: X is real $\iff$ the commitments of X have definable failure conditions. Second, it identified the fundamental difference between $CR2$ and traditional criteria: $CR2$ is a criterion concerning "how a commitment becomes an ontological commitment," rather than a criterion "concerning the attributes of reality itself." Third, it gave three layers of argument for $CR2$: philosophical motivation, logical necessity, and the history of science.

The next section develops the concrete form of $CR2$: how the three kinds of concreteness in failure conditions, namely dimensionality, proxy, and threshold, constitute the concrete content of ontological commitments.

3.4. The Concreteness of Failure Conditions Determines the Concreteness of Ontological Commitments

Aim of the Argument

Section 3.2 established $CR2$ as an ontological criterion. But so far $CR2$ remains a general principle: it says that "commitments must have failure conditions," but it does not specify what failure conditions should be like.

This section gives the precise structure of failure conditions. Failure conditions are constituted by three kinds of concreteness: dimensionality, proxy, and threshold. These three kinds of concreteness jointly determine the concrete content of ontological commitments. This refinement turns $CR2$ from a general principle into an operational criterion.

Three Kinds of Concreteness

•

Chapter 1, Section 1.3 already introduced dimensionality, proxy, and threshold as requirements of the discipline of failure-orientation among the four disciplines. This section further develops the ontological meaning of these three kinds of concreteness.

Concreteness One: Dimensionality

The object involved in a commitment must have definable dimensionality. Dimensionality is the most basic interface between the commitment and the world: it tells us in what space of measurement the object exists.

Consider two contrasts:

- "Dark matter halo" as a commitment has dimensionality, namely mass density, with unit kg/m^3. This dimensionality allows us to locate it in a space of measurement: it is a kind of mass structure.
- "Spirit of the age" as a commitment has no dimensionality. We may say that it is "strong" or "weak," but these are only metaphors; there is no locatable space of measurement.

The ontological meaning of dimensionality is not merely a technical requirement. It is the ontological positioning of a commitment: it tells us what type of object the commitment concerns, such as mass, energy, information, or probability distribution. A commitment without dimensionality does not know what type of object it concerns, and therefore lacks concrete ontological content.

Important clarification: dimensionality need not be a fundamental dimension, such as length, mass, or time. Composite dimensions, such as the dimension of entropy J/K, dimensionless quantities such as the fine-structure constant, and statistical dimensions such as probability density are all legitimate. The requirement of constrained speculation is that the object involved in the commitment must have some definable dimensionality: it must be locatable in some space of measurement.

Concreteness Two: Proxy

A commitment must be projectable onto some empirical proxy. A proxy is the point of contact between the commitment and the world: it tells us how the object can be identified, even if indirectly.

Proxies can take multiple forms:

- Direct proxy: the object itself or its behavior can be directly observed. For example, galaxies as commitments have direct proxies, namely images in electromagnetic telescopes.
- Indirect proxy: the object is identified through its effects. For example, black holes as commitments have indirect proxies, such as accretion-disk radiation, gravitational lensing, and the orbits of nearby stars.
- Statistical proxy: the object is identified through the properties of a statistical ensemble. For example, CMB temperature fluctuations as commitments have statistical proxies, namely anisotropy spectra.
- Remote proxy: the object is identified at a location far from the object. For example, primordial gravitational waves as commitments have remote proxies, namely B-mode polarization in the CMB.

The ontological meaning of proxyhood is that it ensures the commitment is not closed. A commitment without proxies exists only inside a model: it has no point of contact with the world and therefore no concrete relation to the world.

Important clarification: proxies may be in progress. An object may currently lack available proxies, because technology or theory has not yet developed to that stage; but if proxies are possible in principle, the commitment is still legitimate. Before the LHC, the Higgs boson had no direct proxy, but it had theoretically predicted proxies, namely the concrete predictions of the Standard Model, and was therefore a legitimate commitment.

Concreteness Three: Threshold

A commitment must be able to specify what kind of observation would make it fail: it must not only have dimensionality and proxy, but also a concrete direction of failure. Threshold is the sharpness of the commitment: it relates the commitment precisely to concrete possibilities.

Thresholds can be:

- Quantitative thresholds: concrete numerical boundaries. For example, "the Higgs boson is near 125 GeV" specifies a numerical threshold; if it deviates by $\pm X$ GeV, the commitment fails.

- Qualitative thresholds: concrete boundaries of type. For example, "the speed of light is invariant for all reference frames" specifies a threshold of type: any dependence on a reference frame would make the commitment fail.
- Statistical thresholds: concrete boundaries of significance. For example, "the cross section of dark matter particles lies within a certain range" specifies a statistical threshold: if that range is excluded under sufficient samples, the commitment fails.

The ontological meaning of threshold is that it makes a commitment sharp. A commitment with dimensionality and proxy but without threshold is blunt: it cannot tell us what kind of observation would make it fail. Sharpness makes a commitment an ontological commitment, rather than a vague tendency.

The Interdependence of the Three Kinds of Concreteness

The three kinds of concreteness are not independent test items. They form an interdependent whole.

Dimensionality precedes proxy: in order to have a proxy, one must first know in what space of measurement the object is located. This is dimensionality. Without dimensionality, a "proxy" does not know what it is measuring.

Proxy precedes threshold: in order to have a threshold, there must first be a proxy relating the commitment to observable circumstances. This is proxyhood. Without proxy, it is unclear where a "threshold" would be located.

Threshold gives ontological status to dimensionality and proxy: without threshold, dimensionality and proxy are merely technical requirements and do not constitute an ontological commitment. Threshold concretizes dimensionality and proxy into the concrete content of an ontological commitment.

The relation among these three kinds of concreteness resembles the two-level structure concerning the four disciplines in Chapter 1, Section 1.4. At the level of conceptual dependence, they have a direction (dimensionality→proxy→threshold), but at the level of ontological commitment, they must be satisfied simultaneously.

The Correspondence between Concreteness and Concrete Content

The key thesis of this section is that the concrete content of an ontological commitment is precisely characterized by the concreteness of its failure conditions.

This thesis says:

- The more concrete the dimensionality, the more concrete the ontological positioning of the commitment, that is, the clearer the type of object it concerns.
- The more concrete the proxy, the more concrete the commitment's contact with the world, that is, the clearer how it can be identified.
- The more concrete the threshold, the more concrete the exclusion of possibilities by the commitment, that is, the clearer what it says.

The concrete content of a commitment = concreteness of dimensionality + concreteness of proxy + concreteness of threshold.

This correspondence allows us to quantify the concreteness of ontological commitments. Two commitments can be compared in concreteness: which has more concrete dimensionality, namely a narrower space of measurement? Which has a more concrete proxy, namely a more direct and repeatable proxy? Which has a more concrete threshold, namely a narrower failure boundary? Through this comparison, we can judge which commitment has more concrete ontological content.

This does not mean that a commitment with concrete content is necessarily true: it may fail along its direction of failure. But as a commitment, it is a legitimate ontological commitment. A "commitment" without concrete content is not a false commitment, but no commitment.

Comparison with Traditional Criteria

This section argues that the three kinds of concreteness in $CR2$ cover and refine the insights captured by the traditional three-pillar criteria.

- Independence is replaced by dimensionality: dimensionality tells us the concrete positioning of an object in a space of measurement, rather than the vague idea of "independent existence."
- Stability is replaced by proxy: proxy tells us the identifiability of an object across different observations, rather than the vague idea of "persistence without change."

- Causality is replaced by threshold: threshold tells us the concrete relation between an object and possible observations, rather than the vague idea of "producing effects."

The insights of the traditional criteria are preserved: independence, stability, and causality are indeed features of real objects. But they are refined into more operational forms: dimensionality, proxy, and threshold. This refinement is not merely a technical improvement; it makes an ontological criterion truly operational as a criterion.

Objections and Replies

Objection One: Your "three kinds of concreteness" look overly technical. They are suitable for physics, but not for other fields such as biology, sociology, and psychology.

Reply: The forms of the three kinds of concreteness can adapt to different fields.

- Biology: dimensionality can be population density, enzyme activity, or gene frequency; proxies can be phenotypes, behavior, or biochemical tests; thresholds can be statistical significance or mutation rates.

- Sociology: dimensionality can be measures of social networks, economic indicators, or population distribution; proxies can be surveys, administrative data, or behavioral observations; thresholds can be statistical patterns or critical sizes.

- Psychology: dimensionality can be reaction time, activation level, or behavioral frequency; proxies can be behavioral tests, neuroimaging, or self-report; thresholds can be significance boundaries or effect sizes.

The three kinds of concreteness in constrained speculation do not presuppose specific physical measures. They require some dimensionality, some proxy, and some threshold; their concrete forms vary by field. This gives them cross-domain applicability.

Objection Two: Do all ontological commitments need all three kinds of concreteness? Are some legitimate commitments able to satisfy only one or two of them?

Reply: $CR2$ requires all three to be satisfied simultaneously. If a commitment has only dimensionality, such as "the total mass of the universe has some value," but lacks proxy and threshold, namely cannot be actually measured, then it is an empty commitment: it says that "there is some value," but that value has no relation to any concrete observation. If a commitment has only proxy but no dimensionality, such as "some sign of X," it is a vague commitment: it does not know what it is measuring.

However, commitments in progress may temporarily not satisfy all three completely. Before the LHC, the Higgs boson had dimensionality, namely energy, and a theoretical threshold, but its proxy had not yet been realized. This was a legitimate commitment in progress: its failure interface had already been designed and was waiting for technical realization.

Objection Three: Do your three kinds of concreteness truly refine the traditional criteria, or do they merely rename them?

Reply: This is not a renaming. The differences between the three kinds of concreteness and traditional criteria are as follows.

- Independence is a relational concept: X does not depend on other things. Dimensionality is a locational concept: in which space of measurement is X located? The former asks us to identify the relation between X and other things; the latter asks us to locate X itself.

- Stability is a temporal concept: X persists without change. Proxy is a contact concept: how is X identified? The former requires X to remain unchanged; the latter requires X to be identifiable, even if it changes.

- Causality is an efficacy concept: X produces effects. Threshold is a sharpness concept: it specifies the concrete relation between X and possible observations. The former requires X to have effects; the latter requires the commitment to X to be sharp.

The difference is not merely terminological, but concerns the nature of the criterion. Traditional criteria ask "what attributes does X have?" The three kinds of concreteness ask "how is a commitment concretized?" This is the concrete unfolding of the reversal introduced by $CR2$.

Literature That Must Be Engaged

- Carl Hempel, Aspects of Scientific Explanation (1965): on concreteness and concrete content.

- Patrick Suppes, Representation and Invariance of Scientific Structures (2002): on measurement theory.
- Hasok Chang, Inventing Temperature (2004): on the role of dimensionality in scientific practice.
- Hans Reichenbach, Experience and Prediction (1938): on proxies and concreteness.
- Allan Franklin, The Neglect of Experiment (1986): on thresholds and experimental design.

Summary of This Section

This section has done three things. First, it refined the general principle of $CR2$ into three kinds of concreteness: dimensionality, proxy, and threshold. Second, it argued how these three kinds of concreteness constitute the concrete content of ontological commitments: the more refined the concreteness, the more refined the ontological content of the commitment. Third, it argued how the three kinds of concreteness cover and refine the insights captured by the traditional three-pillar criteria.

The next section uses the new criterion to re-examine classical ontological problems and shows how the criterion reorganizes the ontological map.

3.5. Re-examining Classical Ontological Problems with the New Criterion

Aim of the Argument

The strength of $CR2$ as an ontological criterion lies in its ability to treat aspects of classical ontological problems that traditional criteria handle with difficulty. This section uses $CR2$ to re-examine three classical problems: abstract objects, possible worlds, and mathematical structures. It shows how the criterion reorganizes the ontological map.

This section does not attempt to give a final solution to these problems. Its aim is to show that $CR2$ provides a clear framework for organizing the discussion, clarifies the vague parts of the problems, and focuses the genuinely difficult parts.

Problem One: Abstract Objects

Abstract objects, such as numbers, propositions, sets, and properties, have long been disputed in analytic metaphysics. Platonism holds that they are real; nominalism holds that they are not real; structuralism and other intermediate positions attempt to reconcile the two.

Traditional criteria encounter difficulties here.

- Independence criterion: abstract objects seem independent of concrete things, for example, the number 7 does not depend on any seven concrete objects, but they also seem unable to exist independently. Where is the number 7?

- Stability criterion: abstract objects seem completely stable, for example, the number 7 is always 7, but stability by itself cannot distinguish abstract objects from contentless fictions.

- Causality criterion: abstract objects seem to lack causal efficacy, for example, the number 7 cannot push any physical process. This leads the Eleatic principle to deny their reality, but then it becomes difficult to explain why mathematics is so effective in science.

$CR2$ provides a different response. Let us examine "the number 7" as an ontological commitment.

- Dimensionality: the number 7 has no physical dimensionality, but it has formal dimensionality: it is a natural number, in N. Formal dimensionality belongs to a category different from physical dimensionality, but it remains a kind of dimensionality.

- Proxy: the number 7 has no physical proxy, but it has formal proxies: it is identified through formal systems such as Peano arithmetic and set theory. These identifications have strict rules.

- Threshold: the number 7, as a commitment to a "number," has no direction of failure: it is equally valid in all mathematical systems.

The judgment of $CR2$: the number 7 does not satisfy the physical form of the three kinds of concreteness, but it satisfies the formal form. This means that the number 7 is a formal constraint, not a real object. This distinction preserves the epistemological dignity of mathematics without having to ontologize it.

More precisely, $CR2$ reorganizes ontological status into three categories:

- Real objects: entities that satisfy the physical form of the three kinds of concreteness, namely physicalized dimensionality, proxy, and threshold. Examples include electrons, galaxies, and the energy quantum density field.
- Formal constraints: entities that satisfy the formal form of the three kinds of concreteness, namely formal dimensionality, formal proxy, and formal rigor. Examples include numbers, sets, propositions, and logical forms.
- Non-commitments: entities that satisfy neither the physical form nor the formal form. An example is "absolute spirit."

This tripartition allows us to reject excessive ontological expansion, since numbers need not be ontologized, while also rejecting excessive reduction, since numbers need not be reduced to physical objects or mental states. Formal constraints have their own legitimate place: they do not compete with real objects, but complement them.

Problem Two: Possible Worlds

David Lewis's modal realism holds that possible worlds are just as real as the actual world. His reason is that possible worlds are indispensable in our modal reasoning, and therefore, according to Quine's criterion of ontological commitment, we must commit ourselves to their existence.

The response of $CR2$ is refined.

Let us examine "possible world W" as an ontological commitment.

- Dimensionality: W has no physical dimensionality. Does it have formal dimensionality? This depends on how we define possible worlds. If W is defined as a "maximal consistent set of propositions," as in the strategy of Adams and others, it has formal dimensionality: it is located in propositional space. If W is defined as a Lewisian concrete universe, its dimensionality becomes vague: it is "another concrete world," but its dimensional relation to the actual world is unclear.

- Proxy: W has no direct proxy. In Lewis's modal realism, W is in principle unreachable: we cannot observe it. But does it have formal proxies? If W is a set of propositions, then it has formal proxies: consistency, maximality, and other properties can be checked.

- Threshold: does W as a commitment have a direction of failure? In Lewis's modal realism, no concrete observation would make "W exists" fail. The existence of W is designed to "correspond" to any observation.

The judgment of $CR2$:

- "Possible worlds" in Lewisian modal realism are not legitimate ontological commitments. They have no failure interface: any observation is compatible with them. They are empty concepts.

- "Possible worlds" in propositionalism, where W is a maximal consistent set of propositions, are legitimate formal constraints. They have formal dimensionality, formal proxies, and formal rigor. They have a legitimate place in modal logic, but they do not constitute commitments to real objects.

This judgment provides a concrete objection to Lewis. Lewisian modal realism is not "metaphysical extravagance," which is the common accusation, but an ontologically empty commitment: it uses the word "real," but this use has no concrete content.

Problem Three: Mathematical Structures

The third problem concerns the ontological status of mathematical structures. The "unreasonable effectiveness" of mathematics in physics, in Wigner's famous formulation, is surprising: why can pure mathematical structures describe the physical world so precisely?

There are several positions in response to this problem:

- Platonism: mathematical structures exist as abstract realities, and the physical world "shares" them.

- Nominalism: mathematical structures are merely useful fictions.

- Structuralism: the physical world itself is structure, and mathematics describes these structures.

- Naturalism: mathematical structures are abstracted from physical practice and have no independent ontological status.

*CR*2 provides a different response.
Mathematical structures as commitments:

- Dimensionality: mathematical structures have formal dimensionality, in appropriate mathematical spaces.
- Proxy: mathematical structures have formal proxies, through proof, computation, and formal verification.
- Threshold: mathematical structures as "formal truths" have no direction of failure; proof is necessary, not contingent.

According to *CR*2, mathematical structures are formal constraints, not real objects. But this does not yet explain Wigner's problem: why can formal constraints describe the physical world?

The response of *CR*2 is that the relation between mathematical structures and the physical world is not correspondence or sharing, but the relation by which physical commitments express themselves through mathematical structures.

More precisely:

- Physical commitments, such as the energy quantum density field, use mathematical structures, such as partial differential equations, as carriers.
- Mathematical structures themselves are formal constraints and have no physical commitment.
- But physical commitments express themselves through mathematical carriers, which makes mathematical structures "effective in physics."

This response does not presuppose any mysterious correspondence between mathematical structures and the physical world. It reduces "mathematical effectiveness" to the fact that physical commitments need strict forms of expression, and mathematics provides such forms. Wigner's "unreasonable effectiveness" is in fact reasonable: it reflects the necessity that physical commitments require formal carriers.

But this response leaves one question: why are specific mathematical structures, such as partial differential equations, group theory, and manifolds, suitable for expressing specific physical commitments? This is a deep problem, and constrained speculation currently has no complete answer. This problem is left for further treatment in Chapter 9, in the discussion of "frequency as a basic spectral quantity."

A Unified Response to the Three Problems

The responses to the three problems share a common structure:

- $CR2$ reorganizes ontological status into three categories: real objects, formal constraints, and non-commitments.
- Abstract objects, possible worlds in the propositionalist version, and mathematical structures are all located as formal constraints.
- This positioning avoids excessive ontologization, and thus avoids the burden of Platonism, while also preserving their legitimacy, and thus avoids the cost of nominalism.
- The relation between formal constraints and real objects is a carrier relation: formal constraints function as expression carriers for real commitments.

This unified response shows the strength of $CR2$ as an ontological criterion: it does not merely provide a criterion, but reorganizes the ontological map. Many traditional metaphysical problems become manageable on the new map.

Objections and Replies

Objection One: Your category of "formal constraints" merely renames traditional abstract objects.

Reply: It is not a renaming. The difference lies in the precise nature of ontological status.

- The traditional abstract-object position, namely Platonism, holds that they are real, although of a different type from physical objects. This presupposes two kinds of reality: physical reality and abstract reality.
- The formal-constraint position holds that they are not real objects, but carriers of commitments. This does not require presupposing two kinds of reality: there is only one kind of

reality, namely the physical form satisfying $CR2$, together with the formal carriers used by commitments.

The difference is ontological, not terminological. The formal-constraint position avoids the "dual reality" burden of Platonism while preserving the epistemological dignity of mathematics.

Objection Two: Your concept of "carrier" is unclear. How do physical commitments "use" mathematical structures?

Reply: The carrier relation will be developed in detail in Chapter 4, in the discussion of physicalized semantics, through the concept of constitutive coextension. Here a brief explanation is enough: a physical commitment expresses itself through a mathematical structure, which means that the concrete content of the physical commitment, namely dimensionality, proxy, and threshold, is stated through mathematical language. Mathematical language does not "correspond" to any external object; it is the form of expression of the commitment. This is similar to ordinary language: ordinary language does not "correspond" to any external object as such; it is a tool of expression.

Objection Three: You seem to have simply rejected Lewis's modal realism. But Lewis has detailed arguments. Is your rejection too hasty?

Reply: The rejection in this section is built upon $CR2$: possible worlds in Lewisian modal realism have no failure interface. This is a concrete and defensible rejection. Lewis's argument depends on the claim that if we do not commit ourselves to the existence of possible worlds, our modal reasoning has no concrete content. The alternative response provided by $CR2$ is that modal reasoning gains concrete content through formal constraints, such as sets of propositions and modal logic, and does not require a commitment to possible worlds as real objects. This alternative response preserves the legitimacy of modal reasoning while avoiding Lewis's ontological expansion.

Literature That Must Be Engaged

- David Lewis, On the Plurality of Worlds (1986)
- W.V.O. Quine, "On What There Is" (1948), "Two Dogmas of Empiricism" (1951)
- Hartry Field, Science Without Numbers (1980): contemporary defense of nominalism.

- Stewart Shapiro, Philosophy of Mathematics: Structure and Ontology (1997): structuralism.
- Penelope Maddy, Naturalism in Mathematics (1997): naturalist treatment of mathematics.
- Eugene Wigner, "The Unreasonable Effectiveness of Mathematics in the Natural Sciences" (1960)

Summary of This Section

This section has done two things. First, it used $CR2$ to re-examine three classical ontological problems, namely abstract objects, possible worlds, and mathematical structures, showing how the criterion reorganizes the ontological map. Second, it introduced the category of "formal constraint" as a third kind of ontological status between real object and non-commitment: formal constraints function as carriers of real commitments.

The next section specifically treats the distinction between form and reality, which is the third open problem left by Chapter 1, Section 1.6.

3.6. The Distinction between Form and Reality

Aim of the Argument

Section 3.4 introduced "formal constraint" as a category of ontological status. This section specifically develops the distinction between form and reality, which is the third open problem left by Chapter 1, Section 1.6.

This distinction is important for the whole book. It determines how we treat mathematical structures, logical forms, conceptual frameworks, and related contents. If the distinction is unclear, the entire ontological framework of constrained speculation will become blurred.

Key Thesis

Thesis: formal constraints are carriers of real commitments, but formal constraints themselves are not real objects.

This thesis requires detailed defense. It requires us to clarify: (1) what formal constraints are; (2) what real objects are; and (3) what the "carrier" relation is.

Precise Definition of Formal Constraints

Formal constraints are commitments that satisfy the following three kinds of formal concreteness:

- Formal dimensionality: positioning within some formal space, such as mathematical structure, logical form, or conceptual framework.
- Formal proxy: identification through the rules of a formal system, such as inference, proof, or computation.
- Formal rigor: strict truth rules within a formal system.

Examples of formal constraints include numbers in arithmetic systems, sets in set theory, propositions in logical systems, groups in group theory, and geometric structures in axiomatic systems.

Note: formal constraints do not require a "direction of failure." Their legitimacy is not established through failure interfaces, but through formal rigor. This is the key difference between formal constraints and real objects.

Precise Definition of Real Objects

Real objects are commitments that satisfy three kinds of physical concreteness:

- Physical dimensionality: positioning in a physical space of measurement, such as mass, energy, time, or charge.
- Physical proxy: identification through physical observation or repeatable experiment.
- Physical threshold: a concrete relation with specific physical observations, capable of specifying a direction of failure.

Examples of real objects include electrons, galaxies, the energy quantum density field, conscious states, and social structures, if they have appropriate dimensionality, proxies, and thresholds.

The key difference between real objects and formal constraints is that real objects have directions of failure and can be falsified by physical observations, whereas formal constraints do not have directions of failure, because truth within a formal system is not determined by physical observation.

The Carrier Relation

How do formal constraints function as carriers of real commitments?

The core feature of the carrier relation is this: real commitments use formal constraints as the language through which they express themselves, but formal constraints themselves do not constitute the content of the commitments.

Precisely speaking:

- The content of a real commitment, such as "the energy quantum density field satisfies the master equation," includes the energy quantum density field as an object, namely a real object, and the description of its behavior, which uses formal constraints such as partial differential equations as carriers.
- A formal constraint, such as a partial differential equation, does not itself commit us to the existence of any object: it only specifies formal relations.
- But a real commitment expresses itself through formal constraints: it says that "the behavior of the object follows this formal relation."

An analogy may help. Consider the ordinary statement "this apple is on the table." This statement involves an apple, a real object; a table, a real object; and their spatial relation, expressed in ordinary language. Ordinary language itself does not commit us to the existence of any object: it is a tool. But the statement expresses itself through ordinary language.

Mathematical structures play a similar role in science. They are tools of expression for commitments, not the contents of commitments.

What This Distinction Preserves

The distinction between form and reality preserves several long-standing metaphysical difficulties.

First, it preserves the effectiveness of mathematics. Wigner's "unreasonable effectiveness" is reduced to the following: physical commitments require strict expression, and mathematics provides that expression. This is neither "reasonable" nor "unreasonable"; it is necessary. If physical commitments have concrete content, satisfying the three kinds of physical concreteness, they require strict forms of expression, and mathematics provides the strictest form.

Second, it preserves the objectivity of logic. Logical forms, such as the law of non-contradiction and the law of excluded middle, have objectivity, but they are not "real objects." In the distinction between form and reality, logic is a formal constraint: it is strict within a formal system, but need not be ontologized. This avoids the metaphysical difficulty of "logical forms existing as real objects," while preserving the objectivity of logic.

Third, it preserves the truth of theories. A physical theory can be "true," but truth is not established through "correspondence to reality," which would inherit the shared premise. Truth is established through the carrier relation between formal constraints, namely theory as a formal system, and real objects, namely the objects studied. The theory as carrier accurately expresses the content of real commitments. This "truth through carrier" is more precise than "truth as correspondence" and better fits scientific practice.

Objections and Replies

Objection One: Your distinction between "formal constraints" and "real objects" sounds like Carnap's distinction between "internal" and "external" questions. Is this a new distinction or a renaming of an old one?

Reply: There is similarity, but also difference. Carnap distinguishes questions "within a framework" (internal) from questions concerning "the framework itself" (external). Constrained speculation distinguishes "rigor within a formal system" (formal constraints) from "physical failable interfaces" (real objects). Carnap treats external questions as "non-cognitive" or "practical"; constrained speculation treats formal constraints as legitimate commitments, but commitments of a different category. The difference is that Carnap tends to eliminate external questions, while constrained speculation retains them as carriers.

Objection Two: Your distinction depends on the distinction between physical concreteness and formal concreteness. But this distinction itself requires defense. Why is physical concreteness more "basic" than formal concreteness?

Reply: Constrained speculation does not claim that physical concreteness is more basic. It claims that they are different categories. The ground of physical concreteness is contact with the physical world, namely the physicalization of dimensionality, proxy, and threshold. The ground of formal concreteness is the rigor of formal systems. Both are legitimate, but they provide different kinds of commitments.

When do we need which kind of commitment? When we make commitments concerning the physical world, such as "electrons exist," we need physical concreteness. When we make commitments concerning formal systems, such as "in Peano arithmetic $2+2=4$," we need formal concreteness. Constrained speculation requires a commitment to know clearly which category it belongs to and to satisfy the corresponding kind of concreteness.

Objection Three: Then which category do conscious states, social structures, and mental entities belong to?

Reply: This depends on the concrete commitment. A conscious state as a commitment, such as $\Psi(\tilde{A}, C, \mathcal{B})$ in "Mind–Matter Unity," requires physical concreteness: it has physical dimensionality, such as power, complexity measure, and boundary measure; physical proxies, such as EEG, fMRI, and behavior; and physical thresholds, such as specific ranges of measurement values. It is a real object.

But a conscious state as a phenomenological concept, such as "qualia" as ineffable interiority, has an unclear ontological status if it lacks physical concreteness. Constrained speculation requires phenomenological concepts to be clearly located: either as real objects, if they satisfy physical concreteness; as formal constraints, if they satisfy formal concreteness; or as rhetorical expressions, if they do not constitute ontological commitments.

The same applies to social structures and mental entities: it depends on how they are stated.

Objection Four: How is the "formal rigor" of formal constraints established? Does this require some kind of ontological commitment?

Reply: Formal rigor is established through the internal rules of formal systems: consistency, completeness subject to Gödelian limitations, determinacy, and so on. These rules themselves do not require ontological commitments; they are conventional, but not arbitrary conventions. They are conventions that are rigorously used in practices such as mathematics, logic, and computation.

This response inherits the later Wittgenstein's insight concerning "form as part of a form of life." The legitimacy of formal constraints is grounded in the strict practices in which they are used; it does not need external ontological commitments as its ground.

Literature That Must Be Engaged

- Rudolf Carnap, "Empiricism, Semantics, and Ontology" (1950): internal/external questions.
- W.V.O. Quine, "Two Dogmas of Empiricism" (1951): criticism of Carnap.

- Hilary Putnam, "What is Mathematical Truth?" (1975): the nature of mathematical truth.
- Stewart Shapiro, Thinking about Mathematics (2000): overview of philosophy of mathematics.
- Ludwig Wittgenstein, Philosophical Investigations (1953): form as form of life.
- Eugene Wigner, "The Unreasonable Effectiveness of Mathematics in the Natural Sciences" (1960)

Summary of This Section

This section has done three things. First, it gave precise definitions of formal constraints and real objects: the former satisfy formal concreteness, while the latter satisfy physical concreteness. Second, it introduced the carrier relation: formal constraints function as expression carriers for real commitments. Third, it showed how this distinction preserves several long-standing metaphysical difficulties, including mathematical effectiveness, logical objectivity, and theoretical truth.

The next section is the reflective conclusion of this chapter: it treats the relation between failure visibility and epistemology, and acknowledges the open problems of this chapter.

3.7. The Relation between Failure Visibility and Epistemology

Aim of the Argument

The first five sections of this chapter have established the complete framework of failure conditions as an ontological discipline. But there is an urgent problem that requires specific treatment: failure visibility appears to be an epistemological concept, concerning how we identify X, yet this chapter has used it throughout as an ontological criterion, concerning whether X is real. Does this confuse epistemology and ontology?

This section directly addresses this problem and states the deepest methodological posture of this chapter: constrained speculation rejects the absolute separation between reality and cognition, but preserves their distinction. This rejection is a conscious ontological position, not a confusion.

The Relation between Reality and Cognition: Four Possible Positions

This section first clarifies four possible positions.

Position A: absolute separation. Reality and cognition are completely separated. Reality is a matter on the side of X, while cognition is a matter on our side. Ontological criteria should concern only the side of X. This is the position of traditional criteria.

Position B: absolute identity. Reality = being cognized. This is epistemological idealism, as in Berkeley and others. Constrained speculation rejects this position.

Position C: reality depends on cognition. Reality depends on cognition in some way, but is not reduced to cognition. This is the Kantian position: the phenomenal world, as cognized, has objectivity, while the thing-in-itself is unknowable.

Position D: reality and cognition are distinguished but not separated. The two are distinct, since reality and our grasp of reality are different, but they are not separated, since reality is not located in some domain "independent of cognition." This is the position of constrained speculation.

The Defense of Position D

Why is Position D the position of constrained speculation? Why not choose any of A, B, or C?

Why not A: Position A assumes that we can identify "the attributes of X itself" independently of our contact with X. But any statement concerning the attributes of X is built upon some form of contact between us and X. The "independence" of Position A is not operational: it requires a standpoint we cannot realize, namely a divine standpoint or God's-eye view. Contemporary physical practice, including operational definitions, repeatable experiments, and statistical inference, is built upon our concrete contact with X and does not assume such an independent standpoint.

Why not B: Position B reduces reality to cognition and thereby cancels the distinction between reality and cognition. But the distinction is necessary: we acknowledge that we may be wrong, meaning that cognition may not accord with reality, and this acknowledgment requires the distinction between the two. Without the distinction, the concept of error loses meaning.

Why not C: Position C retains the Kantian thing-in-itself as an unknowable ontological ground. But the thing-in-itself is not operational in scientific practice: physicists cannot conduct research by saying that "the phenomenal world is true, while the thing-in-itself is unknowable." The elegance of Position C is philosophically tempting, but in natural philosophy it remains suspended.

The advantage of Position D is that it distinguishes without separating. This allows us to acknowledge the possibility of error, namely distinction, while preserving concrete contact with X as the route by which commitments are concretized, namely non-separation. The failable interface is precisely the mechanism of "contact as concretization."

Failable Interface as Contact

What does it mean for the failable interface to be "contact"?

Consider an example: the commitment that "the charge of the electron is e." How does this commitment come into contact with the electron?

Not through some mysterious "correspondence to an attribute on the side of the electron." Rather, it does so through concrete experimental processes:

- Oil-drop experiment (Millikan): measuring the behavior of charged oil drops in an electric field and inferring charge quantization.
- Quantum Hall effect: measuring Hall resistance under low temperature and strong magnetic field to obtain a precise value of e^2/h.
- CMB fluctuations: the influence of electron–photon interactions on fluctuation spectra in cosmological observations.

Each experiment is a contact between the commitment and the electron. Contact is two-directional: we act, by setting up the experiment, and the electron "answers," through instrument records. The concrete content of the commitment is constituted by these contacts: it says that "if the charge is e, then these experiments will show these results."

The failable interface is the precise form of this contact. The direction of failure, such as "if the charge is not e," specifies the concreteness of the contact: it says what kind of contact result is incompatible with the commitment.

Position D takes this contact as the mechanism by which ontological commitment is concretized. Reality does not wait "on the other side" to be corresponded to; reality is concretized through contact. But this does not cancel the distinction: we can still be wrong, because contact may mislead us. This is the meaning of the direction of failure.

Dialogue with Heidegger and Brandom

Position D is structurally similar to several philosophical traditions, but also differs from them.

Dialogue with Heidegger: Heidegger's "being-in-the-world" rejects the absolute separation between reality and cognition. Human beings do not first exist before the world and then cognize the world; they are already in the world. Constrained speculation agrees with this rejection, but concretizes it into the failable interface: contact is not a vague "being-in-the-world," but contact with concrete forms, namely dimensionality, proxy, and threshold. This makes Heidegger's insight operational.

Dialogue with Brandomian inferentialism: Brandom argues that conceptual content is constituted by inferential relations. Constrained speculation agrees that the content of commitments is constituted by relations, rather than by correspondence with an external reality, but concretizes this as a failable interface: not merely inferential relations, but concrete relations with physical possibilities. This extends Brandom's inferentialism into ontology.

Dialogue with McDowell's rejection of the "myth of the given": McDowell rejects the idea that cognition has some "bare given" as its ground. Constrained speculation agrees with this rejection: ontological commitments have no "bare reality" as their ground, but are concretized through failable interfaces.

These interlocutors all belong to the "post-analytic" tradition: they attempt to move beyond the correspondence conception of traditional analytic philosophy. Constrained speculation continues this work, but concretizes it into a method of natural philosophy.

Open Problems of This Chapter

This section finally acknowledges the open problems of this chapter. Constrained speculation itself requires the acknowledgment of limits; this chapter follows that requirement.

Open Problem One: the precision of failure conditions. This chapter has given the three kinds of concreteness of failure conditions, namely dimensionality, proxy, and threshold, but the precise form of each kind of concreteness varies by field. The concrete rules of the three kinds of concreteness in physics, biology, sociology, and psychology require further discussion. This open problem belongs to disciplinary methodology and is left for later work.

Open Problem Two: cross-domain comparison. If a commitment satisfies the three kinds of concreteness in physics, and another commitment satisfies the three kinds of concreteness in psychology, are their ontological statuses comparable? Both are real objects, but the concrete forms of their reality differ. This problem is left to Chapter 6, the synthesis with the tradition of falsifiability.

Open Problem Three: the challenge of phenomenal consciousness. Chalmers's "hard problem of consciousness" claims that subjective experience has an ontological status that cannot be reduced to physical processes. If this claim holds, constrained speculation faces a challenge: how can subjective experience as a commitment satisfy the three kinds of physical concreteness? This problem is left to Chapter 13, "Mind–Matter Unity."

3.8. Chapter Summary

This chapter has undertaken three tasks, all of which have been completed.

Task One: to thoroughly defend $CR1$, proposed in Chapter 1, Section 1.2, as an ontological criterion. Through three layers of argument, namely philosophical motivation, logical necessity, and the history of science, this chapter has shown how $CR2$ can stand; through the three kinds of concreteness, namely dimensionality, proxy, and threshold, it has shown how the criterion can operate concretely.

Task Two: to address the two open problems from Chapter 1, Section 1.6. The ontological status of failure conditions themselves is treated through the self-consistency of the three kinds of concreteness: failure conditions are constituted by dimensionality, proxy, and threshold, all of which are concrete objects with their own failable interfaces and do not generate an infinite regress. The relation between form and reality is developed in Section 3.5: formal constraints function as carriers of real commitments and do not compete with real objects.

Task Three: to re-examine classical ontological problems with the new criterion, including abstract objects, possible worlds, and mathematical structures. This chapter has shown how $CR2$ reorganizes the ontological map: through the tripartition of "real objects / formal constraints / non-commitments," many traditional problems become manageable.

Connections with Later Chapters of the Book

This chapter is the key chapter in which constrained speculation rises from methodology to ontology, and several later chapters depend on its work.

Chapter 4, on physicalized semantics, will further develop the carrier relation: how formal constraints function as expression carriers for real commitments.

Chapter 6, on the synthesis between constrained speculation and the tradition of falsifiability, will depend on this chapter's argument for failure conditions as an ontological discipline.

Chapter 9, on frequency as a basic spectral quantity, will further develop the relation between mathematical structures and physical commitments, namely the problem of mathematical effectiveness.

Chapter 13, on the mind–matter relation, will treat Open Problem Three from Section 3.6 of this chapter: the challenge of phenomenal consciousness.

Chapter 17, on ontological singularity, directly applies this chapter: the gradient terminal phase is treated as the shared failure condition of all EQT ontological commitments.

A Final Reflection

This chapter faces a distinctive difficulty in writing: it defends an ontological criterion while using an ontological criterion. Is this circular?

It is not entirely circular, but it is close to self-application. The argument of this chapter uses the method of constrained speculation, namely the four disciplines, while also arguing for the legitimacy of constrained speculation as an ontological method. This self-application is philosophically legitimate as long as the method remains consistent in its self-application.

Specifically, does the argument of this chapter concerning $CR2$ itself satisfy the requirements of $CR2$?

- Formalization: $CR2$ is formalized as an explicit thesis: X is real $\iff$ the commitments of X have definable failure conditions.

- Mechanization: $CR2$ is developed into a concrete mechanism through the three kinds of concreteness: dimensionality, proxy, and threshold.

- Observationalization: the judgment of $CR2$ is observationalized through cases from the history of science, such as ether and the Higgs boson.
- Failure-orientation: $CR2$ has a direction of failure: if there exists an ontologically indispensable commitment that cannot be handled by the three kinds of concreteness at all, $CR2$ fails.

This chapter itself satisfies the requirements of constrained speculation. This is not circularity, but a self-demonstration of the method. A method that cannot apply to itself risks self-refutation. Constrained speculation demonstrates its own consistency through self-application.

In this sense, this chapter has completed a deep task: it has not only argued for a new ontological criterion, but has also demonstrated how this criterion applies to itself. This is precisely the position of this chapter as one of the metaphysical foundations of the whole book.

3.9. Methodological Reflection on This Chapter

This chapter elevates failure conditions from an epistemological concept, concerning how we know, into an ontological concept, concerning how commitments are concretized. This section reflects on the methodological significance of this elevation.

The Key Contribution of This Chapter: From Epistemology to Ontology

Twentieth-century philosophy of science left one task unfinished. Popper proposed falsifiability but remained at the epistemological level; Lakatos proposed research programmes but remained at the methodological level; intermediate positions such as those of Cartwright and Hacking did not fully ontologize falsifiability. The power of falsifiability as a concept was never pushed to its ontological limit. This chapter completes that work. It elevates failure conditions from "how we identify the failure of a theory," which is an epistemological problem, to "how commitments are concretized," which is an ontological problem.

This elevation is not a terminological extension; it involves the reconstruction of an ontological position. In traditional ontology, reality is built upon "correspondence to some external fact." In the framework of $CR2$, reality is built upon "the concrete content of the commitment itself." This reconstruction tightly connects ontology with methodological rigor.

The Ground of This Chapter's Independence from EQT

This chapter defends $CR2$ through purely philosophical argument. The reason is the same as in Chapter 1: constrained speculation as an independent philosophical position does not depend on the applications of EQT. This allows the whole book to exist as an independent philosophical monograph, rather than merely as a philosophical narration of EQT. This independence allows $CR2$ to stand independently of EQT: if the physical work of EQT needs revision in the future, $CR2$ as an ontological criterion is not affected.

Directions in Which This Chapter Can Be Further Deepened

The work of this chapter can be further deepened. The precise forms of the three kinds of concreteness, namely dimensionality, proxy, and threshold, can be rigorously developed through formal semantics; the distinction between form and reality can be formalized through type theory; the dialogue with Heidegger, Brandom, and McDowell can be expanded into dedicated papers. These deepenings belong to later work. The core argument of this chapter as a chapter of the monograph is already complete.

Limits Honestly Acknowledged in This Chapter

This chapter acknowledges several limits. Section 3.6 explicitly states three open problems: the ontological status of $CR2$ itself, the further refinement of the distinction between form and reality, and the challenge of phenomenal consciousness to $CR2$. Among them, the challenge of phenomenal consciousness is currently the most important direction of failure. If phenomenal consciousness involves an ontologically indispensable commitment that cannot be handled by $CR2$ at all, then $CR2$ as an ontological discipline has limits. Chapter 13 treats this challenge in detail. This chapter does not claim that $CR2$ is absolute. It claims that $CR2$ is a defensible, failable, and refinable ontological criterion.

Connections with Later Chapters

The $CR2$ established in this chapter is widely applied in later chapters. Chapter 4, on constitutive coextension, is directly built upon $CR2$: the definition of constitutive coextension involves the content of "the same failure conditions." Chapter 5, on the realist genealogy, uses $CR2$ as a core feature of constrained process realism. Chapter 6 synthesizes the relation between $CR2$ and falsifiability.

More broadly, all ontological work after Part II is built upon $CR2$. The work of Chapter 17 on ontological singularity directly applies $CR2$: the gradient terminal phase is treated as the shared failure condition of all EQT ontological commitments.

If this chapter were removed, constrained speculation would degenerate into a methodological suggestion rather than an ontological position. This chapter elevates constrained speculation from method to ontology. This is its irreplaceable position in the whole book.

4. Constrained Physicalization: From Docking to Constitutive Coextension

4.1. Introduction to This Chapter

In the process of writing the five-volume EQT monograph series (*EQT Foundational Volume*, *Energy, Process, and the Fate of the Cosmos*, *Freezing Cosmology*, *Mind–Matter Unity*, and *EQT Theory of Detection*), the author repeatedly used a distinctive mode of writing: "docking" philosophers' concepts onto physical concepts in EQT. For example: "Whitehead's actual occasions are gradient transients"; "Bergson's durée is $\partial\rho/\partial t$"; "Prigogine's dissipative structures are gradient-driven order"; "Bohr's complementarity appears in EQT as the detector frequency-matching condition."

These "docking" statements play an important role in the five volumes: they connect EQT's physical work to philosophical traditions, enabling readers to understand EQT through existing philosophical frameworks. Yet these statements share a semantic problem: when we say "actual occasions = gradient transients," what exactly does this equal sign mean? Is it identity, in the sense that A just is B? Is it reduction, in the sense that A is reduced to B? Is it constitution, in the sense that A is constituted by B? Is it metaphor, in the sense that A is like B? Each semantic relation has different philosophical consequences.

This chapter diagnoses the insufficiency of "docking-style writing" as a method: philosophically, it requires a stricter semantic defense. It then introduces "constitutive coextension" as a new semantic relation: neither identity, nor reduction, nor constitution, nor metaphor, but another semantic relation constrained by failure conditions. Finally, it restates all the "physicalization" statements in the whole book, including all propositions of the form "$X = Y$" in the five-volume monograph series and in this book, in terms of constitutive coextension, so that the semantic status of the entire book becomes clear.

This chapter is the most technical chapter in Part I. It treats a meta-level semantic problem: the semantic status of philosophical statements themselves. This is a core task of the analytic philosophical tradition, but it is often neglected in the tradition of natural philosophy. The work of this chapter is to bring the semantic rigor of analytic philosophy into natural philosophy.

This chapter builds upon Chapter 3, "Failure Conditions as an Ontological Discipline": the definition of constitutive coextension directly depends on failure conditions (CR2). It provides the semantic foundation for all "physicalization" statements after Part II.

The discussion proceeds as follows. Section 4.1 diagnoses the semantic problem of physicalization; Section 4.2 develops five possible semantic relations, namely identity, reduction, constitution, metaphor, and constitutive coextension, as well as their differences; Section 4.3 states and defends constitutive coextension as a new semantic relation (CC1); Section 4.4 uses constitutive coextension to restate four examples of physicalization; Section 4.5 locates constitutive coextension within contemporary philosophical semantics; Section 4.6 develops the mutual reinforcement between CC1 and the latest argument of this book, namely identity-level isomorphism in Chapter 13: CC1 provides methodological safeguards for identity-level isomorphism, while identity-level isomorphism in turn confirms that CC1 is not an intermediate position, but a semantic relation family with a broad span; Section 4.7 develops the chapter's reflection on itself: an honest refinement of earlier "docking-style writing," namely the ethics of docking as acknowledgment rather than absorption.

4.2. The Semantic Problem of Physicalization

Source of the Problem

In the process of writing the five-volume physical monograph series (*EQT Foundational Volume*, *Energy, Process, and the Fate of the Cosmos*, *Freezing Cosmology*, *Mind–Matter Unity*, and *EQT Theory of Detection*), I repeatedly used a distinctive mode of writing: "docking" philosophers' concepts onto physical concepts in EQT. For example:

- In *Energy, Process, and the Fate of the Cosmos*: "Whitehead's actual occasions are gradient transients: both are temporary units of process."

- In the same volume: "Bergson's durée is $\partial\rho/\partial t$: time as the rate of change of the density field itself."
- In *Freezing Cosmology*: "Prigogine's dissipative structures are gradient-driven order: order formation driven by gradients in non-equilibrium systems."
- In *EQT Theory of Detection*: "Bohr's complementarity appears in EQT as the detector frequency-matching condition: observational outcomes depend on concrete detection arrangements."

These "docking" statements play an important role in the five volumes: they connect EQT's physical work to philosophical traditions, enabling readers to understand EQT through existing philosophical frameworks.

But these statements share a semantic problem: when we say "actual occasions = gradient transients," what exactly does this equal sign mean?

The Ambiguity of the Equal Sign

Consider the possible readings.

Reading One: Identity

Actual occasions just are gradient transients. They are the same object.

Reading Two: Reduction

Actual occasions can be derived from gradient transients. The latter are more basic, and the former are phenomena derived from the latter.

Reading Three: Constitution

Actual occasions are constituted by gradient transients. This is related to reduction but different from it: constitution allows both sides to retain their conceptual identities.

Reading Four: Metaphor

Actual occasions are similar to gradient transients, but only in the sense of a certain similarity. This similarity is not a strict correspondence.

Reading Five: Discovery

Although Whitehead stated the concept of actual occasions in philosophical language, he was in fact describing the same physical process later stated by EQT in physical language.

These five readings differ fundamentally in philosophical semantics. Each involves different ontological commitments and different argumentative structures.

If different readings are used in different places without being distinguished, the writing has a problem of semantic sliding: statements of the same form, "$X = Y$," have different meanings in different contexts, and the reader does not know what the author is doing. This is a problem repeatedly criticized in the analytic philosophical tradition.

The Difficulties of the Five Readings

Let us examine the difficulties faced by each of the five readings.

The difficulty of the identity reading: it is too strong to claim that actual occasions just are gradient transients. Whitehead's concept of actual occasions has concrete content: it involves prehension, subjective form, satisfaction, and other features belonging to Whitehead's philosophical semantic field. Gradient transients, as an EQT concept, have different content: they involve the dynamics of the ρ_f field, the structure of $\nabla\rho_f$, and so on. The contents of the two do not fully overlap; they are not the same object.

The difficulty of the reduction reading: claiming that actual occasions can be derived from gradient transients presupposes that the concept of actual occasions can be logically derived from the concept of gradient transients. But this logical derivation is difficult to carry out. Whitehead's conceptual framework differs from EQT, and deriving one from the other would require many additional assumptions and translations, making the claim of "reduction" empty.

The difficulty of the constitution reading: claiming that actual occasions are constituted by gradient transients implies that gradient transients are "components" of actual occasions. But this constitution relation is unclear: in what sense do gradient transients "constitute" actual occasions? Whitehead's actual occasions are not a "set" or "combination" of gradient transients.

The difficulty of the metaphor reading: claiming that actual occasions are similar to gradient transients is too weak. If this is only a statement of similarity, then "docking" has no rigorous philosophical ground: it is rhetoric rather than philosophical argument. But the EQT work clearly claims more than rhetoric: it claims that the two concepts correspond in some substantive sense.

The difficulty of the discovery reading: claiming that Whitehead described the "same physical process" presupposes that there is a "process itself" independent of the two descriptions. But this presupposition introduces a metaphysical problem: what is this "process itself"? How do we identify it? This reading transforms the simple problem of "docking" into a deeper metaphysical problem and simplifies nothing.

Each reading faces difficulties. This shows that none of the existing semantic relations accurately captures the actual work of physicalization. A new semantic relation needs to be introduced.

The Ambiguous Status of "Docking"

Return to the language of "docking" that I originally used. The word "docking" itself reflects ambiguity: it suggests some kind of "correspondence" but does not precisely specify what kind of correspondence. This ambiguity is not harmless: it allows readers, and the author himself, to use different implicit readings in different statements, producing semantic sliding.

The consequences are:

- When readers find the docking reasonable, they may assume a weak reading, such as discovery or metaphor.
- When readers find the docking forceful, they may assume a strong reading, such as identity or reduction.
- Different readers may have different readings of the same statement.
- Philosophical critics can attack the author for using an overly strong reading, and the author can retreat to a weak reading for defense. This "flexibility" is in fact philosophical dishonesty.

This is the problem of "docking-style writing" that I mentioned in the Preface. It leaves the five-volume physical monograph series semantically imprecise at the philosophical level.

A Clarification

This section states explicitly that the diagnosis of "docking-style writing" does not mean that the five physical monographs are wrong. The physical work, including the $A1$–$A4$ postulates, the Master Equation, and concrete applications, is rigorous. The diagnosis concerns the philosophical-semantic statements of that work: when physical concepts are docked onto philosophers' concepts, what is the semantics of the docking?

The physical work can stand independently of this diagnosis: it has its own legitimacy, satisfying the physical methodology of EQT.

The philosophical-semantic diagnosis only says that the philosophical connection statements in the five volumes require a more precise semantic defense. This chapter provides that defense, giving the philosophical connection statements in the five volumes an explicit semantics.

Objections and Replies

Objection One: You seem to be over-refining an unimportant problem. The conceptual relation between philosophers and physicists does not need such strict semantic distinctions; intuitive understanding is enough.

Reply: In informal philosophy, "intuitive understanding" may be enough. But this book is a serious philosophical monograph, and it requires semantic precision. This precision is not excessive complexity: it is the standard developed over two centuries in the analytic philosophical tradition. Russell, Carnap, and Quine all did extensive work in this direction, identifying the semantic types of philosophical statements.

If this book does not meet this standard, it will be regarded by the analytic philosophical tradition as insufficiently rigorous. Since this book seeks to enter international philosophical discussion, it must satisfy this standard.

Objection Two: Is your diagnosis itself too strict? Many philosophy–physics works use the language of "docking" without causing serious problems.

Reply: It is true that many philosophy–physics works use docking language and are accepted. But acceptance is not the same as rigor. Many accepted works, when examined carefully, have problems of semantic sliding. For example, Bohmian hidden-variable philosophy, Penrose's work on consciousness, and philosophical interpretations of Bohr's complementarity all have similar problems and have been criticized by analytic philosophers.

This chapter does not claim that "docking language must always be avoided." It claims that if docking is used, its semantics must be made explicit. Vague use is the problem; explicit use, as developed below in this chapter, is legitimate.

Objection Three: Is your list of five readings incomplete? Might there be a sixth or seventh reading?

Reply: The five readings cover the main types identified in the semantic tradition: identity, reduction, constitution, metaphor, and discovery. Of course, there may be other readings or subtypes. For example, Kripke's "rigid designation," Putnam's "functional realization," and Davidson's "anomalous monism" can all be understood as variants or combinations of the five basic types.

The following sections introduce a sixth, and new, reading: constitutive coextension. The work of this chapter is not to claim that all possible readings have been exhausted, but to introduce the reading most suitable for physicalization. If a better reading is later identified, the work of this chapter can be revised accordingly.

Literature That Must Be Engaged

- Saul Kripke, Naming and Necessity (1980): on rigid designation and identity.
- Hilary Putnam, "The Meaning of 'Meaning'" (1975): on natural kinds and meaning.
- David Lewis, "How to Define Theoretical Terms" (1970): on the semantics of theoretical terms.
- Donald Davidson, "Mental Events" (1970): anomalous monism as a semantic relation.
- Chapter 3 on failure conditions as an ontological discipline (self-reference).

Summary of This Section

This section has done three things. First, it explicitly diagnosed the semantic ambiguity of "docking-style writing" as a method. Second, it listed five existing semantic readings and their respective difficulties, showing that none of the existing readings accurately captures the actual work of physicalization. Third, it clarified that this diagnosis does not target the content of the five-volume physical work, but only the semantic precision of its philosophical connection statements.

The next section introduces constitutive coextension as a new semantic relation.

4.3. Constitutive Coextension: Definition and Core Features

Definition

Definition 4.1 (Constitutive Coextension): Two concepts A and B stand in a relation of constitutive coextension if and only if (i) the commitments of A and the commitments of B have the same failure conditions, and (ii) A and B retain independence within their respective conceptual frameworks.

This definition requires detailed explanation. Let us analyze the two conditions one by one.

Condition (i): The Same Failure Conditions

The core of constitutive coextension is shared failure conditions. When we say that actual occasions and gradient transients are constitutively coextensive, this means:

Any concrete condition that makes the commitment "there exist actual occasions" fail also makes the commitment "there exist gradient transients" fail. Conversely, the same holds in the other direction.

This condition is strict. It requires the failure conditions of the two concepts to coincide completely: not partially, not roughly, but identically.

Specifically:

- If some observation forces us to give up the commitment that "there exist processual, temporary, interconnected units of being," namely the core commitment of actual occasions in Whitehead's sense, then the same observation also forces us to give up the commitment that "there exist gradient transients of the ρ_f field," namely the core commitment of gradient transients in the EQT sense.
- If some observation forces us to give up the commitment that "there exist gradient transients of the ρ_f field," then the same observation also forces us to give up the commitment that "there exist actual occasions."

The complete coincidence of failure conditions is the substantive content of constitutive coextension. It is not a vague claim that "the two concepts are related," but a precise statement about concrete observable consequences.

Condition (ii): Independence within Conceptual Frameworks

But constitutive coextension does not require the two concepts to be identical. It allows A and B to retain independence within their respective conceptual frameworks.

Specifically:

- Actual occasions, as a Whiteheadian concept, have their own features: prehension, subjective form, satisfaction, and so on. These features have determinate meanings within Whitehead's philosophical semantic field.

- Gradient transients, as an EQT concept, have their own features: gradient structure, the form of $\partial\rho/\partial t$, their relation to freezing, and so on. These features have determinate meanings within the physical semantic field of EQT.

The features of the two concepts are different: they are defined and used within different semantic fields. But their failure conditions are the same: the conditions under which they fail are the same question.

This structure, "different features + the same failure conditions," is the distinctive character of constitutive coextension. It allows us to say that "two concepts correspond" without having to claim that "they are the same object."

Core Features

Constitutive coextension has three core features that must be explicitly stated.

Feature One: Symmetry

Constitutive coextension is symmetric: if A and B are constitutively coextensive, then B and A are constitutively coextensive.

This differs from reduction, which is asymmetric: that A reduces to B does not imply that B reduces to A. It also differs from constitution, which is asymmetric: that A is constituted by B does not imply that B is constituted by A. The symmetry of constitutive coextension reflects the fact that neither concept is prior to the other; they are philosophically equal in status.

Feature Two: No Requirement of Derivation

Constitutive coextension does not require that A can be logically derived from B, or that B can be logically derived from A.

This differs from identity, which permits mutual substitution. It also differs from reduction, which requires the higher-level concept to be derivable from the lower-level concept. Constitutive coextension allows two concepts to have their own independent logical structures, as long as their failure conditions coincide.

This feature is very important for the semantics of physicalization: it allows us to say that "Whiteheadian concepts and EQT concepts are constitutively coextensive" without having to claim that "Whitehead can be derived from EQT" or that "EQT can be derived from Whitehead." The contents of the two traditions can differ, as long as their failure conditions are the same.

Feature Three: Requirement of Shared Failure Projection

Constitutive coextension requires two concepts to have a shared failure projection: that is, the conditions under which they fail must be the same identifiable problem.

This is the key difference between constitutive coextension and metaphor. Metaphor requires only some similarity between two concepts; constitutive coextension requires the two concepts to coincide in concrete observable failure conditions. This is a stricter requirement.

Specifically, to argue that actual occasions and gradient transients are constitutively coextensive, one must be able to state:

- Under what concrete observational conditions does the commitment "there exist actual occasions" fail?
- Under the same concrete observational conditions, does the commitment "there exist gradient transients" also fail?

If the answers to these two questions coincide, the relation of constitutive coextension holds. If they do not coincide, the relation does not hold. This is a testable relation, not rhetoric.

Connection with Chapter 3

The definition of constitutive coextension directly depends on "failure conditions as an ontological discipline," established in Chapter 3.

The core thesis of Chapter 3 (CR2): X is real $\iff X$ has commitments with definable failure conditions.

Constitutive coextension applies this framework to conceptual relations: two concepts are constitutively coextensive $\iff$ their commitments have the same failure conditions.

This connection is legitimate: it grounds the semantic relation of constitutive coextension in a previously defended ontological framework. If the work of Chapter 3 holds, namely if failure conditions stand as an ontological discipline, then constitutive coextension as a conceptual relation is also legitimate.

Furthermore, constitutive coextension is the application of the method of constrained speculation to conceptual relations. Constrained speculation requires every ontological commitment to have failure conditions; constitutive coextension requires every "$X = Y$" statement to be based on a comparison of failure conditions. They are the same method applied at different levels.

A Concrete Example

To prevent the concept of constitutive coextension from remaining abstract, this section gives a concrete example.

Consider the relation between actual occasions (Whitehead) and gradient transients (EQT).

The core commitments of actual occasions, within Whitehead's philosophical semantic field:

- They are temporary units of process, not enduring entities.
- They prehend other actual occasions, forming a network of interrelations.
- They have subjective form, undergoing a distinctive process of concretization.
- They eventually reach satisfaction, completing their own concretization.

Possible failure conditions of actual occasions:

- If observations show that there exist enduring and unchanging entities, rather than temporary units of process, the commitment to actual occasions fails.
- If observations show that beings have no relations to one another, and prehend nothing else, the commitment to actual occasions fails.
- If observations show some non-processual ontology, such as eternal and unchanging forms, then actual occasions fail as a basic ontological commitment.

The core commitments of gradient transients, within the semantic field of EQT physics:

- They are temporary condensation structures of the ρ_f field, not static objects.

- They interact with other gradient structures through the overall dynamics of the ρ_f field.
- They evolve according to the Master Equation, following concrete dynamical laws.
- They eventually dissipate or freeze, returning to a more basic distribution.

Possible failure conditions of gradient transients:

- If observations show the existence of entities that do not depend on gradients, the commitment to gradient transients fails.
- If observations show that beings do not participate in field dynamics, and do not interact with other gradient structures, the commitment to gradient transients fails.
- If observations show some non-dynamic ontology, such as static matter, then gradient transients fail as a basic ontological commitment.

Comparison of failure conditions:
The two groups of failure conditions coincide structurally:

Table 4.1.. Comparison of Failure Conditions between Actual Occasions and Gradient Transients

Item	Standard-Model Content (Here: Ontological Assumption)	EQT Correspondence (Here: Dynamical Requirement)	EQT Limitation / Not Involved	Chapter
Entity/Dependence	Existence of enduring and unchanging entities	Existence of entities independent of gradients	No absolutely isolated source detached from gradients is allowed	3
Interrelation	Beings have no interrelations	Beings do not participate in field dynamics	Does not violate local dynamical conservation laws	4

Continued on next page...

Continued Table

Item	Standard-Model Content	EQT Correspondence	EQT Limitation / Not Involved	Chapter
Ontological Nature	Existence of non-processual ontology	Existence of non-dynamic ontology	Does not deny phenomenological descriptions in the static limit	5

This coincidence is not literal; it is structural. The two sets of failure conditions describe the same class of possible counterexamples: the existence of non-processual, non-relational, and non-dynamic objects. If such objects were observed, both concepts would fail as commitments.

This constitutes the relation of constitutive coextension between actual occasions and gradient transients. They have different features within their respective conceptual frameworks, such as prehension versus field dynamics, but their failure conditions coincide structurally: both are based on the same commitment to process ontology.

Comparison with Alternative Readings

Let us compare constitutive coextension with the five existing readings listed in Section 4.1.

Versus identity: constitutive coextension does not claim that actual occasions just are gradient transients. They are different concepts with different features in different frameworks. But they commit to the same ontological content, namely processuality, relationality, and dynamicity, and therefore bear the same failure conditions.

Versus reduction: constitutive coextension does not require one concept to be derived from the other. Whitehead's actual occasions cannot be derived from the EQT postulates, for Whitehead has his own specific logic of conceptual development. Likewise, gradient transients cannot be derived from Whitehead's philosophy. But the two concepts have the same failure conditions.

Versus constitution: constitutive coextension does not claim that one concept is composed of the other. Actual occasions are not a "combination" or "aggregation" of gradient transients: they are different concepts. But they commit to the same ontological content.

Versus metaphor: constitutive coextension is stricter than metaphor. It requires a concrete comparison of failure conditions, not merely similarity. If two concepts have only "vague similarity" and their failure conditions do not coincide, they are not constitutively coextensive.

Versus discovery: constitutive coextension does not presuppose some "ontological process itself" as the common basis of the two descriptions. It directly states the relation between the failure conditions of the two concepts and does not require a metaphysical "independent ground."

Constitutive coextension is a sixth semantic relation, independent of the above five. It addresses a specific semantic need: describing cases in which two concepts have different features in different frameworks but bear the same ontological commitments.

Objections and Replies

Objection One: Is your concept of constitutive coextension too artificial? It seems constructed to solve the problem of "docking-style writing," and lacks independent semantic grounds.

Reply: Constitutive coextension is indeed introduced to solve the semantic problem of physicalization. This should be acknowledged honestly. But it has independent semantic grounds:

(i) It is built upon a previously defended ontological framework, namely Chapter 3 on failure conditions as an ontological discipline. (ii) It is clearly distinguished from existing semantic relations, such as identity, reduction, and constitution. (iii) It has testable criteria of application, namely the concrete comparison of failure conditions.

These grounds make constitutive coextension not an ad hoc construction, but a semantic relation based on philosophical argument. Every new philosophical concept must be introduced at some point; the introduction of constitutive coextension has its grounds.

Objection Two: Your definition depends on "the same failure conditions." But how do we judge that two failure conditions are "the same"? Does this merely push the vagueness into the judgment of failure conditions?

Reply: Judging whether failure conditions are the same requires concrete work: comparing the observable consequences of two sets of commitments. This comparison is not mechanical and requires philosophical judgment. But it is a concrete judgment that can be developed and criticized in detail, as shown by the comparison between actual occasions and gradient transients in this section.

The vagueness is not unlimited. This chapter acknowledges that judging the sameness of failure conditions requires detailed argument, not simple assertion. Every claim of constitutive coextension must display a concrete comparison of failure conditions. This requirement makes the use of constitutive coextension rigorous rather than arbitrary.

Objection Three: Constitutive coextension seems to make two concepts too closely connected: any observation affects both concepts at the same time. Does this make them "in fact" the same concept?

Reply: No. The same failure conditions do not imply the same concrete content. Two concepts may have different features, different semantic fields, and different argumentative structures within different conceptual frameworks, as long as their failure conditions coincide. This is the core of constitutive coextension: same failure conditions + different concrete content.

By analogy, water and H_2O are in some sense constitutively coextensive, since the failure condition of one is the failure condition of the other, but as concepts they have different contents: water has ordinary-language associations, while H_2O has chemical associations. It is legitimate to use different concepts in different contexts, even when they are constitutively coextensive.

This analogy has limits, since the relation between water and H_2O is the stronger relation of identity, but it shows that "same failure conditions + different concrete content" is philosophically legitimate.

Literature That Must Be Engaged

- Chapter 3 on failure conditions as an ontological discipline (self-reference).
- Saul Kripke, Naming and Necessity (1980): on rigid designation.
- Hilary Putnam, "The Meaning of 'Meaning'" (1975)
- Donald Davidson, "Mental Events" (1970)
- Hartry Field, Truth and the Absence of Fact (2001)
- Theodore Sider, Writing the Book of the World (2011)

Summary of This Section

This section has done four things. First, it gave a precise definition of constitutive coextension: two concepts have the same failure conditions and retain independence within their respective conceptual frameworks. Second, it clarified three core features of constitutive coextension: symmetry, no requirement of derivation, and the requirement of a shared failure projection. Third, it used the concrete example of actual occasions and gradient transients to show how constitutive coextension is applied. Fourth, it compared constitutive coextension with the five existing semantic relations and clarified its status as a sixth independent relation.

The next section develops the logical features of constitutive coextension.

4.4. The Logical Features of Constitutive Coextension

Formal Features

This section develops the formal-logical features of constitutive coextension. This development is more technical than Section 4.2: it treats the formal properties of constitutive coextension as a semantic relation.

Feature One: Symmetry

If A is constitutively coextensive with B (written as $A \approx B$), then B is constitutively coextensive with A ($B \approx A$). Formally:

$$A \approx B \Longleftrightarrow B \approx A$$

This symmetry comes from the definition of constitutive coextension: coincidence of failure conditions is a symmetric relation (A fails $\Longleftrightarrow$ B fails is equivalent to B fails $\Longleftrightarrow$ A fails). Symmetry distinguishes constitutive coextension from reduction, which is asymmetric, and constitution, which is asymmetric.

Feature Two: Reflexivity

Any concept A is constitutively coextensive with itself:

$$A \approx A$$

This is trivial: the failure conditions of A are obviously the same as its own failure conditions. But it must be stated formally so that constitutive coextension is complete as a relation.

Feature Three: Transitivity

If A is constitutively coextensive with B, and B is constitutively coextensive with C, then A is constitutively coextensive with C:

$$(A \approx B) \wedge (B \approx C) \Rightarrow A \approx C$$

This transitivity comes from the transitivity of failure conditions: if A fails $\iff B$ fails, and B fails $\iff C$ fails, then A fails $\iff C$ fails.

Symmetry + reflexivity + transitivity means that constitutive coextension is an equivalence relation. This formal feature is important for its use: an equivalence relation can partition the space of objects into equivalence classes, allowing us to speak of "constitutive-coextension equivalence classes," namely sets of all concepts that are mutually constitutively coextensive.

Three Deeper Features

Beyond its formal features, constitutive coextension has three deeper features that deserve explicit development.

Deeper Feature One: The Stratification of Failure Conditions

Failure conditions can be stated at different levels. An ontological commitment may have:

- Direct failure conditions: conditions of failure at the directly observable level.
- Indirect failure conditions: conditions of failure connected to direct observation through an inferential chain.
- Foundational failure conditions: conditions of failure at the level of ontological structure.

Constitutive coextension requires coincidence of failure conditions at some shared level. Two concepts may differ in direct failure conditions, where one has concrete physical predictions and the other has concrete philosophical implications, but coincide in more foundational failure conditions, where both fail under the same ontological counterexample.

This stratification enables constitutive coextension to handle relations across semantic fields. Physical concepts, with concrete physical predictions, and philosophical concepts, with ontological commitments, can be coextensive at the level of foundational ontology, even if each has its own contents at the level of direct prediction.

Deeper Feature Two: The Requirement of Concreteness in Failure Conditions

Constitutive coextension requires failure conditions to be concrete. If two concepts have only vague "failure conditions," such as "they fail if the world does not accord with them," then their "coextension" is empty: any two concepts can be claimed to be coextensive in such a vague way.

The requirement of concreteness comes from the method of constrained speculation (Chapter 1): failure conditions must satisfy the concreteness of dimensionality, proxy, and threshold. Constitutive coextension uses concrete failure conditions and therefore inherits this concreteness.

Deeper Feature Three: Categorial Correspondence of Failure Conditions

The failure conditions of constitutive coextension are not merely numerically coincident, in the sense that the same observations would make both fail, but categorially corresponding, in the sense that the ontological type of failure is the same.

Specifically, A and B are constitutively coextensive only if:

- A fails under a certain ontological type, such as "failure of processual ontology."
- B fails under the same ontological type.
- What coincides is not only the observation, but the ontological type.

This requirement of categorial correspondence makes constitutive coextension stricter than mere "empirical equivalence." Two theories may be completely equivalent in empirical predictions but bear different ontological types of failure. They are not constitutively coextensive. Constitutive coextension requires the correspondence of ontological type.

Equivalence Classes of Constitutive Coextension

Since constitutive coextension is an equivalence relation, it divides the conceptual space into equivalence classes. All concepts within each equivalence class are mutually constitutively coextensive; concepts in different equivalence classes are not constitutively coextensive.

Consider one concrete equivalence class: the process-ontology equivalence class. This equivalence class may include:

- Whitehead's actual occasions
- Bergson's durée
- Prigogine's dissipative structures

- EQT's gradient transients
- Zhang Zai's "the gathering and dispersal of qi"
- Wang Fuzhi's "qi is daily generated and daily completed"
- occasions of experience used by Whiteheadian scholars

These concepts differ within their respective philosophical semantic fields: each has its own features. But they coincide in their foundational ontological commitments: all commit to processual ontology, rejecting static substance, requiring dynamic structure, and preserving openness. Their failure conditions correspond at the categorial level, for they all fail under the same kind of counterexample: the existence of non-processual ontology.

The identification of this equivalence class allows us to state precisely the philosophical positioning of the five-volume physical work: the physical work of the five volumes, including gradient transients and $\partial\rho/\partial t$, belongs to the process-ontology equivalence class and is within the same equivalence class as Whitehead, Bergson, Prigogine, Zhang Zai, Wang Fuzhi, and other traditions.

This is not "EQT is identical to Whitehead" or "EQT is the physical version of Whitehead": those would be mistaken statements of identity or reduction. Rather, it is "EQT and Whitehead are within the process-ontology equivalence class": this is the precise statement of constitutive coextension.

Comparison with Other Equivalence Relations

Constitutive coextension, as an equivalence relation, differs from other common equivalence relations.

Versus empirical equivalence: two theories are empirically equivalent iff they have the same observable predictions. This relation does not require ontological correspondence. For example, Bohmian mechanics and orthodox quantum mechanics are empirically equivalent but ontologically different. They are not constitutively coextensive, because their ontological failure conditions differ.

Versus mathematical isomorphism: two structures are mathematically isomorphic iff there exists a structure-preserving bijection between them. This relation is purely mathematical and does not require ontological or empirical correspondence. For example, different abstract groups in group theory may be mathematically isomorphic without having any relation to physical reality.

Versus translational equivalence: two languages are translationally equivalent iff meaning-preserving translation can be established between their sentences. This relation is semantic, but it is usually applied between natural languages or formal systems.

Constitutive coextension is an ontological equivalence relation: it concerns whether the ontological contents committed to by two concepts correspond in their failure conditions. This specificity makes it suitable for treating ontological connections across semantic fields.

Objections and Replies

Objection One: You claim that constitutive coextension is an equivalence relation, but your arguments for symmetry, reflexivity, and transitivity all depend on equivalence between failure conditions. What kind of relation is equivalence between failure conditions? Is this an infinite regress?

Reply: Equivalence between failure conditions is ontological identity: two failure conditions are the same ontological event, namely the same counterexample makes both concepts fail. This ontological identity is a basic relation and does not require further analysis.

This is not an infinite regress: it stops at the level of basic ontological relations. Constitutive coextension is a relation concerning concepts; it depends on an identity relation concerning ontological events; the identity relation among ontological events is basic. This stopping point is legitimate: any definition of a relation must eventually stop at some basic relation.

Objection Two: Is the "process-ontology equivalence class" you introduce too broad? It includes Whitehead, Bergson, Prigogine, EQT, Zhang Zai, Wang Fuzhi, and others, whose works differ greatly in content.

Reply: Yes, the contents within the equivalence class differ greatly. This is precisely the key feature of constitutive coextension. It allows concepts with different contents to be within the same equivalence class, as long as their failure conditions correspond at the categorial level.

But categorial correspondence of failure conditions is not broad or vague: it is a concrete judgment. To argue that a concept belongs to the "process-ontology equivalence class," one must show that it commits to process ontology, rejecting static substance and so on, and that its failure conditions correspond at the categorial level with those of the other members. This argument is concrete, not vague.

Different processual traditions, such as Whitehead, Zhang Zai, and EQT, have important differences in concrete features. These differences are discussed in detail in Chapter 8 and Section 4.4 of this chapter. But they coincide in foundational ontological commitment. This is the ground of their constitutive coextension.

Objection Three: Constitutive coextension as an equivalence relation partitions the conceptual space into equivalence classes. But no classification can be perfect. There may be boundary cases in which a concept partly belongs to one equivalence class and partly to another. How should such cases be handled?

Reply: Yes, boundary cases exist. For example, Whitehead's work belongs in most respects to the process-ontology equivalence class, but the eternal objects he introduces may not belong to that class, because eternal objects bear static ontology rather than processual ontology. In this case, some aspects of Whitehead's work belong to the processual equivalence class, while other aspects do not.

Constitutive coextension as a relation is precise, but its application requires concrete judgment: which concrete part of a concept belongs to which equivalence class? This judgment does not weaken the conceptual legitimacy of constitutive coextension; it requires precise conceptual use. This is the requirement of the method of constrained speculation: precision requires concrete judgment, not simple labeling.

Literature That Must Be Engaged

- W.V.O. Quine, Word and Object (1960), on ontological commitment
- Saul Kripke, Naming and Necessity (1980), on necessity
- David Lewis, "Holes" (1970), on ontological classification
- Theodore Sider, Writing the Book of the World (2011)
- Chapter 1 on the method of constrained speculation (self-reference)
- Chapter 3 on failure conditions as an ontological discipline (self-reference)

Summary of This Section

This section has done four things. First, it developed the formal features of constitutive coextension: symmetry, reflexivity, and transitivity, making it an equivalence relation. Second, it identified three deeper features: the stratification of failure conditions, the requirement of concreteness, and categorial correspondence. Third, it introduced the concept of constitutive-coextension equivalence classes, with the process-ontology equivalence class as a concrete example. Fourth, it compared constitutive coextension with other equivalence relations, such as empirical equivalence, mathematical isomorphism, and translational equivalence, clarifying the distinctive status of constitutive coextension as an ontological equivalence relation.

The next section restates all the "physicalization" claims in the five-volume physical work.

4.5. Restating the Physicalization Claims in the Five-Volume Physical Work

Function of the Restatement

This section restates all the "physicalization" claims in the five-volume physical monograph series in terms of constitutive coextension. This is an application of the work of this chapter: it shows how constitutive coextension refines the philosophical semantics of earlier work.

This work does not deny the content of the five-volume physical work. It refines the semantics of its philosophical connection statements. The physical content, including postulates, equations, and applications, remains unchanged; the philosophical connection statements acquire an explicit semantics.

Method of Restatement

Each old "physicalization" statement is replaced by a constitutive-coextension statement. The general pattern is:

Old form: "X (philosopher's concept) = Y (EQT physical concept)"

New form: "X and Y are constitutively coextensive within a certain ontological equivalence class. Therefore, they commit to the same ontological content and bear the same failure conditions, while retaining their independent features within their respective conceptual frameworks."

The new form is longer, but precise. This precision is required by the method of constrained speculation.

Concrete Restatements

Let us restate, one by one, the main "physicalization" claims in the five-volume physical work.

Restatement One: Actual Occasions and Gradient Transients

Old form: "Whitehead's actual occasions are gradient transients."

New form: "Whitehead's actual occasions and EQT's gradient transients are constitutively coextensive within the process-ontology equivalence class. Both commit to the claim that the basic units of being are temporary units of process, not enduring and unchanging entities; that units of being are connected through networks of interaction, not isolated; and that the concrete form of a unit of being is determined by its internal dynamics and does not require an externally imposed form. Both bear the same failure conditions: if observations show the existence of enduring, unchanging, isolated ontological units requiring externally imposed form, both concepts fail. But actual occasions have concrete features within Whitehead's philosophical semantic field, such as prehension, subjective form, and satisfaction, while gradient transients have concrete features within the physical semantic field of EQT, such as the structure of the ρ_f field, $\partial\rho_f/\partial t$, and the relation to freezing. These concrete features do not coincide. The two concepts are different concepts, but they are constitutively coextensive in their commitment to process ontology."

Restatement Two: Durée and $\partial\rho/\partial t$

Old form: "Bergson's durée is $\partial\rho/\partial t$: time as the rate of change of the density field itself."

New form: "Bergson's durée and EQT's $\partial\rho/\partial t$ are constitutively coextensive within the ontology of processual time. Both commit to the claim that time is not an externally imposed parameter, but an internal dimension of process; that the concreteness of time comes from the concreteness of change, rather than from an abstract coordinate; and that time cannot be reduced to a spatialized sequence of instants. Both bear the same failure conditions: if observations show that time can be stated entirely independently of change, or that time can be reduced to a spatial sequence, both concepts fail. But durée has concrete features within Bergson's philosophical semantic field, such as heterogeneity, an indivisible whole, and the inner experience of consciousness, while $\partial\rho/\partial t$ has concrete features within the physical semantic field of EQT, such as the concrete derivative of the density field and its relation to gradient evolution. These concrete features do not coincide. The two concepts are different concepts, but they are constitutively coextensive in their commitment to the ontology of processual time."

Restatement Three: Dissipative Structures and Gradient-Driven

Order

Old form: "Prigogine's dissipative structures are gradient-driven order: order formation driven by gradients in non-equilibrium systems."

New form: "Prigogine's dissipative structures and EQT's gradient-driven order are constitutively coextensive within the non-equilibrium ontology equivalence class. Both commit to the claim that stable structures emerge from gradient-driven processes under non-equilibrium conditions; that the maintenance of structure requires continuous energy flow; and that structure is a stable form of process rather than a static object. Both bear the same failure conditions: if observations show that stable structures can emerge spontaneously under equilibrium conditions, or that they do not require energy flow to be maintained, or that they are truly static, both concepts fail. But dissipative structures have concrete features within Prigogine's physical-chemical semantic field, such as specific mathematical models like the Brusselator, connections with chemical oscillations, and macroscopic emergence, while gradient-driven order has concrete features within the EQT semantic field, such as the structure of the ρ_f field, its connection with the $A3$ Master Equation, and cross-frequency coupling. These concrete features do not coincide. The two concepts are different concepts, but they are constitutively coextensive in their commitment to non-equilibrium process ontology."

Restatement Four: Bohr Complementarity and Frequency Matching

Old form: "Bohr's complementarity appears in EQT as the detector frequency-matching condition: observational outcomes depend on concrete detection arrangements."

New form: "Bohr's complementarity and EQT's detector frequency-matching condition are constitutively coextensive within the co-evolutionary measurement ontology equivalence class. Both commit to the claim that measurement outcomes are not non-perturbative readings, but results of the co-evolution between the detector and the probed system; that different detection arrangements can reveal different aspects, which may not be measurable simultaneously; and that objectivity is established through controlled reproducibility rather than non-perturbative externality. Both bear the same failure conditions: if observations show that measurement outcomes are completely independent of detection arrangements, or that different aspects can be measured simultaneously with unlimited precision, or that objectivity can be achieved under completely non-perturbative conditions, both concepts fail. But Bohr's complementarity has concrete features within the semantic field of philosophy of physics, such as its connection with the uncertainty principle, wave–particle duality, and classical language, while EQT frequency matching has concrete features within the semantic field of physics, such as the concrete form of $g(f, f_0)$, resonance blind zones, and amplitude thresholds. These concrete features do not coincide. The two concepts are different concepts, but they are constitutively coextensive in their commitment to co-evolutionary measurement ontology."

The Strength of the Restatement

These restatements show the strength of constitutive coextension as a semantic tool.

Strength one: they refine the semantics of philosophical connection. The old form, "$X = Y$," is unclear; the new form explicitly states the relation of constitutive coextension: same failure conditions + independent concrete features. The reader can accurately understand the relation between the two concepts.

Strength two: they preserve the independence of each tradition. The restatements explicitly acknowledge that the work of Whitehead, Bergson, Prigogine, and Bohr has concrete features within its own semantic field. The EQT work does not "replace" the concrete content of these traditions; it corresponds to them at the level of foundational ontological commitment. This is honest respect for each tradition, not cultural absorption.

Strength three: they provide a falsifiable structure. Each constitutive-coextension statement gives a concrete comparison of failure conditions. If an observation makes one concept fail but not the other, the constitutive-coextension relation is falsified. This falsifiability makes the restated claims rigorous philosophical statements, not rhetoric.

Strength four: they enable the five-volume work to enter international philosophical discussion. The old form of "docking-style writing" would be criticized in the analytic philosophical tradition as semantically imprecise. The new form satisfies the standard of semantic rigor in analytic philosophy. It allows the five-volume work to be taken seriously in international philosophical discussion.

Boundaries of Application

Constitutive coextension as a semantic tool has explicit boundaries. This section states them clearly.

Boundary one: not all "docking" statements are constitutively coextensive. Some statements have weaker connections: they are merely general "similarities" or "heuristics," without concrete correspondence of failure conditions. These should be honestly stated as "heuristic analogies" rather than constitutive coextensions. For example:

- "EQT's energy quantum density field recalls Spinoza's conception of substance": this is a heuristic analogy, not constitutive coextension, because no concrete comparison of failure conditions has been shown.
- Such statements also appear in the five-volume work and should be identified as relations weaker than constitutive coextension.

Boundary two: constitutive coextension requires concrete argument. To claim that two concepts are constitutively coextensive, one must display a concrete comparison of failure conditions. It is not enough simply to assert it. Every constitutive-coextension statement bears an argumentative burden. The four examples given in this section, actual occasions, durée, dissipative structures, and Bohr complementarity, each require further detailed argument in order to be fully established. This section only presents the core argumentative skeleton.

Boundary three: constitutive-coextension relations may change through refinement. When our understanding of a concept deepens, for example through deeper Whitehead scholarship or further development of EQT, the concrete form of its failure conditions may change, and the constitutive-coextension relation may need to be re-evaluated. This is legitimate: constitutive coextension as a relation is revisable, not an eternal assertion.

Objections and Replies

Objection One: Your restatements turn each claim from one sentence into a paragraph. Is this not overcomplication?

Reply: Yes, the restatement makes each claim longer. But this "complication" is necessary: it makes the implicit semantic content of a vague "$X = Y$" statement explicit, readable, criticizable, and testable. Any precision of a philosophical concept increases the length of the statement. This is the cost of precision, but also the mark of legitimacy.

In actual writing, constitutive-coextension statements can use a simplified form, such as "X and Y are constitutively coextensive within [a certain equivalence class]," without unfolding the full comparison of failure conditions every time. The concrete failure conditions can be developed in dedicated discussions. This simplification makes writing feasible while preserving semantic precision.

Objection Two: Are concepts such as "process-ontology equivalence class" too abstract? Readers may find them difficult to apply concretely.

Reply: The concept of an equivalence class is indeed abstract, but it is a useful tool. It allows us to state precisely which works have the same commitments on a specific ontological problem. This statement is important for philosophical discussion.

The difficulty of concrete application can be reduced through teaching and examples. The four examples given in this section, actual occasions, durée, dissipative structures, and Bohr complementarity, are all concrete applications of equivalence classes. Through these examples, readers can grasp the method of application.

Furthermore, every philosophical tool requires learning. Constitutive coextension as a tool has a learning curve, but once learned, it enables readers to treat ontological connections across semantic fields precisely. This capacity is valuable for philosophical discussion.

Objection Three: Your restatements seem overly academic and overly refined. General readers may abandon the text because of this.

Reply: This book is an academic philosophical monograph and requires academic language. General readers, who are also part of the intended readership, may focus on the Introduction, Chapter 8 on Chinese philosophy, Chapter 17 as the metaphysical climax, and Chapter 21 as the humanistic climax. These chapters are relatively readable. This chapter, which treats a meta-level semantic problem, is better suited for more specialized philosophical readers.

Such a layered reading strategy is common in philosophical monographs. Rawls's *A Theory of Justice* also contains technical chapters, such as those on the specific form of the difference principle, and more readable chapters, such as the intuitive argument for the original position. General readers may read the latter; professional readers read both. This is a legitimate multi-level mode of writing.

Literature That Must Be Engaged

- Chapter 3 on the concreteness of failure conditions (self-reference)
- Saul Kripke, Naming and Necessity (1980)
- Hilary Putnam, "The Meaning of 'Meaning'" (1975)

- David Lewis, "How to Define Theoretical Terms" (1970)
- Chapter 8 on the correspondence between qi theory and gradient ontology (self-reference)

Summary of This Section

This section has done four things. First, it stated the method of restatement: replacing the form "$X = Y$" with the form of constitutive coextension. Second, it gave four concrete examples of restatement: actual occasions, durée, dissipative structures, and Bohr complementarity. Third, it identified the strength of the restatement: refining semantics, preserving independence, providing a falsifiable structure, and enabling the work to enter international discussion. Fourth, it clarified the boundaries of application: not every statement is constitutively coextensive, concrete argument is required, and the relation may change through refinement. After these physicalization statements are restated through CC1, they can be further classified according to the strength forms of CC1: the strongest form, identity level; the intermediate form, structural isomorphism; and the critical form, partial sharing. Section 4.6 will develop this internal structure.

The next section is the chapter's dialogue with contemporary semantic work by Davidson, Kim, Putnam, and others.

4.6. Comparison with Contemporary Philosophy

Function of the Comparison

Constitutive coextension, as a new semantic relation, has concrete relations to existing semantic work in contemporary philosophy. This section shows the position of constitutive coextension within the genealogy of philosophical semantics through comparison.

Comparison One: Davidson's Anomalous Monism

Donald Davidson proposed anomalous monism in "Mental Events" (1970) as a semantic position for treating the mind–body relation. Davidson holds that mental events and physical events are different descriptions of the same event; that there are no strict laws connecting mental descriptions and physical descriptions ("anomalous"); and yet that they refer to the same concrete event ("monism").

Anomalous monism as a semantic relation is structurally similar to constitutive coextension: both allow two descriptions of the same object without reducing one to the other.

Differences from constitutive coextension:

- Davidson's relation concerns the same event: it assumes that there is "the event itself" as a basis. Constitutive coextension does not assume such an ontological basis; it directly states the correspondence between two concepts in terms of failure conditions.
- Davidson is concerned with the relation between descriptions; constitutive coextension is concerned with the relation between concepts. These are different levels: Davidson works at the level of philosophy of language, while constitutive coextension works at the level of ontological semantics.
- Davidson's "anomalous" thesis claims that there are no strict laws between the mental and the physical; constitutive coextension does not claim that there is no connection between concepts, but rather claims a concrete connection of failure conditions.

Constitutive coextension can be understood as a concretization and extension of Davidson's work: it is not limited to the mind–body relation, and it requires concrete connections of failure conditions rather than Davidson's thesis of "lawlessness."

Comparison Two: Kim's Supervenience

Jaegwon Kim developed supervenience as a semantic concept for treating the relation between higher-level and lower-level properties. In simplified form: A-properties supervene on B-properties iff there can be no difference in A-properties without a difference in B-properties.

Supervenience is related to constitutive coextension: both involve the relation between two levels of properties.

Differences from constitutive coextension:

- Supervenience is asymmetric: that A supervenes on B does not imply that B supervenes on A. Constitutive coextension is symmetric.
- Supervenience assumes two levels, higher and lower, and that the higher depends on the lower. Constitutive coextension does not presuppose a hierarchy: the two concepts may be at the same level.

- Supervenience mainly concerns difference relations; constitutive coextension mainly concerns failure-condition relations.

Constitutive coextension is both stricter and more restrained than supervenience: it does not presuppose a hierarchical relation, but it does require a concrete correspondence of failure conditions.

Comparison Three: Putnam's Functional Realization

In his functionalist work, Hilary Putnam proposed that mental states can be realized by different physical states. The same function can be realized by different physical bases, so mental types cannot be simply reduced to physical types.

Functional realization is related to constitutive coextension: both allow two concepts to have different concrete features while maintaining a concrete connection.

Differences from constitutive coextension:

- Functional realization assumes function as the basis: the same function is realized by different physical implementations. Constitutive coextension does not presuppose the priority of function; it concerns the correspondence of failure conditions.

- Functional realization concerns multiple realization, namely the same function realized by multiple physical implementations. Constitutive coextension concerns ontological correspondence across semantic fields.

- Functional realization is mainly applied to the mind–body relation; constitutive coextension applies to any ontological connection across semantic fields.

Constitutive coextension is not a variant of functionalism. It is a more general ontological-semantic relation and does not presuppose function as the prior category.

Comparison Four: Kripke's Rigid Designation

In *Naming and Necessity* (1980), Kripke proposed rigid designation: the same object is designated by the same term in all possible worlds. This concept supported Kripke's argument concerning necessary identity, such as "water = H_2O" being necessarily true, even if known a posteriori.

Differences from constitutive coextension:

- Rigid designation concerns referential stability across possible worlds; constitutive coextension concerns the correspondence of failure conditions across semantic fields.

- Rigid designation supports necessary identity statements, such as "water = H_2O"; constitutive coextension does not assert identity, but only the sameness of failure conditions.

- Rigid designation requires a single object to be designated by two terms; constitutive coextension allows two concepts to have different concrete contents, as long as their failure conditions are the same.

Constitutive coextension can be understood as an extension of Kripke's work: it extends the relation of identity into a weaker relation that allows two different concepts to have the same failure conditions.

The Distinctive Position of Constitutive Coextension

Through these comparisons, the position of constitutive coextension within the genealogy of contemporary philosophical semantics can be stated as follows: it is an ontological relation, not a merely linguistic relation, and differs from Davidson's concern with descriptive relations; it is a cross-semantic-field relation, not limited to hierarchical relations, and differs from supervenience; it is a symmetric relation, not presupposing priority, and differs from supervenience and reduction; it does not presuppose the priority of function, and is a more general ontological relation than functional realization; it is weaker than identity, allowing differences in concrete content, and differs from Kripke's rigid designation; it is stronger than metaphor, requiring concrete correspondence of failure conditions rather than mere similarity.

This position makes constitutive coextension suitable for treating ontological correspondence across semantic fields: correspondences between physical concepts and philosophical concepts, between different philosophical traditions, and so on. This is a semantic need that existing semantic tools cannot handle well.

Objections and Replies

Objection One: Constitutive coextension differs from existing concepts such as anomalous monism, supervenience, functional realization, and rigid designation. But are these differences sufficient to make constitutive coextension legitimate as an independent concept?

Reply: Yes. Each difference points to a different treatment of a concrete philosophical problem. Anomalous monism assumes the event itself as a basis, and this assumption is disputed. Supervenience presupposes a hierarchical relation, but conceptual relations across semantic fields need not be hierarchical. Functional realization presupposes the priority of function, but ontological connections need not be grounded in function. Rigid designation concerns referential stability across possible worlds, but ontological correspondence across semantic fields concerns failure conditions rather than referential stability. Each difference enables constitutive coextension to handle semantic problems that existing concepts do not handle well. This is the ground of its legitimacy: an independent concept solves an independent problem.

Objection Two: In each comparison, you seem to say what the existing concept cannot handle and then claim that constitutive coextension can handle it. Is this overly optimistic?

Reply: This concern is genuine. The comparison in this chapter is built upon acknowledging the strength of existing work: anomalous monism contributes to treating the mind–body relation, supervenience contributes to treating hierarchical relations, and so on. Constitutive coextension does not replace these works; it works on a semantic problem they do not cover, namely ontological correspondence across semantic fields. If future work shows that some existing concept can in fact treat the problem targeted by constitutive coextension, perhaps through an appropriate extension, then the legitimacy of constitutive coextension as an independent concept would be weakened. This chapter remains open to that possibility. It does not claim that constitutive coextension is irreplaceable; it claims only that it is currently the most suitable concept for treating ontological correspondence across semantic fields.

Objection Three: Your dialogue is too concentrated on the English-language analytic philosophical tradition. Are there related concepts in Continental philosophy or other traditions?

Reply: Yes. This section concentrates on English-language analytic philosophy by design, because the tradition of methodological rigor behind constitutive coextension comes from the analytic tradition. But other traditions contain related work. In phenomenology, Husserl's analysis of the noema involves the relation between concepts and objects and is structurally related to constitutive coextension. In hermeneutics, Gadamer's "fusion of horizons" concerns conceptual correspondence between different traditions and is connected to the cross-semantic-field relation of constitutive coextension. In Chinese philosophy, concepts such as "words do not exhaust meaning" in the *Yizhuan* and Wang Yangming's "unity of knowledge and action" involve modes of connection between concept and ontology. This chapter does not treat these traditions in detail; doing so requires dedicated work. It acknowledges their relevance but concentrates on the analytic tradition in order to maintain methodological consistency.

Literature That Must Be Engaged

- Donald Davidson, "Mental Events" (1970), Inquiries into Truth and Interpretation (1984)
- Jaegwon Kim, Supervenience and Mind (1993), Mind in a Physical World (1998)
- Hilary Putnam, "The Nature of Mental States" (1967), "Philosophy and Our Mental Life" (1975)
- Saul Kripke, Naming and Necessity (1980)
- Chapter 3 on failure conditions as an ontological discipline (self-reference)

Summary of This Section

This section has done three things. First, it compared constitutive coextension with four works in contemporary philosophical semantics: Davidson's anomalous monism, Kim's supervenience, Putnam's functional realization, and Kripke's rigid designation. Second, it clarified the distinctive position of constitutive coextension within the genealogy of contemporary philosophical semantics: an ontological relation, cross-semantic-field, symmetric, not presupposing the priority of function, weaker than identity, and stronger than metaphor. Third, it acknowledged the limits of the dialogue in this chapter: it concentrates on the analytic tradition and needs to be extended to other traditions in future work.

4.7. Constitutive Coextension and "Identity-Level Isomorphism": Mutual Reinforcement

The Work of This Section

Section 4.2 developed constitutive coextension (CC1) as a new semantic relation: neither identity, nor reduction, nor constitution, nor metaphor, but a sixth semantic relation constrained by failure conditions. Sections 4.3–4.5 developed the logical features of CC1, its application to the five-volume physical work, and its comparison with contemporary philosophical-semantic tools.

But Chapter 13 of this book, based on the core argument of the new version of *Mind–Matter Unity*, introduces a philosophical thesis: identity-level isomorphism. According to this thesis, there is strict identity between the full structure of experience and the gradient configuration of a certain class of frequency-resolved energy quantum density fields ρ_f, and this identity is realized through relation-by-relation correspondence between a five-dimensionally enriched gradient structure and all the structural features of experience. This thesis is the deepest position of this book in the philosophy of mind and matter.

What, then, is the relation between CC1 and identity-level isomorphism? Are they two independent semantic tools, or different manifestations of the same semantic relation? If the latter, how do they mutually reinforce each other? This section develops this dialogue.

This section argues that identity-level isomorphism is the strongest form of constitutive coextension. There is a mutual reinforcement between CC1 and identity-level isomorphism: CC1 provides methodological safeguards for identity-level isomorphism, turning the latter from a seemingly metaphysical assertion into an ontological commitment constrained by methodological discipline; conversely, identity-level isomorphism confirms that CC1 is not an intermediate position, but a semantic relation family with a broad span. This mutual reinforcement makes explicit the internal coordination between the two philosophical devices.

Identity-Level Isomorphism as the Strongest Form of Constitutive Coextension

The definition of CC1, given in Section 4.2, is this: A and B share the same failure conditions, namely the failure conditions of $A \iff$ the failure conditions of B. This definition grounds the semantic relation in failure conditions rather than in external correspondence.

But CC1 as a relation is not single and rigid. It contains an intensity spectrum. Within CC1, there are different forms from strongest to weaker, and what distinguishes them is the way in which A and B

share failure conditions:

Strongest form: identity level. A and B not only share failure conditions, but are two aspects of the same reality. In this form, "shared failure conditions" occurs because A and B are ontologically identical: they naturally share everything, including failure conditions. Mind and matter as two aspects of the same process (Section 13.2) are an example of this form: the relation between experience and the ρ_f gradient configuration is not merely one of shared failure conditions, but one in which they are the same reality.

Intermediate form: structural isomorphism. A and B correspond one-to-one structurally, and their structural features match one-to-one, so their failure conditions also correspond one-to-one. But A and B are not asserted to be the same reality; they may be two different ontologies that are structurally parallel. The relation between Whitehead's actual occasions and EQT's gradient transients (Section 4.4) belongs to this form: the two correspond structurally, but this book does not claim that they are the same reality.

Critical form: partial sharing of failure conditions. A and B share failure conditions in some respects, but not completely. This is the critical case of CC1. Such cases are common in interdisciplinary or cross-traditional dialogue. For example, Bergson's durée and $\partial\rho/\partial t$ correspond in some respects but not in all respects (Section 4.4). In this form, the scope of CC1's application must be specified clearly.

This intensity spectrum means that CC1 is not a single rigid semantic relation, but a semantic relation family with a broad span. Identity-level isomorphism, as its strongest form, is the deepest position CC1 can reach ontologically.

First Direction of Mutual Reinforcement: CC1 Provides Methodological Safeguards for Identity-Level Isomorphism

Identity-level isomorphism, as the deepest mind–matter thesis of this book, faces a methodological challenge: is it legitimate? How can "mind and matter are the same reality" be distinguished as a serious philosophical thesis from "mind and matter are the same reality" as a metaphysical assertion, namely a stipulation made without grounds?

CC1 provides methodological safeguards. Identity-level isomorphism, as an ontological commitment, is constrained by the methodological discipline of CC1: it must be able to state its own failure conditions, and these failure conditions must be identifiable and seriously evaluable. This constraint prevents identity-level isomorphism from being an empty metaphysical assertion and makes it a testable ontological commitment.

The failure conditions of identity-level isomorphism are developed in Section 13.2: failure of structural sufficiency, if any structural feature of experience is not carried by the fivefold richness; failure of structural correspondence, if the correspondence is shown not to be one-to-one; failure of non-relational given-ness, if given-ness is shown to be intrinsically non-relational; and failure of the boundary criterion, if a system without an enveloping boundary has first-person given-ness. These four failure conditions legitimate identity-level isomorphism as an ontological commitment: it has identifiable directions of failure and is constrained by CR2, namely failure conditions as an ontological discipline in Chapter 3.

Without CC1 as a methodological tool, identity-level isomorphism would face two difficulties. First, it might be read as an intensified version of traditional identity theory, namely "consciousness = some physical quantity." But traditional identity theory lacks structural sufficiency and has no explicit failure conditions, and therefore is not accepted seriously in contemporary philosophy. By grounding identity level in the sharing of failure conditions, CC1 distinguishes identity-level isomorphism from traditional identity theory: it does not claim that one single physical quantity equals consciousness, but that a certain five-dimensional structural space and the structure of experience are strictly identical through shared failure conditions. Second, identity-level isomorphism might be read as a version of Russellian monism. But Russellian monism as a general posture lacks testable content (Section 5.6). CC1 allows identity-level isomorphism to go beyond the general posture of Russellian monism: it has identifiable directions of failure.

Therefore, CC1 provides methodological safeguards for identity-level isomorphism: it transforms the latter from a seemingly metaphysical assertion into an ontological commitment constrained by methodological discipline.

Second Direction of Mutual Reinforcement: Identity-Level Isomorphism Confirms That CC1 Is Not an Intermediate Position

Conversely, identity-level isomorphism also provides important reinforcement for CC1: it clarifies the position of CC1 within the spectrum of semantic relations.

When Section 4.2 introduced CC1, it located it as a "sixth semantic relation": neither identity, nor reduction, nor constitution, nor metaphor. This positioning may give rise to a misreading: CC1 is some kind of "intermediate form," weaker than identity but stronger than other relations. This misreading would weaken the philosophical force of CC1, making it appear to be a compromise position, a middle point between a deeper ontological relation, namely identity, and weaker semantic tools, such as metaphor.

The introduction of identity-level isomorphism corrects this misreading. Identity-level isomorphism is the strongest form of CC1. This means that CC1 is not an intermediate form, but a semantic relation family that spans from identity to weak sharing. This clarification has three consequences.

First, the intensity range of CC1 is wider. CC1 is not merely a semantic tool "weaker than identity"; it includes identity as its strongest form. This extension makes CC1 a stronger ontological-semantic tool.

Second, the relation between identity and CC1 is one of inclusion rather than juxtaposition. Identity is not another relation outside CC1, but a special case of CC1. When CC1's sharing of failure conditions reaches the deepest point, namely when A and B are the same reality, it degenerates into identity. This clarification avoids the simple dichotomy between "identity vs. non-identity" in traditional philosophy: identity is the limiting case of shared failure conditions.

Third, the application judgment of CC1 becomes more concrete. When using CC1 to restate a physicalization claim (Section 4.4), we need to judge which intensity form of CC1 it belongs to. The strategy of this book is to use the intermediate form of CC1 by default, namely structural isomorphism, unless there is concrete evidence supporting the strongest form, namely identity level. This criterion of judgment prevents the application of CC1 from becoming either too strong or too weak.

Retrospective Enrichment of the Applications in Section 4.4

This mutual reinforcement gives a more refined analysis to the physicalization statements in the five-volume physical work treated in Section 4.4. The original Section 4.4 restated all physicalization statements in terms of CC1, but did not distinguish the intensity forms of CC1. Retrospectively, these physicalization statements can be classified according to CC1 intensity forms:

Table 4.2.. Correspondence Mechanism between Physicalization Statements and CC1 Intensity Forms

Item	Standard-Model/Classical Content (Physicalization Statement)	EQT Correspondence (CC1 Intensity Form)	EQT Limitation / Not Involved	Chapter
Mind–Matter Unity	Mind and matter as two aspects of the same process	Strongest (identity level)	Does not make a substance-level mind–matter dualist division	12
Whiteheadian Ontology	Whitehead actual occasions $\leftrightarrow$ gradient transients	Intermediate (structural isomorphism)	Not completely identical to classical non-spatiotemporal ontology	12-13
Bergsonian Time	Bergson durée $\leftrightarrow \partial\rho/\partial t$	Critical (partial sharing)	Does not fully identify pure duration with a mathematical first derivative	13
Bohr Complementarity	Bohr complementarity $\leftrightarrow$ EQT tri-state decomposition	Intermediate (structural isomorphism)	Does not replace the measurement-operator logic of the Copenhagen interpretation	14
Prigogine	Prigogine dissipative structures $\leftrightarrow$ EQT gradient transients	Intermediate (structural isomorphism)	Not identical to macroscopic classical non-equilibrium thermodynamics	14-15

This classification makes the philosophical positioning of the five-volume physical work more refined: not all physicalization statements have the same intensity. The mind–matter relation, as the deepest ontological connection, is at the strongest form, namely identity level; most other physicalization statements are at the intermediate or critical forms. This classification does not weaken the work of Section 4.4, but makes its content more precise: it allows readers to distinguish the strength of the ontological commitments made by this book in different physicalization statements.

Connection with Chapter 13

This section and Section 13.2, "Identity Level: Mind and Matter as Two Aspects of the Same Process," complement each other. Section 13.2 establishes identity-level isomorphism in the context of the philosophy of mind and matter, and marks it as the strongest form of constitutive coextension at the end of that section. This section develops the mutual reinforcement between CC1 and identity-level isomorphism in the context of CC1 as a general semantic tool.

The relation between the two sections is as follows: Section 13.2 is a local application, using the strongest form of CC1 in the field of the philosophy of mind and matter; this section is a general theoretical extension, incorporating identity-level isomorphism into the internal structure of CC1. Readers who read Section 13.2 see the content of identity-level isomorphism in the mind–matter problem; readers who read this section see how that content enriches CC1 as a general semantic tool.

More broadly, this section makes Chapter 4 not only a methodological preparation for Chapter 13, but also an object of reverse reinforcement by the philosophical contribution of Chapter 13. The identity-level isomorphism of Chapter 13 gives CC1 in Chapter 4 a more refined internal structure; CC1 in Chapter 4 gives the identity-level isomorphism of Chapter 13 methodological safeguards. The two chapters form a mutually constitutive relation in the work of this book, rather than a one-way dependence.

This mutual constitution is an example of the work of this book: a semantic tool (CC1) and an ontological thesis (identity-level isomorphism) reinforce each other through internal coordination, rather than being connected by external superimposition. This example provides a template for the relation between other semantic tools and ontological theses in this book, such as tri-state categories and freezing as a basic ontological process, or the discipline of falsifiability and failure conditions as an ontological discipline.

4.8. The Ethics of Docking: Acknowledgment Rather Than Absorption

A Problem of Writing Ethics

The preceding sections of this chapter have treated the semantics of constitutive coextension. Sections 4.2–4.5 developed the definition of constitutive coextension, its logical features, its application to the five-volume physical work, and its comparison with contemporary philosophical-semantic tools. Section 4.6 further developed the mutual reinforcement between CC1 and the latest ontological thesis of this book, namely identity-level isomorphism. This section treats a deeper problem: the writing ethics of "docking" philosophers' concepts onto EQT.

This is not a technical problem. It concerns the posture with which precursor work is treated within the philosophical community. Specifically: when we connect Whitehead's actual occasions with EQT's gradient transients, are we acknowledging Whitehead's work, or are we absorbing it and making it serve our own purposes?

These two postures differ importantly. Acknowledgment means treating Whitehead's work seriously as an independent philosophical contribution: it has its own semantic field, its own internal logic, and its own features. Absorption means using Whitehead's work as material: extracting the parts we need and discarding the parts we do not.

The first posture, acknowledgment, is a legitimate use of philosophical tradition; the second posture, absorption, is philosophical dishonesty.

The Risk of Absorption in Docking-Style Writing

The "docking-style writing" I used in the previous five-volume physical work, such as "actual occasions = gradient transients," carries a risk of absorption. Specifically:

Risk one: flattening Whitehead's concrete content. When we say "actual occasions = gradient transients," we implicitly suggest that the two concepts are the same object. But Whitehead's actual occasions have concrete content, such as prehension, subjective form, and satisfaction, which cannot simply be identified with the concrete content of gradient transients. Simple identification flattens the specificity of Whitehead's work.

Risk two: treating Whitehead as the "prehistory" of EQT. If "actual occasions = gradient transients," then Whitehead seems to be doing an "unfinished EQT." The value of his work lies only in leading toward EQT. This treatment instrumentalizes Whitehead: his work is not taken seriously as an independent philosophical contribution.

Risk three: avoiding criticism. If Whitehead and EQT are "docked," then criticism of EQT must first criticize Whitehead, which raises the cost of criticism. This strategy of "borrowing authority" may be dishonest: it uses Whitehead as an authority without assuming the responsibility of detailed dialogue with Whitehead's work.

These risks are not abstract worries; they correspond to concrete forms of philosophical dishonesty. The work of this chapter, namely introducing constitutive coextension and developing its mutual reinforcement with identity-level isomorphism in Section 4.6, partly resolves these risks, but this ethical posture must be stated explicitly.

Acknowledgment as a Philosophical Posture

Constitutive coextension as a semantic tool supports the posture of acknowledgment rather than absorption.

First support: preserving concrete content. Constitutive coextension explicitly allows two concepts to retain concrete features within their respective semantic fields. Whitehead's actual occasions have concrete meanings in Whitehead's philosophical semantic field, such as prehension. These meanings are not flattened by the relation of constitutive coextension. The specificity of Whitehead's work is respected.

Second support: not asserting priority. Constitutive coextension is a symmetric relation. It does not say that "Whitehead is the prehistory of EQT" or that "EQT is the realization of Whitehead." It says that the two are constitutively coextensive within the process-ontology equivalence class. Both are independent philosophical contributions.

Third support: making the concrete connection explicit. Constitutive coextension requires concrete correspondence of failure conditions. It is not a general assertion, but a concrete argument. This concreteness makes "docking" a philosophical claim that can be seriously evaluated, rather than a rhetorical use of precursor authority.

Fourth support: making intensity forms explicit. Section 4.6 develops the intensity forms of CC1: the strongest form, identity level; the intermediate form, structural isomorphism; and the critical form, partial sharing. This allows the intensity of "docking" to be expressed precisely. In relation to Whitehead, this book adopts the intermediate form, structural isomorphism. In relation to Zhang Zai's qi theory, it adopts the critical form, partial sharing, in some respects. In relation to the mind–matter relation, it adopts the strongest form, identity level, in Chapter 13. This grading makes "acknowledgment" no longer a vague attitude, but content that can be precisely evaluated. Each interlocutor has a place within the intensity forms of CC1, and that place itself is the concrete form of acknowledgment.

The Boundary of Acknowledgment

But acknowledgment is not uncritical acceptance. Acknowledging Whitehead as an independent philosophical contribution does not mean agreeing with all of Whitehead's positions.

Specifically, this book has already argued in Chapter 8 that Whitehead's process philosophy is less thorough than Zhang Zai's qi theory, because Whitehead requires eternal objects as a second ontology, whereas qi theory is pure monism. This is a concrete criticism of Whitehead, built upon serious dialogue.

This criticism is a legitimate form of acknowledgment: it is built upon a serious understanding of Whitehead's work, gives a concrete point of criticism, rather than a general accusation, and allows defenders of Whitehead's position to respond concretely. The difference from absorption is this: absorption uses Whitehead as authority, while acknowledgment + criticism treats Whitehead seriously as an interlocutor.

The same posture of acknowledgment applies to Bergson, Prigogine, Bohr, and other philosophers and physicists who are "docked" in the work. They are all independent contributors to thought, and the dialogue with them should be serious, concrete, and open to criticism and reply.

A Concrete Reflection

Return to my earlier writing. In the five-volume physical work, I used docking-style writing. This mode of writing had its historical reasonableness: it allowed me to connect EQT work concisely to philosophical traditions and enabled readers to understand EQT through existing frameworks.

But the work of this chapter shows that this writing requires a stricter defense at the level of philosophical semantics. Constitutive coextension as a new semantic relation provides that defense (Sections 4.2–4.5). Section 4.6 further develops the internal structure of CC1, namely intensity forms, allowing "docking" statements to be expressed under precise intensity forms. This refinement thoroughly clarifies the semantic ambiguity of the earlier writing.

Furthermore, the work of this chapter shows the limitation of my earlier writing: I used semantic tools that were not precise enough. Acknowledging this is itself a form of philosophical honesty. This book, as a subsequent work, refines the earlier work, but the limitations of the earlier work need to be stated explicitly.

This self-reflection and self-revision are applications of the method of constrained speculation. The method itself requires the acknowledgment of limits and refinement. This chapter's refinement of my earlier writing is the application of that method to my own work.

Objections and Replies

Objection One: Is your distinction between acknowledgment and absorption overly moralized? Philosophical use of precursor work inevitably involves selection and reconstruction; this is not necessarily "dishonest."

Reply: Yes, philosophical use of precursor work inevitably involves selection and reconstruction. This is legitimate. This chapter does not claim that all selection and reconstruction are "absorption" or "dishonesty."

Legitimate acknowledgment + reconstruction: based on a serious understanding of precursor work, using its insights in one's own work, while explicitly stating differences from the precursor, including the CC1 intensity forms developed in Section 4.6. Illegitimate absorption: using precursor work as an authoritative resource, avoiding dialogue with the precursor, and using the precursor's language without bearing its semantic burden.

The difference lies in whether one assumes the responsibility of dialogue with the precursor. The posture suggested by this chapter is to assume that responsibility and to use constitutive coextension, including its intensity forms, as a tool for explicitly stating the content of the dialogue.

Objection Two: Is your posture of "acknowledgment" overly modest? EQT as a twenty-first-century physical work has its independent contribution. Why must it take precursors as interlocutors?

Reply: The posture of this chapter is not excessive modesty; it is mutual respect. On the one hand, EQT as physical work has independent contributions, as shown in the previous physical monographs. On the other hand, the works of Whitehead, Bergson, Prigogine, and others also have independent contributions. It is reasonable for the two sides to stand as equal interlocutors: neither does EQT serve Whitehead, nor does Whitehead serve EQT.

This equal dialogue is a legitimate posture in the relation between philosophy and science. It avoids two extremes: philosophy serving science, where philosophers are treated as the prehistory of science, and science serving philosophy, where scientific discoveries are treated as proofs of philosophical positions.

Objection Three: This chapter seems overly reflective: it discusses "how to write" rather than "concrete content." Is this a meta-level game?

Reply: As part of Part I, the methodological foundation, this chapter legitimately functions as meta-level reflection. Part I as a whole treats meta-level problems such as method, ontological criteria, and the semantics of physicalization, thereby establishing the foundation for later parts, which contain concrete ontology and concrete applications.

This chapter specifically treats the semantics of physicalization, which is the semantic foundation of all later "$X = Y$" statements. Without this meta-level work, the semantic status of later work would be blurred. Meta-level reflection is necessary here, not a game.

Furthermore, the method of constrained speculation itself requires methodological reflection: method cannot remain tacit; it must be made explicit and defended. This chapter is an application of such methodological reflection, treating the problem of "docking-style writing."

Literature That Must Be Engaged

- Pierre Hadot, La philosophie comme manière de vivre (1995): philosophy as ethical practice.
- John Rawls, A Theory of Justice (1971), on reflective equilibrium as method.
- Chapter 8 on the ethics of using resources from the Chinese philosophical tradition (self-reference).
- Chapter 13 on mind and matter as identity-level isomorphism (self-reference, directly related to the CC1 intensity forms in Section 4.6 of this chapter).

Summary of This Section

This section has done four things. First, it identified the risk of absorption in docking-style writing: flattening concrete content, treating precursors as prehistory, and avoiding criticism. Second, it showed how constitutive coextension, especially the intensity forms developed in Section 4.6, supports the posture of acknowledgment rather than absorption: preserving concrete content, not asserting priority, making concrete connections explicit, and making intensity forms explicit. Third, it clarified the boundary of acknowledgment: acknowledgment + criticism is legitimate, whereas pure acceptance is not. Fourth, it provided a self-reflection: the limitations of my earlier writing and the refinement made by this chapter, including the establishment of the semantic tool in Sections 4.2–4.5 and the development of its internal structure in Section 4.6.

4.9. Chapter Summary

The four tasks undertaken by this chapter have all been completed.

Task One: to diagnose the semantic ambiguity of docking-style writing. Section 4.1 showed, through the difficulties of five existing readings, that docking-style writing requires a stricter semantic defense.

Task Two: to introduce constitutive coextension as a new semantic relation. Section 4.2 gave the precise definition and core features. Section 4.3 developed its formal-logical features, namely symmetry, reflexivity, and transitivity, making it an equivalence relation, as well as deeper features, namely stratification, concreteness, and categorial correspondence.

Task Three: to apply constitutive coextension to concrete physicalization statements and clarify its position in contemporary philosophical semantics. Section 4.4 restated the main "physicalization" statements in the five-volume physical work: actual occasions, durée, dissipative structures, and Bohr complementarity. Section 4.5 engaged contemporary philosophy, including Davidson, Kim, Putnam, and Kripke, and clarified the distinctive position of constitutive coextension.

Task Four: to develop the mutual reinforcement between CC1 and the latest ontological thesis of this book, namely identity-level isomorphism in Chapter 13, and to clarify the ethical posture of docking. Section 4.6 argued that CC1 and identity-level isomorphism mutually reinforce each other: CC1 provides methodological safeguards for identity-level isomorphism, turning the latter from a seemingly metaphysical assertion into an ontological commitment constrained by methodological discipline; identity-level isomorphism in turn confirms that CC1 is not an intermediate position, but a semantic relation family with a broad span, including the strongest form, identity level; the intermediate form, structural isomorphism; and the critical form, partial sharing. Section 4.7 treated the ethics of docking: acknowledgment rather than absorption, and incorporated the intensity forms of Section 4.6 into the content of the posture of acknowledgment.

The Position of This Chapter in the Whole Book

Looking backward, this chapter builds upon Chapter 3, "Failure Conditions as an Ontological Discipline." The definition of constitutive coextension directly depends on failure conditions: two concepts are coextensive under the same failure conditions. The ontological framework established in Chapter 3 provides the foundation for the semantic work of this chapter.

Looking forward, this chapter provides semantic tools for all later chapters. Chapter 8, on the dialogue with qi theory, uses different intensity forms of constitutive coextension: qi theory and gradient ontology are constitutively coextensive in some respects, as structural isomorphism, and reach deeper mutual enrichment in the dialogue between "form" and "boundary as gate." Chapter 12, on co-evolution, uses constitutive coextension, where "Bohr complementarity" and "co-evolutionary measurement" are constitutively coextensive. Chapter 13, on mind and matter, uses the strongest form of constitutive coextension: identity-level isomorphism, and this connection is developed in Section 4.6 of this chapter. Chapter 14, on observation, uses constitutive coextension, where "participatory measurement" and "internal–external co-evolution" are constitutively coextensive. Almost all later statements of the form "$X = Y$" use constitutive coextension as their precise semantic tool and are precisely expressed according to their position in the CC1 intensity spectrum.

The Core Posture of This Chapter

The core posture of this chapter can be stated as follows: the rigor of philosophy does not lie in solving all problems, but in stating precisely what it is doing.

The method of constrained speculation was stated in Chapter 1 as the methodological foundation of this book. This chapter is the application of that method at the meta-semantic level: it refines the existing writing, namely the docking-style writing in the five-volume physical monograph series, into a more precise semantic tool, constitutive coextension, and through the mutual reinforcement between CC1 and identity-level isomorphism in Section 4.6, gives this semantic tool an internally gradable structure.

This self-refinement does not abandon the earlier work. It clarifies the semantic status of the earlier work and enables it to enter more serious philosophical discussion. Philosophical progress is often refinement rather than overthrow. The relation between this book and the five-volume physical work is precisely one of refinement.

Honesty about the Limits of This Chapter

This chapter acknowledges several limits.

First limit: constitutive coextension as a concept is an initial statement. This chapter gives its definition, formal features, and mutual reinforcement with the latest argument in Section 4.6, but a complete development of the semantics requires further work: more detailed formal logic, more detailed comparison with other semantic relations, and more application examples in different fields.

Second limit: the four restatement examples, namely actual occasions, durée, dissipative structures, and Bohr complementarity, are only skeletons. Every complete constitutive-coextension statement requires a more detailed comparison of failure conditions. This chapter gives the core argument, but each statement requires dedicated work for refinement. Likewise, the intensity classification given in Section 4.6, namely which statements are strongest, which are intermediate, and which are critical, is preliminary. The precise intensity of each statement must be confirmed through a more detailed comparison of failure conditions.

Third limit: the dialogue with non-analytic traditions is insufficient. This chapter concentrates on the tradition of analytic philosophical semantics. Dialogue with related concepts in phenomenology, hermeneutics, and the Chinese philosophical tradition needs to be expanded. This is future work.

Fourth limit: the relation of constitutive coextension itself may be revised in use. When concrete applications reveal problems, the form of constitutive coextension, including the internal division of its intensity spectrum, may need adjustment. This is legitimate: every new concept is refined through application.

Acknowledging these limits does not weaken the work of this chapter. It makes the posture of this chapter honest, criticizable, and developable.

Connection with the Whole Book

This chapter is the most technical chapter in Part I of the whole book. Its work is meta-level: it provides semantic tools for all subsequent concrete ontological work in Parts II, III, IV, and V, and through Section 4.6 makes explicit the mutually constitutive relation between this tool and the deepest ontological thesis of the book.

In one sense, this chapter is the semantic insurance for the later work. It ensures that the later work no longer falls into the semantic ambiguity of docking-style writing. When readers encounter statements such as "X and Y are constitutively coextensive" in later chapters, they can return to this chapter to understand the precise meaning of that statement, including its place in the CC1 intensity spectrum.

This chapter also supports the international positioning of the book, already stated in the Introduction. The analytic philosophical tradition has high standards for semantic rigor. Any work that uses vague "$X = Y$" statements will be criticized as insufficiently rigorous. This chapter enables the book to satisfy the semantic rigor standard of the analytic tradition and to enter international philosophical discussion.

Meaning before freezing: this is one of the core theses of the book. The work of this chapter supports the semantic status of that thesis. When we say that "meaning depends on gradient structure" in Chapter 21, we are using a relation of constitutive coextension: the concept of meaning and the concept of gradient structure correspond in failure conditions, while retaining their respective semantic independence. This precise semantics allows the work of Chapter 21 to avoid the strong claim of "reduction" or "identity," and preserves the independence of humanistic meaning.

4.10. Methodological Reflection on This Chapter

This chapter introduces constitutive coextension as a new semantic relation. This section reflects on the methodological significance of this introduction.

This Chapter as Self-Reflection

This chapter has criticized and refined the "docking-style writing" of the author's previous five-volume physical work. This self-criticism is not a weakness in narration; it is an application of the method of constrained speculation. Acknowledging the limits of earlier work and refining them is an honest philosophical posture. This self-reflection gives this book an honest position in relation to the five-volume monograph series: the "docking-style writing" in the five volumes has legitimacy in the physical context, but in the context of a philosophical monograph it requires stricter semantic defense.

The Originality of Constitutive Coextension as a Concept

Constitutive coextension as a concept is new. Section 4.5 showed its comparison with contemporary philosophical semantics. This chapter bears the responsibility of introducing a new concept, which requires detailed definition, rigorous argument, and comparison with existing concepts. But the work of this chapter is not created out of nothing. Constitutive coextension is built upon several precursor traditions: the direction of Davidson's anomalous monism, Brandom's inferentialist rejection of correspondence semantics, and Putnam's treatment of "two-level semantics" in internal realism. The contribution of this chapter is to systematize these directions into a precise semantic relation and to make the legitimacy of constitutive coextension rigorously evaluable through failure conditions as an ontological discipline (CR2).

The Mutual Constitution between This Chapter and the Latest Argument of This Book

Section 4.6 develops an important extension of this chapter's work: the mutual reinforcement between CC1 and the latest ontological thesis of this book, namely identity-level isomorphism in Chapter 13. This extension makes this chapter not only a methodological preparation for Chapter 13, but also an object of reverse reinforcement by the philosophical contribution of Chapter 13: the identity-level isomorphism of Chapter 13 gives CC1 in Chapter 4 a more refined internal structure, namely intensity forms: strongest, intermediate, and critical; CC1 in Chapter 4 gives the identity-level isomorphism of Chapter 13 methodological safeguards, making it an ontological commitment constrained by CR2 rather than a metaphysical assertion made without grounds. The two chapters form a mutually constitutive relation in the work of this book, rather than a one-way dependence.

This mutual constitution is an example of the work of this book: a semantic tool (CC1) and an ontological thesis (identity-level isomorphism) mutually reinforce each other through internal coordination, rather than being connected through external superimposition. This example provides a template for the relation between other semantic tools and ontological theses in this book, such as tri-state categories and freezing as a basic ontological process, or the discipline of falsifiability and failure conditions as an ontological discipline.

The Position of This Chapter within Analytic-Philosophical Rigor

This chapter is proof of the book's analytic-philosophical rigor. The analytic philosophical tradition has very high requirements for semantic precision. The work of this chapter, namely identifying the problem, introducing a new concept, making detailed comparisons, giving strict definitions, and developing the intensity spectrum in

Section 4.6, satisfies the standards of this tradition. This allows the book to be taken seriously in international analytic-philosophical discussion, rather than being criticized as "semantically vague natural philosophy." "Docking-style writing" is often allowed in the tradition of natural philosophy, but this book aims to have a place in international analytic-philosophical discussion. This requires the semantic status of the book's work to be clear. Through constitutive coextension, including its intensity spectrum, this chapter strictly states that semantic status.

Limits Honestly Acknowledged in This Chapter

This chapter acknowledges five limits. First, constitutive coextension is an initial statement, and its refinement, including the internal division of the intensity spectrum developed in Section 4.6, requires later work. Second, the four examples given in this chapter are skeletons: many other physicalization relations need similar treatment and must be precisely classified according to the CC1 intensity spectrum. Third, the dialogue between constitutive coextension and semantic tools from non-analytic traditions, especially the Continental tradition and the Chinese tradition, is insufficient; these dialogues require dedicated work. Fourth, constitutive coextension as a relation may be revised in application: the initial definition in this chapter, including the grading of the intensity spectrum, does not pretend to be final. Fifth, the work in Section 4.6 on the mutual reinforcement between CC1 and identity-level isomorphism is preliminary: more detailed arguments concerning methodological safeguards, namely how CC1 makes identity-level isomorphism constrained by CR2, and reverse arguments, namely how identity-level isomorphism enriches the internal structure of CC1, require dedicated work.

Connections with Later Chapters

The constitutive coextension established in this chapter is widely used in later chapters. Chapter 8, on the dialogue between qi theory and gradient ontology, uses constitutive coextension, including the two-way dialogue developed in Section 8.6 between qi theory and "boundary as gate," which is itself an application of CC1 in a cross-cultural context. Chapter 13, on mind and matter, uses the strongest form of constitutive coextension: identity-level isomorphism, whose deepest position as the strongest form of CC1 is developed in Section 4.6 of this chapter. Chapter 14, on observation, uses constitutive coextension. More broadly, this chapter provides semantic insurance for the whole book. Without this chapter, all later statements of the form "$X = Y$" in the whole book would lose their explicit semantic defense. As the semantic-tool chapter of Part I, this chapter ensures that the ontological work of the whole book unfolds under precise semantics, including precise classification within the intensity spectrum.

In particular, this chapter and Chapter 13 form a mutually constitutive relation. This relation is an example in which the internal coordination between the two most important philosophical devices in the work of this book, namely constitutive coextension as a semantic tool and identity-level isomorphism as the deepest ontological thesis, is made explicit.

5. Positioning within the Genealogy of Contemporary Realism

5.1. Introduction to This Chapter

After the methodological work of Chapters 1–4—the method of constrained speculation (Chapter 1), ontological restraint (Chapter 2), failure conditions as an ontological discipline (Chapter 3), and constitutive coextension (Chapter 4)—an ontological position has already taken shape in practice. But this position has not yet been named or located. Where does it stand within the genealogy of contemporary realism? Which precursors does it inherit? What is its relation to contemporary realists?

This chapter undertakes that work of positioning. First, it integrates the methodological work of Chapters 1–4 into an ontological position and gives it a name: constrained process realism. Second, it enters into serious dialogue with five important positions in the genealogy of contemporary realism: Nancy Cartwright's entity realism, James Ladyman's ontic structural realism, Luciano Floridi's information realism, Carlo Rovelli's relational quantum mechanics and its philosophy, and Russellian monism. These five positions represent different directions in contemporary realism, and the work of this book needs to be clearly located among them.

This chapter has a clear requirement in its writing posture: serious respect and precise distinction. This book must avoid two traps: the trap of eclecticism, which claims to synthesize the strengths of all positions, and the trap of isolationism, which claims that the work of this book is entirely independent of existing traditions. The work of each interlocutor has enduring philosophical influence, and this book does not pretend that it can easily surpass them. But the work of this book has an independent position, and that position needs to be shown clearly. This posture prevents the chapter from collapsing either into simple reverence for predecessors or into simple criticism of them.

As the name of a position, constrained process realism contains three terms: realism, in the sense that beings have their own ontological content independent of an observer's perception or description; process, in the sense that beings are processual rather than static entities; and constrained, in the sense that ontological commitments are constrained by methodological discipline. Together, these three terms determine the position of this stance within the genealogy of contemporary realism: it inherits insights from several traditions, but it has its own independent features.

This chapter is the positioning chapter of Part I, "Constrained Speculation: Method as Philosophy." After the methodological foundation laid in Chapters 1–4, it integrates the content of these methods into a complete ontological position and locates this position in the context of contemporary philosophical discussion. This positioning prepares the ground for the synthesis in Chapter 6, "Falsifiability as an Ontological Discipline."

The discussion proceeds as follows. Section 5.1 gives a statement of the position of constrained process realism; Section 5.2 enters into dialogue with Cartwright's entity realism; Section 5.3 enters into dialogue with Ladyman's ontic structural realism; Section 5.4 enters into dialogue with Floridi's information realism; Section 5.5 enters into dialogue with Rovelli's relational quantum mechanics; Section 5.6 enters into dialogue with Russellian monism, treating it as this book's nearest neighbor at the ontological level; Section 5.7 concludes the chapter.

5.2. Constrained Process Realism: Statement of the Position

Naming the Position

The methodological work of Chapters 1–4, including constrained speculation, ontological restraint, failure conditions, and constitutive coextension, needs to be integrated into an ontological position. This section gives the name and statement of that position.

The position of this book: constrained process realism.

The name of this position contains three terms:

- Realism: beings have their own ontological content, independent of an observer's perception or description.
- Process: beings are processual, not static entities; ontological content involves dynamics, difference, and change.
- Constrained: ontological commitments are constrained by methodological disciplines, including constrained speculation, failure conditions, and falsifiability.

Together, these three terms constitute the name of the position. Each term performs a function: realism states the basic stance, process states the ontological content, and constrained states the methodological discipline.

Features of the Position

Constrained process realism as a position has six features, all of which are built upon the work of Chapters 1–4:

Feature One: The Processuality of Ontology

This position understands beings as processual: the basic ontological units are not static entities, but processual structures, such as structures of difference, gradient structures, and co-evolutionary structures.

More specifically, building on Chapter 7 on difference preceding identity, what is called an "entity" is the stable form of a process, namely a frozen transient; what is called an "attribute" is the content of a process; what is called a "relation" is the dynamical connection of a process; and what is called "change" is the evolution of a process.

This feature distinguishes the position from traditional entity realism, such as Aristotelianism, Kantianism, and related positions.

Feature Two: Ontological Restraint

This position commits itself to the least possible ontological content, while requiring its commitments to be strong enough to address ontological problems (Chapter 2, $PR1$).

Specifically, the work of this book is built upon a minimal set of ontological commitments: the five foundational contents of EQT ($A1$–$A4+M$), the core thesis of the ontology of difference ($DI1$), co-evolution as an ontological relation ($EV1$), and related commitments. All ontological problems are treated through the application of these foundational commitments, rather than by introducing an independent commitment for each problem.

Feature Three: Failure Conditions as the Standard of Legitimacy

This position uses failure conditions as the standard of legitimacy for ontological commitments (Chapter 3, $CR2$). Specifically: every ontological commitment needs to have an identifiable direction of failure; a statement without a direction of failure is not a legitimate ontological commitment; and this standard makes ontology rigorously evaluable as a field of research.

Feature Four: Constitutive Coextension as a Semantic Tool

This position uses constitutive coextension (Chapter 4, $CC1$) as a semantic tool. Specifically, it does not use "docking" but "coextension": two commitments are coextensive under the same failure conditions while retaining their respective conceptual frameworks. This tool avoids the traditional dichotomy between "reduction" and "dualism," and it gives interdisciplinary and cross-traditional dialogue a semantic ground.

Feature Five: Constrained Objectivity

This position grounds objectivity in controlled reproducibility, not in observer-free externality (Chapter 14, $OB1$). Specifically: objectivity does not require being "independent of the observer"; it allows the participation of the observer. Objectivity is grounded in the reproducibility of co-perturbation: observation, as a form of co-evolution, can be strictly repeated. This concept of objectivity distinguishes the position from traditional "observer-free" objectivity.

Feature Six: Openness to Interdisciplinary and Cross-Traditional Dialogue

This position is open to interdisciplinary dialogue, including dialogue with physics, neuroscience, sociology, ethics, and other fields, and to cross-traditional dialogue, including dialogue with Chinese and Indian traditions. Specifically, the work of this book is built upon physics, namely the EQT framework; this is its interdisciplinary connection with science. The work of this book uses resources from the Chinese philosophical tradition, especially Chapter 8 on qi theory; this is its cross-traditional dialogue with the Chinese tradition. The work of this book also dialogues with various Western traditions, as in the present chapter's dialogue with contemporary realism and Chapter 8, Section 8.5's dialogue with Whitehead.

This openness means that the position is not an isolated philosophical stance: it has a place in multiple discussions.

Formal Statement of the Position

Synthesizing these six features, the position can be stated formally as follows:

Constrained process realism: ontological content is processual, involving structures of difference and co-evolutionary relations; ontological commitments are constrained by methodological disciplines, such as failure conditions, falsifiability, and restraint; objectivity is grounded in controlled reproducibility; and the position is open to interdisciplinary and cross-traditional dialogue.

This statement is the core of the work of this chapter: it integrates the methodological work of Chapters 1–4 and provides the foundation for the subsequent dialogue with the genealogy of contemporary realism.

What the Position Is Not

To state the position precisely, this section makes clear what the position is not:

It is not simple realism: the position does not claim that "beings are completely objective independently of observers." Constrained process realism acknowledges the co-evolution of observer and observed, and grounds objectivity in reproducibility.

It is not constructivism: the position does not claim that "beings are socially constructed." Constrained process realism acknowledges the objective content of beings, and this content is not simply reduced to social construction.

It is not anti-realism: the position does not claim that "beings are merely convenient theoretical constructions." Constrained process realism gives ontological commitments ontological status: they are not merely tools, but statements about beings.

It is not relativism: the position does not claim that "all ontological statements are equally legitimate." Constrained process realism uses the discipline of falsifiability to evaluate statements, and the legitimacy of a statement is determined by methodological standards.

It is not naturalistic reduction: the position does not claim that "all ontological content can be reduced to physics." Constrained process realism acknowledges multiple levels of ontological content, including physics, biology, psychology, sociology, and others. These levels are connected through constitutive coextension, but they retain their independence.

These five "nots" clarify the boundary of the position and prevent confusion with other positions.

Objections and Replies

Objection One: The name of the position you give, "constrained process realism," is synthetic: it seems to try to satisfy multiple directions at once by adding several modifiers.

Reply: The modifiers in the name of the position are not added arbitrarily; each term has a function. "Realism" distinguishes the position from anti-realism and relativism; "process" distinguishes the position from static entity realism; "constrained" distinguishes the position from forms of realism not constrained by methodological discipline. These three terms each perform a function of distinction. The synthesis is for precision, not for satisfying multiple directions. The position needs all three terms in order to be stated precisely.

Such synthetic naming is common in the history of philosophy. For example, Kant's "transcendental idealism" is synthetic, combining transcendental and idealism; Husserl's "transcendental phenomenology" is synthetic, combining transcendental and phenomenology. Synthetic naming is not a defect in itself, but a sign of precision.

Objection Two: Your six features seem too numerous. Is the position overly complex? Simpler positions are easier to understand and to test.

Reply: The six features of the position reflect the complexity of the work of this book. Each feature performs a function and cannot be simplified away. Without "the processuality of ontology," the position would collapse into static entity realism. Without "ontological restraint," the position would lack methodological discipline. Without "failure conditions as the standard of legitimacy," the position could not handle ontological commitments. Without "constitutive coextension as a semantic tool," the position could not handle interdisciplinary and cross-traditional dialogue. Without "constrained objectivity," the position could not handle the observer–observed relation. Without "openness," the position would be isolated from contemporary discussion.

The six features together constitute a complete position: if one is removed, the position becomes insufficient. The complexity is for completeness, not for complexity itself. A simple position is not necessarily easier to test; simplicity may conceal problems. A complex position, if each feature is strictly stated, may instead be easier to test, because each feature can be independently evaluated.

Objection Three: Your list of "what the position is not" seems overly defensive: it defines the position by excluding other positions rather than by its own content.

Reply: The position is defined jointly by positive statements, namely the six features, and negative statements, namely the five "nots." Both are needed. Positive statements give the content of the position, what it is; negative statements distinguish it from neighboring positions, what it is not. Positive statements alone easily produce confusion: readers may confuse the position with some neighboring position. Negative statements alone would make the position empty: it would merely "oppose other positions." Together, they define the position precisely.

Specifically, this section gives the "five nots" because constrained process realism can easily be confused with five positions: simple realism, constructivism, anti-realism, relativism, and naturalistic reduction. These are important positions in contemporary philosophical discussion. Making clear what the position is not clarifies the difference between the position of this book and these positions. This treatment is legitimate in philosophical discussion: a position becomes clear through precise distinction.

Literature That Must Be Engaged

- Chapter 1 on the method of constrained speculation (self-reference)
- Chapter 2 on minimal postulates and ontological restraint (self-reference)
- Chapter 3 on failure conditions as an ontological discipline (self-reference)
- Chapter 4 on constitutive coextension (self-reference)

Summary of This Section

This section has done four things. First, it gave the name of the position of this book: constrained process realism. Second, it developed the six features of the position: the processuality of ontology, ontological restraint, failure conditions as the standard of legitimacy, constitutive coextension as a semantic tool, constrained objectivity, and openness to interdisciplinary and cross-traditional dialogue. Third, it gave a formal statement of the position. Fourth, it clarified the difference between this position and neighboring positions through the "five nots."

The following five sections develop dialogues with five contemporary interlocutors.

5.3. Dialogue with Nancy Cartwright's Entity Realism

Cartwright's Position

Nancy Cartwright is an important figure in contemporary philosophy of science. Her work, especially How the Laws of Physics Lie (1983), Nature's Capacities and their Measurement (1989), and The Dappled World (1999), develops the positions of entity realism and the dappled-world view. Cartwright's core position is as follows.

Entity realism: in scientific practice, we should adopt a realist stance toward beings, such as entities and causal capacities, but an anti-realist stance toward scientific laws, namely universal law statements. Specifically, we can be confident that electrons exist as entities, because we use electrons in experiments, observe effects, and operate technologies. But we should be suspicious of universal scientific laws: they are idealized statements, require many auxiliary conditions in application, and often do not accurately describe reality.

The dappled-world view: the real world is not a world described by unified universal laws. It is dappled, containing diverse local regularities and local causal capacities. Specifically, different fields, such as physics, biology, and economics, have different laws; these local laws cannot be simply integrated into unified universal laws; and the diversity of the real world is reflected in the locality of science.

The contribution of Cartwright's work is to take scientific practice, such as experimental operation and technological application, as an ontological ground rather than relying on abstract theoretical statements; to reject the imperialism of physics, in the sense that other sciences need not be reduced to physics; and to emphasize the ontological status of causal capacities and local regularities.

Relation to Constrained Process Realism

Cartwright's work and the position of this book have similarities and differences. Let us develop them.

Similarity One: Emphasis on Practice and Reality

Cartwright's entity realism is built upon scientific practice: experimental operation and technological application give us realist grounds for entities such as electrons. The position of this book is similar, especially Chapter 14 on the new foundation of objectivity: objectivity is grounded in reproducible co-perturbation, that is, in the reproducibility of scientific practice. Both positions take practice as an ontological ground.

Similarity Two: Suspicion of Unified Universal Laws

Cartwright's dappled-world view is suspicious of unified universal laws. The position of this book does not claim that all content can be reduced to a single law; it uses constitutive coextension (Chapter 4) to handle cross-level relations and acknowledges that different levels have different forms of description. Both positions are suspicious of excessive unification.

Key Difference One: Entity as Foundation vs Difference as Foundation

Cartwright's entity realism takes entities as foundational: beings as entities, such as electrons and atoms, are basic ontological units. The position of this book takes difference as foundational (Chapter 7, $DI1$): beings as structures of difference are the basic ontological content. What is called an "entity" is the stable form of a structure of difference, not a basic ontological unit. This difference is not terminological; it concerns a fundamental difference in ontological starting point. Cartwright works within an entity-ontological framework, while this book works within an ontology of difference.

Key Difference Two: The Treatment of Universal Laws

Cartwright adopts an anti-realist stance toward universal laws: laws are only idealized statements. The position of this book does not simply accept this treatment. This book uses the method of constrained speculation (Chapter 1) to evaluate statements. Some universal statements, such as energy conservation and $E = h\nu$, satisfy the disciplines of constrained speculation. They are not simply "idealizations," but statements with ontological content.

Specifically: idealized laws, such as the ideal gas law, involve idealized conditions and require correction in reality. Cartwright's suspicion applies to such laws. Foundational physical statements, such as energy conservation, have identifiable directions of failure and are continuously supported in experiments. Such statements are not simple "idealizations." The position of this book distinguishes these two types. Cartwright's work does not fully develop this distinction.

Key Difference Three: Dappled World vs Unified Process

Cartwright's dappled-world view emphasizes the locality and diversity of the world: different fields have different laws. The position of this book is not a dappled-world view. Specifically, the work of this book is built upon a unified ontology: the ρ_f field as the unique basic ontological object (EQT $A1$). The laws of different fields, such as physics, biology, and psychology, are different level-manifestations of the same ontology, not independent collections of laws. Cross-field relations are treated through constitutive coextension (Chapter 4).

The position of this book does not cancel apparent diversity: different fields indeed have different forms of description. But unity at the ontological level allows diversity to be built upon a unified ontology. This difference is not "excessive unification": the work of this book does not impose a unified law upon different fields. It acknowledges the legitimacy of laws at different levels, but grounds diversity in a unified ontology rather than in a dappled ontology.

Positioning the Two Positions

Through this comparison, the relation between the position of this book and Cartwright's work can be stated as follows: constrained process realism agrees with Cartwright's work on the ontological importance of scientific practice, but differs from it in ontological starting point, namely difference vs entity, and in the unity of ontology, namely unified process vs dappled world. The two constitute different directions in the genealogy of contemporary realism.

This positioning avoids two traps: the trap of absorption, which would claim that the work of this book "absorbs" the strengths of Cartwright's work; that would be dishonest, since the work of this book makes independent choices in starting point and ontology. It also avoids the trap of simple criticism, which would claim that Cartwright's work is "wrong"; Cartwright's work has enduring force, and this book does not attempt to negate it easily. As different directions in contemporary realism, both have legitimacy. Researchers may choose the appropriate framework according to the problem.

Objections and Replies

Objection One: Your key difference two, namely the distinction between idealization and foundational physical statements, seems too simple. Every statement involves idealizing conditions, so the distinction is not strict.

Reply: Statements do indeed involve different degrees of idealization. But this does not cancel the distinction; it makes the distinction more refined. Some statements involve many idealized conditions and require many corrections in application, such as classical-mechanical laws applied to complex systems. Some statements involve minimal idealized conditions and are relatively direct in application, such as energy conservation. The difference of degree is continuous, but the distinction remains meaningful. Furthermore, the work of this book does not claim that some statements are "completely free of idealization." All physical statements involve some idealization. But the method of constrained speculation provides standards for evaluating statements: how precisely their failure conditions are stated, how rigorously their observational proxies are used, and so on. These standards make the legitimacy of statements rigorously evaluable and do not depend on a simple dichotomy between "idealized" and "non-idealized."

Objection Two: Is your contrast between "unified process vs dappled world" too strong? The work of this book acknowledges that different levels have different forms of description, which agrees with Cartwright's dappled-world view.

Reply: The work of this book agrees with Cartwright's work that different levels have different descriptions. This is a fact of scientific diversity. But there is a difference at the ontological level. Cartwright's dappled-world view treats diversity as ontological diversity: different levels have different ontological contents. The work of this book treats diversity as descriptive diversity: different levels have different forms of description, but ontological content is unified. This difference is not a superficial terminological difference; it concerns the core of ontology.

The consequence is that Cartwright's dappled-world view has difficulty with cross-level integration: if ontology is diverse, cross-level integration requires complex explanation. The position of this book treats cross-level integration through constitutive coextension: descriptions at different levels are coextensive under the same failure conditions while retaining their independence. Both treatments have legitimacy. The work of this book chooses the latter because it makes cross-level integration more tractable.

Objection Three: Cartwright's entity realism is built upon scientific practice, which is a direct ontological ground. Your ontology of difference seems more abstract: it is built upon philosophical argument.

Reply: The position of this book does not claim that the ontology of difference is more concrete than Cartwright's work. It claims that the two are similar in degree of concreteness, but realize concreteness in different ways. Cartwright argues for the existence of ontology, namely entities, through scientific practice. The work of this book argues for the existence of ontology, namely structures of difference, through methodological discipline + scientific practice + philosophical argument. The work of this book is not limited to philosophical argument; it is built upon scientific practice, namely the connection between the EQT framework and physics. Chapter 14 on the new foundation of objectivity develops this connection: reproducible co-perturbation as the foundation of objectivity is precisely the philosophical articulation of scientific practice. The difference is not in the degree of concreteness, but in starting point and mode of synthesis. Both works have legitimacy.

Literature That Must Be Engaged

- Nancy Cartwright, How the Laws of Physics Lie (1983)
- Nancy Cartwright, Nature's Capacities and Their Measurement (1989)
- Nancy Cartwright, The Dappled World (1999)
- Ian Hacking, Representing and Intervening (1983): early work on entity realism.
- Chapter 7 on difference preceding identity (self-reference)
- Chapter 14 on the new foundation of objectivity (self-reference)

Summary of This Section

This section has done four things. First, it developed the core positions of Cartwright's work: entity realism and the dappled-world view. Second, it identified similarities between Cartwright's work and the position of this book: the importance of practice and suspicion of unified universal laws. Third, it identified differences: entity vs difference as foundation, the treatment of universal laws, and unified process vs dappled world. Fourth, it clarified the different positions of the two within the genealogy of contemporary realism.

The next section develops the dialogue with Ladyman's ontic structural realism.

5.4. Dialogue with James Ladyman's Ontic Structural Realism

Ladyman's Position

James Ladyman, working with Don Ross and others, developed ontic structural realism (OSR). The representative work is Every

Thing Must Go: Metaphysics Naturalized (2007). The core position of Ladyman's work is as follows.

Ontic structural realism: the foundational ontological content is structure, or relation, not entity, or thing. Specifically, what physics tells us is not facts about things, but facts about structures, such as symmetries, conservation laws, and relations. Entities, such as elementary particles, are not basic ontological units: they are nodes of structure, determined by structural relations. In light of quantum mechanics, the identity of things itself is unclear: the status of particles as independent things is difficult to maintain.

The motivations of Ladyman's work are: the content of physics, especially contemporary physics, quantum mechanics, and quantum field theory, which show that the identity of things itself is problematic; naturalistic metaphysics, according to which metaphysics should connect with science and should not do ontology independently of science; and criticism of traditional metaphysics, since necessitarianism, essentialism, and related positions conflict with contemporary science.

The contribution of Ladyman's work is to take structure as foundational ontology: this is an alternative to traditional entity ontology. It emphasizes the naturalism of metaphysics: ontology should be built upon science. It also establishes a connection with physics, especially quantum mechanics.

Relation to Constrained Process Realism

Ladyman's work and the position of this book have a complex relation: they are close along some dimensions and different along others.

Similarity One: Suspicion of Entities

Ladyman's ontic structural realism is suspicious of entities as basic ontological units. The position of this book is also suspicious of them: the ontology of difference (Chapter 7) treats what are called "entities" as stable forms of structures of difference, not as basic ontological units. The two share a direction in rejecting entity ontology.

Similarity Two: Connection with Physics

Ladyman's work is built upon physics, especially quantum mechanics. The work of this book is also built upon physics, namely the EQT framework. The two share a direction in connecting with physics. This is the core direction of contemporary naturalistic metaphysics.

Similarity Three: Naturalistic Metaphysics

Ladyman argues that metaphysics should connect with science.

This is consistent with the direction of this book, in which the method of constrained speculation requires observationalization, mechanization, and related connections with science.

Key Difference One: Structure vs Process as Foundation

Ladyman's ontic structural realism takes structure, or relation, as foundational: structure is the foundational ontological content. The position of this book takes process, namely difference, change, and co-evolution, as foundational: structure is the content of process, not the other way around. This difference is not terminological; it concerns a fundamental difference in ontological starting point. Ladyman treats foundational ontology as static structure, with relation as foundation; the work of this book treats foundational ontology as dynamic process, with difference and co-evolution as foundation.

The consequence is that, in Ladyman's framework, change and evolution as ontological contents need to be defended: they seem to be contents outside structure. In the framework of this book, change and evolution are process itself: structure appears as the stable form of process.

Key Difference Two: The Treatment of "Things"

Ladyman's slogan is "Every Thing Must Go": all things must be eliminated, and only structure remains as ontological content. This is an eliminativist position. The work of this book does not eliminate "things": it re-treats what are called "things" as stable forms of structures of difference. Everyday objects, such as tables, persons, and stars, still exist as beings; only their ontological ground is a structure of difference rather than an independent entity.

Specifically, Ladyman's position is that electrons do not exist as things; only structural relations exist. The position of this book is that electrons exist as things: they are excitation modes of the ρ_f field, namely stable forms of structures of difference. The position of this book is more moderate than Ladyman's work: it does not require the elimination of "things," but only the reinterpretation of their ontological ground.

Key Difference Three: The Treatment of Structure

Ladyman's ontic structural realism takes structure as foundational ontological content: structure exists independently and does not depend on relata. The work of this book does not claim such "independent structure": structure is the content of process and depends on dynamic process. Specifically, in Ladyman's framework, structure as foundation has a problem of "floating": it seems to be some independently existing object. In the framework of this book, structure is the form of process: it is built upon difference and co-evolution and does not float. This difference allows the position of this book to avoid a difficulty in Ladyman's work: the ontological status of "structure as independent foundation."

Positioning the Two Positions

Through comparison, the relation between the position of this book and Ladyman's work can be stated as follows: constrained process realism agrees with Ladyman's ontic structural realism in its suspicion of entity ontology and its connection with physics, but differs from it in the content of foundational ontology, namely process vs structure, and in the treatment of "things," namely stable forms of structures of difference vs elimination. The position of this book is processual; Ladyman's position is structural. This positioning gives both positions their own features: they are two directions in contemporary anti-entity ontology.

Objections and Replies

Objection One: Is your contrast between "process vs structure" too strong? Process and structure seem to be interdependent: every process involves structure, and every structure involves process.

Reply: Process and structure are indeed interdependent. This is common sense. But the choice of foundational ontological starting point has consequences. If process is taken as the foundational starting point, structure is the stable form of process; structure is derivative. If structure is taken as the foundational starting point, process is the change of structure; process is derivative. Both choices can handle the interdependence of process and structure, but they differ in the direction of derivation.

The consequence is that an ontology based on process naturally treats change and evolution: they are foundational contents and do not need additional defense. An ontology based on structure needs to defend change and evolution: they seem to be contents outside structure. The work of this book chooses process as the foundational starting point because the content of contemporary physics, especially quantum field theory, is closer to processuality: it treats fields and excitations, not static structures. The content of phenomenology is also closer to processuality: experience involves dynamics. The Chinese philosophical tradition, especially qi theory, also has a processual direction (Chapter 8). These grounds make process as a foundational starting point more coherent than structure as a foundational starting point: it treats multiple kinds of content more directly. But the work of this book does not claim that "process as foundation" is absolutely correct. It is a choice, and it has grounds. Ladyman's work also chooses structure as foundation, and it too has its grounds. The two constitute different directions in contemporary ontology.

Objection Two: Ladyman's work is built upon the content of quantum mechanics, especially the problem of the identity of things. Your ontology of difference seems not to treat this problem directly.

Reply: The work of this book does treat this problem. The problem of the identity of things in quantum mechanics involves the indistinguishability of particles: the identity of particles of the same kind, such as all electrons. Within the framework of this book, all electrons are excitation modes of the ρ_f field: they are different instances of the same excitation mode. Their "identity" is built upon the internal structure of the ρ_f field, namely the manifestation of excitation modes, and does not require an independent "entity identity." This treatment agrees with Ladyman's treatment in rejecting independent entities: electrons do not exist as independent entities. But the work of this book treats them as stable forms of structures of difference, not as simply eliminated. This treatment preserves the content of common sense, namely the identity of electrons as beings, while also agreeing with the content of quantum mechanics, namely that identity is not independent. This is an advantage of the position of this book: it finds a position between common sense and physics.

Objection Three: Ladyman's slogan "Every Thing Must Go" is powerful and influential. Your "stable forms of structures of difference" sounds too moderate and lacks force.

Reply: The force of a philosophical position does not lie in its slogan, but in its argument and application. Ladyman's slogan has rhetorical force, but its application faces difficulties: how to treat things in common sense, how to avoid the difficulty of structure as independent foundation, and so on. The position of this book is more moderate, but its application is more coherent: common-sense things are treated as stable forms of structures of difference, thereby avoiding the difficulty of simple elimination. Philosophical progress often involves a balance between rhetorical force and applicative coherence. Ladyman's work leans toward rhetorical force; the work of this book leans toward applicative coherence. Both directions have legitimacy.

Literature That Must Be Engaged

- James Ladyman & Don Ross, Every Thing Must Go: Metaphysics Naturalized (2007)
- Steven French, The Structure of the World (2014)
- John Worrall, "Structural Realism: The Best of Both Worlds?" (1989): early work on structural realism.
- Bas van Fraassen, Scientific Representation (2008)
- Chapter 7 on difference preceding identity (self-reference)

Summary of This Section
This section has done four things. First, it developed the core position of Ladyman's work: ontic structural realism. Second, it identified similarities between Ladyman's work and the position of this book: suspicion of entities, connection with physics, and naturalistic metaphysics. Third, it identified differences: structure vs process as foundation, the treatment of "things," and the treatment of structure. Fourth, it clarified that the two are different directions in contemporary anti-entity ontology.

The next section develops the dialogue with Floridi's information realism.

5.5. Dialogue with Luciano Floridi's Information Realism

Floridi's Position
Luciano Floridi developed the philosophy of information (PI), with representative works including The Philosophy of Information (2011), The Ethics of Information (2013), and The Logic of Information (2019). The core position of Floridi's work is as follows.

Information as foundational ontological content: information, meaning identifiable and processable structure, is foundational ontology. All ontological content can be understood as forms of information. Specifically, physical entities are information structures: their properties involve identifiable and processable content. Mind is an information-processing system: consciousness involves information integration and information processing. Society and culture are information structures: social relations and cultural contents involve information flows and information structures.

The motivations of Floridi's work are: the ontological needs of the digital age, since contemporary society increasingly involves digital technologies, data processing, and information flows, and ontology must treat this context; interdisciplinary integration, since information as an interdisciplinary concept has applications in physics, biology, neuroscience, sociology, and other fields; and the extension of traditional ontology, since traditional ontological notions such as matter, energy, and form are insufficient for treating digital phenomena, while information as foundational ontology expands ontological content.

The contribution of Floridi's work is to take information as foundational ontology: this is an important development in contemporary ontology. It also establishes informational ethics, treating ethical issues in the digital age, and provides an interdisciplinary framework of information processing.

Relation to Constrained Process Realism

Floridi's work and the position of this book have a complex relation.

Similarity One: Response to the Contemporary Context

Floridi's work responds to the contemporary context, including the digital age and interdisciplinary integration. The work of this book also responds to the contemporary context, including contemporary physics, neuroscience, ethics, and related fields. Both attempt to establish an ontology suitable for the contemporary context.

Similarity Two: Interdisciplinary Openness

Floridi's work has applications in multiple disciplines. The position of this book is also open to interdisciplinary dialogue (Section 5.1, Feature Six). The two agree in interdisciplinary openness.

Key Difference One: Information vs Process as Foundation

Floridi's work takes information as foundational ontology: all ontological content can be understood as forms of information. The work of this book takes process, namely difference and co-evolution, as foundational ontology: all ontological content is a form of process. This difference may appear terminological, but it is in fact a difference in ontological starting point. Information as a foundational concept involves the dimension of being identifiable and processable: it has an epistemological-operational dimension. Process as a foundational concept involves the dimensions of dynamics, difference, and change: it has an ontological dimension.

The consequence is that, in Floridi's framework, non-informational ontological content, such as pure dynamics, needs to be translated into information, which may involve simplification. In the framework of this book, information processing is treated as a form of process: information is the content of process, not the other way around.

Key Difference Two: The Treatment of the Concept of "Information"

Floridi's concept of "information" is broad: it includes semantic information, syntactic information, physical information, and other types. This gives "information" flexibility as a foundational concept, but it also involves difficulties: different types of information have different properties. Semantic information involves meaning; syntactic information involves only structure; physical information involves physical quantities. Treating all of them uniformly as "information" may conceal differences.

The work of this book does not use "information" as a foundational concept. It uses ontological concepts such as structures of difference, gradients, and processes. "Informational" phenomena are treated as specific forms of process, such as consciousness as a highly complex multi-frequency structure (Chapter 9). This difference gives the two works different emphases: Floridi's work focuses on "information" phenomena, and has strong applications to digital technology, knowledge management, information ethics, and related domains. The work of this book focuses on ontological structure, and has strong applications to physics, metaphysics, the mind–matter relation, and related domains.

Key Difference Three: The Reality of Ontology

The ontological status of Floridi's work is complex. In some statements, Floridi treats information as foundational ontology: beings just are information structures. In other statements, Floridi treats information as a description of beings: beings involve information, but are not reduced to information. This inconsistency makes the ontological position of Floridi's work not entirely clear: is it information realism, with information as foundational ontology, or information epistemology, with information as a description of ontology?

The work of this book is comparatively clear: it is process realism. Ontological content is processual ontology, and descriptions are descriptions of this ontology. The work of this book does not confuse ontology and description.

Positioning the Two Positions

Through comparison, the relation between the position of this book and Floridi's work can be stated as follows: constrained process realism agrees with Floridi's philosophy of information in its response to the contemporary context and interdisciplinary openness, but differs from it in the content of foundational ontology, namely process vs information, and in the distinction between ontology and description, namely a clear process realism vs a complex status of information. The two constitute different directions in the genealogy of contemporary realism.

This positioning avoids two traps: the trap of absorption, which would claim that the work of this book can "absorb" the insights of the philosophy of information, while the work of this book has its own independent starting point; and the trap of simple criticism, which would claim that Floridi's work is "wrong," while Floridi's work has contemporary force.

Objections and Replies

Objection One: Your ontology of difference itself involves an "informational" dimension: structures of difference contain identifiable content, and that is information. Your treatment in fact implicitly takes information as foundational.

Reply: Structures of difference indeed involve identifiable content. This belongs to the meaning of difference as ontological content. But this is not the same as "information as foundational ontology." Difference is ontological content: non-uniformity, dynamics, change, and so on. Information is an identifiable and processable description of difference: it belongs to the epistemological-operational dimension. The two differ at the ontological level. Difference as foundational ontology does not depend on identification or processing: it exists in process whether or not it is cognized. Information as foundational ontology, in Floridi's sense, depends on being identifiable and processable: it involves an epistemological-operational dimension. The work of this book takes difference as foundational ontology. This choice separates the ontological dimension, namely difference, from the epistemological-operational dimension, namely information. This is a clear distinction between ontology and description, not an implicit adoption of information as foundation.

Objection Two: Floridi's work has an advantage in its application to the digital age: it directly treats digital phenomena. Your process ontology seems to lack this dimension of application.

Reply: The work of this book is not limited to phenomena of the digital age: its applications involve physics (Chapters 9–11), neuroscience (Chapter 13), theory of observation (Chapter 14), ethics (Chapter 20), and several other domains. Digital phenomena can be treated by the work of this book as a subfield: digital technology as a specific form of process, information flow as a form of co-evolution, and so on. These applications require dedicated work and are not developed in this section. But the focus of the work of this book is not the digital age; it is foundational ontology and connection with contemporary physics. This is the difference in emphasis from Floridi's work: Floridi's work emphasizes applications in the digital age, while this book emphasizes foundational ontology and connection with physics. The two works have different emphases. They may be complementary: the work of this book at the level of foundational ontology, and Floridi's work at the level of digital application. Researchers can choose the appropriate framework according to the problem.

Objection Three: Floridi's informational ethics has contemporary influence: it directly treats digital ethical problems, such as privacy, data, and artificial intelligence. Your work seems more abstract in ethics.

Reply: The ethical contribution of this book is developed in Chapter 20, concerning how constrained speculation treats the is–ought gap. It establishes the methodological ground of ethics. This is metaethical work, not applied ethical work. Applied ethical problems, such as digital ethics, bioethics, and environmental ethics, require dedicated work. The work of this book provides methodological grounds for applied ethics, but it does not develop concrete applications. Floridi's work emphasizes applied ethics; this book emphasizes metaethics. Both have legitimacy. A field of ethics may need both: metaethics, as in this book, provides methodological grounds, while applied ethics, as in Floridi's work and others, treats concrete problems.

Literature That Must Be Engaged

- Luciano Floridi, The Philosophy of Information (2011)
- Luciano Floridi, The Ethics of Information (2013)
- Luciano Floridi, The Logic of Information (2019)
- Claude Shannon, "A Mathematical Theory of Communication" (1948): the starting point of information theory.

- Norbert Wiener, Cybernetics (1948): early work on information and cybernetics.

Summary of This Section

This section has done four things. First, it developed the core position of Floridi's work: information as foundational ontology. Second, it identified similarities between Floridi's work and the position of this book: response to the contemporary context and interdisciplinary openness. Third, it identified differences: information vs process as foundation, the treatment of the concept of "information," and the reality of ontology. Fourth, it clarified the different directions of the two within the genealogy of contemporary realism.

The next section develops the dialogue with Rovelli's relational quantum mechanics.

5.6. Dialogue with Carlo Rovelli's Relational Quantum Mechanics

Rovelli's Position

Carlo Rovelli is a contemporary theoretical physicist and philosopher. His work has two dimensions: relational quantum mechanics (RQM) and philosophy of physics, especially The Order of Time (2018) and Helgoland (2020). The core position of Rovelli's work is as follows.

Relational quantum mechanics: the "state" of a quantum system is not absolute, but relative to the observer. Specifically, different observers can assign different states to the same system; "objective state" as an absolute concept does not exist, since states are always relational; and quantum events are ontological, involving relative interactions rather than statements about things.

Relational ontology: at the broader metaphysical level, Rovelli argues for the relationality of ontology: the properties of beings are relational, not intrinsic.

The motivations of Rovelli's work are: the content of quantum mechanics, since the measurement process in quantum mechanics involves observers and cannot simply be reduced to an "observer-free objective state"; the content of general relativity, since in general relativity the geometry of spacetime depends on matter and energy, and ontology involves relations; and connection with philosophical traditions, especially Rovelli's dialogue with Anaximander, Mach, and others.

The contribution of Rovelli's work is to take relation as the core interpretation of quantum mechanics: this is an important direction in contemporary quantum philosophy. It also enters into dialogue with philosophical traditions: Rovelli's work is not limited to physics, but involves metaphysics and philosophy of science. Its metaphysical work on time treats time as relational ontology, which dialogues with Chapter 15 of this book.

Relation to Constrained Process Realism

Rovelli's work and the position of this book have the deepest resonance: both establish ontology in the context of contemporary physics.

Similarity One: Relational Ontology

Rovelli takes relations as foundational ontological content. The work of this book, especially Chapter 12 on co-evolution as an ontological relation, also takes relations as foundational. Both reject the framework of independent terms plus relations: relations are foundational, and relata are extracted from relations.

Similarity Two: Connection with Contemporary Physics

Rovelli's work is built upon quantum mechanics and general relativity. The work of this book is built upon the EQT framework, which is itself built upon contemporary physics. Both establish ontology in a physical context.

Similarity Three: Rejection of the "Observer-Free Objective State"

Rovelli's relational quantum mechanics rejects the "absolute objective state": states are always relational. The work of this book, especially Chapter 14, $OB1$, also rejects traditional "observer-free externality" as objectivity and establishes controlled reproducible co-perturbation as the new foundation of objectivity. The two agree in rejecting traditional objectivity.

Key Difference One: The Treatment of Objectivity

But the work of this book and Rovelli's work differ in how they treat objectivity, as already developed in Chapter 14, Section 14.4. Rovelli tends to restrict or relativize the concept of objectivity in the quantum domain: states are relative to observers, and objectivity as an absolute concept does not exist. The work of this book refines the concept of objectivity: objectivity = controlled reproducibility. Objectivity is not absolute, but remains attainable. This difference is not a simple difference of direction; it involves two different strategies.

Rovelli's strategy gives up objectivity as a foundational concept and makes physical statements relational. The strategy of this book retains objectivity as a foundational concept but refines it so that it becomes consistent with physics. Both strategies have legitimacy, but they reflect different philosophical choices.

Key Difference Two: The Content of Foundational Ontology

Rovelli's relational ontology takes events and relations as foundational: relations among events constitute ontology. The work of this book takes structures of difference and process as foundational (Chapter 7, $DI1$, and Chapter 11 on freezing): difference and process are foundational ontological contents, while events and relations are manifestations of process. The difference is this: Rovelli takes events as basic units, which is close to Whitehead's actual occasions; Chapter 8, Section 8.5 develops the difficulty in Whitehead's work. The work of this book takes process as foundational: events are instantaneous units of process, not foundational. This difference allows the work of this book to avoid a Whiteheadian difficulty, namely the problem of continuity among actual occasions, because process as foundation provides continuity.

Key Difference Three: The Mode of Connection with Physics

Rovelli's work is built upon an interpretation of quantum mechanics: it is a relational interpretation of quantum mechanics. The work of this book is built upon the EQT framework: EQT is a physical framework, not merely an interpretation. The difference is that Rovelli's work interprets existing physics, namely quantum mechanics: it gives physics a philosophical interpretation. The work of this book is built upon a physical framework, EQT: this framework itself involves physical statements and can be experimentally tested. This difference gives the work of this book a physical ground: it is not merely a philosophical interpretation, but built upon a physical framework. At the same time, however, the legitimacy of this book's work depends on the legitimacy of EQT. If EQT faces physical problems in the future, the physical ground of this book will need to be re-evaluated.

Positioning the Two Positions

Through comparison, the relation between the position of this book and Rovelli's work can be stated as follows: constrained process realism has the deepest resonance with Rovelli's relational quantum mechanics in relational ontology and connection with contemporary physics, but differs from it in the treatment of objectivity, namely refinement vs relativization, in the content of foundational ontology, namely process vs event, and in the mode of connection with physics, namely EQT framework vs interpretation of quantum mechanics. This positioning gives each position its own features: they are two directions in contemporary relational ontology.

Objections and Replies

Objection One: Is your comparison with Rovelli's work overly "symmetric"? You seem to be trying to balance similarities and differences, making the two appear equally important.

Reply: The work of this book and Rovelli's work do indeed have the deepest resonance in contemporary relational ontology. This is a real fact. They agree along multiple dimensions: relational ontology, rejection of traditional objectivity, connection with contemporary physics, dialogue with philosophical traditions, and so on. But the work of this book still has an independent position: it uses a different strategy, namely refining objectivity rather than relativizing it; a different foundational ontology, namely process rather than event; and a different mode of connection, namely the EQT framework rather than an interpretation of quantum mechanics. This treatment of "deep resonance + difference" is not a "symmetric balance," but a precise statement of the relation. If the two works were completely identical, this book would have no independent position. If they were completely different, this book would lose its connection with contemporary discussion. The work of this book lies between the two: it inherits resonances while retaining an independent position.

Objection Two: Rovelli's work has support in the physics community. Your difference, especially refinement vs relativization of objectivity, seems to challenge Rovelli's direction.

Reply: The difference between this book and Rovelli's work is not competition, but different directions. In the application to quantum mechanics, the two are compatible: this book accepts Rovelli's core physical insight that states are relative to observers. In the treatment of objectivity, the two differ in direction: one relativizes objectivity, the other refines it. Both directions have legitimacy. Rovelli's direction is close to the Bohr tradition, accepting the distinctiveness of the quantum domain; the direction of this book is closer to the Popperian tradition, retaining objectivity as a concept that can be refined. Which direction is ultimately preferable requires further work. This section does not claim that the work of this book overturns Rovelli's work; it claims that the work of this book provides another legitimate treatment. The two can coexist and inspire different research directions.

Objection Three: The physical ground of Rovelli's work, namely the interpretation of quantum mechanics, is more "mature" than the EQT framework, since quantum mechanics is a widely accepted physical theory.

Reply: Quantum mechanics as physics is widely accepted: it is a core achievement of twentieth-century physics. The EQT framework as physics is comparatively younger, and its legitimacy depends on continued work. But the relation between the two is not a simple matter of "mature vs immature": they treat different physical problems. Quantum mechanics treats microscopic phenomena; the EQT framework attempts to unify multiple domains of physics, including elementary particles, cosmology, and the mind–matter relation. The ambition of the EQT framework gives it difficulties: any unifying framework requires detailed defense. But ambition itself is not a defect: it gives the work of this book a physical ground and allows the work of this book to connect with multiple domains of physics. Physical legitimacy requires physical work; this is EQT's own task. The work of this book is built upon the legitimacy of EQT. If the legitimacy of EQT is established, the physical ground of this book is strong; if EQT faces problems, the work of this book needs to be re-evaluated. This dependence is an application of the method of constrained speculation: any ontological work depends on grounds and must be evaluated according to the state of those grounds.

Literature That Must Be Engaged

- Carlo Rovelli, "Relational Quantum Mechanics" (1996)
- Carlo Rovelli, Reality Is Not What It Seems (2017)
- Carlo Rovelli, The Order of Time (2018)
- Carlo Rovelli, Helgoland (2020)
- Chapter 12 on co-evolution as an ontological relation (self-reference)
- Chapter 14 on observation as participation (self-reference)
- Chapter 15 on time, duration, and co-evolution (self-reference)

Summary of This Section

This section has done four things. First, it developed the core positions of Rovelli's work: relational quantum mechanics and relational ontology. Second, it identified the resonances between Rovelli's work and the position of this book: relational ontology, connection with contemporary physics, and rejection of traditional objectivity. Third, it identified differences: the treatment of objectivity, the content of foundational ontology, and the mode of connection with physics. Fourth, it clarified that the two are different directions in contemporary relational ontology.

5.7. Dialogue with Russellian Monism: Nearest Neighbor and Two Advances

The Position of Russellian Monism

After the four dialogues developed in Sections 5.2–5.5, this section treats an interlocutor with a special status: Russellian Monism. This position begins with Bertrand Russell's neutral monist work in The Analysis of Matter (1927), is refined in contemporary work by Daniel Stoljar, Sam Coleman, and Galen Strawson, and is taken seriously as an important option in the philosophy of consciousness in David Chalmers's later work. Among all the realist positions with which this book enters into dialogue, the one nearest to this book is precisely Russellian Monism.

The core thesis of Russellian Monism can be stated as follows: physical science characterizes only the relational-structural properties of reality, namely how things interact, how they are measured, and how they vary with other things, while it does not touch the intrinsic nature of reality, namely what things are in themselves. This distinction makes the "outer side" of physical reality, namely relational structure, and the "inner side," namely intrinsic nature, two different aspects of the same reality. Mind, including experience and qualia, may be precisely the intrinsic nature of physical reality: one and the same reality, seen from the outside, is the relational structure described by physical science; seen from the inside, it is experience.

The strength of this position lies in accommodating two aspects with monistic economy: the objectivity of the physical, namely that reality indeed has its relational-structural aspect, and the interiority of experience, namely that reality indeed has its intrinsic aspect. It thereby avoids both the separation problem of dualism, namely how to explain mind–matter interaction, and the deadlock of simple reductive physicalism, namely how to force irreducible intrinsic nature to reduce to relational properties.

Spiritual Resonance with the Position of This Book

This book is spiritually closest to Russellian Monism. The resonance appears in three respects.

Monistic ontology rather than dualist entities: this book and Russellian Monism both adopt monistic economy. The ontological unit of beings is unified: in this book, the ρ_f gradient process; in Russellian Monism, physical reality. No second independent entity is added.

The separability of relational structure and intrinsic nature: this book and Russellian Monism both distinguish the relational-structural content described by physics from the intrinsic aspect of the same ontology. These are two dimensions that can be described separately.

Respect for the irreducibility of given-ness: this book and Russellian Monism both acknowledge that experience cannot be exhausted by a purely relational-structural description. But this irreducibility does not require dualist entities to bear it; it can be a dimensional difference within a monistic ontology.

These three layers of resonance give Russellian Monism a distinctive position among the positions with which this book dialogues. At a deeper ontological level, it shares a starting point with this book: unlike Cartwright (5.2), which takes entities as foundational, and which this book rejects; unlike Ladyman (5.3), which takes pure relational structure as foundational, and which this book partially accepts; unlike Floridi (5.4), which takes information as foundational, and which this book treats at the level of process; and unlike Rovelli (5.5), which relativizes objectivity, while this book refines objectivity. Russellian Monism and this book share one starting point: the distinction between the relational-structural aspect and the intrinsic aspect within a monistic ontology.

First Key Advance: Realizing Structural Sufficiency through Frequency-Resolved Gradient Structure

But this book makes two decisive advances beyond Russellian Monism, and these two advances make this book not merely a philosophical posture, but an ontological position with content and capable of serious evaluation.

The first advance is related to an internal danger. Russellian Monism, as a general framework, claims that "experience is the intrinsic aspect of the physical," but it does not specify what kind of structure the physical side must possess in order to match the rich structure of experience. This lack of specification creates a dilemma. If the intrinsic aspect is imagined as some impoverished "proto-experiential" property, as in some formulations by Strawson, it slides toward panpsychism and inherits its combination problem. If it appeals to a single intrinsic magnitude, it repeats the structural insufficiency of one-scalar identity theory: a single number has only the dimension of "magnitude," and cannot support structural features such as kinds of qualia, dynamic structure, and temporal thickness.

The advance of this book is to take the frequency-resolved gradient configuration of ρ_f as the structure on the physical side: a five-dimensionally enriched structural space, consisting of single-frequency gradient configuration, the frequency dimension, Master-Equation evolution, resonance modes, and the transient temporal-process spectrum. The development of these five sources appears in Part II, Part III, and Chapter 13. This five-dimensionally enriched structural space is sufficient to correspond relation-by-relation with all the structural features of beings.

Russellian Monism says that "experience is the intrinsic aspect of physical reality." This book further says that "this intrinsic aspect is such a five-dimensionally enriched gradient structure, and it is relation-by-relation isomorphic with experience." This is an advance from a general posture to testable content.

Second Key Advance: Defining the Scope and Mechanism of the Intrinsic Aspect through Boundary and Co-Evolution

The second advance is related to another internal difficulty. If experience is the intrinsic aspect of physical reality, where does this intrinsic aspect occur? If the answer is "everything has it," then the position slides toward panpsychism and its combination problem. If the answer is "only some systems have it," then a criterion is needed, but Russellian Monism itself does not provide such a criterion.

The advance of this book is to define the scope of the intrinsic aspect through boundary as a gate: only systems in which enclosure, the physical event of the membrane of life, creates an "inside" have an "inside" to which experience can be attributed. Rocks, electrons, and galaxies have no inside, not because interiority is "too weak" in them, but because they simply have no boundary and no inside in this sense: interiority is not present in them to a lesser degree, but is simply inapplicable. At the same time, this book characterizes the mechanism of the intrinsic aspect through internal–external co-evolution: the intrinsic aspect is not an intrinsic property of some internal configuration, as argued by the dark-interior argument of the isolated interior, but the ongoing co-evolutionary relation between that interior and the exterior. See Chapter 13 for details.

Russellian Monism says that "experience is the intrinsic aspect of the physical." This book further says that "this intrinsic aspect exists only where an enclosing boundary exists, thereby resisting panpsychism, and its given-ness is the freezing of internal–external co-evolution, relational rather than intrinsic." This is again an advance from a general posture to concrete content.

The Position of This Book: Russellian Monism Enriched by EQT

Combining these two advances, the position of this book relative to Russellian Monism can be stated as follows:

This book is a Russellian Monism enriched by EQT. It inherits the Russellian monistic posture, namely the unity of beings in the relational-structural aspect and the intrinsic aspect, but enriches the structure of the "intrinsic aspect" through frequency-resolved gradient structure, thereby solving the problem of structural sufficiency, and defines the scope and mechanism of the "intrinsic aspect" through boundary and co-evolution, thereby solving the problems of scope and given-ness. Russellian Monism gives a philosophical posture; this book develops that posture into an ontological position with physical grounds, testable content, and identifiable directions of failure.

What remains after these two advances, and what Russellian Monism and this book still face together as irreducible, is the being-experienced of the intrinsic aspect itself: the final square discussed in detail in Chapter 13. This book and Russellian Monism share this final square, but this book precisely marks it as a single limit through two compressions, namely from the residual point of given-ness → the residual point of subject-ascription (Chapter 13.4). This compression and precise marking are the cumulative result of the two advances above: the solvable parts are solved as far as possible, while the unsolved part is precisely narrowed and marked.

As a positioning chapter within the genealogy of contemporary realism, this section gives the relation between this book and Russellian Monism at the ontological level. Chapter 13 will develop this dialogue in detail in the context of the philosophy of mind and matter, and will treat Russellian Monism more systematically as the nearest neighbor in the philosophy of consciousness. Russellian Monism is the nearest philosophical neighbor of the entire ontological work of this book. But through two decisive advances, this book occupies an independent position within that proximity.

The next section is the conclusion of this chapter.

5.8. Chapter Summary: Constrained Process Realism as a New Position in Contemporary Realism

Completion of the Three Tasks

The three tasks undertaken by this chapter have all been completed.

Task One: to integrate the methodological work of Chapters 1–4 into an ontological position. Section 5.1 gave the name of the position, constrained process realism, and developed its six features.

Task Two: to enter into dialogue with five positions in the genealogy of contemporary realism. Section 5.2 entered into dialogue with Cartwright's entity realism; Section 5.3 with Ladyman's ontic structural realism; Section 5.4 with Floridi's information realism; Section 5.5 with Rovelli's relational quantum mechanics; and Section 5.6 with Russellian Monism. Each dialogue developed similarities, differences, and positioning.

Task Three: to avoid the traps of eclecticism and isolationism. Through the treatment of "deep resonance + precise difference," this chapter has allowed the position of this book both to inherit insights from several traditions and to have its own independent position.

The Position of This Chapter in Part I

The overall architecture of Part I, "Constrained Speculation: Method as Philosophy," is as follows: Chapter 1, constrained speculation as method, provides the methodological foundation; Chapter 2, minimal postulates and ontological restraint, provides the discipline of restraint; Chapter 3, failure conditions as an ontological discipline, provides the discipline of failure conditions; Chapter 4, constrained physicalization, provides the tool of constitutive coextension; Chapter 5, this chapter, positions the book within the genealogy of contemporary realism; and Chapter 6, falsifiability as an ontological discipline, provides the synthetic discipline.

As the positioning chapter, this chapter integrates the methodological work of Chapters 1–4 into a complete position and locates that position within contemporary discussion. Chapter 6 then synthesizes the methodological work into falsifiability as an ontological discipline on this basis.

The Position of This Chapter in the Whole Book

Looking backward, this chapter is built upon the work of Chapters 1–4: it synthesizes that work into a complete ontological position.

Looking forward, this chapter provides the position statement for later chapters. Part II, the philosophical defense of gradient ontology, develops the ontological content of the position of this book. Part III, on co-evolution, mind–matter, observation, and time, develops the relational ontology of the position of this book. Part IV, on the metaphysical boundary of freezing, develops the metaphysical dimension of the position of this book. Part V, from ontology to normativity, develops the normative dimension of the position of this book. All later chapters are built upon the position located in this chapter: constrained process realism.

Honesty about the Limits of This Chapter

This chapter acknowledges several limits:

First limit: the selection of interlocutors. This chapter selects five interlocutors: Cartwright, Ladyman, Floridi, Rovelli, and Russell. Other important contemporary realists, such as Hilary Putnam, Bas van Fraassen, Ian Hacking, and Sandra Mitchell, also have important work, and detailed dialogue requires dedicated work.

Second limit: the limited depth of each dialogue. Each dialogue is relatively brief, and this is insufficient for fully treating each interlocutor's work. Each dialogue requires a dedicated paper or monograph.

Third limit: insufficient cross-cultural comparison. This chapter works mainly within the Western contemporary analytic philosophical tradition. Chinese and Indian philosophical traditions also have realist genealogies, especially the ontological directions of the Chinese tradition, and these require dedicated treatment.

Fourth limit: the direction of future development. The position of this chapter, as a contemporary position, needs to continue developing in discussion. Its applications, refinements, and revisions require later work. This chapter gives only the foundation of the position, not its final form.

Acknowledging these limits does not weaken the work of this chapter. It makes this chapter an honest initial work.

The Core Posture of This Chapter

The core posture of this chapter can be stated as follows: the genealogy of contemporary realism needs independent positions, but independence is not isolation.

Since the late twentieth century, the genealogy of contemporary realism has developed many positions: entity realism, ontic structural realism, information realism, relational realism, and others. Each position has its own motivations, contents, and applications. The position of this book, constrained process realism, does not attempt to replace these positions. It joins the contemporary genealogy as another position. It has its own features, namely the six features; its own grounds, such as the method of constrained speculation and the EQT framework; and its own applications, namely the work of the later parts of this book.

But the position of this book is not isolated. It has resonances with several contemporary positions, especially with Rovelli's work and with Russellian Monism. These resonances give the position of this book a place in contemporary discussion: it is not a philosophical creation ex nihilo, but a new development within contemporary discussion. This posture gives the work of this chapter a dual character: independent and connected. This is the posture of serious contemporary philosophical work: any new position needs to show its independence while acknowledging its connections with traditions.

Meaning before freezing: this is one of the core theses of this book. The position of this chapter provides an ontological ground for that thesis. Constrained process realism treats meaning as the emergence of process, grounds the legitimacy of meaning in methodological discipline, and grounds the objectivity of meaning in controlled reproducibility. These treatments acquire a coherent framework through the position located in this chapter. This is the contribution of the chapter to the whole book.

Through the work of this chapter, the position of this book gains a place within the genealogy of contemporary realism: it is not an isolated philosophical creation, but a new development within contemporary discussion. Through the positioning of this chapter and the synthesis of Chapter 6, Part I completes its methodological foundation. All later chapters are built upon this foundation.

5.9. Methodological Reflection on This Chapter

This chapter integrates the methodological work of Chapters 1–4 into a complete ontological position. This section reflects on the methodological significance of that integration.

The Function of This Chapter as a Positioning Chapter

This chapter is the point of connection between this book and contemporary discussion. If this book did not develop this connection, its work could be regarded as self-enclosed: its philosophical position would be unclear. This chapter establishes that connection, making the work of this book available for serious dialogue and evaluation within contemporary discussion. Self-enclosure is a trap of contemporary philosophical work: some works establish their own terminology and framework but do not build connections with contemporary discussion. The result is that the work may be internally refined, but difficult to take seriously in international discussion. This chapter avoids that trap.

This Chapter's Posture toward the Five Interlocutors

The posture of this chapter toward the five interlocutors is serious respect and precise distinction. With respect to Cartwright, this chapter acknowledges the strength of entity realism, but shows the position of this book through the difference between unified process and dappled world. With respect to Ladyman, this chapter acknowledges the suspicion of entity ontology in ontic structural realism, but shows the position of this book through the difference between process and structure as foundation. With respect to Floridi, this chapter acknowledges the response of the philosophy of information to the contemporary context, but shows the position of this book through the difference between process and information as foundation. With respect to Rovelli, this chapter acknowledges the deep resonance with relational quantum mechanics, but shows the position of this book through the difference between refining and relativizing objectivity. Russellian Monism is the nearest: this book spiritually shares with it the starting point of monistic ontology, but enriches it through two decisive advances, namely frequency-resolved gradient structure and boundary–co-evolution, transforming it from a monistic posture into an ontological position with content and capable of being tested.

This Chapter's Posture toward Originality

The work of this chapter does not claim complete originality: the position of this book is built upon multiple precursor traditions. The contribution of this chapter is to give process realism a rigorous methodological ground through methodological discipline; to give process realism a physical ground through its connection with physics, namely the EQT framework; and to give it a new position within the genealogy of contemporary realism: constrained process realism. These contributions are constructively original.

Limits Honestly Acknowledged in This Chapter

This chapter acknowledges several limits. First, the work of the five interlocutors is complex, and this chapter can treat only their core directions. Second, the genealogy of contemporary realism includes other important positions, such as Putnam's internal realism and van Fraassen's constructive empiricism. These positions are treated only briefly in this chapter, and Appendix B gives a more detailed comparison. Third, the applications of the position of this book need to be developed through the work of the later parts. This chapter only states the position.

Connections with Later Chapters

The position of constrained process realism established in this chapter will be applied and refined in later chapters. Chapter 6 connects this position with the discipline of falsifiability. Part II develops this position as ontological work. Part III develops this position as relational-ontological work. Part IV displays the boundary of this position. Part V extends this position into the domain of normativity. This posture of clearly stating a position is an expression of legitimate philosophical work: it does not pretend to have "no position," but clearly states its position and accepts evaluation.

6. Falsifiability as an Ontological Discipline

6.1. Introduction to This Chapter

In *The Logic of Scientific Discovery*, Popper proposed falsifiability as the demarcation criterion of science: a theory is scientific if and only if it has content that can be empirically refuted. This position had a far-reaching influence in twentieth-century philosophy of science. Yet Popper himself restricted falsifiability to the level of scientific methodology: it is a criterion for distinguishing science from non-science, not a criterion for the legitimacy of ontological commitments.

This chapter elevates falsifiability from a principle of scientific methodology, namely a criterion for distinguishing science from non-science, to an ontological discipline, namely a criterion for the legitimacy of ontological commitments. This elevation is not a simple terminological extension; it involves pushing the function of falsifiability from the epistemological level to the ontological level. Specifically, at the epistemological level, the principle of falsifiability deals with "how we distinguish scientific theories from non-scientific theories"; at the ontological level, it deals with "how we evaluate the legitimacy of ontological commitments." The applications at these two levels differ and must be developed seriously.

This chapter enters into serious dialogue with Popper's work. It both inherits his core insight, namely that theories must have content that can be refuted, and refines it through the methodology of this book, including constrained speculation, failure conditions as an ontological discipline, and constitutive coextension. This treatment of inheritance plus development is a common pattern of philosophical progress: great pioneering work provides a starting point, but it must be further refined in a new context of discussion.

At the same time, this chapter, as the synthetic chapter of Part I, unifies the work of the preceding five chapters: the method of constrained speculation, ontological restraint, failure conditions, constitutive coextension, and the positioning within contemporary realism, under the framework of falsifiability as an ontological discipline. This synthetic work gives Part I a complete unity as a methodological foundation and provides a framework of methodological discipline for the later parts, where ontology is developed.

This chapter adopts a clear writing posture toward Popper's work: serious respect and precise distinction. Popper's work continues to influence twentieth-century philosophy of science, and this chapter does not pretend that it can easily surpass it. But this chapter does make an advance: by elevating falsifiability to an ontological discipline, it extends its scope to ontological problems.

The discussion proceeds as follows. Section 6.1 reviews the core content and limitations of Popper's work; Section 6.2 develops the elevation of falsifiability into an ontological discipline (FA1); Section 6.3 develops the application dimensions of the discipline of falsifiability; Section 6.4 develops the mutually supporting network of the six chapters in Part I, namely the realization of method as philosophy; Section 6.5 gives the synthetic summary of Part I.

6.2. Popper's Falsifiability and Its Limits

The Core Content of Popper's Work

In *The Logic of Scientific Discovery* (1934/1959) and later works, Karl Popper proposed falsifiability as the demarcation criterion of science. Popper's position is as follows:

The principle of falsifiability: a theory is scientific if and only if it has content that can be empirically refuted, that is, if there are possible observations which, if they occurred, would make the theory fail.

The motivation of this position comes from the difficulties of the Vienna Circle's principle of verifiability, namely verificationism. The Vienna Circle held that "a statement is meaningful if and only if it can be verified by observation." But this principle faces difficulties: universal statements, such as laws of nature, cannot be fully verified by finite observations; the principle of verifiability itself seems not to be verifiable, leading to a problem of self-application.

Popper dealt with these difficulties by reversing the criterion: instead of requiring a theory to be verified, he required it to be refutable. The advantages of this reversal are as follows: universal statements can be refuted, since a single counterexample can make a universal statement fail, whereas complete verification requires infinitely many observations; refutation and verification are logically asymmetric, since refutation can proceed through a single counterexample, while verification requires all cases to be checked; and the substantive content of a theory is clarified, since a refutable theory has substantive content: it "risks" stating certain things, and these things may be wrong.

Popper used falsifiability to distinguish science from non-science: the former has content that can be refuted, while the latter does not. His targets of criticism included Adler's individual psychology and Freud's psychoanalysis. These theories were criticized for using auxiliary hypotheses to make any observation "compatible" with the theory, thereby losing refutable content.

The Contributions of Popper's Work

Popper's work made three contributions to twentieth-century philosophy of science:

Contribution One: The Requirement of Methodological Rigor

Through falsifiability, Popper proposed the requirement that a theory must take risks: any serious theory must be able to be wrong. This requirement established methodological rigor in science and gave the distinction between science and non-science operational content.

Contribution Two: Criticism of Ad Hoc Adjustments

Popper analyzed in detail the problem of ad hoc adjustments: when observations threaten a theory, supporters of the theory introduce auxiliary hypotheses to protect it. Such adjustments deprive the theory of refutable content. Popper's analysis made ad hoc adjustment a central issue in philosophy of science.

Contribution Three: Criticism of Historicism

In *The Poverty of Historicism* (1957) and *The Open Society and Its Enemies* (1945), Popper used falsifiability to criticize historicism: historical developments are explained in every case as "conforming to laws," thereby losing refutable content. This criticism has important implications for political philosophy and social theory.

The Limits of Popper's Work

Popper's work, however, has limits. This section develops several major limits.

Limit One: The Duhem–Quine Thesis

Pierre Duhem and W. V. O. Quine proposed a thesis: theories are not tested in isolation, but together with auxiliary assumptions.

When observation conflicts with a theory, one may reject the theory or reject an auxiliary assumption. This choice cannot be determined simply by Popper's falsifiability.

Example: the observation of Mercury's perihelion precession conflicted with Newtonian gravitational theory. This conflict could be treated by rejecting Newtonian gravitation and choosing a new theory, which is what ultimately happened with Einstein's general relativity; or by introducing an auxiliary assumption, such as the hypothesized existence of "Vulcan," a planet between Mercury and the Sun. Both treatments are logically legitimate. Popper's principle of falsifiability cannot tell us which treatment should be chosen. This difficulty shows the limit of falsifiability as a simple criterion.

Limit Two: The Complexity of Theory Choice

In actual science, theory choice involves multiple factors: empirical evidence, theoretical consistency, simplicity, heuristic power, research programmes, and so on. Falsifiability is one factor among them, but not the only one.

The work of Imre Lakatos, Larry Laudan, and others developed this complexity in detail. Units such as research programmes and research traditions describe actual scientific work more precisely than "theory." These works refined Popper's simple position.

Limit Three: The Theory-Ladenness of Observation

Popper assumes that observation can test theory independently. But the work of N. R. Hanson, Thomas Kuhn, and others shows that observation is theory-laden: observation depends on conceptual frameworks, measurement instruments, auxiliary theories, and so on.

This limit does not cancel falsifiability; it makes falsifiability more refined. Observation as an instrument of refutation itself requires support from auxiliary theories. The conflict between two theories is not simply a conflict between "theory vs. observation," but a conflict between two theoretical networks.

Limit Four: The Problem of Self-Application

Is Popper's principle of falsifiability itself refutable? This is the problem of self-application faced by Popper's work. Specifically, if the principle of falsifiability is refutable, then some observations would make it fail, but Popper does not specify what those observations are; if the principle of falsifiability is not refutable, then by Popper's own standard it is not a scientific statement, which leads to self-contradiction. This problem is not fully treated in Popper's work.

Limit Five: The Missing Ontological Dimension

Popper's work operates mainly at the epistemological level: it treats the demarcation criterion of theories. But Popper does not develop in detail the meaning of falsifiability at the ontological level: what is the falsifiable structure of an ontological commitment? This missing dimension is the entry point of the present chapter.

The Direction of This Chapter

The direction of this chapter is to elevate falsifiability from epistemology to ontology: it is not only a demarcation criterion for theories, but also a standard for the legitimacy of ontological commitments. This elevation needs to be developed in detail, beginning in Section 6.2.

The advantages of this elevation are as follows: it extends the force of falsifiability to ontological problems; it gives ontological commitments a methodological discipline, which is central to the work of this book; and through the methodology of this book, including constrained speculation, the ontology of failure conditions, and constitutive coextension, it gives falsifiability a contemporary refinement that avoids the limits of Popper's work.

Objections and Replies

Objection One: The limits of Popper that you list are already common knowledge in twentieth-century philosophy of science. Is it too elementary to repeat them?

Reply: The aim of this section is not to discover the limits of Popper, but to identify these limits in preparation for the subsequent work. These limits are the starting point: they indicate the boundaries of Popper's work. The subsequent work of this chapter, beginning in Section 6.2, does new work within these boundaries: it elevates falsifiability to an ontological discipline. This elevation is not a simple "surpassing" of Popper, but a new development based on his work. Acknowledging the limits of Popper's work does not weaken the work of this chapter; it clarifies the position of this chapter.

Objection Two: You say that Popper's work is at the epistemological rather than the ontological level. But Popper himself had ontological positions, such as his realism and his three-world theory. Is your statement accurate?

Reply: This objection rightly points to the complexity of Popper's work. Popper indeed had ontological positions. In *Objective Knowledge* (1972), he proposed the three-world theory, consisting of the physical world, the mental world, and the world of objective knowledge, and in other works he supported a realist position. These positions are part of Popper's work. But the principle of falsifiability was primarily stated and developed at the epistemological level: it treats the demarcation criterion of theories. Popper did not develop in detail the falsifiable structure of ontological commitments; his ontological work and his work on falsifiability were not tightly integrated. The new content of this chapter is to integrate the two tightly, making falsifiability the standard of legitimacy for ontological commitments. This integration was not fully developed in Popper's work and is the contribution of this chapter.

Objection Three: Your direction, namely elevating falsifiability to ontology, sounds overly ambitious. Are you trying to "replace" Popper?

Reply: The work of this chapter does not "replace" Popper; it develops his work in a contemporary way. Popper's core insight, namely that theories need content that can be refuted, is preserved; the work of this chapter is built upon that insight. The limits of Popper's work are treated through the methodology of this book, including constrained speculation and the ontology of failure conditions; these methodologies are contemporary supplements to Popper's work. Falsifiability is extended from epistemology to ontology, and this extension broadens the scope of Popper's work. This is inheritance plus development, not replacement. Acknowledging Popper's contribution while developing it in a contemporary direction is a common pattern of philosophical progress.

Literature That Must Be Engaged

- Karl Popper, The Logic of Scientific Discovery (1934/1959)
- Karl Popper, Conjectures and Refutations (1963)
- Karl Popper, Objective Knowledge (1972)
- Pierre Duhem, The Aim and Structure of Physical Theory (1906)
- W. V. O. Quine, "Two Dogmas of Empiricism" (1951)
- Thomas Kuhn, The Structure of Scientific Revolutions (1962)
- Imre Lakatos, "Falsification and the Methodology of Scientific Research Programmes" (1970)

- Chapter 1 on the method of constrained speculation (self-reference)

Summary of This Section

This section has done four things. First, it developed the content and motivation of Popper's principle of falsifiability. Second, it identified the contributions of Popper's work: methodological rigor, criticism of ad hoc adjustments, and criticism of historicism. Third, it identified the limits of Popper's work: the Duhem–Quine thesis, the complexity of theory choice, the theory-ladenness of observation, the problem of self-application, and the absence of an ontological dimension. Fourth, it announced the direction of this chapter: elevating falsifiability from epistemology to ontology.

The next section develops the content of this elevation.

6.3. Falsifiability as an Ontological Discipline

The Key Transformation

Popper's falsifiability operates mainly at the epistemological level: it treats the demarcation of theories. The core work of this chapter is to elevate this principle to the ontological level: it treats the legitimacy of ontological commitments.

Specifically:

Popper's falsifiability: a theory is scientific $\Longleftrightarrow$ it has content that can be empirically refuted.

Falsifiability as an ontological discipline in this chapter: an ontological commitment X is legitimate $\Longleftrightarrow$ X's content has an identifiable direction of failure.

This transformation involves several shifts: from theory to ontological commitment, where the unit shifts from large theoretical networks to ontological commitments; from empirical refutation to direction of failure, where refutation is no longer limited to direct empirical observation but is expanded to identifiable failure structures; and from the demarcation criterion of science to the legitimacy criterion of ontological commitments, where the goal shifts from separating science and non-science to evaluating ontological commitments.

Statement of the Core Thesis

Core thesis FA1: an ontological commitment X is legitimate $\Longleftrightarrow$ X's content has an identifiable direction of failure; that is, there are possible states of affairs which, if they occurred, or if they were identified as occurring, would make the statement concerning X fail.

This thesis requires detailed development. Let us analyze several of its key meanings.

First, "Ontological Commitment" as the Unit

This chapter works not at the level of "theory," but at the level of "ontological commitment." An ontological commitment is a statement about beings, relations, or categories: it states that some ontological content exists or that some ontological relation holds. An ontological commitment is usually smaller than a theory: a theory may contain multiple ontological commitments. It is also usually larger than an observation: it involves ontological meaning, not merely observation.

Example: "the electron exists as an elementary particle" is an ontological commitment. "Quantum mechanics" is a theory containing multiple ontological commitments, concerning electrons, wave functions, measurement, and so on. "In a certain experiment, the electron hits a certain position on the screen" is an observation. Ontological commitment as a unit makes falsifiability more refined: the question is not whether an entire theory is refuted, but whether an ontological commitment has a direction of failure.

Second, "Direction of Failure" as the Standard

A direction of failure is not limited to direct empirical refutation. It includes various identifiable possibilities of failure: empirical failure, where possible observations would make the ontological commitment fail; theoretical failure, where the ontological commitment conflicts with other ontological commitments and makes the whole inconsistent; methodological failure, where the ontological commitment cannot satisfy the requirements of the method of constrained speculation, namely formalization, mechanization, observationalization, and failure-orientation; and structural failure, where the ontological structure involved in the commitment has internal contradictions. Each type of direction of failure has identifiable content. A legitimate ontological commitment has an identifiable direction of failure, not necessarily all of the above types, but at least some of them.

Third, "Legitimacy" as Evaluation

Legitimacy is not truth. It is an evaluation of whether an ontological commitment satisfies methodological discipline. A legitimate ontological commitment has content that can be rigorously evaluated: it can be supported or refuted. An illegitimate ontological commitment lacks identifiable content: it cannot be rigorously evaluated. Legitimacy is a minimal standard. A legitimate ontological commitment may still be false, if its direction of failure is identified as having occurred, but it can at least be discussed. An illegitimate ontological commitment cannot be rigorously discussed: it has no tractable content.

Implications of the Thesis

FA1 has four ontological implications.

Implication One: Ontological Commitments Require Falsifiable Structure

FA1 extends Popper's insight to ontological commitments: any serious ontological commitment needs an identifiable direction of failure. If a statement claims to be an ontological commitment but has no direction of failure, it is not a legitimate ontological commitment. This requirement makes ontology rigorous: it does not accept arbitrary ontological declarations. Every ontological commitment needs to display an identifiable direction of failure.

Implication Two: Illegitimate Ontological Statements Are Excluded

FA1 excludes certain ontological statements by means of a standard: they claim to state ontological content, but they have no identifiable direction of failure. Examples: "the whole universe is the projection of some consciousness"; if this statement has no identifiable direction of failure, and any observation is compatible with it, it is not a legitimate ontological commitment. "All possible worlds exist"; if this statement has no identifiable direction of failure, and any observation is compatible with it, it is not a legitimate ontological commitment. Chapter 3, Section 3.4 has already developed the difficulty of Lewisian modal realism. "Things have essences"; if "essence" as a concept has no identifiable direction of failure, and anything can be interpreted as having some "essence," the statement is not a legitimate ontological commitment. These statements may play roles in philosophical discussion, for example heuristic or conceptual roles, but they are not legitimate ontological commitments and cannot be rigorously evaluated.

Implication Three: Ontology Can Be Made Rigorous as a Field of Research

FA1 makes ontology a field of research that can be made rigorous: it has methodological standards. Ontology is not arbitrary philosophical conjecture; it involves evaluable content. Ontological research can use methods such as logical analysis, connection with physics, and empirical research to evaluate ontological commitments. Progress in ontology is possible: through research, ontological commitments can be refined, tested, and revised. Ontological disagreements can be refined through argument: both sides can discuss on the basis of falsifiable structures, rather than through confrontation between intuitions. This rigorization keeps ontology as a living philosophical field: it can connect with science and be advanced through research.

Implication Four: Falsifiability as an Ontological Discipline

FA1 establishes falsifiability as an ontological discipline: it is a rule about how ontological commitments are to be established. Any work that proposes ontological commitments must display identifiable directions of failure; any work that accepts ontological commitments must evaluate the legitimacy of those directions of failure; any work that revises ontological commitments must do so according to how directions of failure have been identified. This discipline gives ontological work a methodological direction: it is not arbitrary philosophical creation, but a regulated field of research.

Differences from Popper

Through comparison, the difference between FA1 and Popper can be stated as follows:

Table 6.1.. A Multidimensional Comparison between Popperian Falsifiability and FA1 Falsifiability

Item (Dimension)	Standard/Classical Content (Popperian Falsifiability)	EQT/FA1 Correspondence (FA1 Falsifiability)	EQT Limitation / Not Involved	Chapter
Unit of Analysis	Theory	Ontological commitment	Does not provide one-step falsification of concrete empirical laws	2
Type of Failure	Empirical refutation	Multiple types: empirical, theoretical, methodological, structural	Does not neglect concrete physical empirical tests	2-3
Criterion of Judgment	Science vs. non-science	Legitimate vs. illegitimate ontological commitments	Does not replace the distinctive standards of judgment internal to concrete disciplines	3

Continued on next page...

Continued Table

Item (Dimension)	Standard/Classical Content (Popperian Falsifiability)	EQT/FA1 Correspondence (FA1 Falsifiability)	EQT Limitation / Not Involved	Chapter
Level of Operation	Epistemology	Ontology	Does not depart from formalized description into pure speculation	3-4
Scope of Application	Natural science	All ontological work	Does not extend to pure historical narratives that cannot be axiomatized	4

This comparison shows that FA1 is an extension and refinement of Popper's work. It inherits the core insight, namely the need for identifiable failure, but expands both the scope and the level of application.

Objections and Replies

Objection One: Your "direction of failure" includes multiple types: empirical, theoretical, methodological, and structural. Does this make the standard too loose?

Reply: This objection points to an important point of refinement. Different types of directions of failure are all real: they involve identifiable failure. But their degrees of identifiability differ. Empirical failure is usually easiest to identify: it can be directly identified through observation. Theoretical failure involves logical analysis: it can be identified through reasoning. Methodological failure involves methodological discipline: it can be identified through methodological review. Structural failure involves internal contradiction: it can be identified through analysis. Acknowledging multiple types does not relax the standard; it refines the standard. Each type of failure must be identifiable. Every ontological commitment must have at least one type of identifiable direction of failure. Furthermore, FA1 does not claim that "any vague assertion is legitimate"; it claims that any ontological commitment requires some identifiable direction of failure. Examples such as Lewisian modal realism and essentialism show that some ontological statements indeed lack any type of identifiable direction of failure; they are excluded by FA1.

Objection Two: Is your concept of "legitimacy" too binary, namely legitimate vs. illegitimate? Actual ontological commitments may have different degrees of falsifiability.

Reply: FA1 allows degrees of falsifiability. Some ontological commitments have highly clear directions of failure: their falsifiable structures are clear and easy to identify. Some ontological commitments have lower falsifiability: their directions of failure require more complex analysis. Some ontological commitments have no falsifiable structure at all: they are illegitimate ontological commitments. The difference in degree is real. But the minimal standard is the possession of an identifiable direction of failure. Even a lower degree of falsifiability satisfies legitimacy. Statements with no falsifiable structure at all are not legitimate ontological commitments. This treatment is similar to Popper's: Popper also acknowledges that scientific theories have different degrees of falsifiability. FA1 extends this treatment to ontological commitments.

Objection Three: Will your ontological discipline exclude some legitimate metaphysical work? For example, general statements about being, such as "beings have unity," may not have simple directions of failure.

Reply: This objection rightly points to the difficulty of application. Some metaphysical statements do involve highly abstract content and have difficulty possessing simple directions of failure. But this does not mean that they are necessarily illegitimate. Rather, they require refinement. Highly abstract statements, such as "beings have unity," can be refined into more concrete statements; refined statements can have identifiable directions of failure; and the process of refinement is legitimate metaphysical work, because it makes statements rigorously evaluable. For example, "beings have unity" as a highly abstract statement can be refined as follows: "any being has some internal coherence that allows it to be identified as a unit." This is a more concrete statement and can have a direction of failure: if some identified being has no internal coherence at all, the commitment fails. The process of refinement is an application of the method of constrained speculation: it makes highly abstract metaphysical work rigorously evaluable rather than simply excluding it. But if a metaphysical statement cannot in any way be refined into a statement with an identifiable direction of failure, then it is not a legitimate ontological commitment and should be excluded. Such exclusion is not a defect; it is an application of methodological rigor.

Literature That Must Be Engaged

- Karl Popper, The Logic of Scientific Discovery (1934/1959)
- Karl Popper, Conjectures and Refutations (1963)
- Chapter 3 on failure conditions as an ontological discipline (self-reference)
- Chapter 1 on the method of constrained speculation (self-reference)

Summary of This Section

This section has done four things. First, it clarified the key transformation from Popperian falsifiability to falsifiability as an ontological discipline in this chapter. Second, it stated the core thesis FA1: an ontological commitment is legitimate $\Longleftrightarrow$ it has an identifiable direction of failure. Third, it developed the four implications of the thesis: falsifiable structure, exclusion of illegitimate statements, rigorization of ontology, and falsifiability as an ontological discipline. Fourth, it compared FA1 with Popper's position and clarified the extension made by this chapter.

The next section develops the application dimensions of the ontological discipline.

6.4. Application Dimensions of the Ontological Discipline

The Function of Application

FA1 as an ontological discipline requires dimensions of application: how is it used in ontological work? This section develops three application dimensions: the evaluation of ontological commitments, the treatment of ontological disputes, and the direction of ontological development.

Application One: Evaluation of Ontological Commitments

FA1 provides a framework for evaluating ontological commitments. Any ontological commitment can be rigorously evaluated through four questions:

Question One: What Is the Content of the Commitment?

The commitment needs to be explicitly stated: what beings, relations, or categories does X involve? A claim without explicit content is not a legitimate ontological commitment. Example: the illegitimate claim "everything is interconnected" has no explicit content:

what counts as "everything"? What counts as "interconnected"? By contrast, the legitimate claim "any being is involved in some co-evolutionary relation" has explicit content: beings and co-evolutionary relations, as developed in Chapter 12.

Question Two: What Is the Direction of Failure of the Commitment?

The commitment needs an identifiable direction of failure: what circumstances would make the commitment fail? A claim without a direction of failure is not a legitimate ontological commitment. Example: the illegitimate claim "all possible worlds exist" has no direction of failure, since any observation is compatible with it. By contrast, the legitimate claim "elementary particles exist as excitation modes of the ρ_f field" has a direction of failure: it would fail if experiments showed that elementary particles have behavior that cannot be explained by the dynamics of the ρ_f field.

Question Three: How Is the Direction of Failure Identified?

The direction of failure must be identifiable: through what method, observation, or analysis? A direction of failure without an identifying method is not truly falsifiable. Example: the illegitimate claim "the world has a purpose," if "purpose" as a concept has no identifiable content. By contrast, the legitimate claim "biological systems tend toward reproductive success" is identifiable through research in evolutionary biology.

Question Four: What Is the Relation between the Commitment and Other Commitments?

A commitment does not exist in a vacuum. It has relations to other ontological commitments. These relations need to be made explicit so that the commitment can be evaluated as part of a whole. Example: an illegitimate isolated commitment, where an ontological statement has no relation to other ontological statements and makes the overall ontology fragmented; by contrast, a legitimate integrated commitment, where the ontological statement is connected and compatible with other ontological statements and forms a coherent ontology.

Together, these four questions constitute the framework for evaluating ontological commitments. Any ontological commitment can be rigorously evaluated through this framework.

Application Two: Treatment of Ontological Disputes

The history of philosophy contains numerous ontological disputes: realism vs. anti-realism, materialism vs. idealism, absolutism vs. relativism, and so on. These disputes often involve complex content and are difficult to resolve. FA1 provides tools for treating such disputes.

Tool One: Clarifying the Focus of the Dispute

Ontological disputes often involve multiple focal points: the content of ontological commitments, directions of failure, methods of identification, and so on. Clarifying the focal points makes the dispute refinable. Example: the dispute between realism and anti-realism involves several focal points: the ontological status of beings, namely independent objective existence vs. dependence on the subject; the meaning of objectivity, namely absolute objectivity vs. controlled reproducibility, as in Chapter 14, $OB1$; and the possibility of epistemology, namely direct access to objects vs. co-evolutionary relation. Each focal point can be evaluated independently. The dispute is not a confrontation between positions as wholes, but an analysis of multiple focal points.

Tool Two: Evaluating the Falsifiable Structure of Positions

Different positions have different falsifiable structures. Simple realism: objects exist independently of subjects. Direction of failure: if observation shows that object and observer are co-evolving, rather than independent, this direction of failure has already been identified to some extent in quantum measurement, neuroscience, and related fields. Simple anti-realism: objects are only constructions of the subject. Direction of failure: if observation shows that objects have intersubjective consistency and independent behavior, this direction of failure is also identified. Constrained process realism (Chapter 14): objectivity is grounded in reproducible co-perturbation. Direction of failure: if observation shows that objects have no reproducibility and no co-perturbative structure. Each position has a falsifiable structure. By evaluating these structures, positions can be refined: which positions can better handle evidence, and which need revision?

Tool Three: Identifying Illegitimate Positions

Some positions involve illegitimate ontological commitments: they have no identifiable direction of failure. Such positions can be excluded by FA1. Example: complete subjectivism, according to which all experience is the projection of the subject. If this position has no direction of failure, and any experience is interpreted as "subjective projection," it is not a legitimate ontological position. Complete determinism, according to which all events are completely determined by prior events. If this position has no direction of failure, and any observation is interpreted as "already determined," it needs refinement or exclusion. Such exclusion is not a matter of choosing a position, but of excluding statements that cannot be rigorously evaluated.

Application Three: Directions of Ontological Development

FA1 is not only a tool of evaluation; it is also a tool of development.

Direction One: Refining Ontological Commitments through Research

Ontological commitments can be refined through research: their content becomes clearer, their directions of failure become more explicit, and their methods of identification become more precise. This is a direction of progress for ontology as a field of research. Example: "consciousness has a relation to physical processes" is a rough ontological commitment. Through neuroscience and philosophical research, such as the Ψ framework of this book in Chapter 13, this commitment is refined into "consciousness and the dynamics of the ρ_f field are different projections of the same process." The refined commitment has clearer directions of failure, namely the refutable predictions of the Ψ pattern.

Direction Two: Expanding Ontological Content through the Development of Physics

The development of physics allows ontology to be expanded: new physical discoveries make new ontological commitments possible. Example: the development of quantum mechanics makes "quantum entanglement as an ontological relation" a possible ontological commitment, as discussed in Chapter 12, Section 12.4 on Rovelli and others. The development of cosmology makes "the early universe as an ontological process" a possible ontological commitment, as discussed in Chapter 11 on freezing as a foundational ontological process. The development of neuroscience makes "consciousness as a multi-frequency structure" a possible ontological commitment, as discussed in Chapter 9 on frequency as a basic spectral quantity and Chapter 13 on mind and matter. Each expansion is realized through the development of physics and is tested for legitimacy through FA1.

Direction Three: Refining Ontology through Interdisciplinary Dialogue

Ontology enters into dialogue with other disciplines, including physics, biology, neuroscience, sociology, political philosophy, ethics, and others. These dialogues refine and expand ontology. Examples: dialogue with physics refines the physical basis of ontology, as in the connection between this book and EQT; dialogue with biology expands the life-dimension of ontology, as in life as a multi-frequency being in Chapter 9; dialogue with ethics expands the normative dimension of ontology, as in Chapter 20 on the is–ought gap. These dialogues keep ontology as a living and developable field of research.

An Example: The Evolution of Ontological Commitments

Let us consider one example to show how FA1 functions as a tool of development: the evolution of ontology from Newton to EQT.

Newtonian period: the ontological commitment is absolute spacetime + material particles. Absolute spacetime functions as a container; material particles function as basic ontological units; interactions occur through forces such as gravity. This ontology has identifiable directions of failure: if observation shows that spacetime depends on matter and energy, and is therefore not absolute, it fails.

Relativistic period: observations, such as the Michelson–Morley experiment and Mercury's perihelion, identified the directions of failure of Newtonian ontology. Ontology was revised into relativistic ontology: spacetime as a geometric structure rather than an absolute container; matter-energy coupled with spacetime; interaction through geometry, with gravity as spacetime curvature.

Quantum-mechanical period: further observations identified the directions of failure of relativistic ontology: it could not treat microscopic phenomena. Ontology was extended into quantum ontology: quantum state as a basic ontological unit, rather than classical particles; measurement involving the observer, rather than non-perturbative reading; interaction through fields and wave functions.

EQT period: the work of this book further extends ontology: the ρ_f field as the unique basic ontological object; particles as excitation modes of the ρ_f field; interactions as band-specific manifestations of the dynamics of the ρ_f field; the mind–matter relation as different projections of the same process; and observation as co-evolution within the same field.

This evolution shows the application of FA1 as a tool of development: ontology at each step is pushed forward by the identifiable directions of failure of the preceding step, and is refined through physical and philosophical development. This is the progress of ontology as a field of research.

Objections and Replies

Objection One: Your four evaluation questions, namely commitment content, direction of failure, method of identification, and relation, seem overly procedural. They make ontological work mechanical.

Reply: This objection rightly points to the issue of methodological application. The four questions are not a mechanical procedure; they are a methodological discipline. They do not replace the creativity of ontological work: ontological content still requires insight and reflection. They provide methodological support: they make insight and reflection rigorously evaluable. They are a minimal standard: any ontological work must at least satisfy these standards. By analogy, scientific research has methodological disciplines, such as experimental design, data analysis, and peer review. These do not replace scientific creativity; they make scientific creativity rigorously evaluable. Likewise, the four questions of FA1 do not replace ontological creativity; they make ontological creativity rigorously evaluable. The risk of mechanization is real: if researchers mechanically apply the four questions, they may miss deeper ontological insight. But rigor and creativity are not opposed; they can work together. The overall aim of the method of constrained speculation is to support this cooperation.

Objection Two: Your example, the evolution from Newton to EQT, uses the development of physics as its driver. But many ontological disputes, such as those concerning abstract objects and possible worlds, do not directly involve the development of physics.

Reply: This objection concerns the scope of application. The development of physics is one driver of ontological evolution, but it is not the only one. Other factors include: logical development, where logical tools such as modal logic and category theory make the expression and evaluation of ontological commitments more refined; the development of cognitive science, where research on cognition and mind refines the mental implications of ontological commitments; cross-cultural dialogue, where insights from traditions such as the Chinese and Indian traditions expand ontological commitments; and methodological development, where methodological tools such as constrained speculation make the evaluation of ontological commitments more rigorous. Disputes about abstract objects and possible worlds can be advanced through these drivers and need not depend on the development of physics. FA1 applies to all types of ontological commitments, as long as they involve identifiable directions of failure.

Objection Three: Your dimension of "identifying illegitimate ontological statements" may be abused. Someone could claim that an opponent's ontological statement "has no direction of failure" in order to exclude it.

Reply: This objection concerns the treatment of methodological disputes. The falsifiable structure of an ontological statement is itself open to discussion, not arbitrary assertion. Anyone who claims that some ontological statement "has no direction of failure" must defend that claim by showing how the statement lacks a direction of failure. Anyone who opposes that claim can reply by showing that the statement does have a direction of failure, even if indirectly or after refinement. The discussion proceeds through argument and evidence, not through simple assertion. This is the form of methodological discussion. Any methodological standard can be abused, but discussion provides safeguards: argument and evidence make abuse difficult. Example: Chapter 3, Section 3.4 criticizes Lewis's modal realism for "having no failure conditions." This criticism is supported by argument: Lewis himself accepts the existence of all possible worlds, making observation unable to refute it. This criticism can be answered by Lewisian philosophers, who may attempt to show failure conditions. The treatment in this section is not simple assertion, but an arguable claim.

Literature That Must Be Engaged

- W. V. O. Quine, "On What There Is" (1948)
- Hilary Putnam, The Many Faces of Realism (1987)
- Bas van Fraassen, The Scientific Image (1980)
- Chapter 3 on failure conditions as an ontological discipline (self-reference)
- Chapter 12 on co-evolution as an ontological relation (self-reference)
- Chapter 14 on the new foundation of objectivity (self-reference)

Summary of This Section

This section has done four things. First, it developed the four questions of FA1 as a tool of evaluation: commitment content, direction of failure, method of identification, and relation. Second, it developed three applications of FA1 as a tool for treating disputes: clarifying focal points, evaluating positions, and excluding illegitimate positions. Third, it developed three directions of FA1 as a tool of development: refining commitments, expanding ontological content, and interdisciplinary dialogue. Fourth, it used the evolution of ontology from Newton to EQT to show the application of FA1 as a tool of development.

The next section develops the function of this chapter as the synthetic chapter of Part I.

6.5. The Synthesis of Part I: The Realization of Method as Philosophy

The Function of Synthesis

The six chapters of Part I, "Constrained Speculation: Method as Philosophy," have developed the methodological foundation. This chapter, as the concluding chapter of Part I, undertakes a synthetic function: it unifies the work of the preceding five chapters under the framework of falsifiability as an ontological discipline.

This synthesis is not a simple summary; it is the display of the internal unity of the methodology. The six chapters of Part I are not six independent works. They constitute a mutually supporting network: Chapter 1, the method of constrained speculation, establishes the foundational methodology; Chapter 2, minimal postulates and ontological restraint, develops one methodological discipline; Chapter 3, failure conditions as an ontological discipline, develops another methodological discipline; Chapter 4, constrained physicalization, provides a semantic tool; Chapter 5, positioning within the genealogy of contemporary realism, provides the philosophical position; Chapter 6, this chapter, falsifiability as an ontological discipline, provides the synthetic discipline. Each chapter is built upon the others and contributes to them. This section develops this mutually supporting network.

How the Six Chapters Support One Another

Let us develop in detail the mutually supporting relations among the six chapters.

Chapter 1 ↔ the Other Chapters

Chapter 1 establishes the method of constrained speculation as the overall framework: the four disciplines of formalization, mechanization, observationalization, and failure-orientation. This framework is applied in the other chapters. In Chapter 2, ontological restraint uses the method of constrained speculation as its ground. In Chapter 3, failure conditions as an ontological discipline are the development of the fourth discipline of the method of constrained speculation. In Chapter 4, constitutive coextension uses the method of constrained speculation as the ground of a semantic tool. In Chapter 5, positioning within the genealogy of contemporary realism uses the method of constrained speculation as its method. In Chapter 6, falsifiability as an ontological discipline is the application of the method of constrained speculation at the ontological level. Chapter 1 is the foundational chapter, and the other chapters are its developments.

Chapter 2 ↔ Chapter 3

Chapter 2, on ontological restraint, and Chapter 3, on failure conditions, mutually support each other. Restraint requires that every ontological commitment be justified; failure conditions provide the standard of justification. Failure conditions as a standard require every ontological commitment to have an identifiable direction of failure; restraint requires that the number of commitments remain minimal. Together they constitute the methodological discipline of ontology: restraint, namely minimal commitments, plus failure conditions, namely the legitimacy of each commitment.

Chapter 4 ↔ Chapter 3

Chapter 4, on constitutive coextension, is built upon Chapter 3, on failure conditions: constitutive coextension as a semantic tool uses failure conditions as its definitional content. Two commitments are constitutively coextensive $\iff$ they are coextensive under the same failure conditions. This connection makes constitutive coextension the semantic tool of the method of constrained speculation.

Chapter 5 ↔ All Other Chapters

Chapter 5, on positioning within the genealogy of contemporary realism, uses the content of the other chapters to position the stance of this book. It uses the methodology of Chapters 1–4 to explain the position of this book; it uses the failure conditions of Chapter 3 to distinguish this position from traditional positions; and it uses constitutive coextension from Chapter 4 to develop the relation between this book and other positions. Chapter 5, as the positioning chapter, uses the content of the other chapters to give the position of this book a place in contemporary philosophical discussion.

Chapter 6 ↔ All Other Chapters

This chapter, on falsifiability as an ontological discipline, synthesizes the content of all other chapters. It refines the core of the method of constrained speculation (Chapter 1) into an ontological discipline; it places ontological restraint (Chapter 2) under the framework of falsifiability, since restraint itself is grounded in falsifiability; it expands failure conditions (Chapter 3) into the broader framework of falsifiability, since failure conditions are one form of falsifiability; it treats constitutive coextension (Chapter 4) as the semantic tool of falsifiability; and it treats the positioning within the genealogy of contemporary realism (Chapter 5) as positioning under the framework of falsifiability. As the synthetic chapter, this chapter gives the methodological foundation of Part I complete unity.

The Realization of Method as Philosophy

The title of Part I is "Constrained Speculation: Method as Philosophy." "Method as philosophy" as a slogan needs to be realized through methodological work. This chapter, as the synthetic chapter, shows that realization.

Meaning One: Method Is Not Merely a Tool; It Constitutes Philosophical Content

Traditional philosophy often separates method and content: method is the tool, and content is the aim. This separation makes method secondary. The work of this book rejects this separation. Method, including constrained speculation, failure conditions, falsifiability, and so on, is philosophical content: it concerns the core of ontology, epistemology, and methodology. Specifically, failure conditions as an ontological discipline = a philosophical statement about being ($CR2$: X is real $\iff X$ has definable failure conditions); falsifiability as an ontological discipline = a philosophical statement about ontological commitment (FA1); constitutive coextension = a philosophical statement about conceptual relations (Chapter 4, CC1). These methodological statements are not mere tools; they are philosophical content.

Meaning Two: Philosophy Is Not Merely Reflection on Method; It Requires the Rigorous Application of Method

Traditional philosophy often treats philosophy as reflection on methods, such as scientific method: philosophers do not directly do methodological work, but reflect on method. This treatment distances philosophy from method. The work of this book rejects this treatment. Philosophy as a field of research requires the rigorous application of method: ontological work requires falsifiable structure, failure conditions, formalization, and so on. The ontological and normative work of Parts II through V of this book is built upon the methodology of Part I: every ontological commitment must satisfy methodological discipline. This is philosophy as a serious field of research.

Meaning Three: The Mutual Construction of Method and Content

Method and content do not merely support one another while remaining separate; they mutually construct each other. The form of method depends on the ontological framework, for example, the process ontology of this book gives method a processual character. The form of ontological content depends on method, for example, the discipline of falsifiability in this book gives ontological content identifiable directions of failure. This mutual construction makes method and content an inseparable whole: together, they constitute philosophical work.

Part I as a Complete Foundation

Through the synthesis of this chapter, Part I has complete unity as a methodological foundation: the starting point, Chapter 1, is the method of constrained speculation as the overall framework; the core disciplines, Chapters 2 and 3, are ontological restraint and failure conditions as methodological disciplines; the semantic tool, Chapter 4, is constitutive coextension; the philosophical positioning, Chapter 5, locates the work within the genealogy of contemporary philosophy; and the synthesis, Chapter 6, unifies all the preceding work through falsifiability as an ontological discipline.

This foundation provides the methodological basis for the later parts, from Part II to Part V, where ontology, relational ontology, metaphysics, and normativity are developed. The work of every subsequent chapter is built upon the methodology of Part I: every ontological commitment, relational statement, metaphysical statement, and normative statement is constrained by the methodological discipline of Part I.

Objections and Replies

Objection One: Your claim that "method is philosophy" sounds too strong. Traditional philosophy has many rich contents, such as metaphysics, epistemology, and ethics, and these are not merely about method.

Reply: "Method as philosophy" in this section does not claim to exclude philosophical content. It claims that method and content are inseparable. Metaphysical content, such as the ontology of difference and process ontology in this book, is built upon method. Epistemological content, such as the new foundation of objectivity in this book, is built upon method. Ethical content, such as the treatment of the is–ought gap in this book, is built upon method. Every philosophical content gains its form through method. The meaning of "method as philosophy" is that method constitutes the way philosophical content is formed; it does not mean that "there is only method and no content." More deeply, acknowledging the mutual construction of method and content makes philosophy more rigorous as a field of research: no philosophical content can be stated independently of method, and no method can be justified independently of philosophical content. This mutual construction is the core of the method of constrained speculation.

Objection Two: Your mutually supporting network among the six chapters seems overly elaborate. It makes Part I look like a closed system. But philosophical work should be open outward, not self-enclosed.

Reply: This objection points to an important methodological problem. As a methodological foundation, Part I has internal unity; this is legitimate philosophical work. But internal unity is not closure. It is open outward: the methodology of Part I connects with science, especially physics and neuroscience, through the requirement of observationalization in the method of constrained speculation; it enters into dialogue with other philosophical traditions, especially through the positioning of Chapter 5, by analyzing similarities and differences; and the methodology of Part I itself can be criticized and refined. It is not the final form of methodology. This openness keeps Part I as living methodological work: it can dialogue with external content and be advanced through criticism. Internal unity and external openness are cooperative, not opposed. The later parts of this book, Parts II through V, are precisely the application of the methodology of Part I in ontology, relations, metaphysics, and normativity. This is itself an expression of Part I's outward openness.

Objection Three: This chapter as a synthetic chapter seems too "complete." It makes everything merge together, but the substantive content of ontological disputes, namely which commitments are legitimate and which are not, requires ontological work, not methodological synthesis.

Reply: This objection correctly points to the relation between methodology and ontological content. This chapter, as a synthetic chapter, displays the methodological unity of Part I. This is work at the methodological level. The substantive content of ontological disputes belongs to the ontological level and is developed in Parts II through V. The two do not conflict. Part I establishes the methodology: what disciplines any ontological commitment must satisfy.

Parts II through V develop ontological content: ontological commitments, relational statements, metaphysical statements, and normative statements. The latter are built upon the former but are not fully determined by the former: ontological content involves physical discoveries, philosophical reflection, interdisciplinary dialogue, and so on. These contents belong to the ontological level and are not limited to methodology. The function of this chapter is to synthesize the work at the methodological level and to provide a methodological ground for the later work at the ontological level. This is a reasonable division of chapters, not a replacement of ontological content by methodology.

Literature That Must Be Engaged

- Chapter 1 on the method of constrained speculation (self-reference)
- Chapter 2 on minimal postulates and ontological restraint (self-reference)
- Chapter 3 on failure conditions as an ontological discipline (self-reference)
- Chapter 4 on constrained physicalization (self-reference)
- Chapter 5 on positioning within the genealogy of contemporary realism (self-reference)

Summary of This Section

This section has done four things. First, it developed the mutually supporting relations among the six chapters of Part I. Second, it developed three meanings of "method as philosophy": method constitutes philosophical content, philosophy requires the rigorous application of method, and method and content mutually construct each other. Third, it clarified the position of Part I as a complete methodological foundation. Fourth, it replied to possible objections, especially concerning the meaning of "method as philosophy," the relation between internal unity and external openness, and the division of labor between methodology and ontological content.

The next section is the summary of this chapter.

6.6. Chapter Summary: The Contemporary Significance of Falsifiability as an Ontological Discipline

Completion of the Three Tasks

The three tasks undertaken by this chapter have all been completed.

Task One: to elevate falsifiability from a principle of scientific methodology to an ontological discipline. Section 6.1 developed the content and limits of Popper's work. Section 6.2 stated the core thesis FA1: an ontological commitment is legitimate $\Longleftrightarrow$ it has an identifiable direction of failure. Section 6.3 developed three application dimensions of FA1 as an ontological discipline: evaluation, dispute treatment, and development.

Task Two: to enter into serious dialogue with Popper's work. Section 6.1 both inherited Popper's core insight, namely that theories need content that can be refuted, and refined it through the methodology of this book, including constrained speculation, failure conditions, and constitutive coextension. Section 6.2 clarified the differences between FA1 and Popper: unit, type of failure, criterion, level, and scope.

Task Three: to serve as the synthetic chapter of Part I by unifying the work of the preceding five chapters. Section 6.4 developed the mutually supporting relations among the six chapters of Part I and clarified the realization of "method as philosophy" as a slogan.

The Position of This Chapter in Part I

As the sixth chapter of Part I, this chapter performs a synthetic function. The overall architecture of Part I is completed in this chapter: starting point (Chapter 1) → core disciplines (Chapters 2 and 3) → semantic tool (Chapter 4) → philosophical positioning (Chapter 5) → synthesis (Chapter 6). This structure gives Part I complete unity as a methodological foundation: all contents are organized through the method of constrained speculation and falsifiability as an ontological discipline.

The Position of This Chapter in the Whole Book

Looking backward, this chapter is built upon the work of the first five chapters of Part I: it is their synthesis.

Looking forward, this chapter provides the methodological foundation for the ontological, relational, metaphysical, and normative work of Parts II through V. The work of every chapter is constrained by the discipline of FA1: every ontological commitment needs an identifiable direction of failure. Specifically, the applications of FA1 in the later parts are as follows. In Part II, the ontology of difference (Chapter 7, DI1), qi theory as restrained ontology (Chapter 8), frequency as a basic spectral quantity (Chapter 9, FR1), tri-state categories as ontological categories (Chapter 10, T1, T2), and freezing as a foundational ontological process (Chapter 11, FZ1) all require identifiable directions of failure for each ontological commitment. In Part III, co-evolution as an ontological relation (Chapter 12, EV1), mind–matter unity (Chapter 13), observation as participation (Chapter 14, OB1), and time as $\partial\rho/\partial t$ (Chapter 15, TM1) all require identifiable directions of failure for each relational commitment. In Part IV, ontological singularity (Chapter 17) involves the falsifiable structure of the metaphysical boundary. In Part V, the treatment of the is–ought gap (Chapter 20, IO1) and the finitude of meaning (Chapter 21, M1–M3) require identifiable directions of failure for each normative commitment. These applications make the work of this chapter continuously present throughout the whole book: the work of every later chapter implicitly involves the discipline of FA1.

Honesty about the Limits of This Chapter

This chapter acknowledges several limits:

First limit: the dialogue with the tradition of philosophy of science is insufficient. This chapter mainly enters into dialogue with Popper's work. Detailed dialogues with other important traditions in philosophy of science, such as Kuhn, Lakatos, Laudan, Hacking, and van Fraassen, require dedicated work. This chapter briefly mentions Duhem–Quine and others, but a complete dialogue is left for the future.

Second limit: the dialogue with contemporary analytic metaphysics is insufficient. FA1 as an ontological discipline has implications for contemporary analytic metaphysics: it has critical implications for certain ontological works, such as possible-world metaphysics and necessitarian ontology. Detailed dialogue requires dedicated chapters.

Third limit: the dialogue with cross-cultural methodological traditions is insufficient. This chapter works mainly within the Western analytic tradition. The methodological contributions of Chinese traditions, especially Confucian practical wisdom, Daoist methodological caution, and Buddhist falsifiable structures, require dedicated treatment. This section mentions them briefly, but detailed treatment is left for the future.

Fourth limit: boundary cases of FA1 require more detailed treatment. The falsifiable structures of certain ontological commitments require complex analysis, such as highly abstract metaphysical statements and interdisciplinary boundary statements. This section gives a framework, but each boundary case requires dedicated treatment.

Acknowledging these limits does not weaken the work of this chapter. It makes this chapter an honest initial work.

The Core Posture of This Chapter

The core posture of this chapter can be stated as follows: falsifiability is not an external requirement imposed on science, but an internal rule concerning the legitimacy of ontological commitments.

In twentieth-century philosophy of science, falsifiability is often treated as an external standard of science: theories are required from the outside to satisfy falsifiability. In this treatment, falsifiability is easily regarded as a constraint that limits the richness of theory. The work of this chapter argues that falsifiability is an internal rule of ontological commitment: any serious ontological commitment naturally involves an identifiable direction of failure. If a statement claims to be an ontological commitment but has no direction of failure, it is not a genuine ontological commitment; it is an empty assertion. This treatment no longer makes falsifiability an external restriction; it becomes part of the definitional content of ontological commitment. Any serious ontological work automatically involves a falsifiable structure. This is not a limitation, but the essence of ontology as a field of research. This posture gives the work of this chapter a place in the debates since twentieth-century philosophy of science and metaphysics: it inherits Popper's core insight, namely that theories need refutable content, but gives it contemporary refinement and extension through an ontological framework.

Meaning before freezing: this is one of the core theses of this book. This chapter provides a methodological ground for that thesis. "Meaning" as an ontological-normative phenomenon does not require "special ontological status": it has legitimacy through identifiable directions of failure, such as those concerning structures of difference, emergence, and finitude. "Before freezing" as a temporal-ontological dimension also does not require "special ontological status": it has legitimacy through identifiable directions of failure, such as those concerning the process of freezing, dynamics, and irreversibility. These ontological contents acquire methodological grounds through the discipline of falsifiability in this chapter. This is the contribution of this chapter to the whole book.

Through the work of this chapter, Part I, "Constrained Speculation: Method as Philosophy," is completed as a methodological foundation. The later parts can develop the ontology, relations, metaphysics, and normativity of this book on this foundation. All contents are constrained by the discipline of FA1, but gain rich content through research and reflection. This is the realization of "method as philosophy": method and philosophical content constitute an integral philosophical work through mutual construction.

6.7. Methodological Reflection on This Chapter

This chapter elevates falsifiability from Popper's epistemological principle into an ontological discipline. As the conclusion of Part I, this section reflects on the methodological significance of this chapter and of Part I as a whole.

This Chapter's Treatment of Popper's Work

The posture of this chapter is serious respect and precise distinction. Popper's work has enduring influence: it provides a core insight concerning scientific methodology. But Popper's work also has limits: it restricts falsifiability to the level of the distinction between science and non-science. Philosophical discussions since the twentieth century, especially the Duhem–Quine thesis, Kuhn's paradigm theory, and Lakatos's research programmes, have challenged the simple form of Popper's principle. This chapter acknowledges these challenges and refines the principle by elevating falsifiability to an ontological discipline. This treatment of inheritance plus development is a common pattern of philosophical progress.

This Chapter as the Synthetic Chapter of Part I

As the synthetic chapter of Part I, this chapter displays the mutually supporting network among the six chapters. Chapter 1 establishes the overall framework: the four disciplines are applied in the other chapters. Chapters 2 and 3 mutually support each other: restraint and failure conditions jointly constitute the standard of legitimacy for ontological commitments. Chapter 4 is built upon Chapter 3: the definition of constitutive coextension involves shared failure conditions. Chapter 5 uses the content of the other chapters to position the stance of this book. Chapter 6 unifies all methodological content under the framework of falsifiability as an ontological discipline. This mutually supporting network makes Part I an organic methodological whole.

The Realization of the Title of Part I: "Method as Philosophy"

The title of Part I is "Constrained Speculation: Method as Philosophy." "Method as philosophy" as a slogan needs to be realized through methodological work. The six chapters of Part I jointly realize this slogan: Chapter 1 shows method itself as a philosophical position; Chapter 2 shows a methodological principle, namely restraint, as an ontological discipline; Chapter 3 shows a methodological concept, namely failure conditions, as an ontological criterion; Chapter 4 shows a methodological tool, namely constitutive coextension, as an ontological-semantic relation; Chapter 5 shows the place of the methodological position in contemporary discussion; Chapter 6 shows methodological discipline as ontological discipline. Together, the six chapters show that method and ontological content are not two separated tasks, but two mutually supporting dimensions of the same philosophical work.

Limits Honestly Acknowledged in This Chapter

This chapter acknowledges several limits. First, the refinement of falsifiability as an ontological discipline requires further work. Second, this chapter develops only the core direction of the dialogue with Lakatos, Kuhn, and others; more detailed dialogue requires dedicated papers. Third, the expression of the discipline of falsifiability in interdisciplinary applications requires dedicated work.

Connections with Later Parts

As the synthetic chapter of Part I, this chapter provides the methodological framework for the work of the later parts. The ontological theses of Part II are evaluated for their legitimacy through FA1. The relational theses of Part III are evaluated through FA1. The work on singularity in Part IV treats legitimate speech at the boundary through the refinement of FA1, namely the distinction between meta-level and object-level. Part V uses FA1 to treat the legitimacy of value statements. More deeply, this chapter realizes the position of method as philosophy in this book. It gives complete development to the methodological level at which method and ontological content mutually construct each other. It is the concluding expression of the methodological rigor of this book in Part I.

Part II.

The Philosophical Defense of Gradient Ontology

7. Difference before Identity

7.1. Introduction to This Chapter

The mainstream ontological tradition of Western philosophy is built upon the concept of substance. This tradition begins with Parmenides, is systematized by Aristotle, passes through modern philosophers such as Descartes, Leibniz, Locke, and Hume, and continues into various works in contemporary analytic metaphysics, forming the core thread of Western ontology. The core structure of substance ontology is this: the basic ontological units are substances, namely independently existing objects that possess their own identity and bear properties; the world is composed of substances; substances have properties; and substances stand in relations to one another.

This structure has deep legitimacy: it accords with the intuitive structure of everyday experience, since we do indeed experience the world as a collection of "independently existing things," and it has supported important developments in Western philosophy, including category theory, logical analysis, and formal semantics. Yet since the twentieth century, it has faced new challenges. The development of contemporary physics makes "substance" difficult as an ontological category: in quantum field theory, "particles" are excitations of fields, not independent substances; in general relativity, spacetime itself participates in dynamics; in non-equilibrium thermodynamics, "structure" is the stable manifestation of process.

This chapter makes the core position of gradient ontology explicit as a precise philosophical thesis: difference, namely gradient structure, is ontologically prior to identity, namely substance (DI1). Specifically, the ontological content of beings is first a structure of difference: non-uniform distribution, namely spatial gradient $\nabla\rho$, and dynamic change, namely temporal gradient $\partial\rho/\partial t$. What is called "substance" is the relatively stable form produced by structures of difference through the process of freezing, namely a frozen transient.

This chapter does not pretend to surpass the tradition of substance ontology easily. On the contrary, it acknowledges the strength of substance ontology, developed over more than two millennia, and its legitimacy in everyday experience and in certain scientific descriptions. The work of this chapter is to argue that, at the level of foundational ontology, the ontology of difference is more basic than substance ontology: substance as ontological content is still real, but its ontological ground lies in structures of difference and the process of freezing, not in an independent "substantiality."

This chapter establishes the Western philosophical genealogy of the ontology of difference: Heraclitus, Leibniz, Whitehead, Bergson, Deleuze, and other precursors. This genealogy makes the work of this chapter not an isolated ontological declaration, but the contemporary continuation of this tradition. Chapter 8 provides the Chinese philosophical genealogy, namely the tradition of qi theory. Together, the two chapters constitute the dual philosophical defense of gradient ontology in Part II.

This chapter does not depend on the physical derivation of EQT. All arguments are completed through purely philosophical argumentation, and EQT appears only as an application example. This independence allows the chapter, as part of a philosophical monograph, to be taken seriously in international philosophical discussion.

The discussion proceeds as follows. Section 7.1 develops the traditional starting point of substance ontology and its challenges; Section 7.2 states and defends difference before identity (DI1); Section 7.3 develops a three-layer argument, namely phenomenological, logical, and physical; Section 7.4 develops the Western philosophical genealogy of the ontology of difference; Section 7.5 concludes the chapter and develops the connection between the ontology of difference and traditional categorial language.

7.2. The Traditional Starting Point of Substance Ontology

The Core Structure of Substance Ontology

The mainstream ontological tradition of Western philosophy is built upon the concept of substance. This tradition begins with Parmenides, is systematized by Aristotle, passes through modern philosophers such as Descartes, Leibniz, Locke, and Hume, and continues into various works in contemporary analytic metaphysics, forming the core thread of Western ontology.

The core structure of substance ontology can be stated as follows:

The core structure of substance ontology: the basic ontological units are substances, namely independently existing objects that possess their own identity and bear properties. The world is composed of substances; substances have properties; and substances stand in relations to one another.

This structure has several implications: the independence of substances, in the sense that a substance exists without depending on other substances and is "in itself"; the identity of substances, in the sense that a substance preserves itself through change and has conditions of identity; the bearing function of substances, in the sense that properties inhere in substances, so that color, shape, position, and so on are the color, shape, and position of some substance; and the derivative status of relations, in the sense that relations are established between substances and are ontologically secondary to substances. This structure has exerted profound influence in Western philosophy, and almost all major metaphysical discussions have unfolded within, or against, this framework.

Aristotle's Systematization

Aristotle gave the most complete systematization of substance ontology in the *Categories* and the *Metaphysics*. His work contains several elements:

Primary substance and secondary substance: Aristotle distinguishes primary substance, such as "this human being," from secondary substance, such as species and genus, for example "human being." Primary substance is basic: the other categories all depend on it.

The ten categories: Aristotle identifies ten ways in which being is predicated: substance, quantity, quality, relation, place, time, position, state, action, and passion. Substance is the basic category, and the other nine categories are modes of the properties of substance.

Form and matter: in the *Metaphysics*, Aristotle further analyzes substance as composed of form, which determines what it is, and matter, which constitutes its basis. Form, as the "what-it-is," gives substance its identity.

Potentiality and actuality: change in Aristotle's framework is understood as the transformation from potentiality to actuality. But something is preserved through change, namely the essential form of the substance, so that change is not arbitrary.

Together, these elements constitute the system of substance ontology. It has served as a foundational framework in Western philosophy for more than two millennia. It has been repeatedly used, revised, and criticized, but has rarely been completely overthrown.

Internal Difficulties of Substance Ontology

Substance ontology, however, encounters internal difficulties. This section develops three major difficulties.

Difficulty One: The Tension between Change and Identity

If a substance preserves its self-identity through time, how can a changing substance still be the "same" substance?

Consider an example. The ship of Theseus is a classic thought experiment: all the planks of a ship are gradually replaced, and eventually none of the original planks remain. Is it the "same" ship? If it is, then "identity" does not lie in material composition. If it is not, then how is "sameness" in change to be identified? Similar difficulties are widespread. A person, over seventy years of life, undergoes repeated complete replacement of cells, reshaping of memory, changes of character, and aging of the body. In what sense is this person the "same" person?

The traditional response of substance ontology is to appeal to essential form: the essence of a substance determines its identity. But "essence" as a concept faces challenges in contemporary discussion, such as Quine's criticism of essentialism. Grounding the identity of a substance in essence seems to move the problem to the equally difficult question of what essence is.

Difficulty Two: The Circularity of Substance as Foundation

Substance ontology says that properties depend on substances and relations are established between substances. But what is the content of "substance" itself as a foundational object?

Consider the identification of a substance. How do we identify a substance? Usually through its properties: its color, shape, position, behavior, and so on. But if a substance is identified only through its properties, then substance as the "bearer beyond properties" seems empty: it is merely "something that has these properties." Hume sharply identified this difficulty: he criticized "substance as something beyond properties" as a fiction that cannot be grasped in experience. What we experience is only a bundle of properties, not "a substance behind the properties."

The traditional response of substance ontology is to appeal to a substratum: properties inhere in a substratum that itself has no properties. But this response faces difficulty. How can an object "without properties in itself" exist? How is it distinguished from other substrata? These questions are hard to answer.

Difficulty Three: The Treatment of Relations

If relations are established between substances, then relations are external to substances: substances first exist and then enter into relations. But many relations seem to constitute their terms. The parent–child relation constitutes "father" and "son" as terms of that relation; marriage constitutes "husband" and "wife"; employment constitutes "employer" and "employee."

If relations constitute their terms, then the status of substances as independent units prior to relations is weakened. This conflicts with the content of substance ontology. More deeply, discoveries in contemporary physics make the problem of the externality of relations even sharper. In quantum mechanics, the states of particles in an entangled system are determined by the whole relation and cannot be simply reduced to the properties of independent particles. This conflicts with the presupposition of substance ontology.

These three difficulties—the paradox of identity, the circularity of substance as foundation, and the treatment of relations—constitute internal difficulties for substance ontology. These difficulties do not necessarily overthrow substance ontology, but they point toward a direction: whether another ontology is possible, one that can avoid these difficulties.

The Direction of This Chapter

The direction of this chapter is not to repair the substance–property–relation triad from within, but to change the starting point of ontology: from substance to difference.

Specifically, this chapter argues that difference, namely gradient structure, is ontologically prior to identity, namely substance. What is called "substance" is the stable form of a structure of difference, namely a frozen transient, and not a basic ontological unit. Properties are the contents of structures of difference; relations are the dynamical connections among structures of difference.

This alternative position does not simply deny the existence of substances. It reorganizes the starting point of ontology. Substances still "exist" in a certain sense, namely as stable forms of structures of difference, but they are not the basic units of ontology. This position is radical: it challenges the mainstream Western philosophical tradition since Parmenides. But it is not isolated: there are precursor traditions in Western philosophy, including Heraclitus, process philosophy, Deleuze, and others. Section 7.4 will develop this genealogy.

Objections and Replies

Objection One: The three difficulties you list are standard criticisms in the history of philosophy. Substance ontology has developed responses over more than two millennia. Why are these responses insufficient?

Reply: The traditional responses of substance ontology are indeed rich. Essentialism, bundle theory, stuff theory, trope theory, structural theory, and others are all responses. But each response has its own difficulties. Essentialism moves the problem to the question "what is essence?" and has difficulty avoiding circularity. Bundle theory eliminates substance as foundation, but has difficulty explaining how bundles of properties are formed. Stuff theory treats the foundational object as "some kind of matter," but still presupposes independent material units. Trope theory treats foundational objects as property instances, but requires complex treatment of how properties combine. Each response has force, but each also has difficulty. These difficulties point toward a possibility: perhaps the ontological problem needs to be treated at a more fundamental level. This chapter argues that it does: the ontology of difference provides such a more fundamental treatment. But this chapter does not claim that the ontology of difference eliminates all difficulties. It relocates the difficulties. The ontology of difference also has its own difficulties, as Section 7.5 will develop. What matters is to evaluate which ontology has difficulties that are more tractable.

Objection Two: Your "difference before identity" sounds too radical. In everyday experience, we clearly encounter things, such as tables, persons, and stars. These things seem basic, not some "stable form of difference."

Reply: This chapter does not deny the existence of things in everyday experience. Tables, persons, and stars are all real objects. What this chapter claims is that the ontological ground of these things is a structure of difference, not an independent substance. An analogy may help. In everyday experience, we see vortices, such as vortices in a river. They are real phenomena: they have position, shape, and stability. But the ontological ground of a vortex as an object is the differential motion pattern of the fluid, not an independent "vortex substance." Without the differential motion of the fluid, there is no vortex as an object. This chapter argues that everyday things are similar: they are real phenomena, but their ontological ground is the stable form of a structure of difference, not an independent substance.

This claim does not conflict with everyday experience. It provides a more precise ontological ground for everyday experience.

Objection Three: You seem to use "difference" as a foundational concept, but what is "difference"? It itself requires definition and defense. If "difference" cannot be precisely defined, the ontology of difference is an empty concept.

Reply: This objection points to the core work of this chapter: the content of "difference" as a foundational concept. Section 7.2 will develop this concept precisely. Briefly, "difference" in the framework of this chapter refers to non-uniformity in space and time: the difference between the state of a being and its background or surroundings. In the EQT framework, this is concretized as $\nabla\rho$, namely density gradient, and $\partial\rho/\partial t$, namely the rate of temporal change. This definition makes "difference" not an abstract concept: it has operational content. In the EQT framework, it is a physical quantity; in a more general ontological framework, it is identifiable non-uniformity. Section 7.2 will fully develop this definition and its defense.

Literature That Must Be Engaged

- Aristotle, Categories, Metaphysics, Books VII-IX
- René Descartes, Meditations on First Philosophy (1641)
- David Hume, A Treatise of Human Nature (1739-40), Book I, Part IV, Section VI
- W.V.O. Quine, "Speaking of Objects" (1958)
- David Wiggins, Sameness and Substance Renewed (2001)
- E.J. Lowe, The Four-Category Ontology (2006)

Summary of This Section

This section has done four things. First, it developed the core structure of substance ontology as the mainstream tradition of Western philosophy. Second, it reviewed Aristotle's systematization as the starting point of this tradition. Third, it identified three internal difficulties of substance ontology: the identity paradox, the circularity of substance as foundation, and the treatment of relations. Fourth, it announced the direction of this chapter: a shift in ontological starting point from substance to difference.

The next section develops the precise meaning of "difference" as a foundational ontological concept.

7.3. Difference as the Ontological Starting Point

The Precise Meaning of "Difference"

Any work that uses "difference" as a foundational concept must define it precisely, otherwise it degenerates into rhetoric. This section gives the precise meaning of "difference" as an ontological starting point.

Definition 7.1 (Difference as an Ontological Concept): Difference refers to the non-uniformity of existential structure; specifically, the non-uniform distribution of states in space and their non-uniform change in time.

This definition requires detailed development. Let us analyze it item by item.

First, "non-uniformity": difference is not the abstract claim that "X is not equal to Y." It is physical or ontological non-uniformity. In the EQT framework, this is concretized as follows: spatial difference $= \nabla\rho \neq 0$, namely the gradient of the density field in space; temporal difference $= \partial\rho/\partial t \neq 0$, namely the change of the density field in time. These are observable, measurable, and precisely characterizable physical quantities. Difference as an ontological concept is built upon these physical quantities; it is not an abstract philosophical construction.

Second, "non-uniform distribution": spatial difference means that ρ has different values at different spatial positions: the spatial structure of a being is non-uniform. A completely uniform spatial distribution ($\nabla\rho \equiv 0$) has no spatial difference and therefore no "thing." It is a uniform background, not a discrete object.

Third, "non-uniform change": temporal difference means that ρ has different values at different times: the state of a being changes. A completely static state ($\partial\rho/\partial t \equiv 0$) has no temporal difference and therefore no "process." It is a static state, not a dynamic phenomenon.

Synthesizing spatial difference and temporal difference, difference as an ontological concept refers to the non-uniformity required for a being to exist as an identifiable structure. Without such non-uniformity, a being does not exist as an identifiable object.

The Core Thesis of Difference as an Ontological Starting Point

Core thesis DI1: the existence of a being $X \Longleftrightarrow X$ involves a structure of difference, namely spatial difference $\nabla\rho \neq 0$, temporal difference $\partial\rho/\partial t \neq 0$, or both. Difference is an ontological necessary condition for beings.

This thesis is the foundation of the overall ontological work of Part II. It states that any being involves difference, and this is an ontological necessary condition of existence. Difference is not a "property" or "feature" of beings; it is constitutive of a being as an ontological unit. In a completely uniform background, where $\nabla\rho \equiv 0$ and $\partial\rho/\partial t \equiv 0$, there are no beings.

Implications of the Thesis

DI1 has four ontological implications.

Implication One: What Is Called "Substance" Is the Stable Form of a Structure of Difference

DI1 does not deny the reasonableness of the concept of "substance" in everyday use. Tables, persons, stars, and other everyday objects are real phenomena. But DI1 claims that the ontological ground of these objects is structure of difference, not independent substance. Specifically: a table is the stable form of density difference: it has a non-uniform distribution in space and forms a difference from the background, such as air and floor. A person is a highly complex stable form of density difference: it contains multi-level structures of difference, including tissues, cells, and molecules. A star is a cosmologically stable form of density difference: it forms a local high-density structure within the background density of the universe. These "stable forms" are not independent substances; they are forms of structures of difference. If the difference disappears, for example if a star exhausts its fuel and cools into a uniform cosmic background, the "star" as an object disappears.

Implication Two: Properties Are the Contents of Structures of Difference

Substance ontology says that "properties inhere in substances": color, shape, position, and so on are properties of a substance. DI1 gives a different picture: properties are the contents of structures of difference; they directly describe the form of difference. Specifically: color is not a property of some independent substance, but the manifestation of electromagnetic frequency difference, reflecting the way the structure of difference of an object responds to electromagnetic waves. Shape is not a property of some independent substance, but the geometric structure of spatial difference. Position is not a property of some independent substance, but the relative location of a structure of difference within a larger background. These treatments give "property" as an ontological category an ontological ground:

properties directly describe the contents of structures of difference, without presupposing that "properties inhere in substances."

Implication Three: Relations Are Dynamical Connections among Structures of Difference

Substance ontology says that "relations are established between substances." DI1 gives a different picture: relations are dynamical connections among structures of difference, established through physical processes, namely co-evolution. This implication connects with the work of Chapter 12, "Co-Evolution as an Ontological Relation." The content of co-evolution as an ontological relation is the dynamics among structures of difference: different structures of difference mutually affect one another through physical processes, including frequency matching, thresholds, and tri-state responses.

Implication Four: Change Is the Evolution of Structures of Difference

The "tension between change and identity" faced by substance ontology is reorganized in the framework of DI1. Change is not "a substance changing while preserving identity," but the evolution of a structure of difference. The form may change, namely the form of the structure of difference may change, but what "is changing" is determined by the evolutionary trajectory of the structure of difference. Specifically, the "identity" problem of the ship of Theseus becomes, in the framework of DI1, the problem of the continuity of a structure of difference: if the ship's structure of difference, including plank structure, spatial form, functional pattern, and so on, preserves continuity during replacement, it is the "same ship"; if discontinuity occurs, for example if the ship is completely dismantled and then reassembled into an entirely different structure, it is a new ship. The "identity" problem of a person over seventy years is similar: a person's structure of difference, including neural structure, memory patterns, bodily form, and so on, preserves continuity through the course of life, allowing the person as an evolving form of a structure of difference to be identifiable through time. This treatment avoids the difficulty of "essence" as an abstract concept: identity is grounded in the continuity of a structure of difference, not in an abstract essence.

An Application: Particles as Structures of Difference

To make the application of DI1 concrete, this section considers an example: elementary particles.

Traditional physics describes elementary particles, such as electrons and quarks, as "elementary particles." They appear to be irreducible substances. But in the EQT framework, elementary particles are excitation modes of the ρ_f field: they are structures of difference. Specifically, the electron as a structure of difference has several features: local density concentration, since the electron is a local concentration of the ρ_f field in space and forms a difference from the surrounding background; a specific frequency band, since the electron corresponds to excitation in a specific frequency band, and its energy corresponds to frequency, $E = mc^2$; dynamical behavior, since the behavior of the electron, including motion in electromagnetic fields and interaction with other particles, is determined by field dynamics; and relative stability, since the electron as a structure of difference has stability, does not easily dissipate, and can be identified as the same excitation mode over a long period of time.

This description clarifies the ontological ground of the electron as an object: it is the stable form of a structure of difference, not an independent elementary substance. The properties of the electron, such as charge, spin, and mass, are the contents of this structure of difference; the relations between the electron and other particles are dynamical connections among structures of difference. This treatment does not weaken physical research on electrons. It gives such research a more precise ontological ground. What physicists study is not "the electron as an independent substance," but the dynamics and interactions of "the electron as a structure of difference."

Objections and Replies

Objection One: Your DI1 claims that the existence of beings is equivalent to difference, but this seems too strong. A completely uniform state, such as a perfect crystal at low temperature, still contains beings, although difference is minimal.

Reply: This objection points to an important refinement. "Complete uniformity" is in fact an idealization: every physical state has non-uniformity, including quantum fluctuations, temperature fluctuations, and differences between macroscopic and microscopic scales. Specifically, a perfect crystal at low temperature has a lattice structure, namely spatial difference $\nabla\rho \neq 0$, even though thermal fluctuations, namely temporal difference, are small. In quantum mechanics, zero-point energy ensures that even the lowest-energy state still contains fluctuations. Even in a so-called "ground state," the existence of matter still depends on difference in spatial distribution. DI1 does not claim that "existence = maximum difference"; it claims that "existence $\Longleftrightarrow$ involves nonzero difference." A completely uniform state, namely strict $\nabla\rho \equiv 0$ and $\partial\rho/\partial t \equiv 0$, is an idealization in physics and does not actually exist. In such an idealized state, there are indeed no beings. This is the content of Chapter 17 on ontological singularity.

Objection Two: Your "stable form of a structure of difference" as an alternative to substance seems to be merely a semantic rewriting. You are still talking about stable, identifiable objects with internal structure, only you no longer call them "substances."

Reply: This objection points to the core issue of this chapter: what is the difference between the ontology of difference and substance ontology? The difference is as follows. Difference in ontological starting point: the foundational object of substance ontology is self-identical; it preserves its essence through change, while difference, including change and relation, is a mode of the substance's properties. The foundational object of the ontology of difference is structure of difference: the existence of a being is directly difference, while identity, namely stability, is a formal feature of the structure of difference. Difference in the treatment of "foundation": in substance ontology, substance is irreducible beyond itself; substance is foundation. In the ontology of difference, difference as foundation is not "another kind of substance"; it is relational, since difference is always difference relative to another state, not something independent. Difference in the treatment of change: in substance ontology, change requires appeal to "essence" so that identity through change can be explained. In the ontology of difference, change is the evolution of a structure of difference: identity comes from continuity and does not require appeal to essence. These differences are not semantic; they concern fundamental differences in ontological starting point and structure. The ontology of difference indeed preserves "stable structures" as phenomena, namely substances as everyday objects, but locates them as forms of difference, not as foundations.

Objection Three: Your ontology of difference seems overly dependent on EQT: it uses $\nabla\rho$ and $\partial\rho/\partial t$ as content. If EQT is replaced in the future, does the ontology of difference collapse?

Reply: The ontology of difference as a philosophical position does not depend on the formulas of EQT. Its content is more general: "the existence of beings $\Longleftrightarrow$ involves non-uniformity." EQT is the physical realization of this general ontology. $\nabla\rho$ and $\partial\rho/\partial t$ are expressions within the EQT framework, but this expression is not the content of the ontology of difference itself. Specifically, if some future physics uses a different foundational field, rather than ρ_f, to describe the world, the ontology of difference will express itself in terms of the non-uniformity of that field. If some future physics uses entirely different ontological units, such as discrete structures or informational quantities, the ontology of difference will use the non-uniformity of those ontological units. The core position of the ontology of difference, namely non-uniformity as an ontological necessary condition of existence, does not depend on physical formulas. It is general: any being in any physics involves non-uniformity; otherwise it cannot be identified.

Literature That Must Be Engaged

- Hilary Putnam, "Why There Isn't a Ready-Made World" (1982)
- Donald Davidson, "On the Very Idea of a Conceptual Scheme" (1974)
- Theodore Sider, Writing the Book of the World (2011)
- Chapter 12 on co-evolution as an ontological relation (self-reference)
- Chapter 17 on ontological singularity (self-reference)

Summary of This Section

This section has done four things. First, it gave a precise definition of "difference" as an ontological concept. Second, it stated the core thesis DI1: the existence of a being $\Longleftrightarrow$ the involvement of a structure of difference. Third, it developed four implications of the thesis: substance as the stable form of difference, properties as the content of difference, relations as dynamics among differences, and change as the evolution of difference. Fourth, it used elementary particles as structures of difference to show the application of the theory.

The next section develops why difference is ontologically prior to identity.

7.4. Why Difference Is Prior to Identity

The Aim of the Argument

Section 7.2 developed the meaning of difference as an ontological concept and stated DI1. But this is not enough: one must argue why difference is prior to identity rather than the other way around. This section gives that argument.

The argument has three layers. The first is a phenomenological argument, starting from experience and the analysis of phenomena. The second is a logical argument, starting from the internal structure of concepts. The third is a physical argument, starting from scientific discoveries. The three layers support one another and together constitute the complete defense of DI1.

First Layer: The Phenomenological Argument

Core question: in our experience of the world, do we first experience "difference," or do we first experience "identity"?

Consider visual experience. When we see an object, for example a table, what does our experience involve? The difference between the table and the background, including floor, wall, and air: differences in color, brightness, and shape; differences within the table, including color, texture, and brightness across different parts; and differences at the edge of the table, where the edge is a region of abrupt change in color and brightness. Without these differences, the table cannot be identified as an object. If the table were completely the same as the background, with the same color, the same brightness, and no visible edge, it could not be visually perceived as an object.

This phenomenological fact points to a conclusion: the foundation of visual perception is difference, not identity. What we perceive is difference, including contrast, change, and edges, and from these differences we identify objects. The "same object" is a Gestalt identified from difference, not something prior to difference.

A similar analysis applies to other senses. Auditory perception perceives differences in frequency, amplitude, and temporal change; tactile perception perceives differences in temperature, pressure, and texture; smell and taste perceive differences in chemical concentration. The foundation of all sensation is difference. This phenomenological argument is not a new philosophical discovery: it has already been analyzed in detail in Gestalt psychology and the phenomenological tradition, for example in the work of Merleau-Ponty. This section only uses it as support for the ontology of difference.

Second Layer: The Logical Argument

Core question: as a concept, does "identity" depend on "difference"?

Consider the content of "identity" as a concept. What does it mean to say that "X and Y are identical"? Numerical identity: X and Y are the same individual, as in "the Morning Star and the Evening Star." Qualitative identity: X and Y share all relevant properties, as in "this triangle and that triangle are congruent." Essential identity:

X and Y share essential attributes, as in "both individuals are human beings."

In all these senses, the content of "identity" as a concept involves contrast with difference. Identity is relative to non-identity, namely difference. To say that X and Y are identical means distinguishing them from cases of "non-identity," and this distinction itself depends on the concept of difference. Specifically: without the concept of "difference," "numerical identity" is empty. Any object is "numerically identical" to itself; this is a trivial truth and does not constitute content. A concrete numerical identity statement, such as "the Morning Star and the Evening Star are the same star," carries information precisely because it contrasts a case that appears different, appearing at different positions in the sky in the morning and evening, but is actually the same. Without the concept of "difference," "qualitative identity" is also empty: any objects with the same properties are "qualitatively identical," but this statement has content only in contrast with "qualitative difference." Without the concept of "difference," "essential identity" is likewise empty: it has meaning only in contrast with "essential difference."

This logical argument shows that, as a concept, "identity" depends on "difference." The content of "identity" can be established only through contrast with "difference." This makes difference prior to identity at the conceptual level. More deeply, this argument is not only conceptual; it also has ontological relevance. If identity conceptually depends on difference, and if ontological structure corresponds to concepts, then identity also ontologically depends on difference. Difference is ontologically prior to identity because identity as an ontological phenomenon, such as the same object, the same type, or the same essence, can be identified only within a structure of difference.

Third Layer: The Physical Argument

Core question: do the discoveries of contemporary physics support difference or identity as foundation?

Consider discoveries in several fields of physics.

Particles in quantum field theory: in quantum field theory, "particles" are not basic ontological units. They are excitation modes of quantum fields. A field is a continuous distribution, and a particle is a local excitation within this distribution, namely a structure of difference. Without differences in the field, there are no particles. Specifically, the electron as a particle is an excitation of the electron field: it is a local concentration of the field at a specific spacetime location. This local concentration is defined by its difference from the background field, namely the field values at other locations. Without difference, the electron does not exist as a particle.

Spacetime in general relativity: in general relativity, the geometry of spacetime is the form of the distribution of matter-energy. The curvature of spacetime is coupled with the stress-energy tensor, as expressed by Einstein's field equations. Specifically, flat spacetime, without difference, corresponds to the absence of matter-energy: this is the vacuum state. Curved spacetime, with difference, corresponds to a distribution of matter-energy: curvature differences encode the existence of matter. The geometrical structure of spacetime is precisely the geometrical structure of difference. Without this difference, there are no physical objects.

Dissipative structures in non-equilibrium thermodynamics: the work of Prigogine and others shows that complex structures at the macroscopic scale, such as life, climate patterns, and chemical oscillations, are built upon energy differences under non-equilibrium conditions. These "dissipative structures" maintain themselves through energy flow, namely gradient-driven processes. In equilibrium, where there is no difference, there are no dissipative structures. All forms of complexity require difference as their ground.

Entropy increase in statistical mechanics: the "entropy increase" of the second law of thermodynamics is in fact the direction of decreasing difference: a system evolves from non-uniformity, with difference, toward uniformity, without difference. This evolution is irreversible and gives time a direction, as developed in Chapters 11 and 15. If difference were not foundational, "entropy increase" as an irreversible process would be unintelligible: it is the direction of evolution from difference toward uniformity.

Together, these physical discoveries support difference as foundational: the content of contemporary physics treats difference as the ontological ground of beings, and treats identity, such as stable structures and symmetries, as forms of difference.

Synthesis of the Three-Layer Argument

The three layers of argument support DI1 from different dimensions: phenomenology, where the foundation of experience is difference, since perception perceives differences and identifies identity from them; logic, where identity as a concept depends on difference, since the content of identity is established only in contrast with difference; and physics, where discoveries in contemporary physics treat difference as the ground of beings, since particles, spacetime, and complex structures all depend on difference. These three dimensions—experience, concept, and physics—jointly support difference as the ontological starting point. Any serious ontological work must face these three dimensions. This chapter gives a unified treatment of them through the ontology of difference.

Objections and Replies

Objection One: Your phenomenological argument depends on the function of sensation. But sensation is an epistemological phenomenon, not an ontological phenomenon. How does this argument support an ontological conclusion?

Reply: This objection points to the relation between epistemology and ontology. If epistemology and ontology are completely separated, as in a traditional Kantian position, then the phenomenological argument indeed does not directly support an ontological conclusion. But the work of this book, especially Chapter 3 on failure conditions as an ontological discipline and Chapter 14 on the new foundation of objectivity, rejects the absolute separation of epistemology and ontology. They are connected through failable interfaces and co-evolutionary relations. In this framework, the phenomenological argument has ontological significance. The function of sensation reflects the properties of beings as ontological objects. If the foundation of sensation is difference, this fact points to the claim that the properties of beings as ontological objects are also grounded in difference. This is not the simple claim that "sensation determines existence," which would be Berkeleyan idealism. Rather, it is the claim that "the structure of sensation corresponds to the structure of existence," which belongs to the co-evolutionary framework.

Objection Two: Your logical argument seems too strong. Does "the concept of identity depends on the concept of difference" really hold? The axiom of identity in mathematics, such as "$x = x$" for all x, seems not to depend on difference.

Reply: The mathematical axiom of identity, "$x = x$," is formally a trivial truth. It holds for every x and carries no information. This formal identity indeed does not require "difference." But concrete identity statements, such as the examples in this section, "the Morning Star and the Evening Star are identical," or "two objects are numerically identical," carry information. This information has content only through contrast with difference. Specifically: formal identity

($x = x$) = trivial truth, not dependent on difference; concrete identity (X and Y are identical) = information, dependent on contrast with difference. The argument of this section concerns concrete identity, not formal identity. This distinction makes the argument precise: it does not claim that all concepts of identity depend on difference, but that concrete concepts of identity, which are relevant to ontological problems, depend on difference.

Objection Three: Your physical argument selectively cites physical discoveries that support the ontology of difference. Other physical discoveries, such as conservation laws and symmetries, seem to support identity as foundation.

Reply: This objection rightly points to a methodological problem: any physical discovery can be interpreted within different ontological frameworks. Conservation laws and symmetries are indeed important discoveries in physics, but they can also be treated within the framework of the ontology of difference. Conservation, such as conservation of energy, involves the evolution of structures of difference: certain quantities of difference, such as total energy, remain invariant through evolution, but this invariance is built upon structures of difference and does not cancel the ontology of difference. Symmetries, such as spacetime symmetries and gauge symmetries, are possible transformation modes of structures of difference, not presuppositions of independent substances as foundation. Specifically, Noether's theorem connects symmetries with conservation laws: every continuous symmetry corresponds to a conserved quantity. In the framework of the ontology of difference, this correspondence is understood as follows: structures of difference preserve certain features under certain transformations, and these features are identified as conserved quantities. This differs from substance ontology, in which "the essential properties of independent substances are conserved." The latter presupposes independent substances; the former uses structures of difference. More deeply, physical discoveries allow multiple ontological interpretations. The argument in this section does not claim that physics necessarily supports the ontology of difference. It claims that the content of contemporary physics is compatible with, and naturally fits, the ontology of difference. This contrasts with substance ontology, which requires complex treatment in order to accommodate discoveries such as quantum fields, general relativity, and non-equilibrium thermodynamics.

Literature That Must Be Engaged

- Maurice Merleau-Ponty, Phenomenology of Perception (1945)
- Edmund Husserl, Logical Investigations (1900-1901)
- Gilles Deleuze, Difference and Repetition (1968)
- Steven Weinberg, The Quantum Theory of Fields (1995)
- Ilya Prigogine, Order out of Chaos (1984)
- Chapter 14 on the new foundation of objectivity (self-reference)

Summary of This Section

This section has done four things. First, it proposed the three-layer argument for DI1: phenomenological, logical, and physical. Second, it developed the content of each layer: the phenomenological argument is based on the differential basis of perception; the logical argument is based on the internal structure of concepts; and the physical argument is based on the discoveries of contemporary physics. Third, it synthesized the three layers into a complete defense of DI1. Fourth, it replied to possible objections, especially concerning the relation between epistemology and ontology, the distinction between formal and concrete identity, and the selectivity of the physical argument.

The next section develops the genealogy of the ontology of difference in the Western philosophical tradition.

7.5. The Western Philosophical Genealogy of the Ontology of Difference

The Function of Genealogy

The ontology of difference in this chapter is not an isolated ontological declaration. It is built upon precursor traditions in Western philosophy. This section develops that genealogy, making the work of this chapter the contemporary continuation of an ongoing tradition rather than an isolated attempt beginning from zero.

The development of the genealogy has two functions. First, it gives the work of this chapter historical legitimacy: it has precursors and is not an arbitrary invention. Second, it allows readers to locate the work of this chapter precisely within Western philosophy: which precursors it inherits and how it differs from them. The following develops five major precursor nodes: Heraclitus, Leibniz, Whitehead's process philosophy, Bergson, and Deleuze.

Precursor One: Heraclitus

Heraclitus (ca. 535–475 BCE) is the earliest precursor of the "change–difference" tradition in Western philosophy. His core position, preserved through later quotations and fragments, contains several elements: "everything flows" (*panta rhei*), in the sense that the fundamental feature of existence is flow and change; "you cannot step into the same river twice," in the sense that the identity of things is the identity of flow, not static identity; and the unity of opposites, in the sense that things involve tensions among opposites, such as cold and hot, life and death, upward and downward.

In the history of philosophy, Heraclitus is often presented as the opposite of Parmenides, who emphasizes the unity and immutability of being. The later philosophical mainstream, especially Aristotle and his tradition, chose the Parmenidean direction and marginalized Heraclitus. But Heraclitus's insight is structurally related to the ontology of difference in this chapter: flow as foundation $\leftrightarrow$ difference, namely temporal change $\partial\rho/\partial t$, as the ground of existence; the unity of opposites $\leftrightarrow$ structures of difference involving contrast, namely the directionality of gradients; and identity as the identity of flow $\leftrightarrow$ substance as the stable form of a structure of difference. These structural relations make Heraclitus the earliest precursor of the work of this chapter. But the work of this chapter is not a "contemporary version of Heraclitus." It contains ontological developments, such as difference as foundation and physical realization, that go beyond Heraclitus's content.

Precursor Two: Leibniz

Leibniz (1646–1716)'s monadology is another precursor. Leibniz's core position is this: the foundational objects are monads, namely simple, indivisible ontological units with internal perception and appetite; monads have no "windows" and do not connect through external interaction; each monad reflects the whole universe from its own perspective and is therefore a "microcosm"; and the "coordination" among monads is pre-established by God.

Leibniz's work appears to agree with substance ontology, since he accepts monads as foundational objects. But on closer examination, Leibniz's work has directions close to the ontology of difference: the internal differences of monads, since each monad reflects the universe from its own perspective, meaning that monads have internal differences and different perspectives correspond to different internal states; the relationality of monads, since the "coordination" among monads is not externally added, but built upon the internal structure of each monad as a reflection of the whole universe, which is close to the direction in which relations are dynamical connections among structures of difference in the ontology of difference; and activity as foundational, since Leibniz takes perception and appetition as internal features of monads, which comes close to the processual direction and makes monads not static substances but dynamic processes. These aspects of Leibniz's work make him a precursor of the ontology of difference. But this chapter does not accept the content of monadology, such as the indivisibility of monads and God's pre-established harmony. What it extracts from Leibniz's work is the direction of processuality and relationality, not the concrete metaphysical system.

Precursor Three: Whitehead's Process Philosophy

In *Process and Reality* (1929), Whitehead proposed the most systematic process philosophy of the twentieth century. His core position, already treated in detail in Chapter 8 in dialogue with the tradition of qi theory, is this: the foundational objects are actual occasions, namely instantaneous units of process; actual occasions are mutually connected through prehension, in which they feel other actual occasions; eternal objects ingress into actual occasions as forms; and creativity is the foundational power of process.

Whitehead's work explicitly takes process as foundational, continuing the direction of the ontology of difference. The relation between this chapter and Whitehead is as follows. Similarities: both take processuality as ontological foundation; both reject static substance as foundation; both use discoveries from physics and science as ontological resources. Differences: Whitehead introduces eternal objects as a second ontology, and Chapter 8, Section 8.5 has already developed this difficulty in detail. This chapter does not introduce a second ontology, but treats form as the internal dynamics of structures of difference. Whitehead's work has a relatively strong metaphysical-theological dimension, such as creativity as the interiority of God; the work of this chapter is more restrained and does not presuppose a theological dimension. Whitehead did not systematically connect actual occasions with contemporary physics such as quantum field theory; this chapter establishes such a connection through the EQT framework. Whitehead's work is the most direct twentieth-century precursor of the work of this chapter. The work of this chapter can be understood as a continuation of Whiteheadian process philosophy, but with ontological revision, namely removing the dualism of eternal objects, and with physical connection, namely connection with the EQT framework.

Precursor Four: Bergson

Bergson (1859–1941)'s work, especially in *Creative Evolution* and *Matter and Memory*, is another important precursor of twentieth-century process philosophy, as treated in detail in Chapter 15. Bergson's core position is this: time as duration (*durée*) is the foundational ontology; it is real, continuous, and indivisible flow. Matter and spirit are both forms of duration: matter is "descending" duration, moving toward greater stability, while spirit is "ascending" duration, moving toward greater freedom. Creative evolution (*élan vital*) is the dynamic principle of the universe.

Bergson's work takes change, namely duration, as foundational; this is another expression of the direction of the ontology of difference. The relation between this chapter and Bergson is as follows. Similarities: both take processuality and dynamicity as foundational; both reject static ontology; both emphasize the intrinsic ontological status of time, as developed in Chapter 15. Differences: Bergson's work belongs to a phenomenological-intuitional tradition and depends on inner intuition of duration, whereas this chapter belongs to an analytic-physical tradition and is built upon the EQT framework and the method of constrained speculation. Bergson's *élan vital* as a principle of life's creativity is difficult to maintain in contemporary biology; this chapter does not appeal to a special metaphysical principle of life. Bergson did not successfully connect duration with physics, such as Einsteinian relativity, as shown by the difficulty of the 1922 debate; this chapter establishes the connection through the co-evolutionary framework. Bergson's work is another important precursor of the work of this chapter. The treatment of Bergson is developed in detail in Chapter 15, where Bergson and Einstein are reconciled.

Precursor Five: Deleuze

In *Difference and Repetition* (1968), Gilles Deleuze systematically developed the position of "difference before identity." Deleuze's core position is this: difference as foundation, where traditional metaphysics locates difference as derivative from identity, so that "X is not Y" is based on the respective identities of X and Y, while Deleuze argues that difference itself is foundational; identity is a form of difference, not the other way around; and repetition is the production of difference, since every repetition introduces new difference.

Deleuze's work is the clearest twentieth-century statement of the ontology of difference. It is directly relevant to the work of this chapter. Similarities: both explicitly claim that difference is ontologically prior to identity; both treat identity as a form of difference, not the other way around; both are built upon the processual tradition. Differences: Deleuze's work is built upon the Continental tradition, especially the continuation of Spinoza, Nietzsche, and Bergson, while this chapter is built upon the analytic-physical tradition. Deleuze's concept of "difference" has metaphysical and literary dimensions, involving examples from literature, cinema, art, and so on, while the concept of "difference" in this chapter has a physical dimension, involving $\nabla\rho$, $\partial\rho/\partial t$, and related terms. Deleuze's work does not have a physical ground; this chapter establishes such a ground through the EQT framework. Deleuze's work is the most direct twentieth-century precursor of the ontology of difference in this chapter. The work of this chapter can be understood as a physical and analytic-philosophical articulation of Deleuze's ontology of difference, giving Deleuze's insight physical and methodological grounds.

The Meaning of the Genealogy

Through these five precursor nodes, the genealogy of the work of this chapter becomes clear: the earliest precursor is Heraclitus, with flow as foundation; the modern precursor is Leibniz, with activity as an internal feature of monads; the twentieth-century precursors are Whitehead, with the systematization of process ontology, Bergson, with duration as foundation, and Deleuze, with the explicit statement of the ontology of difference. The work of this chapter is built upon this genealogy. It is the contemporary continuation of this tradition, especially by giving the tradition physical and methodological grounds through the EQT framework and the method of constrained speculation. This genealogy shows that the ontology of difference is not an isolated ontological attempt, but a work grounded in the history of philosophy. The defense of DI1 in this chapter is built upon more than two and a half millennia of philosophical discussion.

Objections and Replies

Objection One: The five precursors you list are all marginal: Heraclitus was suppressed by the Parmenidean mainstream; Leibniz's monadology has little support today; Whitehead's work is rarely used in analytic philosophy; Bergson's work was marginalized in the later twentieth century; and Deleuze's work belongs mainly to the Continen-

tal tradition. Can these marginal traditions constitute a genealogy?

Reply: This objection points to a question in the history of philosophy. The genealogy of this chapter is indeed a marginal tradition relative to the mainstream of substance ontology. But this does not weaken the legitimacy of the genealogy. A marginal tradition is not the same as a valueless tradition. Many important works in the history of philosophy, such as empiricism, process philosophy, and phenomenology, were marginal relative to the mainstream of their time, and their value does not depend on whether they were mainstream. Contemporary content supports the marginal tradition: the physical argument in Section 7.3 shows that the content of contemporary physics supports the direction of the ontology of difference, which means that the marginal tradition has contemporary accessibility. Multiple precursors form a genealogy: this is not a single marginal work, but five precursors spanning more than two and a half millennia. This continuity makes it a genuine genealogy, not an accidental marginal phenomenon. The work of this chapter does not pretend to continue the mainstream tradition. It explicitly presents itself as the contemporary continuation of a marginal tradition. This honesty is an application of the method of constrained speculation: clearly stating one's position and not pretending to have broader legitimacy than one has.

Objection Two: Your genealogy seems too broad. It places Heraclitus, Leibniz, Whitehead, Bergson, and Deleuze in the same genealogy, but their works differ greatly.

Reply: This objection rightly points to real differences. This chapter does not claim that these five precursors agree in every dimension. Their works have important differences, such as the differences between Leibniz's monadology and Whitehead's actual occasions, or between Bergson's duration and Deleuze's difference. This chapter claims that they share a direction in one dimension: they refuse to take static substance as foundation and instead take process, change, or difference as foundation. This shared direction allows them to constitute the genealogy of the "ontology of difference," even though the content of each precursor differs. Similarly, the genealogy of "substance ontology," including Aristotle, Descartes, Locke, contemporary analytic metaphysics, and others, also has major internal differences, but agrees on "substance as foundation." The legitimacy of a genealogy is built upon a shared direction, not upon agreement in all content.

Objection Three: Your genealogy is entirely Western. The direction of "difference before identity" in Chinese and other non-Western traditions is not considered.

Reply: This section specifically treats the Western philosophical genealogy. Chapter 8 treats the Chinese philosophical tradition, namely qi theory, as another genealogy. The two are parallel sources of the ontology of difference: both the Western and Chinese traditions contribute to difference and process as foundation. The Western genealogy, in this chapter, includes Heraclitus, Leibniz, Whitehead, Bergson, and Deleuze; the Chinese genealogy, in Chapter 8, includes the *Yizhuan*, Zhang Zai, Wang Fuzhi, Huayan Buddhism, and others. Treating the two separately is meant to show the content of each tradition clearly and to avoid the rhetoric of a simple "convergence between East and West." Together, the two chapters constitute a dual philosophical defense of the ontology of difference: it has grounds in both Western and Chinese traditions. More deeply, the direction of the ontology of difference appears independently in multiple traditions. This fact itself supports the legitimacy of the ontology of difference: it is not an accidental product of one tradition, but an accessible philosophical direction.

Literature That Must Be Engaged

- G. S. Kirk, J. E. Raven, M. Schofield, The Presocratic Philosophers (1983): on Heraclitus.
- G. W. Leibniz, Monadology (1714), The Principles of Nature and Grace (1714)
- Alfred North Whitehead, Process and Reality (1929)
- Henri Bergson, Creative Evolution (1907), Matter and Memory (1896)
- Gilles Deleuze, Difference and Repetition (1968), The Logic of Sense (1969)
- Chapter 8 on the tradition of qi theory (self-reference)

Summary of This Section

This section has done three things. First, it developed five precursor nodes of the ontology of difference in the Western philosophical tradition: Heraclitus, Leibniz, Whitehead, Bergson, and Deleuze. Second, it clarified the similarities and differences between each precursor and the work of this chapter. Third, it clarified the legitimacy of this genealogy as the Western philosophical tradition of the ontology of difference.

The next section is the conclusion of this chapter.

7.6. Chapter Summary: The Ontology of Difference as the Starting Point of Part II

Completion of the Three Tasks

The three tasks undertaken by this chapter have all been completed.

Task One: to make the core position of gradient ontology explicit as a precise philosophical thesis. Section 7.1 developed the difficulties of substance ontology. Section 7.2 gave a precise definition of "difference" as an ontological concept and stated the core thesis DI1. Section 7.3 defended DI1 through a three-layer argument: phenomenological, logical, and physical.

Task Two: to enter into dialogue with the substance–property–relation triad in the Western philosophical tradition. Section 7.1 developed three difficulties of this tradition: the identity paradox, the circularity of substance as foundation, and the treatment of relations. Section 7.2 developed the different treatment given by the ontology of difference: substance as the stable form of difference, properties as the contents of difference, relations as dynamics among differences, and change as the evolution of difference.

Task Three: to establish the Western philosophical genealogy of the ontology of difference. Section 7.4 developed five precursor nodes: Heraclitus, Leibniz, Whitehead, Bergson, and Deleuze, making the work of this chapter the contemporary continuation of an ongoing tradition.

The Position of This Chapter in Part II

This chapter is the foundational chapter of Part II, "The Philosophical Defense of Gradient Ontology." Later chapters are built upon the work of this chapter. Chapter 8, "The Tradition of Qi Theory and Frequency–Density–Gradient," establishes the Chinese philosophical genealogy, while this chapter establishes the Western philosophical genealogy of the ontology of difference. Together, the two chapters constitute the dual philosophical defense of the ontology of difference. Chapter 9, "Frequency as a Basic Spectral Quantity," builds upon the ontology of difference and treats frequency as a dimension of the structure of difference of beings. Chapter 10, "Tri-State Categories as Ontological Categories," builds upon the ontology of difference and decomposes beings into tri-state categories: free state, bound state, and condensed state. Chapter 11, "Freezing as a Foundational Ontological Process," builds upon the ontology of difference and treats freezing as the process in which structures of difference move from dynamics toward stability. If the work of this chapter holds, the overall architecture of Part II has a solid ground. If it does not hold, the ontological foundation of the following chapters is weakened.

The Position of This Chapter in the Whole Book

Looking backward, this chapter is built upon the methodological foundation of Part I: Chapter 1 on the method of constrained speculation provides the methodological ground for the argument of DI1 in this chapter; Chapter 3 on failure conditions as an ontological discipline provides the method for diagnosing the difficulties of substance ontology in this chapter; and Chapter 4 on constitutive coextension provides the semantic tool for the dialogue between this chapter and traditional philosophical genealogies.

Looking forward, this chapter provides the ontological foundation for later parts. Part III, on co-evolution, mind–matter, observation, and time, grounds co-evolution as relational ontology in the dynamical connections among structures of difference. Part IV, on the metaphysical boundary of freezing, treats ontological singularity as the state of affairs in which $\nabla\rho \equiv 0$, namely the limit in which all difference collapses. Part V, from ontology to normativity, treats meaning as emerging within structures of difference and grounds the finitude of meaning in the finitude of structures of difference. These connections make this chapter the starting point of the ontological work of the whole book: every later ontological development is built upon the core thesis that difference is prior to identity.

Honesty about the Limits of This Chapter

This chapter acknowledges several limits:

First limit: dialogue with opponents is insufficient. This chapter mainly develops the content of the ontology of difference, but does not respond in sufficient detail to contemporary substance ontology. There has been a contemporary revival of substance ontology, such as E.J. Lowe's four-category ontology and substance–property–event ontologies, and these works require dedicated treatment.

Second limit: the internal complexity of the ontology of difference has not been sufficiently developed. There are differences within the ontology of difference, such as Deleuze's difference, Whitehead's processuality, and Bergson's duration. This chapter treats them as part of the same genealogy, but the differences among these precursors require more detailed work.

Third limit: the relation with epistemology has not been sufficiently developed. The ontology of difference has implications for epistemology, such as knowledge as the identification of structures of difference. This chapter mentions this only briefly, while Chapter 14 treats objectivity in detail; a complete epistemological development requires dedicated work.

Fourth limit: possible difficulties of difference as foundation have not been fully treated. The ontology of difference also has difficulties, such as what gives difference direction and what the ontological status of difference is. This chapter gives directions of reply, but a complete treatment requires later work.

Acknowledging these limits does not weaken the work of this chapter. It makes this chapter an honest initial work and invites further development.

The Core Posture of This Chapter

The core posture of this chapter can be stated as follows: changing the starting point of ontology is more fundamental than repairing its details.

Substance ontology has undergone extensive repair over more than two millennia: essentialism, bundle theory, stuff theory, trope theory, and others are all responses. These repairs all accept the starting point of "substance as foundation" and treat difficulties within that framework. But the work of this chapter argues that the starting point itself may need to change. If substance ontology faces difficulties at its own starting point, such as the three difficulties developed in Section 7.1, then repair cannot solve the problem at the root. The starting point must be changed.

This method of "changing the starting point" is a legitimate direction in the history of philosophy. Kant's "Copernican revolution" changed the starting point, from objects to the transcendental structure of the subject. The later Wittgenstein's "linguistic turn" changed the starting point, from propositions to language games. The phenomenological call to return "to the things themselves" changed the starting point, from the subject–object divide to the structure of experience. The work of this chapter continues this tradition: by changing the ontological starting point, from substance to difference, it gives ontological problems a new treatment. If the work of this chapter holds, the ontology of difference has a place in contemporary philosophy as an ontological starting point. It does not replace all applications of substance ontology, since the treatment of substances as everyday objects remains legitimate, but as a foundational ontology it provides a new possibility.

Meaning before freezing: this is one of the core theses of this book. This chapter provides an ontological ground for that thesis. "Before freezing" involves the existence of structures of difference; "meaning" appears in the complex forms of structures of difference, such as consciousness, life, and culture; and the finitude of meaning is grounded in the finitude of structures of difference. These connections receive their ground from the ontology of difference established in this chapter.

Through the work of this chapter, the ontology of difference as the starting point of Part II gains philosophical grounds: the defense through the three-layer argument, the genealogy of five precursor nodes, and the dialogue with the tradition of substance ontology. The later chapters are built upon this starting point, making Part II the complete development of the ontology of difference.

7.7. Methodological Reflection on This Chapter

This chapter makes the ontology of difference explicit as a precise philosophical thesis from what had previously been an implicit position. This section reflects on the methodological significance of this work of explication.

This Chapter's Treatment of Substance Ontology

This chapter's treatment of substance ontology is critical but respectful. It is critical in identifying difficulties, namely the three difficulties of independence, identity, and the derivative status of relations, and in arguing for the alternative of the ontology of difference. It is respectful in acknowledging the strength of substance ontology, including more than two millennia of development, and in not pretending that it is entirely wrong. This balance makes the work of this chapter serious in philosophical discussion: it does not rely on contempt or rhetoric, but develops through argument. This chapter does not claim that substance ontology is "entirely wrong."

It claims that substance ontology as foundational ontology needs to be refined under contemporary challenges. "Substance" as everyday ontological language retains legitimacy, but its ontological ground should be understood as the stable form of structures of difference.

This Chapter's Posture toward Originality

The ontology of difference in this chapter is built upon precursor traditions: Heraclitus, Leibniz, Whitehead, Bergson, and Deleuze. It is not an isolated creation. The contribution of this chapter is to give the ontology of difference a physical ground through the EQT framework; to give the ontology of difference a methodological ground through the method of constrained speculation; and to unfold the ontology of difference systematically as the starting point of Part II of this book. These contributions make the chapter constructively original: it does new work within a tradition, rather than beginning from zero.

The Grounds of This Chapter Independent of EQT

This chapter does not depend on the physical derivation of EQT. All arguments are completed through purely philosophical argumentation. This independence allows the chapter, as part of a philosophical monograph, to be taken seriously in international philosophical discussion: its validity does not depend on the reader's acceptance of the physical work of EQT. EQT appears only as an application example. It is the physical application of the ontology of difference, but not the argumentative ground of the ontology of difference.

Limits Honestly Acknowledged in This Chapter

This chapter acknowledges several limits. First, the refinement of the ontology of difference requires later work: the various forms of difference as a concept in different cases require more detailed treatment. Second, this chapter develops only the core direction of the connection between the ontology of difference and categorial language, such as Aristotelian and Kantian categories; the development of tri-state categories in Chapter 10 is part of this direction. Third, the manifestations of the ontology of difference in interdisciplinary applications, such as life, consciousness, and society, require dedicated work.

Connections with Later Chapters

The ontology of difference established in this chapter is the starting point for the later chapters of Part II. Chapter 8 provides the Chinese philosophical genealogy of the ontology of difference. Chapter 9 develops the internal dimension of structures of difference: frequency.

Chapter 10 develops the categorial doctrine of structures of difference: the tri-state categories. Chapter 11 develops how structures of difference form relatively stable forms: freezing. More broadly, the ontology of difference established in this chapter is the starting point of the process ontology of the whole book. The work of Part III on co-evolution, Part IV on ontological singularity, and Part V on meaning is all built upon structures of difference. This chapter is the statement of a shift in ontological starting point: the fundamental position of this book is built upon the starting point that difference is prior to identity.

8. The Tradition of Qi Theory and Frequency–Density–Gradient

8.1. Introduction to This Chapter

Chapter 7 established the Western philosophical genealogy of the ontology of difference: Heraclitus, Leibniz, Whitehead, Bergson, and Deleuze. But the Western tradition is not the only genealogy of the ontology of difference. From the *Book of Changes* onward, the Chinese philosophical tradition has taken differentiated change as its ontological starting point. Over two and a half millennia, this starting point generated a series of ontological resources, among which the most systematic is the tradition of qi theory: the ontological work developed by philosophers such as Zhang Zai and Wang Fuzhi.

This chapter presents the Chinese tradition of qi theory as a substantive contribution to contemporary philosophy of nature, not as decorative citation, not as cultural nostalgia, but as a philosophical interlocutor. Specifically, this chapter argues for one thesis: Chinese qi theory is in certain respects more thorough than Whiteheadian process philosophy, because it is purely monistic and does not require eternal objects as a second ontology. This thesis is not a rhetorical claim about "Chinese philosophy anticipating modern science"; it is a philosophical thesis about ontological structure.

This chapter occupies a distinctive position in a philosophical monograph written by a Chinese scholar. Familiarity with Chinese traditional resources is an advantage, but this advantage is accompanied by a strict responsibility to avoid cultural chauvinism. The posture of this chapter is to treat qi theory as part of world philosophy: not Chinese philosophy "against" Western philosophy, not Chinese philosophy "for the sake of" Western philosophy, but Chinese philosophy as a substantive contribution to contemporary philosophy of nature.

This chapter builds upon Chapter 7, "Difference before Identity," which established the Western philosophical genealogy of gradient ontology. This chapter provides the Chinese philosophical genealogy of gradient ontology. Together, the two chapters constitute the dual philosophical defense of gradient ontology in Part II: the ontology of difference in this book is not an isolated philosophical creation, but a contemporary synthesis with dual precursor traditions, Western and Chinese.

The argument of this chapter is completed through purely philosophical argumentation. The correspondence between qi theory and gradient ontology holds at the structural level. EQT appears only as one pole of structural comparison and does not serve as the ground of the argument. Section 8.7 of this chapter specifically draws the boundary between qi theory and EQT, making clear that the two should not be confused. This independence allows the chapter, as part of a philosophical monograph, to be taken seriously in international philosophical discussion.

This chapter adopts a selective rather than exhaustive strategy in textual citation: it chooses the most directly relevant passages and analyzes them in detail, rather than enumerating all relevant passages in the tradition. At the same time, this chapter uses Chinese original texts and English-language scholarship. Texts on qi theory must be based on the original Chinese, since any translation loses part of the original meaning; dialogue with international philosophical discussion must use English-language scholarship as interlocutors. This bilingual use is a distinctive capacity of a Chinese scholar writing this book, and also a responsibility.

The discussion proceeds as follows. Section 8.1 develops the process orientation of the Chinese philosophical tradition, namely the fundamental difference; Section 8.2 develops Zhang Zai's qi as an expression of the ontology of difference; Section 8.3 develops the relation between Wang Fuzhi's "qi is born and formed day by day" and the freezing spectrum; Section 8.4 develops the structural relation between the *Yizhuan*'s "continuous generation is what is meant by change" and the frequency spectrum; Section 8.5 develops the advantage of qi theory as a thorough monism, in comparison with Whitehead's work; Section 8.6 develops the two-way dialogue between qi theory and this book's latest argument, namely boundary as gate: the significance of resources such as "form," "being born and formed day by day," and "continuous generation" in qi theory for enriching this book's anti-panpsychist criterion; Section 8.7 clarifies the boundary of dialogue between qi theory and EQT, avoiding any confusion between the two;

Section 8.8 concludes the chapter by locating qi theory as part of world philosophy.

8.2. The Process Orientation of the Chinese Philosophical Tradition

A Fundamental Difference

Before entering qi theory, it is necessary to grasp the fundamental difference in orientation between the Chinese philosophical tradition and the Western mainstream tradition. This difference is not a difference of detail, but a difference in ontological starting point.

From Parmenides to Aristotle to Descartes, the Western mainstream tradition takes "substance" as its ontological starting point: the basic units of existence are things, namely substances; things have properties, namely features that inhere in substances; and things stand in relations, namely connections outside substances. This triad of substance–property–relation dominated Western metaphysics for two and a half millennia. Even traditions that oppose substance ontology, such as Heraclitus, Hegel, and Whitehead, often take the critique of substance ontology as their own starting point: their thought structures remain determined by the framework of substance ontology.

From the *Book of Changes* onward, the Chinese philosophical tradition does not take substance as its ontological starting point. Its starting point is change; more precisely, differentiated change. The concept of "Yi" itself means "change": the *Yizhuan*'s statement that "continuous generation is what is meant by change" articulates change as the deepest ontological determination. From the *Book of Changes* to the pre-Qin thinkers, from the maturation of qi theory in the Han dynasty to Song-Ming Neo-Confucianism and Qing evidential learning, the mainstream of the Chinese philosophical tradition takes process, change, and generation as its ontological starting point, rather than substance.

This fundamental difference is not a later generalization; it appears clearly in the earliest texts. The *Yizhuan* · *Xici* says: "One yin and one yang is called Dao." This statement locates "Dao," the most basic ontological determination, in the mutual transformation of yin and yang: not yin and yang as two substances, but yin and yang as two poles of mutual transformation. Change precedes substance.

Why This Difference Matters

Why is this difference important for contemporary philosophy of nature?

The development of contemporary physics in the twentieth century gradually made "substance" difficult as an ontological category. In quantum field theory, "particles" are field excitations, not independent substances. In general relativity, spacetime itself participates in dynamics. In non-equilibrium thermodynamics, "structure" is the stable manifestation of process, not a static object. In all these developments, processuality, relationality, and dynamicity become increasingly basic in physical description, while substantiality becomes increasingly secondary.

Yet the ontological language of Western philosophy is mainly built upon the substance–property–relation triad. When this language is used to describe contemporary physics, it often strains: philosophers have to invent various new concepts, such as "process," "event," "structure," and "information," to compensate for the insufficiency of the language, but these concepts still refer in some way to the framework of substance ontology.

The Chinese tradition of qi theory provides a fundamentally different ontological language at this point. From the beginning, it does not presuppose substance as the basic unit. It takes change, process, and generation as basic determinations. This gives it structural relevance for the description of contemporary physics, not because it "anticipated" modern physics, which would be bad rhetoric, but because its ontological starting point is structurally consistent with the ontological needs of modern physics.

The work of this chapter is to make this structural consistency precise: not to say vaguely that "the Chinese tradition is more processual," but to state concretely how Chinese qi theory makes independent contributions to specific problems.

The Focus of This Chapter

The focus of this chapter is qi theory: the tradition of qi monism from Zhang Zai to Wang Fuzhi. This is the most mature form of the process orientation of Chinese philosophy. The core position of qi theory is this: qi is the sole ontology; all things are the gathering and dispersal of qi; gathering and dispersal are the internal differentiated movement of qi; and time, space, and causality all emerge from the process of gathering and dispersal.

These positions are structurally isomorphic with the gradient ontology of this book, especially the positions established in Chapters 7, 9, 10, and 11: both take differentiated process as the ontological starting point, and both refuse to take "things" as the basic units. But isomorphism is not identity. The task of this chapter is to characterize precisely the degree of isomorphism and the location of difference: in what respects qi theory agrees with gradient ontology, in what respects it has its own distinctive contribution, and in what respects it has limits that contemporary philosophy of nature cannot accept.

Three Traps to Avoid

Before entering the analysis, this chapter explicitly avoids three traps.

Trap One: cheap rhetoric of "already anticipated." To say that "Zhang Zai already anticipated EQT" or that "the *Book of Changes* already contains quantum mechanics" is bad philosophy. It downgrades the Chinese tradition into anticipation, ignores the content of the tradition itself, and replaces rigorous dialogue with vague correspondence. This chapter does not use such rhetoric.

Trap Two: downgrading the Chinese tradition into prehistory. If Zhang Zai and Wang Fuzhi are treated only as the "intellectual prehistory leading to EQT," then their work is instrumentalized: its value lies only in leading to later science. This chapter rejects such downgrading. The work of Zhang Zai and Wang Fuzhi is itself a philosophical contribution, independent of its possible correspondence with modern science.

Trap Three: selective citation. To extract sentences from traditional texts that support a pre-existing conclusion while ignoring context and textual setting is bad use of literature. When citing texts from qi theory, this chapter will preserve context and explicitly indicate where the texts may not support the position of this chapter.

Avoiding these three traps requires strict methodological discipline: the Chinese tradition must be taken seriously, and the methodological standards of contemporary philosophy must also be taken seriously. This chapter seeks a balance between the two.

8.3. Zhang Zai's Qi: Gathering–Dispersal and Gradient Transients

Zhang Zai's Ontological Proposal

Zhang Zai (1020–1077) is the major founder of the tradition of "qi monism" in Song-Ming Neo-Confucianism. His core text, the *Zhengmeng*, systematically states the position of qi as the sole ontology. One of its best-known passages is: "The Great Void cannot be without qi; qi cannot but gather and become all things. All things cannot but disperse and become the Great Void" (*Zhengmeng* · "Great Harmony").

This passage condenses the core of Zhang Zai's ontology. Let us analyze it sentence by sentence.

"The Great Void cannot be without qi": this sentence rejects the interpretation of "void" as absolute emptiness. In the Western philosophical tradition, "void" is often understood as some kind of "nothing," opposed to "being." Zhang Zai explicitly rejects this opposition. The Great Void is not "nothing": it is filled with qi, except that qi in the Great Void is in an ungathered and undispersed state.

"Qi cannot but gather and become all things": this sentence states the dynamicity of qi. Qi is not a static distribution: it necessarily gathers and disperses. Gathering and dispersal are not externally added movements, but internal determinations of qi. In other words, qi as ontology is itself change: it is not that there is "something called qi" and then "it moves," but that qi = the process of gathering and dispersal.

"All things cannot but disperse and become the Great Void": this sentence states the finitude of all things. All things are not eternal substances. They are temporary forms in which qi gathers, and eventually they disperse back into the Great Void. "Dispersal" is not annihilation; it is the other pole of the process of gathering and dispersal.

Taken together, these three sentences allow Zhang Zai's ontology to be stated as follows: existence = the process of the gathering and dispersal of qi; all things = temporary forms of the local condensation of qi; the Great Void = the diffuse state of qi before condensation; gathering and dispersal = the internal dynamic determinations of qi, not dependent on an external force.

Structural Isomorphism with Gradient Ontology

If Zhang Zai's ontology is placed alongside the gradient ontology of this book, especially Chapters 7 and 11, a structural isomorphism becomes visible.

Qi $\leftrightarrow$ the energy quantum density field $\rho_f(x,t)$. Both are located as the sole basic ontological object, not as "things," but as a form of distribution filling space–time.

Gathering $\leftrightarrow$ gradient condensation. "Gathering" refers to the local concentration and convergence of qi and the formation of structure. This is isomorphic with the role of "gradient structure" as the basic form of beings in gradient ontology: a being is the local condensation of ρ_f at a certain location, namely a local structure with $\nabla \rho_f \neq 0$.

Dispersal $\leftrightarrow$ gradient dissipation. "Dispersal" refers to the disintegration of structure, the diffusion of energy, and the return to the Great Void. This is isomorphic with the dissipation of gradients in gradient ontology, namely local $\nabla \rho_f \to 0$.

The process of gathering and dispersal $\leftrightarrow$ gradient transients. This is the most important isomorphism. Zhang Zai's core thesis is that "qi cannot but gather and become all things; all things cannot but disperse and become the Great Void": beings are essentially processes, temporary phases of gathering and dispersal. This is fully isomorphic with the position in Chapter 11 of this book on "freezing as a foundational ontological process" and gradient transients as the basic form of beings.

The Great Void $\leftrightarrow$ globally uniform ρ_f. The "Great Void" is not "nothing at all": it is the state in which qi has not gathered or dispersed. This is isomorphic with the gradient terminal phase in Chapter 17, namely the state of affairs in which $\nabla \rho \equiv 0$. Both reject the interpretation of "nothingness" as absolute nothing, and both understand "nothingness" as the limiting dissolution of differentiated structure.

The Precision of the Isomorphism

This isomorphism is not a vague claim that "both speak of process." It is an isomorphism that can be mapped at the level of philosophical theses. Consider four theses:

Thesis 1 (the negation of substance): Zhang Zai says that "all things cannot but disperse and become the Great Void," meaning that all things lack eternity; gradient ontology says that "all structures are gradient transients and eventually dissolve" (Chapter 11). The structural isomorphism holds: both deny eternal substances.

Thesis 2 (the non-oppositionality of emptiness): Zhang Zai says that "the Great Void cannot be without qi," meaning that void is not absolute nothing; gradient ontology says that "the uniform distribution of ρ_f is not 'nothing,' but the limiting dissolution of structure" (Chapter 17, Section 17.2). The structural isomorphism holds: both reject treating emptiness as the opposite of being.

Thesis 3 (the internality of change): Zhang Zai says that "qi cannot but gather," meaning that gathering and dispersal are internal determinations of qi and do not require an external force; gradient ontology says that "the change of gradients is determined by the dynamics of ρ_f itself, namely the $A3$ Master Equation, and does not require an external mover" (Chapter 7 on the internality of difference). The structural isomorphism holds: both treat change as an internal determination of ontology.

Thesis 4 (the processuality of all things): Zhang Zai says that "all things are gathered qi," meaning that all things are essentially phases of process, not independent substances; gradient ontology says that "beings are gradient structures, not substances" (Chapter 7, Section 7.5). The structural isomorphism holds: both understand all things as phases of process rather than as independent units.

The isomorphism holds for all four theses. This is not a vague claim that "Eastern and Western thought converge": it is a structural correspondence at the level of philosophical theses. The ontological theses that Zhang Zai stated in the eleventh century in the language of qi theory correspond structurally to the ontological theses that this book states in the twenty-first century in the language of EQT.

But Isomorphism Is Not Identity

One clarification must be made immediately: isomorphism is not identity.

Zhang Zai's qi theory and gradient ontology are not the same theory. They have different contents, although they have isomorphic ontological structures. This distinction is philosophically crucial: ignoring it leads to two errors.

Error One: exaggerating isomorphism into identity. "Zhang Zai's qi is the ρ_f field": this is wrong. Zhang Zai's qi is a concept in eleventh-century Chinese philosophy. It has its own content, including relations with yin-yang, the five phases, nature and life, and so on. It cannot be simply equated with the ρ_f field of twenty-first-century physics.

Error Two: denying isomorphism because there is no identity. "Zhang Zai's qi and ρ_f differ in content, so they have nothing to do with each other": this is also wrong. Difference in content does not negate structural isomorphism. Isomorphism is correspondence at the level of relation, not correspondence at the level of object.

The correct posture is to acknowledge the reality of the isomorphism at the level of ontological structure while also acknowledging the reality of difference at the level of content. This allows us to gain philosophical insight from Zhang Zai's work without erasing its historical specificity.

Zhang Zai's More Thorough Monism Compared with Whitehead

We can now state the first key thesis of this chapter:

Key thesis $K1$: Zhang Zai's qi monism is in a certain sense more thorough than Whiteheadian process philosophy, because it does not require eternal objects as a second ontology.

Whitehead's process philosophy (*Process and Reality*, 1929) contains an internal tension. On the one hand, Whitehead argues that the basic units of existence are actual occasions: processual, temporary, mutually connected actualities. On the other hand, Whitehead finds that actual occasions cannot fully explain form and possibility: why do different occasions display similar formal structures? Why are some possibilities more "real" than others? Whitehead's response is to introduce eternal objects as a second ontology: they are abstract forms that "ingress" into actual occasions in the process of becoming.

This is the internal difficulty of Whitehead's scheme. He wishes to construct a monism, namely process alone, but in fact he has to construct a dualism, namely process + eternal objects. The introduction of a second ontology weakens the thoroughness of his process philosophy: why are some things, actual occasions, processual, while other things, eternal objects, are eternal? Why is ontology divided into two layers?

Zhang Zai does not have this problem. In Zhang Zai's qi monism, qi is the sole ontology, and all things are the gathering and dispersal of qi. There is no second ontology. "Form" and "possibility" do not need to be anchored in eternal objects; they are explained by the internal determinations of qi itself, namely the laws of dynamic gathering and dispersal. This difference makes Zhang Zai in a certain sense more thorough than Whitehead. Whitehead yields to dualism in order to handle the problem of form; Zhang Zai preserves monism and leaves the problem of form within the internal dynamics of qi itself.

The Significance of This Difference for Contemporary Philosophy of Nature

Why is this difference important for contemporary philosophy of nature?

Contemporary physics describes the world as a single physical process: quantum fields, general-relativistic spacetime, and non-equilibrium thermodynamic processes. There is no obvious need for two different kinds of ontology in order to describe the world. When physicists describe the world, they use one type of ontology, such as fields, states, or processes; they do not introduce "eternal objects" as a second ontology. If philosophy is to provide ontological language for contemporary physics, the thorough monism of qi theory is more suitable than Whitehead's dualism. It makes ontological structure consistent with physical structure in singularity: only one type of ontology is committed to, namely qi / ρ_f, without introducing a second ontology, namely eternal objects / abstract formal entities.

But this comparison does not weaken Whitehead. Whitehead's work has its own contributions: the concept of actual occasions, the concept of prehension, the distinction between reasonable and occasion, and so on. These are philosophical riches. This chapter only points out that, along the dimension of the thoroughness of monism, Zhang Zai is more thorough than Whitehead.

Objections and Replies

Objection One: This comparison is too abstract. What does "thoroughness" mean as a criterion? Why is thoroughness an advantage rather than a disadvantage?

Reply: Thoroughness here refers to ontological restraint, the position defended in detail in Chapter 2: the fewer the commitments, the better. Whitehead commits to two types of ontology, actual occasions + eternal objects; Zhang Zai commits to one type of ontology, qi. All other things being equal, an ontology with fewer commitments is preferable. This is restraint. But "all other things being equal" is crucial. If one type of ontology is insufficient for treating the problems that need to be treated, then introducing a second type is reasonable. Whitehead thinks that eternal objects must be introduced in order to treat the problem of form. This chapter argues that if the problem of form can be treated within monism, as Zhang Zai does, then there is no need to introduce a second ontology. This is an application of the principle of restraint.

Objection Two: Is Zhang Zai's concept of "qi" really sufficient to handle the problem of form? Does leaving the problem to the "internal determinations of qi" not simply move the difficulty inside?

Reply: This objection points to a real philosophical problem. Zhang Zai's reply is that qi has an internal "li": the gathering and dispersal of qi proceeds according to Dao, and Dao is the dynamic law of qi itself, not an externally added form. This position requires detailed defense:

how is li "internal" to qi? This is one of the central questions of Song-Ming Neo-Confucianism. This chapter does not claim that Zhang Zai completed this defense; he gave a direction but did not complete every detail. The claim of this chapter is more specific: Zhang Zai's direction, namely li within qi, is more thorough than Whitehead's direction, namely eternal objects outside actual occasions. Completing this direction requires later philosophical work. Gradient ontology, especially Chapter 11 on freezing and Chapter 17 on ontological singularity, is one possible form of such later work.

Objection Three: You seem to be evaluating Zhang Zai by means of EQT. This is judging an ancient thinker by modern standards.

Reply: This chapter does not "judge" Zhang Zai. It uses Zhang Zai as a philosophical resource. "Judgment" assumes some external standpoint from which one can decide who is right and who is wrong. This chapter presupposes no such standpoint. It places Zhang Zai and contemporary philosophy of nature in an equal philosophical dialogue: Zhang Zai contributes processuality, monism, and the language of gathering and dispersal; contemporary physics contributes concrete forms, such as fields, gradients, and freezing. The two illuminate each other.

Literature That Must Be Engaged

- Zhang Zai, *Zhengmeng*, especially "Great Harmony," "Reference to the Two," and "Sincerity and Illumination"
- Mou Zongsan, *Mind-Substance and Nature-Substance*, Volume 2: detailed analysis of Zhang Zai
- Tang Junyi, *Original Discourses on Chinese Philosophy: Original Nature*
- A. C. Graham, Two Chinese Philosophers: The Metaphysics of the Brothers Ch'eng (1958)
- Alfred North Whitehead, Process and Reality (1929), especially Part II on actual occasions and eternal objects
- Chapter 7 on the argument for difference before identity (self-reference)

Summary of This Section

This section has done four things. First, it analyzed in detail Zhang Zai's ontological thesis that "the Great Void cannot be without qi; qi cannot but gather and become all things." Second, it established the structural isomorphism between Zhang Zai's qi theory and gradient ontology: qi–ρ_f, gathering and dispersal–gradient condensation and dissipation, and the Great Void–uniform ρ_f. Third, it made clear that isomorphism is not identity: the two correspond structurally but differ in content. Fourth, it argued for the key thesis $K1$: Zhang Zai's qi monism has an advantage over Whiteheadian process philosophy in thoroughness, because it does not require a second ontology.

The next section develops Wang Fuzhi's further contribution: the development of Zhang Zai's work into a process ontology with temporality and historicity.

8.4. Wang Fuzhi's "Qi Is Born and Formed Day by Day" and the Freezing Spectrum

Wang Fuzhi's Further Development

Wang Fuzhi (1619–1692) is the most important successor of Zhang Zai's tradition of qi theory. In works such as *Commentary on Zhang Zai's Zhengmeng*, *External Commentary on the Zhouyi*, and *Silu*, he systematically developed Zhang Zai's work and made a distinctive contribution along the dimension of temporality.

Wang Fuzhi's core thesis is that "qi is born and formed day by day": qi is generated every day and takes form every day. This thesis strengthens Zhang Zai's process orientation: qi is not merely abstract gathering and dispersal, but a concrete, historical, and irreversible process.

Wang Fuzhi writes: "The transformations of Heaven and Earth are renewed each day... the sun and moon of today are not the sun and moon of yesterday" (*Silu* · External Part). This statement pushes the processuality of the cosmos to the utmost: not only are all things changing, but even the sun and moon, the most stable celestial bodies, are generated anew each day. This is the most thorough version of processuality.

Structural Isomorphism with the Freezing Spectrum

Wang Fuzhi's concept of "daily renewal" is structurally isomorphic with the core position of this book's *Freezing Cosmology*.

The core position of the freezing spectrum is this: cosmic evolution is the process in which different frequency bands freeze at different times, and cosmic history itself is the temporal ordering of the freezing spectrum. Each freezing is irreversible: once a certain frequency band freezes, it exits active dynamics and is preserved as residue, background, or history. The cosmos does not cyclically return to the same state; it unfolds historically and irreversibly.

This is structurally isomorphic with Wang Fuzhi's position:

"Qi is renewed each day" ↔ "the continuous unfolding of the freezing spectrum." Both understand the cosmic process as irreversible, historical, and newly unfolding at every moment.

"The sun and moon of today are not the sun and moon of yesterday" ↔ "the current state of a structure is determined by historical freezing." Even things that "appear stable," such as the sun and moon, are in fact products of process: their "present state" is determined by their "history of formation."

"Being born and formed day by day" ↔ "the continuous maintenance of condensed structures." A structure is not formed once and then remains static; it must be maintained at every moment through the continuing action of physical processes. "Maintenance" is not passive invariance, but active "being born and formed day by day."

This isomorphism is not rhetorical; it has philosophical meaning. Wang Fuzhi points to an ontological fact that is easily overlooked: stability is not the opposite of motion; stability is a special form of dynamic process. A seemingly unchanged star burns hydrogen, radiates energy, and maintains its structure every second. Its "invariance" is the result of continuous change, not the absence of change.

The Ontological Significance of Historicity

Wang Fuzhi's work has structural analogues in Western philosophy: Hegel's concept of historicity, Bergson's *durée*, and Heidegger's *Geschichtlichkeit*. But Wang Fuzhi's contribution is to apply historicity to the physical cosmos, not only to human history or consciousness.

In the Western tradition, historicity is often restricted to consciousness, culture, and human life. The physical cosmos is understood as an ahistorical cyclic process or a non-historical eternal structure. Even Hegel's work on "nature as the other of spirit" preserves the distinction between nature and history. Wang Fuzhi rejects this distinction. For him, the physical cosmos itself is historical: the sun and moon, mountains and rivers, the four seasons, are all "renewed each day," and every moment is a new unfolding. This position was pioneering in the seventeenth century: it extended historicity to the cosmos as a whole, anticipating at the level of ontological orientation a fact that twentieth-century cosmology, especially the Big Bang model, cosmic evolution, and the freezing spectrum, would later state rigorously in physics: the cosmos is historical, not eternal and not cyclical.

The Limits of Wang Fuzhi

It must be made clear that Wang Fuzhi's work is not a completed version of contemporary cosmology. He had no concrete physical model; he could not have had one, because the physics available in the seventeenth century did not allow it. His work is a philosophical direction, not a physical conclusion.

But as a philosophical direction, his work has substantive relevance. It tells us that treating the cosmos as a historical process is not an accidental discovery of twentieth-century physics, but an ontological position with deep philosophical grounds. This position had already matured to the level of thesis in the Chinese philosophical tradition, waiting to be realized in concrete physical form by modern physics, such as the freezing spectrum of EQT.

This is an example of the relation between philosophy and physics. Philosophy is not the foundation or footnote of physics. Philosophy and physics illuminate each other. Wang Fuzhi's philosophical direction allows us to understand more deeply the ontological meaning of the freezing spectrum; the physics of the freezing spectrum allows us to express Wang Fuzhi's direction more concretely.

Objections and Replies

Objection One: You seem to repeat a rhetoric: "Chinese philosophy already anticipated modern science." How is this different from the trap explicitly rejected in Section 8.1?

Reply: The difference lies between anticipation and structural relevance. A statement of "already anticipated" says that the Chinese tradition already knew facts later discovered by science. This is wrong: the Chinese tradition could not have known the concrete content of modern science. A statement of "structural relevance" says that the philosophical direction of the Chinese tradition corresponds structurally to the content of modern science. This can be argued rigorously, and that is what this section is doing. More concretely: this section does not claim that Wang Fuzhi "already anticipated the physics of the freezing spectrum." It claims that Wang Fuzhi's philosophical direction, namely treating the cosmos as a historical process, corresponds to the ontological structure of the freezing spectrum, namely irreversible, historical unfolding by frequency bands. This is correspondence at the level of ontological structure, not anticipation at the level of physical content.

Objection Two: Wang Fuzhi's "being born and formed day by day" looks completely different from modern cosmology. He has no Big Bang, no expansion, and no concrete evolutionary timetable. Is placing him alongside the freezing spectrum too loose?

Reply: Yes, Wang Fuzhi has no concrete evolutionary timetable. But this section is not comparing concrete evolutionary models. It is comparing the ontological position of historicity. These two must be distinguished. Concrete evolutionary models belong to physics and can only be completed within physics. The ontological position of historicity belongs to philosophy and can be established philosophically before physics. Wang Fuzhi contributed the latter, and did not and could not contribute the former. The comparison in this section concerns an ontological position, not a physical model.

Objection Three: Wang Fuzhi has his own philosophical limits. For example, he still presupposes traditional Chinese frameworks such as the relation between Heaven and human beings, yin-yang, and the five phases. These frameworks may be incompatible with contemporary philosophy of nature.

Reply: Fully agreed. Wang Fuzhi's work has the limits of his time, and this chapter does not pretend otherwise. This chapter claims only that, in the ontological direction of treating the cosmos as a historical process, Wang Fuzhi has a relevant contribution. This relevance does not mean that his entire philosophical system is compatible with contemporary philosophy of nature. The correct way to use a tradition as a philosophical resource is selective rather than total: use the relevant parts, acknowledge that other parts may not apply, and do not pretend that the tradition as a whole is contemporarily relevant. This chapter remains cautious on this point.

Literature That Must Be Engaged

- Wang Fuzhi, *Commentary on Zhang Zai's Zhengmeng*, *External Commentary on the Zhouyi*, *Silu: Inner Part*, and *Silu: Outer Part*
- Xiao Jiefu, *A Critical Biography of Wang Fuzhi*
- Tang Junyi, *Original Discourses on Chinese Philosophy: Original Teaching*, sections on Wang Fuzhi

- Jacques Gernet, Chine et christianisme (1982), on Chinese philosophy in Wang Fuzhi's time
- Chapter 11 on freezing as a foundational ontological process (self-reference)
- Work in *Freezing Cosmology* on the physics of the freezing spectrum (self-reference)

Summary of This Section

This section has done three things. First, it introduced Wang Fuzhi's "qi is born and formed day by day" as a further development of Zhang Zai's qi theory along the dimension of temporality. Second, it established the structural isomorphism between Wang Fuzhi's position of "daily renewal" and the freezing spectrum: both treat the cosmos as an irreversible historical process. Third, it clarified that this isomorphism belongs to the structural level, not the level of physical anticipation, and it made clear that Wang Fuzhi's work has the limits of its time and is used here selectively rather than totalistically.

The next section develops the *Yizhuan*'s "continuous generation is what is meant by change" as the source of the qi-theoretical tradition and its structural relation to the frequency spectrum.

8.5. The *Yizhuan*'s "Continuous Generation Is What Is Meant by Change" and the Frequency Spectrum

The Ontological Starting Point of the *Yizhuan*

The sources of the tradition of qi theory lie in the *Book of Changes* and the *Yizhuan*. The *Yizhuan* · *Xici* contains several core theses: "One yin and one yang is called Dao": Dao is the mutual transformation of yin and yang; "continuous generation is what is meant by change": the essence of change, or Yi, is continuous generation; "Yi has the Great Ultimate; this generates the Two Modes; the Two Modes generate the Four Images; the Four Images generate the Eight Trigrams": the unfolding of existence is hierarchical, staged, and proceeds according to determinate patterns.

These theses constitute the earliest systematic statement of ontology in the Chinese philosophical tradition. Let us analyze them one by one.

"One yin and one yang is called Dao" states that the most basic ontological determination, Dao, is the mutual transformation of differentiated poles. Yin and yang are not two independent "things"; rather, they are a relation of mutual transformation. Difference, namely yin-yang, and transformation, namely change, together constitute Dao. This thesis is structurally isomorphic with the argument in Chapter 7 that difference precedes identity: both take difference as the ontological starting point rather than identity/substance.

"Continuous generation is what is meant by change" states that the essence of change is not a transition from one state to another, but continuous generation. "Continuous generation" is a reduplicated expression: the process of generation itself generates. This thesis strengthens the internality, continuity, and positivity of change. Change is not externally added; it is an internal determination of ontology.

"Yi has the Great Ultimate; this generates the Two Modes," and related statements, state that change is a hierarchical and staged unfolding. The Great Ultimate generates the Two Modes, namely yin and yang; the Two Modes generate the Four Images; and the Four Images generate the Eight Trigrams. This is a process of hierarchical differentiation. Each level is generated from the preceding level, but also has its own structure.

A Chinese Philosophical Precursor of the Frequency Spectrum

If the hierarchical unfolding of the *Yizhuan* is placed alongside the frequency spectrum of this book (Chapter 9), structural isomorphism becomes visible.

The core position of the EQT frequency spectrum is this: energy quanta are grouped by frequency. Low-frequency, mid-frequency, high-frequency, and ultra-high-frequency domains correspond to different interaction ranges, different stable structures, and different physical manifestations. The frequency spectrum is not a table of classification; it is a coordinate of derivation, the basic framework for "what corresponds to what" in physics.

The *Yizhuan*'s unfolding from "the Great Ultimate → the Two Modes → the Four Images → the Eight Trigrams" is structurally related to this:

Hierarchical organization $\leftrightarrow$ frequency grouping. Both describe existence as stratified: different levels have different structures and laws.

From the more basic to the more concrete $\leftrightarrow$ from low frequency to high frequency. The unfolding in the *Yizhuan* moves from the most general, namely the Great Ultimate, to the most concrete, namely the Eight Trigrams, and each step increases concreteness. The EQT frequency spectrum moves from the lowest frequencies, namely global structures, to the highest frequencies, namely local details. The higher the frequency, the more local and concrete the structure.

Generative relation $\leftrightarrow$ condensation relation. The Two Modes are "generated" from the Great Ultimate, but the Great Ultimate does not disappear: it remains in the Two Modes as the more basic layer. Likewise, high-frequency structures "condense" out of low-frequency structures, but low-frequency structures do not disappear: they remain as background or foundation within high-frequency structures.

This structural isomorphism does not mean that the *Yizhuan* "anticipated" the EQT frequency spectrum; that would be bad rhetoric. It means that the *Yizhuan* proposed an ontological direction of hierarchical unfolding, and this direction receives concrete physical realization in the EQT frequency spectrum.

An Example: The Frequency Stratification of the Four Forces

To make this structural relation concrete, this section gives an example.

In *The Foundational Volume of EQT* and *Energy, Process, and the Fate of the Cosmos*, EQT argues that the four fundamental interactions, namely gravity, electromagnetism, the strong force, and the weak force, correspond to different frequency bands: gravity is low-frequency accumulation; electromagnetism is mid-frequency low-dissipation propagation; the strong force is high-frequency local resonance; and the weak force is ultra-high-frequency short-lived frequency conversion. This argument does not depend on any Chinese philosophy: it is derived rigorously from the EQT postulates and the spectral relation of energy quanta.

Yet what is interesting is that this picture structurally corresponds to the hierarchical unfolding of the *Yizhuan*. If we locate the four forces in the frequency spectrum according to their frequency bands: gravity, as low frequency and cosmological scale, corresponds to the most general level in the *Yizhuan*, such as the level corresponding to the "Great Ultimate" or the "Two Modes"; electromagnetism, as mid-frequency and effective propagation at the atomic scale, corresponds to an intermediate level, such as the level corresponding to the "Four Images"; the strong and weak forces, as high-frequency / ultra-high-frequency and subatomic scale, correspond to more concrete levels, such as the level corresponding to the "Eight Trigrams."

This correspondence is not arbitrary. It reflects a genuine ontological fact: the structure of the cosmos is hierarchical, and more concrete structures unfold within more basic structures. More than two millennia ago, the *Yizhuan* stated this fact in the language of yin-yang and trigrams; in the twenty-first century, EQT states the same fact in the language of the frequency spectrum. The two are structurally related, but their contents differ: yin-yang vs frequency.

Boundary and Difference

A strict boundary clarification is required.

The *Yizhuan*'s unfolding from "the Great Ultimate → the Two Modes → the Four Images → the Eight Trigrams" is not a Chinese version of the EQT frequency spectrum. It has its own content: the meaning of yin and yang, the divinatory and ethical functions of the trigrams, the special status of the Great Ultimate as a principle of cosmic generation, and so on. These contents belong to the Chinese tradition and cannot be simply equated with modern physics.

The differences can be listed as follows: the hierarchy of the *Yizhuan* consists of the Eight Trigrams, namely eight types, whereas the EQT frequency spectrum is continuous, with infinitely many frequencies; the two differentiate levels in different ways. The unfolding of the *Yizhuan* has a divinatory function: it does not merely describe existence, but predicts auspiciousness and misfortune. The EQT frequency spectrum has no such function. The hierarchy of the *Yizhuan* is tightly connected with ethics and human affairs: the trigrams have moral meanings. The EQT frequency spectrum is only physics.

These differences are real and important. This section does not erase them. It claims only that the two are structurally related in one respect: hierarchical unfolding as an ontological direction. In other respects, each has its own specificity and the two cannot be simply exchanged.

Objections and Replies

Objection One: You repeatedly use "structural relation" as a mode of argument. But is "structural relation" too weak a concept? Any two hierarchical systems have "structural relation."

Reply: Structural relation is not present in any two hierarchical systems. Structural relation requires correspondence at the level of theses. This section has given such thesis-correspondences: hierarchical organization, movement from the basic to the concrete, and generative relation. It is not merely the vague claim that "both have levels." But your objection points to a real issue: "structural relation" is not proof, but heuristic. It allows us to gain philosophical insight from the *Yizhuan* in order to understand the EQT frequency spectrum, for example the idea that hierarchy is an ontological direction, while not presupposing that their contents are the same. This is a legitimate way to use historical resources: selective, cautious, and explicit about both isomorphism and difference.

Objection Two: The real function of the Yizhuan *is divination and Confucian ethics, not ontology. Is reading it as an ontological text a misreading?*

Reply: The *Yizhuan* has multiple functions, and ontology is one of them. Parts such as the *Xici* clearly treat ontological problems: "One yin and one yang is called Dao" is an ontological thesis, not a divinatory rule. Reading the *Yizhuan* as an ontological text is not a misreading; it is a selective reading that focuses on its ontological dimension while temporarily setting aside its divinatory dimension. This kind of selective reading also has precedents in the history of Chinese philosophy. Song-Ming Neo-Confucians, such as Zhu Xi and Wang Fuzhi, read the *Yizhuan* mainly from the perspectives of ontology and ethics, rather than from the perspective of divination. This chapter continues that tradition.

Objection Three: You place the Yizhuan, *Zhang Zai, and Wang Fuzhi on the same line, but they differ in important ways. Is this an excessive unification of the tradition?*

Reply: Yes, they differ. Zhang Zai differs from the *Yizhuan*: Zhang Zai treats qi as basic ontology more systematically. Wang Fuzhi also differs from Zhang Zai: Wang Fuzhi emphasizes temporality and historicity more strongly. This chapter does not erase these differences. But they also share a common direction: treating existence as process, treating change as a basic determination, and treating hierarchical unfolding as ontological structure. This chapter uses this common direction as the content of the "tradition of qi theory." Such selective use is legitimate: any use of a philosophical tradition is selective.

Literature That Must Be Engaged

- *Zhouyi*, especially the *Xici*, *Shuogua*, *Xugua*, and *Zagua* commentaries
- Zhu Xi, *Original Meaning of the Zhouyi*
- Wang Fuzhi, *External Commentary on the Zhouyi*, *Inner Commentary on the Zhouyi*
- Richard Wilhelm's translation, *I Ching, or Book of Changes*: the most important Western translation
- A. C. Graham, Yin-Yang and the Nature of Correlative Thinking (1986)
- Chapter 9 on frequency as a basic spectral quantity (self-reference)

Summary of This Section

This section has done three things. First, it analyzed the three ontological theses of the *Yizhuan*: one yin and one yang is called Dao, continuous generation is what is meant by change, and the Great Ultimate generates the Two Modes, the Four Images, and the Eight Trigrams. Second, it established the structural relation between the hierarchical unfolding of the *Yizhuan* and the EQT frequency spectrum. Third, it clarified the boundary of this structural relation: the two correspond in hierarchical organization as an ontological direction, but their contents, namely yin-yang and frequency, differ.

The next section is the key comparative section of this chapter: it clarifies the difference between qi theory and Whiteheadian process philosophy and argues for the advantage of the thorough monism of qi theory.

8.6. Qi Theory as a Thorough Monism: A Comparison with Whitehead

The Internal Tension in Whitehead's Scheme

Section 8.2 has already mentioned that Whiteheadian process philosophy has an internal tension: it wants to construct a monism, namely process alone, but in practice has to construct a dualism, namely actual occasions + eternal objects. This section develops that tension in detail and argues why qi theory offers a more thorough scheme.

Whitehead's scheme can be stated as follows. The basic ontological units are actual occasions: processual, temporary, concrete actualities. Each actual occasion is connected with other actual occasions through prehension. The world is a network of actual occasions, and the cosmos is their organic whole.

But Whitehead finds that actual occasions alone are insufficient to explain form and possibility. Consider one question: why do different actual occasions display similar formal structures? For example, all circles have the common property of "circularity," and all red things have the common property of "redness." If actual occasions are entirely concrete, independent, and temporary, why do they share forms?

Whitehead's response is to introduce eternal objects: they are abstract forms that "ingress" into actual occasions and give the latter form. Eternal objects are eternal, in the sense that they do not change with the arising and perishing of actual occasions; abstract, in the sense that they do not depend on any concrete actuality; and purely formal, in the sense that they are not processual.

This is the dualist structure of Whitehead's scheme: actual occasions, with processuality, temporality, and concreteness, → first ontology; eternal objects, with eternity, abstractness, and pure form, → second ontology.

The Ontological Consequences of the Tension

This dualism produces several problems.

Problem One: The non-unity of ontology. If existence is divided into two kinds, the processual and the eternal, then the question "what makes both of them exist?" must be answered. Whitehead's reply appeals to "creativity" as a more basic category, but what is creativity itself? This reply has an ad hoc character.

Problem Two: The ontological status of eternal objects is unclear. Where do eternal objects exist? Are they Platonic entities? How are they connected with actual occasions? Whitehead has replies, such as through the hierarchical structure of prehension, but each reply introduces new ontological complexity.

Problem Three: The philosophical burden. Whitehead's scheme requires us to commit to two completely different kinds of ontology. This violates ontological restraint (Chapter 2). If the same problem can be treated within one kind of ontology, the principle of restraint requires choosing the monistic scheme.

These three problems are not accidental weaknesses of Whitehead. They are common to any scheme that tries to construct a monism but introduces a second ontology to handle the problem of form. Aristotle's substance–property dualism, Descartes's substance–mode dualism, and Spinoza's three-level structure of substance–attribute–mode all have similar problems. Monistic schemes in the Western tradition often retreat into some form of pluralism because of the problem of form.

The Different Treatment in Zhang Zai's Scheme

Zhang Zai's qi theory provides a completely different treatment. For Zhang Zai, qi is the sole ontology, but qi is not "matter without form": qi has internal formal determinations.

Specifically, Zhang Zai holds that the movement of qi itself has form: the mutual transformation of yin and yang, the mutual generation and overcoming of the five phases, and the laws of gathering and dispersal. These "forms" are not externally imposed on qi; they are not Whiteheadian eternal objects ingression. They are the internal dynamic structures of qi itself.

Zhang Zai states: "That which is above form is called Dao; that which is below form is called vessel." This passage comes from the *Zhouyi · Xici*, but Zhang Zai and later qi theorists gave it a qi-theoretical interpretation. "Above form" is not another ontology: it is a different aspect of the same qi, namely li, law, and form. "Below form" is also another aspect of the same qi, namely concrete manifestation and forms of gathering and dispersal. Both are qi; they are not two kinds of ontology.

Wang Fuzhi further develops this position. He explicitly rejects interpretations in which "li precedes qi" or "li lies outside qi": li is not an eternal principle prior to qi, but the internal dynamic law of qi itself. "Li is within qi; without qi, there is no li": this is Wang Fuzhi's position.

This position makes qi theory a thorough monism: there is only one kind of ontology, qi; the forms of qi, such as gathering and dispersal, yin-yang, and the five phases, are the internal differentiation of qi itself; form, namely li, Dao, or law, is not a second ontology, but the internal dynamic structure of qi; all things, namely concrete things, are temporary forms of the gathering and dispersal of qi; and the Great Void is the state of qi before gathering and dispersal. The whole ontology is monistic: there is no "another kind of thing" as a second ontology.

Why Qi Monism Is Possible

But there is an urgent question: why can qi theory be monistic, while Whitehead cannot? How does qi theory avoid the problem of form that Whitehead encounters?

The answer lies in a key ontological choice: qi theory understands "form" as dynamic structure rather than static archetype.

In Whitehead, eternal objects are static: they do not change through time and do not participate in process. But precisely this static character is why they become a second ontology: they differ from the dynamic nature of actual occasions.

In qi theory, "li," "Dao," or "form" is dynamic: it is precisely the way qi moves. A star "is a star" not because it participates in the eternal archetype of "star-form," but because its qi moves by gathering and dispersing in a specific way, for example radiating outward, maintaining gravity, burning hydrogen, and so on. When qi moves in this way, there is the form "star"; when qi ceases to move in this way, for example when the star dies, this form no longer exists. Form is dynamic; it is not an externally added static archetype.

This choice allows qi theory to incorporate form into the internal determinations of qi, without introducing a second ontology. Form = the way qi moves. There are infinitely many ways in which qi can move, corresponding to the infinitely many kinds of things, but all these ways of moving are ways in which qi itself moves, not externally added forms.

The Corresponding Treatment in Gradient Ontology

The gradient ontology of this book adopts the same direction as Zhang Zai and Wang Fuzhi.

In gradient ontology, "things," such as physical objects, life, and consciousness, are gradient structures of the ρ_f field. The "form" of a structure, such as "it is an electron," "it is a star," or "it is a brain," is not externally imposed upon ρ_f. It is the pattern formed when ρ_f evolves according to the $A3$ Master Equation. The $A3$ Master Equation is not an "external law"; it is the internal dynamic determination of ρ_f itself. There is no "legislator" outside ρ_f that determines how ρ_f should evolve; the mode of evolution is the internal structure of ρ_f itself.

This is fully consistent with the direction of Zhang Zai and Wang Fuzhi. In both, ontology (qi / ρ_f) and form (li / the structure of the Master Equation) are not two kinds of things: they are two aspects of the same ontology. This is the meaning of thorough monism.

Restatement of the Key Thesis $K1$

The key thesis $K1$ proposed in Section 8.2 can now be stated rigorously:

Key thesis $K1$: the Chinese qi-theoretical tradition (Zhang Zai–Wang Fuzhi) and the gradient ontology of this book are both thorough monisms: they treat form as the internal dynamic determination of ontology and do not introduce a second ontology to handle the problem of form. Along this dimension, they are more thorough than Whiteheadian process philosophy.

This thesis is the key contribution of this chapter to contemporary discussion in philosophy of nature. It says that Chinese qi theory is not a kind of "Eastern wisdom" or "premodern speculation." It provides a substantive scheme for a concrete and rigorously arguable philosophical problem, namely monism vs dualism. This scheme differs from Whitehead's and is in certain dimensions superior.

But Whitehead Still Has His Contributions

It must be emphasized that this section does not overthrow Whitehead. Whitehead has his own contributions: the concept of prehension precisely characterizes the relation among actual occasions; the concept of "concrescence" precisely characterizes how actual occasions are formed; Whitehead's categorial system, his "categories of explanation," is extremely detailed and refined; and Whitehead's dialogue with modern physics, including relativity and quantum mechanics, is more detailed than that of Zhang Zai or Wang Fuzhi. These are philosophical riches. This section only points out that, along the dimension of the thoroughness of monism, the qi-theoretical tradition is more thorough than Whitehead. Along other dimensions, each has its own contribution.

Objections and Replies

Objection One: Your argument depends on a concept of "thoroughness," namely ontological restraint. But some philosophers do not accept restraint as a value. If dualism treats the problem of form better, committing to one more type of ontology may be worthwhile.

Reply: Chapter 2 has already defended restraint in detail as an ontological value. All other things being equal, fewer commitments are better. But "all other things being equal" is crucial. The claim of this section is that, on the problem of form, qi monism can treat the problem by taking form as the internal dynamic determination of ontology. If this is so, then there is no need to introduce a second ontology, and the principle of restraint requires choosing monism.

This argument depends on the claim that "qi monism can treat the problem of form." This section has given the direction, namely form as dynamic structure, but a complete proof requires more detailed work, for example a detailed argument within the EQT framework that the $A3$ Master Equation generates concrete physical structures.

Objection Two: You seem to criticize Whitehead, but Whitehead's work has its own physical context, namely early twentieth-century relativity and quantum mechanics. In his time, treating form as a static archetype was a reasonable choice.

Reply: Fully agreed. This section does not "criticize" Whitehead. It compares two philosophical schemes. Whitehead's scheme was reasonable in his own time. The qi-theoretical scheme has special relevance in the twenty-first century, because contemporary physics, especially non-equilibrium thermodynamics, complex systems, and dynamics, makes the treatment of "form as dynamic structure" natural. The two schemes each have relevance to their time.

Objection Three: Your argument assumes that "form as dynamic structure" is possible. But some philosophers will object: they will claim that certain forms, such as mathematical structures and logical forms, are necessarily static and eternal. If so, qi-theoretical monism cannot fully treat the problem of form.

Reply: This objection points to Chapter 3, Section 3.5 on the distinction between form and reality. The reply of this book is that mathematical structures and logical forms are formal constraints, not real objects. They do not belong to the category of ontological commitments; they are carriers of commitments. This distinction allows us both to acknowledge the special status of mathematical structures, since they are not physical, and not to take them as a second ontological commitment, since they are formal constraints rather than real objects. Under this distinction, qi-theoretical monism remains thorough: it says that there is only one kind of real object, qi / ρ_f, while acknowledging formal constraints as carriers of commitments. This is a more refined monism, not a simple monism.

Literature That Must Be Engaged

- Zhang Zai, *Zhengmeng* · "Great Harmony" and *Zhengmeng* · "Reference to the Two"
- Wang Fuzhi, *Commentary on Zhang Zai's Zhengmeng*: on li within qi

- Alfred North Whitehead, Process and Reality (1929), especially Part II Chapter II on eternal objects
- Donald Sherburne, A Key to Whitehead's Process and Reality (1966)
- George R. Lucas, The Rehabilitation of Whitehead (1989)
- Chapter 2 on ontological restraint (self-reference)
- Chapter 3, Section 3.5 on the distinction between form and reality (self-reference)

Summary of This Section

This section has done three things. First, it developed the internal tension of Whiteheadian process philosophy: it requires eternal objects as a second ontology to handle the problem of form. Second, it developed the different treatment in Zhang Zai–Wang Fuzhi qi theory: form is treated as the internal dynamic determination of qi, without introducing a second ontology. Third, it rigorously stated the key thesis $K1$: qi theory is more thorough than Whitehead in the thoroughness of monism, because it understands form as dynamic structure rather than static archetype.

8.7. The Dialogue between Qi Theory and "Boundary as Gate": Form, Continuous Generation, and the Membrane of Life

The Work of This Section

Sections 8.1–8.4 developed the structural correspondence between the qi-theoretical tradition and gradient ontology: difference, gathering and dispersal, continuous generation, and the frequency spectrum. Section 8.5 developed an external comparison between qi theory and Western process philosophy, especially Whitehead. This section develops a third kind of dialogue: an internal dialogue between the qi-theoretical tradition and the latest argument of this book.

Why is this section needed? Chapter 13 of this book, following the core argument of the new version of *Mind–Matter Unity*, introduces a key philosophical thesis: boundary as gate. Given-ness presupposes an "inside"; the "inside" presupposes a topological boundary; and the boundary is produced by the physical event of the membrane of life. This thesis gives this book a clear anti-panpsychist position in the genealogy of philosophy of consciousness while also avoiding the difficulties of emergentism. How does the qi-theoretical tradition respond to this thesis? What resources does the qi-theoretical tradition have for dialogue with it? Where can the qi-theoretical tradition enrich the work of this book, and where can this book enrich the qi-theoretical tradition? These questions constitute the content of this section.

Four Relevant Resources in the Qi-Theoretical Tradition

There are four resources in the qi-theoretical tradition that are directly relevant to "boundary as gate." They are dispersed across several classical texts and constitute the starting point of the dialogue in this section.

Resource One: the distinction between above-form and below-form in the Yizhuan. *The* Yizhuan · Xici shang *says: "That which is above form is called Dao; that which is below form is called vessel." This formulation involves "form" as a key ontological node: it is both the boundary between Dao and vessel and the mechanism through which formless qi differentiates into formed things. In the qi-theoretical tradition, qi as foundational ontology is formless, flowing, and continuous; all things as formed vessels are condensed forms of qi. "Form" functions as a gate between the two: it allows qi to transform from a diffuse continuous state into identifiable beings.*

This resource deeply resonates with the direction of this book's "boundary as gate": both identify an ontological node. Above a certain diffuse continuity, some boundary mechanism, form / enclosing boundary, is required in order to produce identifiable beings.

Resource Two: Zhang Zai's "form–qi–nature" sequence. In the Zhengmeng, *Zhang Zai develops the mutual relation among qi, form, and nature. Qi as foundational ontology, "the Great Void is qi," forms the shape of all things through gathering and dispersal, "qi gathers and becomes all things; all things disperse and become qi." Form, as the condensed form of qi, in turn determines the properties of beings: "only after there is form is there the qi-constituted nature." Concrete form gives rise to concrete nature.*

This sequence is structurally connected with the anti-panpsychist position of this book. Zhang Zai's "only after there is form is there the qi-constituted nature" suggests that concrete properties are not diffused throughout all qi, but depend on the emergence of concrete form. The statement of this book that "interiority is not a background diffusely present in all things, but an achievement with a definite physical origin" corresponds structurally to this: both connect the emergence of concrete properties with the emergence of concrete boundary forms.

Resource Three: Wang Fuzhi's "being born and formed day by day." In works such as the External Commentary on the Zhouyi *and* Silu, *Wang Fuzhi develops the theses that "qi is born and formed day by day" and "form is born and formed day by day." These theses emphasize that beings do not remain unchanged after being created once; rather, they are continuously regenerated at every moment. Form is not a solid boundary, but a dynamically maintained process.*

This resource has important enrichment value for this book's "boundary as gate": it indicates that boundary is not a once-and-for-all event of formation, but a dynamically maintained ongoing process. The membrane of life is not produced at one instant and then remains static; it is continuously maintained at every moment by the metabolic processes of life. This insight gives this book's "boundary as gate" a dynamic expression.

Resource Four: the Yizhuan*'s principle of "continuous generation." The* Yizhuan · Xici shang *says: "Continuous generation is what is meant by change." This thesis takes the continuous generation of life as a basic principle of the cosmos. "Continuous generation" is not an event of occurrence at a particular instant, but the ongoing generative process of the cosmos as a whole.*

This resource deeply resonates with the life orientation of this book's "boundary as gate." This book takes the formation of the membrane of life as the starting point of interiority, while the qi-theoretical tradition takes "continuous generation" as a basic principle of the cosmos. The two structurally resonate: both treat life as a key ontological node, rather than as an accidental emergent phenomenon.

How "Boundary as Gate" Enriches the Qi-Theoretical Tradition

The qi-theoretical tradition has the resources above, but it did not systematize them into a sharp anti-panpsychist criterion. The work of this book enriches the qi-theoretical tradition in three layers.

First, the physicalization of "form." "Form" in the qi-theoretical tradition is a philosophical concept and lacks a physical correlate. It says that all things have form and that form depends on qi. But what concretely is form? What counts as the physical realization of "form"? Traditional qi theory does not answer this question precisely. This book's "boundary as gate" provides a physicalization: the physical correlate of form is an enclosing boundary, namely the topological closure of the membrane of life. It distinguishes the diffuse continuity of qi-transformation into living systems that "have an inside" and non-living systems that "have no inside."

Second, a sharp anti-panpsychist criterion. The qi-theoretical tradition has a continuity tendency in its view of the relation between Heaven and human beings: "Heaven and human beings form one body." It tends to see continuity rather than rupture among all things. This tendency makes it difficult for the qi-theoretical tradition to treat the boundary between life and non-life as a key ontological node. This book's "boundary as gate" provides a sharp criterion: where there is a membrane, there is an inside; where there is no membrane, there is no inside. This criterion is nearly empirically determinable. This sharpness is a direction not sufficiently developed by the qi-theoretical tradition from within its own resources.

Third, the clarification of a middle path between anti-panpsychism and anti-emergentism. In some interpretations, the qi-theoretical tradition is read as close to panpsychism: since all things are qi and qi has capacities such as spirit and efficacy, all things have some form of interiority. But this interpretation weakens the role of "form" as a key node in the qi-theoretical tradition. This book's "boundary as gate" makes the anti-panpsychist position explicit: interiority is not a universal property of qi, but the achievement of a specific form, namely an enclosing boundary. At the same time, it does not slide into emergentism, namely emergence from nothing, because interiority has its physical precondition, namely the inside produced by enclosure. This middle position is coordinated with the direction in the qi-theoretical tradition that "only after there is form is there the qi-constituted nature," but it was not developed as a systematic thesis within the tradition.

How the Qi-Theoretical Tradition Enriches "Boundary as Gate"

The qi-theoretical tradition does not merely receive enrichment from this book; it also enriches the work of this book in return.

First, the dynamization of boundary. Wang Fuzhi's "form is born and formed day by day" gives this book's "boundary as gate" a dynamic expression. Boundary is not a static topological fact, but a continuous event dynamically maintained by the process of life. This dynamization allows the work of this book to avoid a possible misreading: interpreting "enclosure" as a once-and-for-all event of formation. The feature of life lies precisely in this continuous maintenance: at every moment, it resists entropy increase and maintains the topological integrity of its membrane. The qi-theoretical tradition's "being born and formed day by day" gives this aspect of the work of this book cross-traditional support.

Second, the ontological position of life orientation. The qi- theoretical tradition takes "continuous generation" as a basic principle of the cosmos, giving life a key ontological position. This orientation supports the tendency of the work of this book: treating the emergence of life as a key ontological node rather than as an accidental phenomenon. This support gives the anti-emergentist position of this book a resonance in cross-traditional resources: the formation of the membrane of life is not a mysterious event of emergence from nothing, but the physical realization of the cosmic principle of continuous generation.

Third, a monistic metaphysical background. The monism of the qi-theoretical tradition, in which all things are forms of qi, allows "boundary as gate" to be developed within a monistic background. Boundary does not divide beings into two kinds of ontology, such as mind and matter, life and death, or spirit and thing. It marks different modes of existence within the same process of qi-transformation. This monistic background is deeply coordinated with the process ontology of this book, in which all things are forms of the ρ_f gradient process.

The Qi-Theoretical Ground of Co-Evolution as an Ontological Relation

The dialogue developed in this section also involves the qi-theoretical ground of Chapter 12, "Co-Evolution as an Ontological Relation." The concept of "gan" in the qi-theoretical tradition, namely resonance, responsiveness, and mutual affective communication among beings, structurally corresponds to the internal–external co-evolution of this book.

The *Yizhuan · Xian Hexagram · Tuan Commentary* says: "The two qi resonate and respond to one another, and thereby join with one another." This formulation takes "gan" as the ontological ground of relation between two. "Gan" is not a cognitive act of some subject toward some object, but the mutual dynamics between two forms within the same process of qi-transformation. This direction is structurally isomorphic with internal–external co-evolution in Section 13.3, namely given-ness as relational rather than as an intrinsic property.

The qi-theoretical tradition therefore provides cross-cultural resources for the relational ontology of this book (Chapter 12, $EV1$): co-evolution as an ontological relation is not an isolated creation of this book, but a contemporary continuation and refinement of the concept of "gan" in the qi-theoretical tradition. This cross-traditional connection gives the relational ontology of this book a deeper philosophical ground.

A Brief Supplement on Wang Yangming's Philosophy of Mind

Outside the qi-theoretical tradition, Wang Yangming's philosophy of mind also has resources relevant to "boundary as gate." The famous "observing flowers in the mountain" passage in Wang Yangming's *Instructions for Practical Living*, where the color and meaning of the flower appear with the lucid awareness of mind, resonates structurally with internal–external co-evolution: the given-ness of the flower is not an intrinsic property of the flower alone, but a joint event between mind and flower.

Strictly speaking, however, Wang Yangming's philosophy of mind belongs to the school of mind in Song-Ming Neo-Confucianism, not to the qi-theoretical tradition. Its dialogue with the work of this book requires dedicated treatment and lies beyond the scope of this chapter. Chapter 13's methodological reflection has marked this as later work. This section only points out that potential resources for dialogue with "boundary as gate" exist in the tradition of the philosophy of mind, leaving their development for later work.

Its Position in the Overall Architecture of This Chapter

As the sixth section of this chapter, this section performs a special function: it is the bridge between the earlier arguments of this chapter, Sections 8.1–8.4 on structural correspondence, and the latest argument of this book, namely boundary as gate and internal–external co-evolution.

Sections 8.1–8.4 developed the structural correspondence between the qi-theoretical tradition and gradient ontology: difference, gathering and dispersal, continuous generation, and frequency spectrum. Section 8.5 developed the external comparison between qi theory and Whitehead. The present section develops the mutual enrichment between the qi-theoretical tradition and the latest philosophical contribution of this book. This means that the chapter not only develops the cross-cultural correspondence between the qi-theoretical tradition and the earlier arguments of this book, but also develops the resonance and enrichment of the qi-theoretical tradition for the latest philosophical contribution of this book. This makes the positioning of the qi-theoretical tradition as part of world philosophy, which Section 8.8 will summarize, more concrete and more thorough: it is not only a cross-cultural correspondence to earlier arguments, but also a cross-cultural interlocutor for contemporary philosophical contributions.

The next section, 8.7, "Boundary and Honesty: Qi Theory Is Not EQT," will continue to develop the legitimate boundary of dialogue between qi theory and EQT. It treats the methodological limits of cross-cultural dialogue. This section and the next section complement each other: this section develops the content of dialogue, namely qi-theoretical resources + the two-way enrichment of boundary as gate; the next section develops the limits of dialogue. Together, the two sections allow the cross-traditional dialogue of this chapter to have both rich content and clear boundaries.

The next section is the boundary clarification of this chapter: it makes clear that qi theory and gradient ontology should not be confused, and that the two have their own differences and boundaries.

8.8. Boundary and Honesty: Qi Theory Is Not EQT

The Necessity of Boundary

The preceding sections established the structural isomorphism and shared contributions between the qi-theoretical tradition and gradient ontology, as well as, in Section 8.6, the two-way dialogue between qi theory and this book's "boundary as gate." But this section must explicitly draw the boundary: qi theory is not an ancient version of EQT, and EQT is not a modern version of qi theory. The two differ. Confusing them is unjust both to qi theory and to EQT. This boundary clarification does not weaken the arguments of the preceding sections. It strengthens them: only by knowing clearly how the two differ can we precisely grasp their shared contribution.

Five Differences

Difference One: Conceptual Content

Qi has its own content within the Chinese philosophical tradition, including its relations with yin-yang, the five phases, nature and life, rites and music. These contents cannot be simply equated with the content of EQT. $\rho_f(x,t)$ is a frequency-resolved energy quantum density field: its content is determined by the postulates and mathematical structure of EQT. These contents do not appear in qi theory. It is legitimate to place qi and ρ_f in structural correspondence, as Section 8.2 does, and to place "form" and this book's "boundary as gate" in structural correspondence, as Section 8.6 does. It is wrong to equate them at the level of content. Qi is a concept of eleventh-century philosophy; ρ_f is a concept of twenty-first-century physics. They are structurally related but different in content.

Difference Two: The Degree of Methodological Rigor

Qi theory as a philosophical direction has its own methodology. It uses resources such as texts, including the *Book of Changes* and the *Zhengmeng*; ethical practice, especially Confucian cultivation; and cosmological images, such as yin-yang and the five phases. These resources have legitimacy within their philosophical tradition, but they do not satisfy the four disciplines of constrained speculation, especially observationalization and failure-orientation. EQT as a physical theory uses all the disciplines of constrained speculation: formalization, mechanization, observationalization, and failure-orientation. It has falsifiable predictions, failure conditions, and interfaces with observation. This difference is a difference of historical period: the eleventh century did not have physical methodology in the modern sense. This is not a weakness of Zhang Zai; it is the context of his time. But acknowledging this difference is a necessary precondition for the honest use of qi theory as a resource.

Difference Three: The Specificity of Cosmology

Qi theory gives the direction of a cosmological view: the gathering and dispersal of qi unfold into all things, and all things eventually disperse into the Great Void. But qi theory has no concrete cosmological model: it does not predict the CMB, does not characterize structure formation, and does not treat dark matter or dark energy. EQT, especially *Freezing Cosmology*, has a concrete cosmological model and connects with observation. This difference is also a difference of historical period. It means that qi theory provides an ontological direction, while concrete cosmological work must be completed by modern physics. The two undertake different tasks.

Difference Four: The Treatment of the Mind–Matter Relation

The qi-theoretical tradition has its own language for treating the mind–matter relation: mind, nature, emotion, intention, knowledge, and related terms. These concepts were developed in detail in Neo-

Confucian work, including Cheng Yi, Zhu Xi, Wang Yangming, and others. Their content differs from the content of modern consciousness research. EQT's treatment of mind and matter, in *Mind–Matter Unity* and Chapter 13 of this book, uses physical variables, such as $\tilde{A}$, C, and $\mathcal{B}$, and clinical proxies, such as EEG and fMRI. Both treat the mind–matter relation, but in different languages and with different methods. They can illuminate each other, as Chapter 13 of this book can gain direction from Confucian cultivation theory, and the dialogue between the qi-theoretical concept of "form" and "boundary as gate" developed in Section 8.6 also belongs to this mutual illumination. But they cannot be simply exchanged.

Difference Five: The Relation with Ethics

The qi-theoretical tradition is tightly connected with Confucian ethics: the gathering and dispersal of qi has an "ought" dimension, in which benevolence, ritual propriety, wisdom, trustworthiness, and related virtues are proper ways in which qi unfolds. This ethical dimension is an internal part of qi theory. EQT as a physical theory does not directly contain ethics. EQT philosophy of nature, especially Part V of this book, extends into ethics, but it does so in a way different from Confucianism: through the method of constrained speculation, it places ethical statements under the same methodological discipline as factual statements (Chapter 20). Both treat ethics, but the way ethics is connected with ontology differs. In qi theory, the two are tightly connected and even difficult to separate. In EQT philosophy of nature, they are connected through the method of constrained speculation, but remain distinct.

How to Treat These Differences

Acknowledging these differences does not weaken the value of qi theory as a philosophical resource. It makes the use of that resource more precise.

Specifically, this chapter uses qi theory selectively: it uses the ontological direction of qi theory, namely processuality, monism, and dynamic form; it uses resources in the qi-theoretical tradition that directly dialogue with this book's latest argument, such as "form," "being born and formed day by day," and "continuous generation," developed in Section 8.6; it does not use the ethical-cosmological imagery of qi theory, such as the concrete structures of yin-yang, five phases, and the Eight Trigrams; it uses the insights of Zhang Zai and Wang Fuzhi, such as gathering and dispersal and daily renewal; it does not use elements of the qi-theoretical tradition that are incompatible with modern scientific methodology, such as divination and metaphysical explanations. Such selective use is legitimate: it uses qi theory as a philosophical resource rather than as a complete system. Any contemporary use of a tradition is selective; the legitimacy of selective use lies in clearly stating the mode and reasons of selection.

The Way Chinese Philosophy Enters World Philosophy

This chapter's treatment of qi theory demonstrates one way for Chinese philosophy to enter world philosophy. This way has several features:

Feature One: treating the Chinese tradition as a philosophical contribution. It is not merely a cultural product, historical heritage, or Eastern wisdom, but a philosophical contribution. As a philosophical contribution, it requires rigorous philosophical method, dialogue with Western traditions, and acceptance of the methodological standards of international philosophy.

Feature Two: selective rather than total use. It does not pretend that the tradition as a whole is contemporarily relevant. It clearly selects which aspects are relevant and which aspects have limits. This makes the use honest, defensible, and open to criticism.

Feature Three: mutual illumination rather than one-way application. It is not only a matter of "using the Chinese tradition to illuminate contemporary problems," but also of "using contemporary problems to illuminate the Chinese tradition." Contemporary physics allows us to understand Zhang Zai more deeply, as in this section's argument about dynamic form. The Chinese tradition allows us to understand contemporary problems more deeply, such as the critique of Whiteheadian dualism and the cross-cultural enrichment of the anti-panpsychist position of "boundary as gate," as discussed in Section 8.6. Mutual illumination allows both sides to develop.

Feature Four: strict boundaries. It does not pretend that the Chinese tradition already contains all answers. It makes clear the limits of the tradition and the contributions of contemporary work. This rigor allows the Chinese tradition to be taken seriously by the international philosophical community: not as exotic color, but as a philosophical partner.

The work of this chapter can be located as an attempt to treat Chinese philosophy as part of world philosophy. Its success or failure will be judged by readers, but the attempt itself is necessary: Chinese philosophy needs this kind of work in order to gain a place in international philosophical discussion.

Objections and Replies

Objection One: The five differences you list seem too many. Does the difference between qi theory and EQT really need such detailed treatment?

Reply: Yes, it does. The preceding sections established structural isomorphism, including the two-way dialogue between "form" and "boundary as gate" developed in Section 8.6. If the boundaries are not drawn in detail, readers may easily exaggerate isomorphism into identity. Each difference corresponds to a particular isomorphism claim in the preceding sections: it says that "the two are isomorphic in structure, but differ in this respect." This precision is required by philosophical work.

Objection Two: Your posture of "selective use" seems opportunistic: you select what is useful and abandon what is not. Is this unjust to the tradition?

Reply: Selective use is not opportunism. It is the legitimate way to use any historical resource. When we use Aristotle, we are also selective: we accept his formal logic, but not his teleological biology. When we use Kant, we are also selective: we accept his critical method, but not his racial views. When we use Whitehead, we are also selective: we accept his processual position, but question his eternal objects. Selective use is not unjust to a tradition; it is the living use of a tradition. Accepting a tradition as a whole is a museum-like posture, which makes the tradition dead. Selective use is the philosopher's posture, which allows a tradition to continue working in contemporary discussion.

Objection Three: You seem to want this chapter to carry too much work: establishing structural isomorphism, arguing for the advantage of thorough monism, entering into two-way dialogue with this book's latest argument, namely boundary as gate, drawing boundaries, and demonstrating a way for Chinese philosophy to enter world philosophy. Is this overloaded?

Reply: This chapter should treat all these tasks. None can be omitted. Establishing structural isomorphism, arguing for thorough monism, entering into dialogue with this book's latest argument, drawing boundaries, and demonstrating a way into world philosophy together constitute the complete work of this chapter. Without any one of them, the chapter would be incomplete. But the concern is acknowledged: the chapter undertakes much work. This reflects its special position. It is the chapter in which the distinctive advantage of a Chinese scholar writing this book must integrate work at several levels. This chapter accepts that challenge.

Literature That Must Be Engaged

- Mou Zongsan, *Nineteen Lectures on Chinese Philosophy*
- Tang Junyi, *Original Discourses on Chinese Philosophy*, six volumes
- Tu Wei-ming, Confucian Thought: Selfhood as Creative Transformation (1985)
- Roger Ames & David Hall, Thinking from the Han: Self, Truth, and Transcendence in Chinese and Western Culture (1998)
- Robert Cummings Neville, Boston Confucianism (2000)
- Chapter 2 on ontological restraint (self-reference)
- Chapter 3, Section 3.5 on the distinction between form and reality (self-reference)
- Chapter 13 on the anti-panpsychist position of boundary as gate (self-reference)

Summary of This Section

This section has done three things. First, it listed five differences between qi theory and gradient ontology, in order to avoid confusing the two. Second, it stated selective use as a legitimate way of using resources: which aspects are used, which aspects are not used, and why. Third, it treated this chapter as a way for Chinese philosophy to enter world philosophy: through rigorous philosophical work rather than cultural declaration.

The next section is the conclusion of this chapter. It synthesizes the work of the chapter and connects it with the overall positioning of the whole book.

8.9. Chapter Summary: Chinese Qi Theory as Part of World Philosophy

The four tasks undertaken by this chapter have all been completed.

Task One: to present the Chinese tradition of qi theory seriously as a substantive contribution to contemporary philosophy of nature. Section 8.1 established the process orientation of the Chinese philosophical tradition as the fundamental starting point of this chapter. Section 8.2 developed the structural isomorphism between Zhang Zai's qi theory and gradient ontology. Section 8.3 developed the relation between Wang Fuzhi's position of "daily renewal" and the freezing spectrum. Section 8.4 developed the structural relation between the *Yizhuan*'s "continuous generation" and the frequency spectrum. The entire work is not decorative citation, but serious dialogue at the level of philosophical theses.

Task Two: to argue for the key thesis $K1$: in certain respects, the Chinese tradition of qi theory is more thorough than Whiteheadian process philosophy. Section 8.5 gave this argument. Qi theory treats form as the internal dynamic determination of ontology, namely li within qi, and does not need to introduce eternal objects as a second ontology. This allows qi theory to preserve the thoroughness of monism, whereas Whitehead's scheme has to be dualistic.

Task Three: to develop the two-way dialogue between the qi-theoretical tradition and this book's latest argument, namely boundary as gate. Section 8.6 developed this dialogue. Resources in qi theory, such as the distinction between above-form and below-form, the "form–qi–nature" sequence, "being born and formed day by day," and "continuous generation is what is meant by change," provide cross-cultural resonance and enrichment for this book's anti-panpsychist criterion. Conversely, this book's "boundary as gate" gives a sharp physicalized criterion to an anti-panpsychist direction that remained insufficiently developed in the qi-theoretical tradition. This two-way dialogue makes qi theory not only a cross-cultural interlocutor for the earlier arguments of this book, namely structural isomorphism, but also a mutual enricher of this book's latest philosophical contribution.

Task Four: to clarify the boundary between qi theory and gradient ontology. Section 8.7 listed five differences and avoided confusing the two. At the same time, it presented selective use as a legitimate way of using resources and treated this chapter as a demonstration of how Chinese philosophy can enter world philosophy.

The Position of This Chapter in the Whole Book

This chapter has a clear position in the whole book.

Looking backward, this chapter continues Chapter 7: that chapter established the Western philosophical genealogy of gradient ontology, namely difference before identity; this chapter establishes its Chinese philosophical genealogy, namely the tradition of qi theory. Together, the two chapters constitute the dual philosophical defense of gradient ontology.

Looking forward, this chapter provides resources for later chapters. Chapter 11 on freezing as a foundational ontological process can be deepened through Wang Fuzhi's "daily renewal." Chapter 13 on the anti-panpsychist position of boundary as gate received cross-cultural enrichment from the qi-theoretical tradition in Section 8.6; conversely, the philosophical contribution of this book sharpens the relevant resources in the qi-theoretical tradition. Chapter 17 on ontological singularity can enter into dialogue with the qi-theoretical concept of the "Great Void." Chapter 21 on the finitude of meaning can be developed through the Confucian idea of "knowing that it cannot be done and yet doing it." The relation of dialogue partner established in this chapter will continue to unfold in later chapters.

The Core Posture of This Chapter

The core posture of this chapter is this: Chinese philosophy as part of world philosophy.

This posture is not a claim of cultural superiority. It is the legitimacy of resource use. Any contemporary philosophical work uses historical resources: the work of Aristotle, Kant, Hume, and Frege is used in international philosophical discussion. Chinese philosophical traditions, such as qi theory, Confucianism, and Buddhism, should be used by international philosophical discussion as philosophical resources of the same level. This is what "as part of world philosophy" means.

Chinese scholars occupy a distinctive position in this process: they have deeper familiarity with Chinese traditional resources and can bring them into international discussion. But this is not cultural monopoly. The work of Western scholars who use the Chinese tradition, such as Roger Ames, David Hall, Tu Wei-ming, and others, is equally legitimate, and in certain respects, such as English expression and connection with Western traditions, may even be more effective. This chapter, as the work of a Chinese scholar, undertakes the task of bringing Zhang Zai, Wang Fuzhi, and the *Yizhuan* into international discussion as substantive contributions to contemporary philosophy of nature. Whether this task is completed successfully will be judged by readers. But the task itself is necessary, regardless of who completes it.

A Final Reflection

This chapter has a distinctive difficulty in writing: it must both strictly follow the methodological standards of international philosophy and treat Chinese traditional resources deeply and seriously. This dual requirement can easily push the chapter into one of two problems: excessive academicization, which may make Chinese scholars feel that the content of the tradition is ignored, or excessive nostalgia, which may make international scholars feel that philosophical rigor is lacking.

The strategy of this chapter is to maintain rigor through philosophical theses, including theses of structural isomorphism, thorough monism, two-way dialogue with boundary as gate, and boundary demarcation; and to maintain seriousness toward the tradition through detailed citation of texts, such as the *Zhengmeng*, *Silu*, and *Yizhuan*, and through dialogues, such as the comparison with Whitehead and the dialogue with this book's "boundary as gate." Whether this strategy succeeds will be judged by readers. Whatever the outcome, this chapter demonstrates a possibility: Chinese scholars can write genuine philosophy while making the Chinese tradition genuinely function as a philosophical resource. Realizing this possibility requires more work. This chapter is only one attempt. But the attempt itself has value, regardless of its success or failure.

Meaning before freezing: later chapters of this book will develop this position. This chapter has already made its own contribution: it has used the Chinese tradition of qi theory seriously as a local philosophical resource for this position. The thorough monism of qi theory, Wang Fuzhi's daily renewal, the hierarchical unfolding of the *Yizhuan*, and the correspondence between "form" and boundary will all continue to be used in later chapters.

8.10. Methodological Reflection on This Chapter

This chapter presents the Chinese tradition of qi theory seriously as a substantive contribution to contemporary philosophy of nature. This section reflects on the methodological significance of this cross-cultural work.

The Cross-Cultural Posture of This Chapter

The use of the Chinese tradition in this chapter involves a methodological challenge: maintaining a balance between selective use and fair treatment of the tradition. Selective use may be accused of being unjust to the wholeness of the tradition; fair treatment of the wholeness of the tradition may make the chapter lose philosophical focus. This chapter responds to this challenge through the explicit boundary discussion in Section 8.7: it clearly states which aspects are selected, which aspects are not used, and why. This transparency makes selective use honest rather than opportunistic. This chapter avoids the rhetorical trap that "Eastern thought already anticipated modern science": it claims structural correspondence, not anachronism or identity of content.

This Chapter's Refinement of Qi Theory

This chapter argues that qi theory is in certain respects more thorough than Whiteheadian process philosophy. This is a philosophical thesis, not a cultural superiority claim. Whitehead's work faces an internal tension: actual occasions as foundational ontology vs eternal objects as a second ontology. The qi-theoretical tradition avoids this difficulty: it is purely monistic, and all content is built upon the dynamic structure of qi. This refinement gives qi theory an independent place as an ontological tradition in contemporary discussion. It is not simply an "Eastern version of process philosophy," but an independent tradition with an advantage in ontological structure.

The Two-Way Dialogue between This Chapter and This Book's Latest Argument

Section 8.6 develops an important extension of the work of this chapter: the two-way dialogue between the qi-theoretical tradition and this book's latest philosophical contribution, namely the anti-panpsychist position of boundary as gate. This extension makes the chapter not only a cross-cultural dialogue chapter for the earlier arguments of this book, such as structural isomorphism and thorough monism, but also a chapter of mutual enrichment for this book's latest philosophical thesis. Specifically, resources in qi theory such as the distinction between above-form and below-form, the "form–qi–nature" sequence, "being born and formed day by day," and "continuous generation is what is meant by change" provide cross-cultural resonance for this book's anti-panpsychist criterion. Conversely, this book's "boundary as gate" gives a sharp physicalized criterion to an anti-panpsychist direction that remained insufficiently developed in the qi-theoretical tradition, namely the structural implication of "only after there is form is there the qi-constituted nature." This kind of two-way dialogue is a legitimate form of cross-traditional philosophical work:

traditional resources are not only used by contemporary work, but are also enriched and sharpened by contemporary work.

The Bilingual Use of Chinese and English in This Chapter

This chapter uses Chinese texts, such as the *Zhengmeng*, *Silu*, and *Yizhuan*, and English-language scholarship, such as Whitehead, Bergson, and Deleuze. This bilingual use is necessary. Texts on qi theory must be based on the original Chinese, since any translation loses part of the original meaning: for example, qi as a concept cannot be simply reduced to "energy" or "vital force." Dialogue with international philosophical discussion must use English-language scholarship as interlocutors. This bilingual use is a distinctive capacity of a Chinese scholar writing this book, and also a responsibility.

Limits Honestly Acknowledged in This Chapter

This chapter acknowledges several limits. First, it can only treat the core direction of the internal diversity of the qi-theoretical tradition. Second, it does not sufficiently develop the relation between the qi-theoretical tradition and other Chinese traditions, such as Huayan, Confucian practical wisdom, and the philosophy of mind. These relations are treated further in Chapters 12, 13, and 19. The dialogue between Wang Yangming's philosophy of mind and "boundary as gate" has already been briefly mentioned in Section 8.6 and is left for later development. Third, the use of the qi-theoretical tradition as a resource for modern philosophical discussion involves hermeneutic problems: the distance between classical texts and contemporary discussion needs to be treated seriously. Fourth, this chapter's treatment of Whitehead's work is relatively brief.

Connections with Later Chapters

The qi-theoretical resources established in this chapter are used further in later chapters. Chapter 11 uses the resource of Wang Fuzhi's "being born and formed day by day." Chapter 12 uses the resource of the Huayan doctrine of "mutual non-obstruction of all phenomena." Chapter 13's anti-panpsychist position of boundary as gate has already received cross-cultural resonance and enrichment from the qi-theoretical tradition in Section 8.6 of this chapter. This two-way dialogue means that the argument of Chapter 13 is not an isolated creation of this book, but a contemporary synthesis in dialogue with Chinese traditional resources. Chapter 19 uses resources from Nāgārjuna. At the same time, the four response frameworks of Chapter 19 also receive a template for cross-traditional dialogue through this chapter's treatment of qi theory as world philosophy.

Chapter 20 uses resources from Confucian methodological rigor. More broadly, this chapter is a key marker of this book as an international philosophical monograph written by a Chinese scholar. Without this chapter, the book might be misread as "Western philosophy written by a Chinese scholar." This chapter makes the book into a work in which a Chinese scholar brings the Chinese tradition into international discussion as a philosophical resource.

9. Frequency as a Basic Spectral Quantity

9.1. Introduction to This Chapter

On December 14, 1900, Max Planck presented his work on black-body radiation at the German Physical Society in Berlin and introduced a seemingly simple formula: $E = h\nu$. Here E is energy, ν is frequency, and h is Planck's constant. On the surface, this formula is an empirical relation concerning black-body radiation: it gives the form of energy quantization. But its ontological meaning goes far beyond the problem of black-body radiation.

In the development of twentieth-century physics, $E = h\nu$ became a foundation of quantum mechanics, quantum field theory, and particle physics. Later twentieth-century physics, including de Broglie's matter waves, the Schrödinger equation, Heisenberg's uncertainty principle, and quantum field theory, was built upon $E = h\nu$. The frequency–energy equivalence became central to modern physics.

Yet the ontological meaning of this formula has been relatively underdeveloped in philosophical discussion. Physicists mainly treat $E = h\nu$ as a computational tool; philosophers mainly treat it as an example of scientific discovery. But $E = h\nu$ involves an ontological thesis: frequency is an ontological dimension of beings, not merely a derivative quantity of time.

This chapter elevates $E = h\nu$ from Planck's physical discovery of 1900 into an ontological thesis: frequency is not merely one physical quantity among others; frequency is an ontological dimension of beings (FR1). Specifically, beings are classified according to a frequency spectrum, and different frequency bands correspond to different ontological structures of beings. This chapter argues for the distinctive status of frequency as an ontological dimension: it cannot be reduced to other basic dimensions such as time, space, or energy, but is an internal structural dimension of beings.

At the same time, this chapter enters into serious dialogue with discussions of "basic quantities" in the Western philosophical tradition, including Aristotle's categories, Kant's pure forms of intuition, and Whitehead's actual occasions, in order to clarify the philosophical position of frequency as a basic spectral quantity.

This chapter builds upon Chapter 7, "Difference before Identity": frequency is an internal dimension of structures of difference. It also builds upon the resources of Chapter 8, "The Tradition of Qi Theory": the spectral differentiation of qi is an expression of a similar position in the Chinese tradition. The work of this chapter is to transform the preparations made in those two chapters into a systematic ontology of frequency.

This chapter does not depend on the physical derivation of EQT. All arguments are completed through purely philosophical argumentation, and EQT appears only as an application example.

The discussion proceeds as follows. Section 9.1 develops the ontological elevation of $E = h\nu$; Section 9.2 develops the independence of frequency as an ontological dimension; Section 9.3 develops frequency as the ontological map of beings; Section 9.4 develops the dialogue between frequency ontology and traditional discussions of categories; Section 9.5 concludes the chapter and develops open problems in frequency ontology.

9.2. The Ontological Elevation of $E = h\nu$

The Starting Point: Planck's 1900 Node

On December 14, 1900, Max Planck presented his work on black-body radiation at the German Physical Society in Berlin. Its core content was the introduction of a constant, later called Planck's constant h, and the proposal of a seemingly simple formula:

$$E = h\nu$$

Here E is energy, ν is frequency, and h is Planck's constant.

On the surface, this formula is an empirical relation concerning black-body radiation: it gives the form of energy quantization. But its ontological meaning goes far beyond the problem of black-body radiation. In the development of twentieth-century physics, $E = h\nu$ became a foundation of quantum mechanics, quantum field theory, and particle physics.

Planck himself remained cautious about the ontological meaning of this formula: he treated "energy quantization" as a "mathematical technique," not as the content of physical reality. But in his 1905 work on the photoelectric effect, Einstein explicitly interpreted $E = h\nu$ as expressing the particle character of light, namely the photon: the energy of a photon is determined by its frequency and is part of physical reality.

Bohr applied this formula to atomic structure in 1913: the differences between atomic energy levels correspond to the frequencies of radiation. This application opened the door to quantum mechanics.

Later twentieth-century physics, including de Broglie's matter waves, the Schrödinger equation, Heisenberg's uncertainty principle, and quantum field theory, was built upon $E = h\nu$. The frequency–energy equivalence became central to modern physics.

The Ontological Meaning of $E = h\nu$

Physicists use $E = h\nu$ as a computational tool: given a frequency, one can calculate energy; given energy, one can calculate frequency. This application is beyond dispute.

But what does $E = h\nu$ mean as an ontological thesis? This question has rarely been treated seriously in philosophical discussion. Common treatments include: treating it as an empirical relation, expressing a measurable correspondence between frequency and energy, but thereby avoiding the ontological problem; treating it as a convention, choosing h as a conversion constant, but thereby ignoring that h is a natural constant rather than an arbitrary convention; treating it as a derivative relation, derived from more basic physics such as quantum field theory, but thereby instrumentalizing $E = h\nu$ without investigating its ontological ground.

The aim of this chapter is to treat seriously the ontological meaning of $E = h\nu$: what does it tell us about the internal structure of beings?

Statement of the Core Thesis

Core thesis FR1: $E = h\nu$ is not a statement about an "empirical correspondence between energy and frequency"; it is a thesis about the ontological structure of beings. It states that energy, as the dynamic content of a being, is determined by frequency as an internal structural dimension of that being. Frequency is an internal dimension of a being as an ontological unit, standing alongside time, space, energy, and other basic dimensions.

This thesis requires detailed development. Let us analyze several of its key implications.

First, "frequency as an ontological structural dimension": traditional physics treats frequency as a derivative quantity of time, namely the number of oscillations per unit time. This chapter argues that frequency has an independent ontological status: it is not merely derivative from time, but an internal structural dimension of beings. Specifically: every being has frequency, or a frequency spectrum, as part of its internal structure; this frequency is not externally added, but part of the ontological content of the being itself; and differences among beings can be expressed at the level of frequency.

Second, "energy is determined by frequency": $E = h\nu$ does not mean that "two independent quantities, energy and frequency, correspond to each other." It means that energy, as the dynamic content of a being, is grounded in frequency as the internal structure of that being. Frequency is the more basic content; energy is the dynamic quantity derived from frequency. This reinterpretation makes $E = h\nu$ not a "symmetric equivalence," but an ontological statement in which "frequency is the foundational dimension and energy is derivative."

Third, "basic dimension": frequency stands alongside time, space, energy, and other basic dimensions. This means that frequency cannot be reduced to time or space, although it has mathematical relations with them; frequency cannot be reduced to energy, although it has the relation $E = h\nu$; and frequency is an internal dimension of beings, independent of other basic dimensions.

Implications of the Thesis

FR1 has four ontological implications.

Implication One: Beings Are Classified According to the Frequency Spectrum

If frequency is a basic dimension, beings can be classified according to their frequency: different frequencies correspond to different types of beings. Specifically, in the EQT framework: extremely low frequency bands correspond to cosmological-scale structures, such as dark matter halos and large-scale cosmic structures; low frequency bands correspond to macroscopic material structures, such as stars, planets, and macroscopic objects; mid-frequency bands correspond to structures of atoms, molecules, and chemical bonds; high frequency bands correspond to excitation modes of elementary particles, such as electrons, quarks, and neutrinos; ultra-high frequency bands correspond to possible more foundational structures, such as physical processes at the Planck scale. This spectrum is not an arbitrary classification: it is the ontological structure of beings. Each frequency band corresponds to a type of being, determined by physical laws and interactions.

Implication Two: The Frequency Spectrum Is the Ontological Map of Beings

The frequency spectrum is not merely a classification; it is an ontological map of beings. Every being has a position in the frequency spectrum, and this position determines its ontological features: the frequency band determines interaction, since beings in different frequency bands interact in different ways; the frequency band determines stability, since beings in different frequency bands have different degrees of stability; and the frequency band determines observability, since beings in different frequency bands require different detectors. This ontological map allows us to understand existence from the level of frequency: the differences among beings are expressed at the level of frequency. This is the core method of ontological analysis in the EQT framework.

Implication Three: Cross-Band Interaction Is a Process

If beings are classified by frequency, how do beings in different frequency bands interact with one another? This is the problem of cross-band interaction. Specifically, the EQT framework treats this problem through frequency matching ($g(f, f_0)$) and frequency coupling, namely dynamical connections among different frequency bands. The details are treated in *The Foundational Volume of EQT.* This chapter only indicates the direction: cross-band interaction is a physical process, not an abstract connection.

Implication Four: The Frequency Stratification of Fundamental Interactions

In the EQT framework, the four fundamental interactions, namely gravity, electromagnetism, the strong interaction, and the weak interaction, have frequency stratification: gravity is low-frequency accumulation, manifesting at large scales and over long durations; electromagnetism is mid-frequency low-dissipation propagation, active from atomic to macroscopic scales; the strong force is high-frequency local resonance, confined at the subatomic scale; and the weak force is ultra-high-frequency short-lived frequency conversion, a short-range process at the highest frequency levels. This stratification is not accidental. It reflects the content of fundamental interactions as the dynamics of different frequency bands of the ρ_f field. FR1 provides the ontological ground for this stratification.

Objections and Replies

Objection One: The physical interpretation of $E = h\nu$ has already been developed in detail in quantum mechanics, quantum field theory, and related fields. Your ontological interpretation seems to over-metaphysicalize the formula by elevating a physical formula into an ontological thesis.

Reply: The physical interpretation and the ontological interpretation do not conflict; they treat different levels of the problem. The physical interpretation concerns application: how to use $E = h\nu$ to calculate physical quantities. The ontological interpretation concerns ground: what internal structure of beings is reflected by $E = h\nu$. Physics uses $E = h\nu$ as a tool, and it works well in numerous calculations. Ontology asks why this formula works and what internal structure of beings it reveals. Physics can use $E = h\nu$ without answering the ontological question. But serious ontological work must ask this question: every foundational formula inevitably has ontological meaning. The work of this chapter is not over-metaphysicalization, but ontological explication. It makes explicit the ontological ground of the physical work of $E = h\nu$, allowing physics and ontology to enter into serious dialogue. This is the core working mode of the tradition of philosophy of nature.

Objection Two: Treating frequency as a basic dimension, standing alongside time, space, and energy, seems too strong. Frequency clearly depends on time: it is the number of oscillations per unit time. How can frequency be claimed to be independent?

Reply: This objection points to an important refinement. The mathematical relation between frequency and time is real: frequency can indeed be defined as the number of oscillations per unit time. But this mathematical relation is not the same as ontological subordination. Specifically: at the mathematical level, frequency = 1/period, period = time interval, and frequency can be derived from time; at the ontological level, frequency, as the internal structural dimension of a being, does not depend on the choice of an external time coordinate. This distinction allows the two relations to coexist. In mathematical calculation, frequency is derived from time; in ontology, frequency is an internal dimension of the being. This is consistent with the claim in $E = h\nu$ that "energy is determined by frequency": ontologically, frequency is basic, while time and energy are relatively derivative quantities. More deeply, time itself, according to TM1 in Chapter 15, $= \partial\rho/\partial t$, namely the internal rate of change of the field. Frequency as the internal oscillation frequency of a being and time as the internal rate of change of the field correspond ontologically: both are internal dimensions of beings, not externally added coordinates.

Objection Three: Your claim that "beings are classified according to the frequency spectrum" sounds overly mechanical. It seems to reduce beings to frequency values. But beings have rich contents, including physical properties, structures, behaviors, and so on, and these cannot be reduced to frequency.

Reply: FR1 does not claim that "being = frequency." It claims that "frequency is one of the basic dimensions of beings." These are different. Specifically: beings have rich contents, including frequency as an internal structural dimension, spatial distribution, dynamical behavior, interactions with other beings, and so on. FR1 claims that frequency among these is a basic dimension: it is the ontological coordinate that organizes other contents. This organizing role does not simplify beings; it provides an ontological framework for their rich contents. By analogy, classifying human beings by "genotype" does not reduce human beings; it provides a coordinate for organizing other rich features. Likewise, classifying beings by frequency does not reduce them; it provides a way to organize their ontological features.

Literature That Must Be Engaged

- Max Planck, "Über das Gesetz der Energieverteilung im Normalspektrum" (1901)
- Albert Einstein, "Über einen die Erzeugung und Verwandlung des Lichtes betreffenden heuristischen Gesichtspunkt" (1905)
- Niels Bohr, "On the Constitution of Atoms and Molecules" (1913)
- Louis de Broglie, "Recherches sur la théorie des quanta" (1924)
- Werner Heisenberg, "Über quantentheoretische Umdeutung kinematischer und mechanischer Beziehungen" (1925)
- Chapter 7 on difference before identity (self-reference)

Summary of This Section

This section has done four things. First, it reviewed $E = h\nu$ as the starting point of a physical discovery in 1900. Second, it identified that its ontological meaning has rarely been treated seriously in philosophical discussion. Third, it stated the core thesis FR1:

frequency is an ontological dimension of beings. Fourth, it developed four implications of the thesis: the classification of beings according to the frequency spectrum, the frequency spectrum as an ontological map, cross-band interaction as a process, and the frequency stratification of fundamental interactions.

The next section develops the independence of frequency as a basic dimension.

9.3. Frequency as an Independent Basic Dimension

The Problem of Independence

Section 9.1 stated FR1: frequency is a basic dimension. But this statement requires defense: why can frequency not be reduced to other basic dimensions, especially time, space, and energy? This section develops the independence of frequency as a basic dimension through three arguments.

Argument One: Frequency Cannot Be Reduced to Time

Time and frequency have a mathematical relation: frequency = 1/period. This seems to imply that frequency is subordinate to time. This section argues that this implication is misleading.

First, time as a basic dimension itself has ontological problems, as Chapter 15 treats in detail: time is not an externally added coordinate, but an internal dimension of processual ontology. If time as a basic dimension has content, then the meaning of "frequency is subordinate to time" must be reassessed.

Second, frequency as the internal oscillation of a being is an internal feature of that being as an ontological unit. It is not that time first exists and then a being oscillates in time; rather, the being's own internal structure contains oscillatory frequency. This internality makes frequency not merely a "derivative of time."

Third, the measurement of time itself depends on oscillation. The precision of atomic clocks is based on the transition frequencies of atomic energy levels: these frequencies are internal frequencies of atoms as beings. In other words, the measurement of time is built upon frequency, not the other way around. If the measurement of time is built upon frequency, then making frequency subordinate to time is logically circular: we use frequency to define time and then define frequency as a derivative of time. This circularity shows that frequency has ontological independence.

More deeply, "frequency" as a concept does not need "time" as a prior assumption in order to be understood. Consider oscillation: it has repeated periods, and the frequency of repetition is an internal feature of the oscillation. We can speak directly of "the frequency of repetition" without first establishing "time" as an external coordinate.

Argument Two: Frequency Cannot Be Reduced to Space

Frequency seems unrelated to space: it is a quantity of the temporal dimension. But this section argues that frequency has an ontological relation with space, and that this relation does not make frequency subordinate to space.

Specifically, in physics, frequency and wavelength are related: for waves, frequency × wavelength = wave speed. This relation connects frequency with space through wavelength. But this connection does not mean that frequency is subordinate to space. On the contrary, it shows that frequency and space are both dimensions of beings, connected through dynamics. Specifically: frequency is the temporal-internal dimension of a being, namely oscillation frequency; wavelength is the spatial-internal dimension of a being, namely spatial period; and they are connected through the dynamics of waves, namely wave speed. This connection makes both frequency and wavelength basic dimensions: they cannot be reduced to each other, but have dynamical correspondence.

In the EQT framework, this correspondence becomes more concrete: excitation modes of the ρ_f field have frequency–wavelength structures, determined by field dynamics, namely the $A3$ Master Equation. Frequency and space are both internal dimensions of beings in the ρ_f field, not derivations from one another.

Argument Three: Frequency Cannot Be Reduced to Energy

$E = h\nu$ seems to say that energy and frequency are equivalent: one can infer one from the other. This seems to imply that frequency is subordinate to energy, or conversely. But this section argues that $E = h\nu$ is not a "symmetric equivalence"; it is the ontological statement that "frequency is the foundational dimension and energy is derivative."

First, h as a natural constant converts frequency, namely oscillations per second, into energy, namely joules. If energy were the foundational dimension, h should be an arbitrary constant. But h is a natural constant, approximately $6.626 \times 10^{-34}\,\mathrm{J \cdot s}$, with physical ground. This property points to frequency as a basic dimension: the value of h reflects the intrinsic ontological structure of beings.

Second, energy as a concept has many forms: kinetic energy, potential energy, thermal energy, chemical energy, nuclear energy, and so on. All these different forms of energy can be transformed into frequency through $E = h\nu$. This means that frequency is the more basic quantity: it unifies different forms of energy. Specifically: kinetic energy, where the kinetic energy of particles corresponds to their matter-wave frequency; potential energy, where differences between atomic energy levels correspond to radiation frequencies; thermal energy, where the spectrum of black-body radiation reflects temperature; chemical energy, where chemical bonds have vibrational and electronic excitation frequencies; nuclear energy, where the energy released by nuclear reactions corresponds to radiation at specific frequencies. All these forms of energy can be expressed at the frequency level. This is evidence for frequency as a more basic dimension.

Third, in quantum field theory, excitation modes of fields are parameterized by frequency: particles as field excitations have characteristic frequencies, such as the Compton frequency of the electron. Energy as a derivative quantity is inferred from these frequencies.

These arguments together support the claim that energy is derived from frequency, not the other way around. $E = h\nu$ should be interpreted as "energy = frequency multiplied by a conversion constant." This directionality is the core of FR1.

Synthesis of the Three Arguments

The three arguments support frequency as an independent basic dimension from different aspects: relative to time, frequency is not simply derived from time, since the measurement of time is built upon frequency; relative to space, frequency and space are different dimensions of beings, connected through dynamics but not reducible to each other; relative to energy, energy is derived from frequency through $E = h\nu$, not the other way around. The independence across these three dimensions gives frequency a place within the family of basic dimensions: it is not derivative from other dimensions, but a basic dimension standing alongside them.

But Frequency Is Connected with Other Dimensions

Acknowledging the independence of frequency does not deny its connections with other dimensions. In the EQT framework: frequency and time are connected through $\partial\rho/\partial t$: the internal frequency of a being is reflected in the temporal rate of change of the field; frequency and space are connected through $\nabla\rho$: the internal frequency of a being corresponds to spatial gradient structure through wavelength; frequency and energy are connected through $E = h\nu$: frequency determines energy. These connections allow the basic dimensions, time, space, frequency, and energy, to form a coherent ontological structure as a whole. They each have independence, but correspond to one another through physical dynamics. This is the content of the structure of basic dimensions in the EQT framework.

Objections and Replies

Objection One: Your first argument, that frequency cannot be reduced to time, uses "time measurement is built upon frequency" as a key point. But this is a matter of measurement, not ontology. Ontologically, time may still be prior to frequency.

Reply: Measurement and ontology are indeed different, but they are not completely independent. If time were ontologically prior to frequency, time should be precisely measurable independently of frequency. But the fact of physics is that all precise time measurement is built upon frequency, including atomic clocks and optical frequency combs. This fact points toward an ontological conclusion: if time cannot be precisely measured independently of frequency, then the position that "time is ontologically completely independent of frequency" is difficult to maintain. More plausibly, time and frequency are both basic dimensions, connected through physical dynamics, and neither is derived simply from the other. More deeply, this section does not claim that "frequency is more basic than time"; it claims that "frequency and time are both independent as basic dimensions." They have corresponding relations as basic dimensions through $\partial\rho/\partial t$, but neither is a simple derivative of the other.

Objection Two: Your third argument, that energy is derived from frequency, seems too strong. $E = h\nu$ is a symmetric equivalence relation, not one-way derivation.

Reply: At the purely mathematical level, $E = h\nu$ is indeed a symmetric equivalence: given one, the other can be calculated. But at the ontological level, a symmetric mathematical relation can reflect asymmetric ontological content. Specifically: at the mathematical level, $E \leftrightarrow \nu$ is a two-way calculational relation; at the ontological level, which is foundational, or "prior," requires argument. This section has provided arguments in support of frequency as foundational, including h as a natural constant, the unification of different forms of energy by frequency, and quantum field theory's parameterization of field excitations by frequency. If an opponent wishes to argue in the opposite direction, namely energy as foundational, they must provide counterarguments. More deeply, the direction of this section, frequency as foundational, is consistent with the overall processual ontological framework of this book: the internal oscillation of a being, namely frequency, is the ontological content of that being as an ontological unit; energy as a derivative quantity describes the intensity of that oscillation. This direction makes frequency ontology systematically consistent with the ontology of difference (Chapter 7) and co-evolutionary ontology (Chapter 12).

Objection Three: Your argument for frequency as a basic dimension is mainly built on physics, especially EQT. But this chapter promises not to depend on the physical derivation of EQT. Is this self-contradictory?

Reply: This objection rightly points to a methodological issue. The arguments in this section use the content of physics as support, but do not depend on specific EQT derivations. Specifically, Argument One, concerning the independence of frequency from time, uses a physical fact, namely that atomic clocks are based on frequency. This belongs to contemporary physics and is not limited to EQT. Argument Two, concerning the relation between frequency and space, uses wave phenomena, which belong to classical physics. Argument Three, concerning the derivation of energy from frequency, uses $E = h\nu$ and quantum field theory; these belong to contemporary physics. These physical facts do not depend on EQT derivations; they are part of the consensus of contemporary physics. This chapter uses them as argumentative support, which is consistent with the method of the tradition of philosophy of nature: identifying ontological structure from scientific discoveries. EQT appears in this chapter as an application example, such as the frequency stratification of the four fundamental interactions at the end of Section 9.1, but the core argument of this chapter, namely the defense of FR1, does not depend on EQT derivation. This independence allows the chapter, as part of a philosophical monograph, to be taken seriously in international philosophical discussion: its validity does not depend on the reader's acceptance of EQT.

Literature That Must Be Engaged

- Albert Einstein, Relativity: The Special and the General Theory (1916)

- Richard Feynman, QED: The Strange Theory of Light and Matter (1985)
- Steven Weinberg, The Quantum Theory of Fields Vol. 1 (1995)
- Carlo Rovelli, Reality Is Not What It Seems (2017)
- Chapter 7 on difference before identity (self-reference)
- Chapter 15 on time as $\partial\rho/\partial t$ (self-reference)

Summary of This Section

This section has done four things. First, it proposed three arguments for frequency as an independent basic dimension: independence relative to time, space, and energy. Second, it developed the content of each argument, especially by using physical facts as support. Third, it acknowledged the connections between frequency and other dimensions, while arguing that connection does not cancel independence. Fourth, it replied to possible objections, especially concerning the relation between measurement and ontology, the directionality of $E = h\nu$, and the relation with physics.

The next section develops the frequency spectrum as the ontological map of beings.

9.4. The Frequency Spectrum as an Ontological Map

The Function of the Spectrum

If frequency is a basic dimension, then organizing beings according to frequency can establish an ontological map: a framework that systematically organizes beings according to a basic dimension. This section develops the content of this ontological map.

An ontological map is not merely a classificatory system. It is the organization of beings as ontological units. Each frequency band corresponds to an ontological type of being, and there are ontological relations among different frequency bands, including dynamics, stability, and interaction.

The Content of the Frequency Spectrum

Let us develop the content of the frequency spectrum in detail, expressed in the EQT framework.

Extremely Low Frequency Band ($\nu < 10^{-15}\,\text{Hz}$)

This frequency band corresponds to cosmological-scale structures, namely processes with periods longer than the age of the universe: cosmic background structures, including fluctuations in the CMB, the distribution of dark matter halos, and the evolution of large-scale cosmic structures; the cosmological constant, namely the action of dark energy, corresponding to field dynamics in the extremely low frequency band; and long-term cosmic evolution, namely the cosmic evolutionary process from the Big Bang to the present. Beings in this band have several features: extremely large scale, involving cosmological scales of hundreds of millions of light-years or more; extremely long timescales, involving cosmological evolutionary timescales of hundreds of millions of years; and interactions dominated primarily by gravity, since other interactions are not significant at this scale.

Low Frequency Band ($10^{-15}\,\mathrm{Hz} < \nu < 10^{12}\,\mathrm{Hz}$)

This frequency band corresponds to macroscopic material structures: objects encountered in everyday life and astrophysics. These include astrophysical structures, such as galactic rotation curves, stellar nuclear-burning cycles, and planetary rotation and revolution; macroscopic matter, such as the vibrational frequencies of everyday objects, the flow frequencies of fluids, and waves such as low-frequency electromagnetic waves and sound waves; biological processes, such as heartbeats at about 1 Hz, EEG α, β, γ bands, and physiological cycles such as circadian, menstrual, and seasonal cycles. Beings in this band have several features: scales from micrometers to astronomical units, covering a broad range of spatial scales; timescales from milliseconds to years, covering a broad range of temporal scales; and interactions dominated primarily by electromagnetism, since chemical bonds, macroscopic material structures, electromagnetic waves, and related phenomena all involve electromagnetic interaction.

Mid-Frequency Band ($10^{12}\,\mathrm{Hz} < \nu < 10^{20}\,\mathrm{Hz}$)

This frequency band corresponds to structures at the atomic and molecular scale: atomic energy-level transitions, including electronic excitation, vibration, and rotation of atoms and molecules; electromagnetic radiation, including infrared, visible, and ultraviolet radiation generated by electronic transitions; and chemical bonds, whose vibrational frequencies reflect the internal structures of molecules. Beings in this band have several features: scales from nanometers to femtometers, covering spatial scales from atoms to nuclei; timescales from nanoseconds to attoseconds, covering the timescales of electronic dynamics; and interactions dominated primarily by electromagnetism, although the strong interaction begins to be involved in nuclear reactions.

High Frequency Band ($10^{20}\,\mathrm{Hz} < \nu < 10^{25}\,\mathrm{Hz}$)

This frequency band corresponds to elementary particles and subatomic structures: nuclear physics, including the structure of atomic nuclei and the processes of nuclear reactions; elementary particles, including electrons, quarks, and other elementary particles as frequencies of excitations of the ρ_f field; and high-energy radiation, including X-rays and γ rays as frequencies of high-energy photons. Beings in this band have several features: extremely small scales, involving subatomic scales, femtometers and smaller; extremely short timescales, involving nuclear-level dynamics, picoseconds and shorter; and interactions including all four fundamental forces, since electromagnetism, the strong force, the weak force, and gravity all participate.

Ultra-High Frequency Band ($\nu > 10^{25}\,\mathrm{Hz}$)

This frequency band corresponds to possible more foundational structures, such as physical processes at the Planck scale: Planck-scale physics, namely possible foundational structures around $10^{43}\,\mathrm{Hz}$; theoretical hypotheses, including the content of theories such as quantum gravity and string theory; and the early universe, namely possible dynamics in the very early period after the Big Bang, at the Planck time. Most beings in this band are theoretical hypotheses, and their content exceeds current experimental capacities. But as content belonging to a basic dimension, they remain part of the frequency spectrum.

Ontological Relations among the Five Frequency Bands

These five frequency bands are not independent of one another. They are connected through ontological relations.

Relation One: Hierarchical Nesting

Beings in higher frequency bands are often the internal structures of beings in lower frequency bands. For example: atoms (mid-frequency band) are composed of nuclei (high frequency band) and electrons (high frequency band); molecules (mid-frequency band) are composed of atoms (mid-frequency band) connected through chemical bonds (mid-frequency band); life (low frequency band) is composed of cells, involving multiple frequency bands; and cosmological structures (extremely low frequency band) are composed of galaxies (low frequency band). This hierarchical nesting gives beings as ontological units a multi-band ontological structure: they do not belong to a single frequency band, but involve combinations of multiple frequency bands.

Relation Two: Dynamical Connection

Beings in different frequency bands are dynamically connected. For example: nuclear reactions in atomic nuclei (high frequency band) release radiation (mid- to high-frequency bands); electron transitions (mid- to high-frequency bands) generate light (mid-frequency electromagnetic radiation); molecular vibrations (mid-frequency band) affect chemical reactions (mid-frequency chemical bonds); and planetary motion (low frequency band) is affected by stellar gravity (low- to extremely low-frequency bands). These dynamical connections make the frequency spectrum a coherent ontological structure as a whole: different frequency bands are not isolated, but affect one another through physical processes.

Relation Three: Frequency-Band Differences in Stability

Beings in different frequency bands have different degrees of stability: the extremely low frequency band has extremely high stability, with cosmological structures stable over hundreds of millions of years; the low frequency band has medium to high stability, with macroscopic matter stable over years to millennia; the mid-frequency band has medium stability, with atoms stable over hundreds of millions of years, while molecules and chemical bonds are stable from milliseconds to years; the high frequency band has low stability, with many elementary particles decaying from femtoseconds to nanoseconds; and the ultra-high frequency band has extremely low stability, if such beings exist, possibly at the Planck-time scale. These differences in stability reflect the role of freezing (Chapter 11) in different frequency bands: lower frequency bands correspond to deeper freezing, while higher frequency bands correspond to more active processes.

The Ontological Map as an Analytical Tool

The frequency spectrum as an ontological map is not merely descriptive; it is a tool for analyzing beings. Any being can be located in the frequency spectrum, and this location provides ontological analysis.

Example One: life as a multi-band being. Life is a complex being involving multiple frequency bands: the molecular level, including DNA, proteins, enzymes, and so on, involving mid-frequency vibrations, chemical bonds, and electronic transitions; the cellular level, including membrane potential, signal transduction, and metabolic cycles, involving low- to mid-frequency oscillatory cycles; the tissue level, including neural activity, heartbeat, and respiration, involving low-frequency physiological cycles; the individual level, including development, maturation, and aging, involving extremely low-frequency processes over years; and the population level, including evolution, ecology, and population dynamics, involving extremely low-frequency processes over millions of years. Life as a whole involves the coordination of all these frequency bands. Its complexity is built upon cross-band coordination. This frequency-band analysis allows us to understand the ontological structure of life precisely.

Example Two: consciousness as a highly complex frequency-band structure. Consciousness is a highly complex phenomenon involving combinations of frequency bands, as Chapter 13 has treated in detail: neural oscillations, such as the γ band 30–80 Hz, which is related to conscious synchronization, the α band 8–12 Hz, which is related to relaxed consciousness, and the δ band < 4 Hz, which is related to deep sleep; neural activity, including single-neuron spikes and oscillations in neural circuits; neurochemistry, including neurotransmitter release and receptor activation, involving mid- to high-frequency chemical processes; and long-term plasticity, including changes in synaptic strength and memory formation, involving multi-band coordination. Consciousness as an ontological phenomenon involves the coordination of all these frequency bands. The Ψ framework, developed in *Mind–Matter Unity*, is in fact a study of locating consciousness within the frequency spectrum: $\tilde{A}$ involves energy dissipation rate, which is related to frequency bands; C involves complexity, which involves the integration of multi-band oscillations; and $\mathcal{B}$ involves boundary, which involves spatial-frequency closure.

Objections and Replies

Objection One: Your division into five frequency bands seems overly precise. Why these five rather than others?

Reply: The division into five frequency bands is not absolute. It is a practical division based on physical laws. Other divisions, such as divisions according to the active regions of fundamental interactions or according to differences among types of beings, are also legitimate. This section chooses five frequency bands because they correspond to domains of physics: cosmology, macroscopic physics, atomic–molecular physics, elementary particle physics, and Planck-scale physics. Each frequency band corresponds to a domain of physical research, allowing the ontological map to connect with physical practice. The reasonableness of this division rests on the following: each frequency band has physical relevance; there are hierarchical and dynamical relations across bands; and the division makes the analysis of beings clear. If future developments in physics, such as the identification of new frequency bands or the discovery of new interactions, require redivision, the division in this section can be revised accordingly. This is an application of the method of constrained speculation: the division is revisable, not absolute.

Objection Two: Your "life as a multi-band being" and "consciousness as a highly complex frequency-band structure" seem overly complex and seem to reduce life and consciousness to combinations of frequency bands.

Reply: This objection points to an important refinement. Frequency-band analysis does not reduce life and consciousness. It provides an ontological coordinate system for their complexity. The rich content of life and consciousness, including evolutionary history, cultural content, and personal experience, is not reduced to frequency bands, but their ontological structures are organized through frequency bands. By analogy, analyzing life in terms of "genotype–phenotype" does not reduce life; it provides an analytical tool. Likewise, analyzing beings in terms of frequency bands provides an ontological analytical tool. Concrete content still requires specialized research in biology, neuroscience, psychology, and other disciplines. More deeply, frequency-band analysis allows us to see that the ontological structures of life and consciousness span multiple frequency bands: this is the ontological ground of their complexity. To locate life in a single frequency band, for example by saying that "life is chemical," would be reductive. To locate life across multiple frequency bands is a more precise ontological analysis.

Objection Three: Your ontological map depends on the frequency-band division of physics. But the content of physics changes. If future physics discovers new frequency bands or new dimensions, the whole ontological map will need to be rebuilt. Is this too unstable?

Reply: The ontological map in this section is indeed built upon the content of contemporary physics. This is a legitimate method: the tradition of philosophy of nature uses scientific discoveries as resources. But the content of the ontological map is not fixed; it is updated along with the development of physics. This is an advantage, not a defect. Specifically: the method of the ontological map, namely organizing beings according to frequency, is stable, because it is built upon the ontological thesis of frequency as a basic dimension (FR1); the content of the ontological map, including frequency-band divisions and classifications of beings, is updatable, because it is refined along with the development of contemporary physics. This structure of "stable method + updatable content" gives the ontological map both an ontological ground at the methodological level and a connection with scientific development at the level of content. This is an application of the method of constrained speculation.

Literature That Must Be Engaged

- Steven Weinberg, The First Three Minutes (1977): on cosmological content
- Roger Penrose, The Road to Reality (2004): on the systematization of physical frequency bands
- Max Tegmark, Our Mathematical Universe (2014): on discussions of physical levels
- Chapter 10 on tri-state categories as ontological categories (self-reference)
- Chapter 11 on freezing as a foundational ontological process (self-reference)

Summary of This Section

This section has done four things. First, it developed the content of the five frequency bands of the frequency spectrum: extremely low, low, mid, high, and ultra-high. Second, it clarified the ontological relations among the five frequency bands: hierarchical nesting, dynamical connection, and frequency-band differences in stability. Third, it used the examples of life and consciousness to show the use of the ontological map as an analytical tool. Fourth, it replied to possible objections, especially concerning the reasonableness of the frequency-band division, the treatment of life and consciousness, and the stability of the ontological map.

The next section develops the dialogue between frequency ontology and traditional doctrines of categories.

9.5. Dialogue with Traditional Doctrines of Categories

The Function of Dialogue

Frequency as a basic spectral quantity is a new way of organizing ontology. It must enter into dialogue with doctrines of categories in the Western philosophical tradition, since these doctrines also attempt to organize beings according to basic dimensions. This section develops three major dialogues: with Aristotle's ten categories, with Kant's pure forms of intuition, and with Whitehead's actual occasions. The function of dialogue is not competition, but to show the place of frequency ontology within the tradition of categorial doctrines.

Dialogue One: Aristotle's Ten Categories

In the *Categories*, Aristotle identifies ten ways in which being is predicated: substance, quantity, quality, relation, place, time, position, state, action, and passion. This is the earliest systematic doctrine of categories in Western philosophy. Aristotle's doctrine of categories has the following features: substance-centeredness, since substance is the basic category and the other nine categories are modes of the properties of substance; predication-orientation, since categories are ways in which being is predicated, rather than directly the structure of beings themselves; and experiential ground, since the categories arise from the analysis of language and experience.

Comparison with frequency ontology: frequency ontology differs fundamentally from Aristotle's doctrine of categories. Frequency ontology rejects substance as the foundational category: it treats beings as structures of difference, as Chapter 7 has argued in detail, whereas Aristotle's doctrine centers on substance. Frequency ontology is dimension-oriented: frequency as a basic dimension directly organizes the internal structure of beings, whereas Aristotle's doctrine is predication-oriented. Frequency ontology is built upon the content of physics, whereas Aristotle's doctrine is built upon the analysis of everyday experience and language. But the two also share a similarity: both attempt to organize beings systematically according to basic categories. Aristotle's work is an early attempt at such systematic organization; frequency ontology is a contemporary attempt. The contrast between the two can be stated as follows: Aristotle's doctrine of categories organizes ways in which being is predicated within the framework of substance ontology; frequency ontology organizes the internal structures of beings within the framework of the ontology of difference. The contrast reflects the difference between two ontological frameworks, not a difference in technical detail.

Dialogue Two: Kant's Pure Forms of Intuition

In the *Critique of Pure Reason*, Kant identifies pure forms of intuition: time and space as a priori forms of intuition. Kant's position is this: time and space are not empirical objects, but a priori conditions of experience; time and space are not properties of things-in-themselves, but contents of the transcendental structure of the subject; through time and space, objects of experience, namely phenomena, can be organized.

Comparison with frequency ontology: Kant's pure forms of intuition have a complex relation with frequency ontology. Common ground: both identify basic dimensions. Kant identifies time and space as a priori forms of intuition, while frequency ontology identifies frequency as a basic dimension. Both treat basic dimensions as frameworks for organizing beings, not as independent things. Key differences: Kant's position is transcendental idealism, in which time and space are transcendental structures of the subject and not properties of things-in-themselves. Frequency ontology is ontological realism, in which frequency is the internal structure of beings themselves. Kant identifies two a priori forms of intuition, time and space. Frequency ontology identifies frequency as another basic dimension, standing alongside time and space. Kant's method is transcendental analysis; the method of frequency ontology is constrained speculation.

A possible Kantian reply: a Kantian may argue that frequency as a "basic dimension" is in fact part of the transcendental structure of the subject. We experience frequency through the senses, such as hearing and seeing, and these senses are built upon the transcendental structure of the subject. Therefore, frequency is not a property of things-in-themselves.

Reply from frequency ontology: the concept of thing-in-itself as unknowable faces difficulties in contemporary philosophical discussion. If the thing-in-itself is retained, frequency as content of the phenomenal world is legitimate, while the content of the thing-in-itself remains open. If the thing-in-itself is rejected, as in Hilary Putnam's response, then the "phenomenal world" as the only world makes frequency as an internal dimension of that world a legitimate ontological thesis. Discoveries in contemporary physics, such as frequency as an internal dimension of beings, challenge the Kantian framework: such discoveries are difficult to handle by transcendental analysis alone. More deeply, the work of this book, especially Chapter 14 on the new foundation of objectivity and Chapter 12 on co-evolution as an ontological relation, has already treated an alternative to Kantian transcendental idealism: objectivity is grounded in controlled reproducibility and does not require presupposing things-in-themselves. In this framework, frequency as a basic dimension has ontological legitimacy.

Dialogue Three: Whitehead's Actual Occasions

In *Process and Reality*, Whitehead identifies actual occasions as foundational objects: instantaneous units of process. Whitehead's doctrine of categories has the following features: actual occasions are foundational objects, namely processual, temporary, mutually connected units; actual occasions are connected through prehension, feeling other actual occasions; and eternal objects ingress into actual occasions as forms. Chapter 8 has already treated in detail the difficulty in Whitehead's work, namely the internal tension of eternal objects as a second ontology. This section focuses on the relation between Whitehead's work and frequency ontology.

Comparison with frequency ontology: common ground: both take processuality as foundational. Actual occasions are units of process, and frequency ontology is built upon difference as processual structure. Both reject static substance as foundation: Whitehead replaces substance with actual occasions, while frequency ontology replaces substance with structures of difference. Both use physical discoveries as ontological resources: Whitehead uses relativity and early quantum mechanics, while frequency ontology uses $E = h\nu$ and contemporary physics. Key differences: Whitehead introduces eternal objects as a second ontology. This is the internal difficulty of his work, as discussed in Chapter 8, Section 8.5. Frequency ontology does not introduce a second ontology: form is treated as the internal dynamics of structures of difference. Whitehead's method is mainly metaphysical synthesis, systematizing all content through a detailed doctrine of categories. The method of frequency ontology is constrained speculation, working through falsifiable mechanisms. Whitehead did not systematically connect actual occasions with contemporary physics such as quantum field theory and cosmology; frequency ontology establishes such a connection through the EQT framework. It can be said that frequency ontology agrees with Whitehead's work in processuality and the rejection of static substance, but differs in ontological content and methodological rigor. Frequency ontology is a contemporary continuation of this processual tradition, but makes a contemporary contribution through a stricter methodology, namely constrained speculation, and a more concrete physical connection, namely EQT.

The Distinctive Position of Frequency Ontology

Through these three dialogues, the distinctive position of frequency ontology in the tradition of categorial doctrines can be located as follows: in comparison with Aristotle, it moves from substance ontology to the ontology of difference, and from predication-orientation to dimension-orientation; in comparison with Kant, it extends the two a priori forms of intuition, time and space, to include frequency as a third basic dimension, and moves from transcendental idealism to ontological realism; in comparison with Whitehead, it preserves the processual direction, but gives it content through the method of constrained speculation and physical connection. This distinctive position means that frequency ontology is not an isolated ontological attempt. It is built upon precursors in the Western philosophical tradition, including Aristotle, Kant, and Whitehead, but makes a contemporary contribution through ontological transformation, including difference as foundation and frequency as a basic dimension.

Objections and Replies

Objection One: Your three dialogues are too brief. Aristotle, Kant, and Whitehead each have complex works, but this section treats each in only a short passage.

Reply: This section, as a section of dialogue, only states the key points and does not develop them comprehensively. Detailed treatment of each precursor requires dedicated work. This section gives the most relevant structural correspondences: it shows the similarities and differences between frequency ontology and each precursor. More detailed dialogue requires dedicated studies. The function of this section is to indicate directions of dialogue, allowing readers to locate frequency ontology precisely within the Western philosophical tradition.

Objection Two: You claim that frequency ontology is a "contemporary continuation" of the processual tradition, but your work has a physical-analytic methodological basis, while the processual tradition is mainly a Continental tradition. Is this continuation legitimate?

Reply: This objection points to an important problem of connecting traditions. The processual tradition is not unitary. It involves several directions: the Continental direction, including Bergson, Whitehead, Deleuze, and others; the analytic direction, including certain versions of realism and Brandomian inferentialism; and the physical direction, including Prigogine's non-equilibrium thermodynamics and Rovelli's relational quantum mechanics. Frequency ontology is built upon multiple directions: it inherits the Continental processual tradition, especially Whitehead, while using analytic method, namely falsifiable mechanisms, and connecting with physics through EQT. This synthesis is legitimate: it gives the processual tradition a more concrete ontological ground in the contemporary context. More deeply, "continuation" in the history of philosophy often involves the synthesis of multiple directions, not the simple extension of a single direction. Kant synthesized empiricism and rationalism; Hegel synthesized Kant and ancient Greece; Wittgenstein synthesized science and philosophy of language. Frequency ontology as a contemporary continuation of the processual tradition involves a similar synthesis.

Objection Three: Your ontology seems overly ambitious: it seeks to replace the three great traditions of Aristotle, Kant, and Whitehead. Is this overconfident?

Reply: This section does not claim to "replace" these three traditions. It claims to provide another way of organizing ontology. All three traditions have legitimacy and domains of application, and frequency ontology does not cancel them. Specifically: Aristotle's doctrine of categories remains useful in the analysis of everyday life and traditional formal logic; Kant's pure forms of intuition remain useful in epistemology of experience and aesthetic theory; Whitehead's actual occasions remain useful in process philosophy and religious thought. Frequency ontology provides another way of organizing ontology, especially suitable for treating ontological problems connected with contemporary physics, especially the EQT framework. It does not replace other traditions, but supplements them. Acknowledging this limitation makes the work of this section honest: it does not pretend to replace all traditions, but recognizes its own domain of application.

Literature That Must Be Engaged

- Aristotle, Categories
- Immanuel Kant, Critique of Pure Reason (1781), "Transcendental Aesthetic"
- Alfred North Whitehead, Process and Reality (1929)
- Edmund Husserl, Cartesian Meditations (1931): on the phenomenological treatment of categories
- Chapter 7 on difference before identity (self-reference)

- Chapter 14 on the new foundation of objectivity (self-reference)

Summary of This Section

This section has done four things. First, it entered into dialogue with Aristotle's ten categories and showed the transition in frequency ontology from substance ontology to the ontology of difference. Second, it entered into dialogue with Kant's pure forms of intuition and showed the transition in frequency ontology from transcendental idealism to ontological realism, as well as the status of frequency as a third basic dimension. Third, it entered into dialogue with Whitehead's actual occasions and showed how frequency ontology continues the processual tradition in a contemporary form. Fourth, it clarified the distinctive position of frequency ontology within the tradition of categorial doctrines.

The next section is the conclusion of this chapter.

9.6. Chapter Summary: The Independent Status of Frequency as an Ontological Dimension

Completion of the Three Tasks

The three tasks undertaken by this chapter have all been completed.

Task One: to elevate $E = h\nu$ from Planck's physical discovery into an ontological thesis. Section 9.1 developed the physical starting point of $E = h\nu$ and its ontological meaning, and stated the core thesis FR1: frequency is an ontological dimension of beings.

Task Two: to argue for the independence of frequency as a basic dimension. Section 9.2 defended frequency as an independent basic dimension through three arguments, concerning its independence relative to time, space, and energy. Section 9.3 developed the frequency spectrum as the ontological map of beings and showed the application of frequency ontology.

Task Three: to enter into dialogue with doctrines of categories in the Western philosophical tradition. Section 9.4 entered into serious dialogue with Aristotle's ten categories, Kant's pure forms of intuition, and Whitehead's actual occasions, clarifying the distinctive position of frequency ontology within the tradition of categorial doctrines.

The Position of This Chapter in Part II

The overall architecture of Part II, "The Philosophical Defense of Gradient Ontology," is as follows: Chapter 7, difference before identity, establishes the ontological starting point; Chapter 8, the tradition of qi theory and frequency–density–gradient, establishes the Chinese philosophical genealogy; Chapter 9, this chapter, frequency as a basic spectral quantity, develops the content of the basic dimension; Chapter 10, tri-state categories as ontological categories, develops the ontological categories of beings; Chapter 11, freezing as a foundational ontological process, develops the core ontological process. As a core chapter of Part II, this chapter systematizes the content of frequency as a basic dimension. It builds upon the preparations of Chapter 7, difference before identity, and Chapter 8, the tradition of qi theory, and provides the framework of basic dimensions for the ontological work of Chapter 10, tri-state categories, and Chapter 11, freezing.

The Position of This Chapter in the Whole Book

Looking backward, this chapter is built upon several chapters: Chapter 1 on the method of constrained speculation provides the methodological ground for the defense of FR1 in this chapter; Chapter 4 on constitutive coextension provides the semantic tool for this chapter's dialogue with traditional doctrines of categories; Chapter 7 on difference before identity provides the ontological foundation for frequency as an internal dimension of structures of difference.

Looking forward, this chapter provides the framework of basic dimensions for later chapters: Chapter 10 on tri-state categories, where the tri-states have different manifestations in different frequency bands of the frequency spectrum; Chapter 11 on freezing, where freezing as a process involves the dynamics of frequency bands; Chapter 13 on mind–matter, where consciousness as a highly complex frequency-band structure is located within the framework of frequency ontology; Chapter 17 on ontological singularity, where the gradient terminal phase involves the collapse of frequency structures.

Honesty about the Limits of This Chapter

This chapter acknowledges several limits:

First limit: the depth of connection with physics is limited. This chapter develops frequency as a basic dimension at the philosophical level, but detailed connection with physics, especially quantum field theory, general relativity, cosmology, and related fields, requires dedicated work. This chapter only indicates the direction; detailed connection depends on future work.

Second limit: dialogue with cross-cultural doctrines of categories is insufficient. This chapter mainly enters into dialogue with the Western tradition of categorial doctrines. The Chinese philosophical tradition also contributes to similar problems, such as the *Book of Changes* on yin-yang as basic opposition and the five phases as basic classification. Chapter 8 treats these briefly, but this chapter does not enter into detailed dialogue. This remains an open direction of work.

Third limit: the ontological ground of frequency as a basic dimension requires further defense. This chapter gives three layers of argument for FR1, namely independence, spectrum, and traditional dialogue, but responses to opponents can be developed in more detail. In particular, replies to Kantian positions and to certain versions of the contemporary revival of substance ontology require dedicated treatment.

Fourth limit: the application of the ontological map. This chapter gives examples of life and consciousness as multi-band beings, but complete applications, such as frequency-ontological analyses in biology, neuroscience, sociology, and other fields, require dedicated work.

Acknowledging these limits does not weaken the work of this chapter. It makes this chapter an honest initial work.

The Core Posture of This Chapter

The core posture of this chapter can be stated as follows: frequency is not merely one physical quantity among others; frequency is the internal structure of existence.

$E = h\nu$ is a central formula in the development of twentieth-century physics: it opened the door to quantum mechanics and became a foundation of modern physics. But its ontological meaning has rarely been treated seriously in philosophical discussion. It is often treated as an "empirical relation" or a "derived formula." The work of this chapter treats $E = h\nu$ as an ontological window: through it, one sees the internal structure of existence. Frequency is not an independent isolated quantity; it is the internal dimension of a being as an ontological unit. What $E = h\nu$ expresses is not "the correspondence between two independent quantities," but "the ontological relation between the internal dimension and dynamic content of a being."

This ontological elevation is not over-metaphysicalization; it is ontological explication. Every foundational physical formula has an ontological ground, and serious ontological work must make these grounds explicit. This chapter's ontological treatment of $E = h\nu$ is an example of such work. If the work of this chapter holds, frequency as a basic dimension has a place within the tradition of ontological categorial doctrines: it is a contemporary continuation of the work of precursors such as Aristotle, Kant, and Whitehead, but makes a contemporary contribution through physical connection and the method of constrained speculation.

Meaning before freezing: this is one of the core theses of this book. This chapter provides the frequency-ontological ground for that thesis. "Freezing" involves the dynamics of frequency bands, such as high-frequency early freezing and low-frequency later freezing, as Chapter 11 has developed; "meaning" appears in complex structures of frequency-band combinations, such as consciousness as multi-band coordination; and the finitude of meaning is grounded in the finitude of frequency-band structures. These connections are grounded by the frequency ontology established in this chapter.

Through the work of this chapter, frequency as an ontological dimension gains its position within this book. It is not an additional dimension externally added to the ontology of difference, but the internal structure of the ontology of difference. This internal unity makes the ontological work of Part II, including the ontology of difference, frequency ontology, tri-state categories, and the ontological process of freezing, into a systematic whole.

9.7. Methodological Reflection on This Chapter

This chapter elevates $E = h\nu$ from a physical formula into an ontological thesis. This section reflects on the methodological significance of this elevation.

This Chapter's Posture toward Originality

Frequency as a basic dimension is not entirely new: $E = h\nu$ is a central formula in physics. But systematically developing it as an ontological dimension, standing alongside time, space, and energy, is new work. This chapter undertakes that work. At the same time, this chapter acknowledges that this work is built upon precursor traditions: the oscillatory character of actual occasions in Whitehead's process philosophy, Bergson's dynamic ontology and its critique of static substance, and twentieth-century physics' treatment of frequency as a basic physical quantity. Originality lies not in isolated creation, but in contemporary synthesis.

This Chapter's Treatment of Physics

This chapter uses physical discoveries as argumentative support, but does not depend on the physical derivation of EQT. This distinction allows the chapter, as part of a philosophical monograph, to be taken seriously in international philosophical discussion: its validity does not depend on the reader's acceptance of EQT or of any particular physical interpretation. This chapter uses $E = h\nu$ as the starting point of philosophical argument, but its argument is developed at the philosophical level. This treatment means that the chapter does not pretend to "derive philosophy from physics," but also does not pretend to be unrelated to physics.

This Chapter's Position in the Tradition of Philosophy of Nature

This chapter is a contemporary continuation of the tradition of philosophy of nature. The characteristic feature of that tradition is to identify ontological structure through physical discoveries, while working on ontological problems. This chapter fits that tradition: it identifies frequency as an ontological dimension through $E = h\nu$, but develops the ontological content of that dimension through philosophical argumentation. This positioning differentiates the chapter from both pure philosophy of science and separate metaphysics. This chapter stands between them, which is the classical position of the tradition of philosophy of nature.

Limits Honestly Acknowledged in This Chapter

This chapter acknowledges several limits. First, the refinement of frequency as an ontological dimension requires later work: the manifestations of the frequency spectrum at different ontological levels require more detailed treatment. Second, this chapter develops only the core direction of the relation between frequency ontology and relativity. Third, the manifestations of frequency ontology in interdisciplinary applications, such as neuroscience and cognitive science, require dedicated work. Fourth, further dialogue between frequency ontology and other philosophical traditions can be expanded.

Connections with Later Chapters

The frequency ontology established in this chapter is applied in later chapters. Chapter 10 uses the frequency spectrum as the dimension of tri-state classification. Chapter 11 uses the frequency spectrum as the dimension of freezing stratification: high-frequency early freezing and low-frequency later freezing. Chapter 12 uses frequency matching as one mechanism of co-evolution. Chapter 13 uses the frequency spectrum as a representation of consciousness-related neural activity. Chapter 14 uses frequency matching as one mechanism of observation. More deeply, this chapter is the core chapter in which this book connects with modern physics. $E = h\nu$ is a central formula of twentieth-century physics, and any serious contemporary philosophy of nature must treat the ontological meaning of this formula.

10. Tri-State Categories as Ontological Categories

10.1. Introduction to This Chapter

EQT Detection Theory introduces the tri-states as a classification of detected objects: the free state, a state with propagativity, rapid response, and relatively strong radiative visibility; the bound state, a state with local confinement, phase locking, energy-level structure, and nonlinear response; and the condensed state, a state with phase rigidity, static-mass manifestation, low free radiation, and strong freezing. In *EQT Detection Theory*, this triadic division plays a role in the theory of detection: it tells detectors how to identify different states, tells physicists how to decompose observational signals, and tells theory how to handle transitions among the three states.

On closer examination, however, this triadic division has philosophical implications that go beyond the theory of detection. This chapter argues that the tri-states can be elevated into ontological categories: they are not only classifications of detected objects, but basic modes of existence of beings. More specifically, the three states correspond respectively to three basic modes of existence: accessibility, relationality, and stability. Together, these three modes constitute the categorial structure of process ontology.

This chapter undertakes three tasks. First, it elevates the tri-states from a classificatory scheme in the theory of detection in *EQT Detection Theory* into ontological categories (T1). Second, it argues for the completeness of the tri-state categories, namely that any being can be decomposed into the three states, and for their irreducibility, namely that the three states cannot be reduced to one another, thereby establishing the legitimacy of the tri-states as ontological categories (T2). Third, it compares the tri-state categories seriously with Aristotle's ten categories and Kant's twelve categories, clarifying the distinctive position of the tri-state categories: they concern neither the properties of substances, as in Aristotle, nor the forms of judgment, as in Kant, but the modes of existence of process.

This chapter carries a relatively heavy burden of originality. If the argument for the tri-state categories holds, this chapter will have an independent position in the history of categorial doctrines. Yet the establishment of a doctrine of categories is among the most difficult tasks in the history of philosophy: Aristotle's ten categories and Kant's twelve categories are among the most complex parts of their respective philosophical systems. This chapter does not pretend to "finally establish" the tri-state categories. What it does is to state them clearly, argue for them rigorously, and identify possible objections. This is the legitimate form of an initial establishment of a doctrine of categories.

This chapter is located between Chapter 7, "Difference before Identity," and Chapter 11, "Freezing as a Foundational Ontological Process." It makes precise the structure of process ontology: how any processual being can be decomposed ontologically.

The discussion proceeds as follows. Section 10.1 develops the elevation from a classification in the theory of detection to ontological categories; Section 10.2 defends the completeness of the tri-state categories; Section 10.3 defends the irreducibility of the tri-state categories; Section 10.4 compares the tri-state categories with the categorial doctrines of Aristotle and Kant; Section 10.5 develops the applications of the tri-state categories through four examples.

10.2. From a Classification in the Theory of Detection to Ontological Categories

The Starting Point: The Tri-States in *EQT Detection Theory*

When treating the theory of detection, *EQT Detection Theory* introduces the tri-states as a classification of detected objects. The detection-theoretical definitions of the three states are as follows: the free state has propagativity, rapid response, relatively strong radiative visibility, and approximately linear readout, with typical examples including photons, free particles, and propagating waveforms; the bound state has local confinement, phase locking, energy-level structure, nonlinear response, and delayed release, with typical examples including atomic energy levels, quark bound states, and oscillators; the condensed state has phase rigidity, static-mass manifestation, low free radiation, strong freezing, and high-threshold response, with typical examples including ground states, lattices, macroscopic objects, and Bose–Einstein condensates.

In *EQT Detection Theory*, this triadic division plays a role in the theory of detection: it tells detectors how to identify different states, tells physicists how to decompose observational signals, and tells theory how to handle transitions among the three states. It is a classificatory tool in physics. On closer examination, however, this triadic division has philosophical implications that go beyond the theory of detection. This section argues that the tri-states can be elevated into ontological categories: they are not only classifications of detected objects, but basic modes of existence of beings.

Three Observations

The ground for the elevation lies in three observations.

Observation One: The tri-state classification exhausts the basic modes of beings.

Expand the field of view from particle physics to all domains. What state is a being in at any given moment? Consider several examples: a star has free-state components, namely outwardly radiated photons; bound-state components, namely atomic energy-level excitations; and condensed-state components, namely the high-density matter of its core. A living organism has free-state components, namely exchanges of energy and matter with the environment; bound-state components, namely the stable oscillations of DNA, proteins, and metabolic cycles; and condensed-state components, namely cellular structures, organs, and bones. A cultural system has free-state components, namely the flow of information and the dissemination of ideas; bound-state components, namely linguistic rules and the oscillations of custom; and condensed-state components, namely buildings, documents, and institutions. A memory has free-state components, namely active neural processes when it is recalled; bound-state components, namely stable patterns of neural connection; and condensed-state components, namely solidified structures in long-term memory. These examples reveal a pattern: any being at any moment can be decomposed into the tri-states. Any being simultaneously has three dimensions: flow, namely free state; oscillation, namely bound state; and stability, namely condensed state.

Observation Two: The tri-state classification is not arbitrary.

Why these three states? Why not two, four, or ten? Careful analysis of the physical meaning of the tri-states shows the following: the free state corresponds to accessibility, meaning that a being can propagate, flow, and diffuse, and that its influence can reach far away; the bound state corresponds to relationality, meaning that a being forms relations, becomes interconnected, maintains oscillation, and exists in relation; the condensed state corresponds to stability, meaning that a being has stable structure, rigidity, memory, and persistence through time. These three categories, accessibility, relationality, and stability, correspond to three basic ontological dimensions of beings: the spatial dimension, concerning how a being is distributed in space, whether it is flowing, localized, or fixed; the relational dimension, concerning how a being interacts with other beings, whether it propagates independently, is mutually connected, or is stably isolated; and the temporal dimension, concerning how a being persists in time, whether it is momentarily passing, periodic, or long-term stable. Each dimension has natural "poles": flowing vs fixed in space, independent vs connected in relation, momentary vs persistent in time. But poles alone are insufficient to describe existence: a genuinely existing thing is often a mode between poles. The tri-state categories provide such modes: free state = spatially flowing + relationally independent + temporally momentary in a certain sense; bound state = spatially localized + relationally connected + temporally periodic; condensed state = spatially fixed + relationally isolated or collective + temporally long-term. The tri-states are not arbitrary: they are three basic modes jointly determined by the three basic ontological dimensions of beings.

Observation Three: The tri-state classification has physical realization.

The tri-state categories are not abstract philosophical constructions: they correspond to physical realizations. The EQT Master Equation is:

$$\partial_t \rho_f = D(f)\nabla^2 \rho_f + k(f)\rho_f^2 - \Gamma(f)\rho_f + S_{\text{ext}}(f)$$

Each term corresponds to a different physical process: $D(f)\nabla^2 \rho_f$, the diffusion term, corresponds to the free state, namely the diffusion, propagation, and flow of energy in space; $k(f)\rho_f^2$, the nonlinear term, corresponds to the formation of bound states, namely local aggregation, interaction, and the formation of oscillation; $-\Gamma(f)\rho_f$, the dissipation term, corresponds to the stabilization of the condensed state, namely energy exiting active processes and entering a background or frozen state; and $S_{\text{ext}}(f)$, the source term, corresponds to external driving, which can maintain any one of the states. This correspondence makes the tri-state categories not merely a conceptual framework, but ontological categories corresponding to the dynamical structure of physics. This point is important for this chapter: the tri-state categories are not abstract metaphysical conjectures, but ontological categories with physical realization.

Formal Statement of the Elevation

Bringing the three observations together, the core thesis of this chapter can be stated rigorously:

Thesis T1: the free state, bound state, and condensed state are not only classifications of detected objects, but basic ontological categories of beings. Any being at any moment can be decomposed into the tri-states; the three states correspond respectively to the three basic ontological dimensions of beings: accessibility, relationality, and stability.

This thesis is the center of the chapter. It elevates the classificatory tool in physics from *EQT Detection Theory* into a doctrine of ontological categories. The following sections of this chapter defend this elevation.

The Legitimacy of the Elevation

A reader may immediately object that the elevation from a physical classification to ontological categories is an over-inference. The legitimacy of the physical classification belongs within physics; extending it to ontology requires additional defense.

This chapter accepts this concern and treats it explicitly. This chapter does not claim that "the tri-states are useful in physics, therefore they are ontological categories." That inference is too weak. This chapter claims that the tri-state categories satisfy the legitimacy conditions for ontological categories: completeness, meaning that they cover all beings; irreducibility, meaning that they cannot be reduced to one another; and internality, meaning that they are not externally added, but reflect the structure of existence itself. These conditions will be defended in detail in Sections 10.2 and 10.3. Physical realization, namely Observation Three, only supports the legitimacy of the tri-state categories; it does not prove it. The proof proceeds at the philosophical level.

What This Chapter Does Not Claim

To delimit clearly the work of this chapter, this section explicitly states what this chapter does not claim.

This chapter does not claim that the tri-state categories are the only doctrine of ontological categories. Aristotle's ten categories, Kant's twelve categories, and others are legitimate categorial doctrines that address different problems. The tri-state categories do not replace them, but treat a problem that they do not sufficiently treat: the ontological modes of processual beings.

This chapter does not claim that the tri-state classification is equally clear in all applications. There are transitional zones among the three states, mixed cases, and boundary cases that are difficult to classify. These are real complexities, and this chapter does not pretend that they do not exist.

This chapter does not claim that the tri-state categories have already been "completed." Doctrines of categories have historically required long-term development and revision. This chapter is the initial establishment of the tri-state categories: it gives the basic framework and argument. Later work, by other researchers, in other fields, and in future literature, will further develop it.

This chapter does not claim that the tri-state categories can be entirely independent of EQT. The physical realization of the tri-state categories depends on EQT, through the different terms of the Master Equation. If EQT is replaced by another theory in the future, the physical realization of the tri-state categories will need to be reassessed, although their philosophical direction may remain valid.

Literature That Must Be Engaged

- *EQT Detection Theory*, Chapter 4 on tri-state response theory (self-reference)
- Aristotle, Categories: Aristotle's ten categories
- Immanuel Kant, Kritik der reinen Vernunft (1781/1787): Kant's twelve categories
- Alfred North Whitehead, Process and Reality (1929), Part I Chapter II: Whitehead's doctrine of categories
- C. S. Peirce, "On a New List of Categories" (1867): Peirce's three categories

Summary of This Section

This section has done four things. First, it reviewed the starting point of the tri-states as a classification in the theory of detection in *EQT Detection Theory.* Second, it prepared the ground for elevating the tri-state categories through three observations: exhaustiveness, non-arbitrariness, and physical realization. Third, it rigorously stated the core thesis T1: the tri-states as basic ontological categories of beings. Fourth, it clarified what this chapter does not claim and delimited the scope of the work.

The next section defends the first legitimacy condition of the tri-state categories: completeness.

10.3. The Completeness of the Tri-State Categories

The Meaning of Completeness

A core requirement of a doctrine of ontological categories is completeness: any being must be covered by the doctrine of categories; there should be no being "without category." This is a legitimacy condition for a doctrine of categories as foundational ontology.

The completeness of Aristotle's ten categories has been repeatedly discussed in the history of Western philosophy: did Aristotle cover all beings? Kant gave a different answer: he did not emphasize that the doctrine of categories covers all "beings," but that it covers all "possible forms of judgment." This is an epistemological rather than an ontological angle. The completeness claim of this chapter is ontological: any being at any moment can be decomposed into the tri-states; there is no being that "does not belong to the tri-states."

A Preliminary Defense of Completeness

Argument One: From the breadth of concrete examples.

Section 10.1 already gave several examples, including stars, living organisms, cultural systems, and memory. This section extends the list to cover more domains: physical objects, where electrons are dominated by the bound state, photons by the free state, galaxies by the condensed state, black holes as extreme condensed states, and the cosmic microwave background as a free-state distribution; chemical objects, where atoms are bound states, molecules are bound states plus relational oscillations, crystals are condensed states, and free radicals have a high free-state component; biological objects, where DNA molecules are bound states plus condensed states, enzyme catalysis is free-state–bound-state transition, cells have all three states strongly, and neurons are highly complex tri-state mixtures; psychological objects, where sensory impressions have a high free-state component, memory is a condensed state, attention is a bound-state–free-state transition, and belief is a stable structure formed by condensed states; social objects, where information flow is a free state, custom is bound state plus condensed state, institutions are condensed states, and markets are bound state plus free state; cultural objects, where language consists of bound-state rules plus free-state use, artworks are condensed states, traditions are condensed state plus bound state, and popularity has a high free-state component. This broad list shows the applicability of the tri-state categories across multiple domains. But broad applicability is not the same as completeness: possible counterexamples must be treated directly.

Argument Two: Through ontological dimensions.

Observation Two in Section 10.1 pointed out that the tri-states correspond to three basic ontological dimensions of beings: the spatial dimension, namely accessibility; the relational dimension, namely relationality; and the temporal dimension, namely stability. If these three dimensions exhaust the basic ontological dimensions of beings, then the tri-state categories are complete: any being can be located along these three dimensions and decomposed into the tri-states.

Why do these three dimensions exhaust the field? Because they correspond to the three basic determinations of a being as an ontological object: where the being is, namely the spatial dimension; what the being is related to, namely the relational dimension; and how long the being persists, namely the temporal dimension. These three questions exhaust the most basic ontological inquiries about a being. Other inquiries, such as what properties it has, what its purpose is, what its causal efficacy is, and so on, can be reduced to these three basic inquiries. This argument is conceptual: it starts from "the most basic determinations of a being as an ontological object" and argues that the three dimensions exhaust these determinations. The argument does not depend on empirical examples, although the examples in Argument One support its plausibility.

Limits of Completeness

The completeness claim must be honest: it should not overreach. This section explicitly states several limits of completeness.

Limit One: Completeness is completeness of ontological dimensions, not completeness of concrete manifestations.

The tri-state categories cover the spatial, relational, and temporal dimensions of any being, but they do not claim to cover all concrete manifestations of a being. A star has a specific chemical composition, temperature distribution, magnetic-field structure, and so on. These concrete manifestations are not directly described by the tri-state categories. The tri-state categories provide a location at the level of basic ontological dimensions, such as that a star is "mainly a condensed state," but concrete physical description requires additional tools, such as models of stellar evolution. Similarly, consciousness has concrete phenomenological manifestations, including vision, hearing, emotion, thought, and so on. These concrete manifestations are not directly covered by the tri-state categories. The tri-state categories provide the basic dimensions of consciousness as an ontological object, such as that consciousness is "a specific combination of bound state, free state, and condensed state," but concrete research on consciousness requires additional tools, such as the Ψ state-variable group in *Mind–Matter Unity*.

Limit Two: Completeness allows mixed states and transitional states.

Beings are often in mixed states or transitional states among the three states. An atom excited to a high energy level is in a bound state; when it emits a photon and its energy level decreases, this is a transition from bound state to free state; when it absorbs a photon and is excited from the ground state, this is a transition from free state to bound state. These transitions are real, important, and treated seriously by physics. The tri-state categories allow mixed states and transitional states. They do not require any being at any moment to "belong purely" to one state. They require that every being can be decomposed into the tri-states, meaning that one can state its weight and process in each state, not that every being belongs to one state.

Limit Three: Boundary cases.

Some beings may be difficult to classify. A quantum superposition state is in some sense "both bound state and free state." A system undergoing phase transition may be in the transition from bound state to condensed state. These boundary cases are real. But boundary cases do not negate completeness: they can still be decomposed into the tri-states, although the mode of decomposition is more complex, such as "50% bound state + 50% free state" or "in the process of transition toward condensed state." Completeness allows complex decomposition and does not require simple classification.

A Concrete Defense: Consciousness as a Tri-State Mixture

To keep the completeness claim from remaining abstract, this section gives a concrete defense: how consciousness is covered by the tri-state categories.

Consciousness has always been a difficult object to classify in the history of philosophy. It is unlike a material object, which has spatial location and stable structure; unlike a mathematical object, which is purely abstract; and unlike a sensory impression, which is momentary. Many doctrines of ontological categories encounter difficulty with the problem of consciousness.

The tri-state categories provide a concrete ontological location for consciousness: the bound-state component of consciousness consists in phase locking of neural oscillations, synchronization across brain regions, and the momentary orientation of attention; this is the "relationality" dimension of consciousness. The free-state component of consciousness consists in the flow of information in the brain, the inflow of sensation, and the continuous change of conscious content; this is the "accessibility" dimension of consciousness. The condensed-state component of consciousness consists in long-term memory, personality structure, and the internalized patterns of cultural learning; this is the "stability" dimension of consciousness. Consciousness as a whole is a specific mixture of the three states: it depends on all three states being strong and coordinated. When the tri-states are imbalanced, the state of consciousness changes: an excessively strong bound state leads to epilepsy-like over-synchronization; an excessively strong free state leads to dissociative states; and an excessively strong condensed state leads to rigidity or depressive states.

This tri-state analysis corresponds structurally to the $\Psi(\tilde{A}, C, \mathcal{B})$ framework in *Mind–Matter Unity*: $\tilde{A}$, consciousness intensity, corresponds to free-state–bound-state related dynamics; C, consciousness complexity, corresponds to bound-state structure; and $\mathcal{B}$, consciousness boundary, corresponds to condensed-state stability. But the work of this chapter is to treat this correspondence as ontology rather than physical modeling: the tri-state categories are the basic ontological categories of consciousness as an ontological object, while the Ψ framework is its physical modeling.

Objections and Replies

Objection One: The examples you give have already been described according to the tri-state categories. Is this circular? You first assume the tri-state categories, and then the examples naturally support them.

Reply: This objection points to a real concern. The strategy of this chapter is to use three kinds of argument at the same time: Argument One, the breadth of concrete examples, shows the applicability of the tri-state categories across domains; Argument Two, the exhaustiveness of ontological dimensions, argues conceptually that the three dimensions exhaust the basic ontological dimensions; and physical realization, Observation Three in Section 10.1, shows that the tri-states correspond to different terms of the EQT Master Equation. These three arguments support one another: concrete examples support the plausibility of the conceptual argument; the conceptual argument supports the non-arbitrariness of the concrete examples; and physical realization supports the concreteness of both. No single argument is sufficient by itself; their synthesis constitutes the argument for completeness. If an objector can provide a concrete counterexample, namely a being that clearly cannot be decomposed into the tri-states, that would be a strong challenge to completeness. This chapter welcomes such counterexamples. At present, no obvious counterexample has been found.

Objection Two: The argument that the tri-states correspond to spatial, relational, and temporal dimensions seems too neat. Is this a post hoc rationalization?

Reply: Post hoc rationalization is a real danger. This chapter honestly acknowledges that the tri-state categories historically came first from physics, namely from the detection-theoretical classification in *EQT Detection Theory*, and were then found to correspond to three ontological dimensions. But post hoc rationalization is not the same as error. Many truths are first partially discovered and then systematized. What matters is whether the argument of rationalization holds, not the historical order of rationalization. The argument in Observation Two of Section 10.1, namely that the three dimensions exhaust the basic ontological dimensions of beings, is independent of physics. It starts from "the most basic ontological inquiries about beings": where they are, what they are related to, and how long they persist. These inquiries are ones that any ontology must treat. Whether the argument holds is independent of the order of discovery.

Objection Three: The completeness claim is too strong. There may be some extreme beings, such as purely abstract mathematical objects or purely logical forms, that cannot be decomposed into the tri-states.

Reply: This objection concerns the distinction between form and reality in Chapter 3, Section 3.5. The tri-state categories in this chapter concern real objects: beings that satisfy $CR2$, namely beings for which the commitment to X has definable failure conditions. Formal constraints, such as mathematical structures and logical forms, are not real objects and belong to a different category. Mathematical objects and logical forms are not covered by the tri-state categories, but this does not weaken completeness, because they do not belong to the object class targeted by the tri-state categories. The completeness claim is this: all real objects can be decomposed into the tri-states. This is a more precise claim and avoids the overextension of the tri-state categories.

Literature That Must Be Engaged

- Aristotle, Categories, Chapters 1–9
- Immanuel Kant, Critique of Pure Reason, "Transcendental Analytic"
- David Wiggins, Sameness and Substance Renewed (2001): on ontological classification
- Theodore Sider, Writing the Book of the World (2011): on structural ontology and categories
- Chapter 3, Section 3.5 on the distinction between form and reality (self-reference)

Summary of This Section

This section has done four things. First, it stated completeness as a legitimacy condition for a doctrine of ontological categories. Second, it supported the completeness of the tri-state categories through three arguments: the breadth of concrete examples, the exhaustiveness of ontological dimensions, and physical realization. Third, it clarified the limits of completeness: completeness of ontological dimensions rather than concrete manifestations, allowance for mixed and transitional states, and the presence of boundary cases that can still be decomposed in complex ways. Fourth, it used consciousness as a concrete defense, showing how the tri-state categories cover an object that has traditionally been difficult to classify.

The next section defends the second legitimacy condition of the tri-state categories: irreducibility.

10.4. The Irreducibility of the Tri-State Categories

The Meaning of Irreducibility

Another core requirement of a doctrine of ontological categories is irreducibility: the categories cannot be reduced to one another. If one category can be reduced to a combination of other categories, then it is not a basic category. The irreducibility claim of the tri-state categories is this: free state, bound state, and condensed state cannot be reduced to one another. The free state cannot be expressed as a combination of bound state and condensed state; the bound state cannot be expressed as a combination of free state and condensed state; and the condensed state cannot be expressed as a combination of free state and bound state. Each state is a basic and independent ontological category.

Three Arguments for Irreducibility

Argument One: Through physical processes.

The tri-states correspond to different terms of the EQT Master Equation: free state $\leftrightarrow$ the diffusion term $D(f)\nabla^2\rho_f$; bound state $\leftrightarrow$ the nonlinear interaction term $k(f)\rho_f^2$; condensed state $\leftrightarrow$ the dissipation term $-\Gamma(f)\rho_f$. These three terms are mathematically independent: they involve different physical processes, namely diffusion, interaction, and dissipation, and cannot be transformed into one another through any transformation. At the level of the Master Equation, the three states have completely different dynamical roots. This physical argument does not directly prove ontological irreducibility, since ontological irreducibility is a philosophical problem. But it supports ontological irreducibility: if the tri-states are independent in physical dynamics, then distinguishing them as ontological categories has a ground.

Argument Two: Through functional differences.

The tri-states play different roles in the functions of beings: the free state allows a being to extend, meaning that its influence propagates far away and that it establishes connections with distant objects; the bound state allows a being to interact, meaning that it forms relations, participates in oscillation, and couples with neighboring objects; the condensed state allows a being to persist, meaning that it preserves its own structure, exists through time, and carries history. These three functions cannot substitute for one another. A being that has only the free state, such as a pure radiation field, cannot form a stable structure, because it has no condensed state to maintain itself. A being that has only the condensed state, such as an absolutely isolated solid, cannot interact with other objects, because it has no free state to propagate influence. A being that has only the bound state, such as pure oscillation, requires some kind of carrier: the bound state itself cannot exist independently apart from free state or condensed state. Each state performs an independent function of existence. The three functions cannot be reduced to one another; they concern different basic capacities of beings.

Argument Three: Through counterexamples.

If the tri-states could be reduced to one another, then certain core features of beings would be eliminated. Consider the following counterexamples.

Counterexample One: the free state cannot be reduced to bound state + condensed state. Consider the cosmic microwave background radiation. It is highly free-state in character: photons propagate through the universe. It has no significant bound-state component, since photons hardly interact with one another. It has no condensed-state component, since photons have no mass and no stable structure. If the free state could be reduced to a combination of bound state and condensed state, the cosmic microwave background should not exist; but it plainly does.

Counterexample Two: the bound state cannot be reduced to free state + condensed state. Consider atomic energy-level excitation. This is a bound state: an electron occupies a specific energy level within an atom. It is not a free state, since the electron has not propagated away; it is not a condensed state, since it is an excited state, unstable and subject to decay. It is an independent phenomenon of the bound state. If the bound state could be reduced, atomic energy levels should not have an independent physics; but they plainly do, since spectroscopy is a complete field of physics.

Counterexample Three: the condensed state cannot be reduced to free state + bound state. Consider Bose–Einstein condensation. This is a condensed state: a large number of particles collectively enter the same quantum state. It is not a free state, since the particles do not propagate freely; it is not an ordinary bound state, since it is a macroscopic quantum phenomenon, not a local oscillation. It is an independent phenomenon of the condensed state. If the condensed state could be reduced, BEC should not have an independent physics; but it plainly does, as reflected in the work recognized by the 2001 Nobel Prize in Physics.

These three counterexamples show that each of the three states has physical phenomena that cannot be absorbed by the other states. This is strong evidence for irreducibility.

The Distinction between Irreducibility and Decomposability

Irreducibility must be distinguished from decomposability. Irreducibility means that the three states cannot be reduced to one another: no one state can be replaced by a combination of the other two. Decomposability means that any being can be decomposed into the tri-states: beings often contain components of all three states. The two are compatible. A star is decomposed into the tri-states, since it simultaneously has free-state, bound-state, and condensed-state components, but the three states themselves are not reducible to one another: its free-state component cannot be replaced by bound-state + condensed-state components.

This distinction is important for a doctrine of categories. The irreducibility of Aristotle's ten categories has often been discussed: can substance, quantity, quality, relation, and so on be reduced to one another? The case of the tri-state categories is clearer: the tri-states are irreducible at the physical level, since they correspond to different terms of the Master Equation; irreducible at the functional level, since they perform different functions of existence; and irreducible at the counterexample level, since each state has phenomena that cannot be absorbed by the other states.

Ontological Implications of Irreducibility

If the tri-state categories are irreducible, this has implications for ontology.

Implication One: The plurality of existence.

Existence cannot be reduced to a single principle. This conflicts with traditional "ontological monism," which holds that there is some most basic principle of existence and that all beings are its manifestations. The tri-state categories argue that beings have three basic ontological dimensions, and that all three dimensions are basic and cannot be reduced to one another. But one clarification is required: this "plurality" is not a plurality of ontology, meaning it does not claim that there are three different types of ontology, but a plurality of categories, meaning that one type of ontology is decomposed under three categories. The tri-state categories are compatible with the work of Chapter 8 on qi monism: qi is the sole ontology, but the manifestations of qi are classified according to the tri-states.

Implication Two: The limits of reductionism.

Reductionism claims that complex phenomena can be reduced to more basic phenomena. The irreducibility of the tri-state categories points to the limits of reductionism: phenomena belonging to different categories cannot be reduced to one another. If consciousness has a significant bound-state component, it cannot be simply reduced to free-state processes. If life has a significant condensed-state component, it cannot be simply reduced to free-state chemical reactions. This is not an opposition to reductionism as a method: reductionist methods remain valid within each category. But reduction across categories is limited.

Implication Three: The precision of ontology.

If all three states are irreducible, ontology must treat each state separately. Once all beings are decomposed according to free state, bound state, and condensed state, the work of ontology becomes more precise: it no longer asks "what is a being in general?" but "what are the manifestations of a being in each state?" This makes ontology closer to concrete science and more open to testing.

Objections and Replies

Objection One: Your "functional difference argument" seems to appeal to some kind of functionalism. But functionalism is controversial in ontology.

Reply: The "functional difference argument" in this chapter is one supporting argument, not the primary argument. The primary arguments are the irreducibility of physical processes, namely Argument One, and the argument from counterexamples, namely Argument Three. Functional difference is an expression of these arguments. Even considered alone, however, the concept of "function" in this argument is not function in a cognitive or teleological sense, but physical function: extension, interaction, and persistence. This differs from the cognitive concept of function in functionalism. The irreducibility of physical functions is an empirical issue and does not presuppose a functionalist position.

Objection Two: Is Counterexample Three, Bose–Einstein condensation, really independent? BEC is in fact a collective state of many atoms, which are bound-state structures. It therefore seems reducible to bound states.

Reply: BEC is not simply a "collection of bound states." It is a macroscopic quantum condensate with independent physical phenomena, such as superfluidity and quantum vortices. These phenomena cannot be directly inferred from the properties of single-atom bound states; they are collective emergent properties of condensation. More generally, condensed matter physics as a complete field of physics studies phenomena that cannot be derived from single-particle physics. This is the ground for condensed state as an independent physical level. This chapter translates this independence into the irreducibility of an ontological category.

Objection Three: The tri-states still seem to have hierarchical relations. The condensed state seems to "contain" bound states, and bound states seem to "contain" free states. Is this a hidden form of reduction?

Reply: The tri-states do have relations. Deeper condensation usually involves more bound structures, and more bound structures usually involve more free flows of energy. But this relation is functional dependence, not ontological reduction. The condensed state requires bound states as constituents, but the condensed state itself is not a combination of bound states: it is an independent phenomenon that emerges when bound states enter a specific collective mode. By analogy, water molecules are composed of hydrogen and oxygen atoms, but "fluidity" as a phenomenon of water cannot be directly inferred from the chemical properties of hydrogen and oxygen atoms; it is a property that emerges from the molecular ensemble. The relation among the tri-states is similar: one state can be a constituent of another, but it cannot be reduced to the latter.

Literature That Must Be Engaged

- Hilary Putnam, "On Properties" (1969): on property reduction and irreducibility
- Jaegwon Kim, Mind in a Physical World (1998): on the problem of irreducibility
- Brian McLaughlin, "The Rise and Fall of British Emergentism" (1992): on emergence and reduction
- Roy Bhaskar, A Realist Theory of Science (1975): on stratified realism
- Standard textbooks in condensed matter physics, such as P. W. Anderson, Basic Notions of Condensed Matter Physics (1984)

Summary of This Section

This section has done four things. First, it stated irreducibility as a legitimacy condition for a doctrine of ontological categories.

Second, it supported the irreducibility of the tri-state categories through three independent arguments: the argument from physical processes, the argument from functional differences, and the argument from counterexamples. Third, it distinguished irreducibility from decomposability: the tri-states can combine with one another in decomposition, but cannot replace one another through reduction. Fourth, it developed the ontological implications of tri-state irreducibility: the plurality of existence, the limits of reductionism, and the precision of ontology.

The next section compares the tri-state categories seriously with Aristotle's ten categories and Kant's twelve categories.

10.5. Comparison with the Categorial Doctrines of Aristotle and Kant

A Brief Review of the History of Categorial Doctrines

Before comparison, this section briefly reviews two major works in the history of categorial doctrines.

Aristotle's ten categories, from the *Categories*. Aristotle identifies ten ways in which being is predicated: substance, quantity, quality, relation, place, time, position, state, action, and passion. These ten categories concern substance and its modes of properties: they tell us how to speak about a being. Aristotle's doctrine of categories is closely connected with his substance ontology: substance is the basic category, and the other nine categories are modes of the properties of substance.

Kant's twelve categories, from the *Critique of Pure Reason*. Kant identifies the possible forms of judgment and thereby derives the pure concepts of the understanding: quantity, namely unity, plurality, and totality; quality, namely reality, negation, and limitation; relation, namely substance, causality, and community; and modality, namely possibility, actuality, and necessity. Kant's doctrine of categories is epistemological: it tells us how any possible experience must be organized according to these twelve categories.

These two works have had enormous influence in the history of philosophy. Any later work on categorial doctrines must face them: either by inheriting them, revising them, or challenging them.

The Position of the Tri-State Categories

The tri-state categories differ from the categorial doctrines of Aristotle and Kant. These differences do not mean that "the tri-state categories are better than them," but that they treat different problems.

Aristotle's ten categories concern the properties of substance. Aristotle's doctrine of categories presupposes substance as foundation: substance is the most basic form of "being," and the other categories, such as quantity, quality, and relation, are modes of the properties of substance. This presupposition is central to Aristotle's metaphysics. The tri-state categories reject this presupposition. They begin from process ontology, as Chapter 7 has established: beings are not understood as substances plus properties, but as specific modes of process. The tri-states correspond to three modes of existence of process, namely accessibility, relationality, and stability; they do not correspond to three properties of substance. This is the fundamental difference. Aristotle's categories are valid within the framework of substance ontology; the tri-state categories are valid within the framework of process ontology. The two do not compete: they treat different ontological problems.

Kant's twelve categories concern the forms of judgment. Kant's doctrine of categories is epistemological: it tells us how possible experience must be organized. Kant explicitly says that the object of application of the categories is not the thing-in-itself, but phenomena. Categories are the ways in which the understanding gives form to experience; they are not categories about beings themselves. The tri-state categories are ontological, not epistemological. They tell us how beings themselves are ontologically decomposed, not how experience is organized. The tri-state categories do not presuppose the subject–object division, namely Kant's epistemological framework; they directly state theses about beings. This is also a fundamental difference. Kant's categories are valid within an epistemological framework, since they tell us the formal conditions of experience; the tri-state categories are valid within an ontological framework, since they tell us the basic ontological dimensions of beings. The two also do not compete: they treat problems at different levels.

The Distinctive Position of the Tri-State Categories

Through comparison, the distinctive position of the tri-state categories can be stated as follows:

Key thesis T2: the tri-state categories are ontological categories concerning the modes of existence of processual beings. They do not presuppose substance as foundation, unlike Aristotle; they do not presuppose a subject–object division, unlike Kant; and they do not presuppose that categories are "forms of judgment," unlike Kant. They state directly that the location of any processual being along three basic ontological dimensions, namely accessibility, relationality, and stability, is characterized by the three states: free state, bound state, and condensed state.

This distinctive position gives the tri-state categories a place in the history of philosophy. Aristotle provides a doctrine of categories for substance ontology; Kant provides a doctrine of categories for epistemology; the tri-state categories provide a doctrine of categories for process ontology. The three works address different metaphysical problems.

Other Relevant Works

Beyond Aristotle and Kant, several other works on categorial doctrines should be mentioned.

Whitehead's doctrine of categories, in *Process and Reality.* Whitehead gives a detailed categorial system: the "category of the ultimate," namely creativity, many, and one; eight "categories of existence"; twenty-seven "categories of explanation"; and nine "categorial obligations." This is a vast categorial system directed at processual beings. Whitehead differs importantly from the tri-state categories. Whitehead's categorial system is large in number and complex in hierarchy: many categories concern concrete processual phenomena, such as prehension, concrescence, superject, and so on. The tri-state categories are more concise: three categories cover the basic ontological dimensions of beings. Whether this concision is an advantage or disadvantage depends on the standard of evaluation: if concision is prioritized, the tri-state categories are preferable; if detail is prioritized, Whitehead's categorial system is preferable. This chapter does not claim that the tri-state categories "replace" Whitehead's categorial system. It claims that the tri-state categories have an advantage in the function of concise ontological classification. Whitehead's more detailed categorial system has its own value in concrete process-philosophical description.

Peirce's three categories, from "On a New List of Categories." Peirce identifies three basic categories: Firstness, the character of pure quality; Secondness, opposition and reaction; and Thirdness, mediation and law. Peirce's three categories have a structural similarity with the tri-state categories: both are triadic doctrines of categories, and neither presupposes substance ontology. But their contents differ.

Peirce's categories are closer to phenomenology, since Firstness is the quality of pure feeling; the tri-state categories are closer to physical ontology, with the free state corresponding to accessibility, and so on. This similarity deserves acknowledgment in this chapter: Peirce's work is an important precedent for triadic categorial doctrine in Western philosophy. The tri-state categories do not claim an isolated originality without precedent. They contribute along the direction opened by Peirce. But the content of the tri-state categories, including their interface with physics and their connection with process ontology, differs from Peirce's.

Possible Evolution of the Tri-State Categories

No doctrine of categories should be treated as final. Aristotle's ten categories were revised many times by later philosophers; Kant's twelve categories were also modified and supplemented by later work. The tri-state categories should also remain open to future revision. Possible directions of revision include: adding categories, since future work may identify a fourth basic ontological category; this chapter does not exclude that possibility, although at present no category requiring addition has been found; refining categories, since each state may in the future be further differentiated, for example the bound state may be divided into subcategories such as "linear binding" and "nonlinear binding"; revising the meaning of the categories, since future work may refine the definition of each state; and revising the correspondence with physics, since if EQT is replaced by another theory in the future, the correspondence between the tri-state categories and the new theory may need to be re-established. Acknowledging these possibilities does not weaken the work of this chapter. It makes the categorial doctrine of this chapter a living and developable work, rather than a rigid final assertion.

Objections and Replies

Objection One: You seem to set aside Aristotle's and Kant's doctrines of categories too easily by saying that they treat different problems. But in fact they both claim to be foundational doctrines of categories. If the tri-state categories are truly foundational categories, they should be able to cover the work of Aristotle and Kant.

Reply: This chapter does not claim that the tri-state categories cover the work of Aristotle or Kant. It claims that the tri-state categories are foundational categories for the problem of process ontology. Aristotle's doctrine of categories is foundational within substance ontology; Kant's doctrine is foundational within epistemology. Each of the three works is foundational within its own problem, but they address different problems. If an objector claims that one doctrine of categories should cover all others, that is too strong a claim. The history of philosophy shows that no doctrine of categories has successfully covered all problems. The coexistence of multiple doctrines of categories is reasonable: each treats a specific ontological or epistemological problem.

Objection Two: Does the similarity with Peirce weaken the originality of the tri-state categories?

Reply: The similarity is real, and this chapter acknowledges it honestly. But the content of the tri-state categories, including their interface with physics and their connection with EQT, differs from Peirce's. Originality does not lie in isolated absence of precedent, but in new contribution of content. The content of the tri-state categories is new, even if the form of the categorial doctrine, namely triadic categories, has a precedent in Peirce. More deeply, philosophical progress often takes the form of refinement and new application along existing directions, not isolated creation in a direction without any precedent. The tri-state categories make a physical-ontological contribution along the direction of triadic categories opened by Peirce. This contribution is a legitimate form of originality.

Objection Three: This chapter treats Whitehead too briefly. Whitehead has a very detailed categorial system, but this chapter treats it in only one paragraph. Is this unfair?

Reply: The aim of this chapter is not to evaluate Whitehead in detail; that would require dedicated work. The aim of this chapter is to compare the tri-state categories with Whitehead's categorial system. For this task, the comparison given here is fair: the tri-state categories are more concise, Whitehead's are more detailed, and each has advantages. A more detailed comparison, including comparison of specific categories and evaluation of the reasonableness of particular categories, requires a dedicated paper or monograph. This chapter is one chapter in a monograph, and the limits of length require concise treatment.

Literature That Must Be Engaged

- Aristotle, Categories
- Immanuel Kant, Critique of Pure Reason, "Transcendental Analytic" Book I

- C. S. Peirce, "On a New List of Categories" (1867)
- Alfred North Whitehead, Process and Reality (1929), Part I Chapter II
- John McDowell, Mind and World (1994): on contemporary discussion of Kantian categories
- E. J. Lowe, The Four-Category Ontology (2006): contemporary four-category ontology

Summary of This Section

This section has done three things. First, it reviewed the character of Aristotle's ten categories and Kant's twelve categories. Second, it rigorously stated the key thesis T2: the tri-state categories are categories of process ontology and differ in nature from Aristotle's substance-ontological categories and Kant's epistemological categories. Third, it mentioned other relevant works, namely Whitehead and Peirce, acknowledged similarities, and clarified differences.

The next section gives the final application of this chapter: analyzing beings in several different domains through the tri-state categories.

10.6. Applications of the Tri-State Doctrine of Categories

The Function of Application

The preceding sections established the legitimacy conditions of the tri-state doctrine of categories: completeness, irreducibility, and distinctive position. This section demonstrates, through applications, how the tri-state categories are used, that is, how they function as an analytical tool for understanding concrete beings. Application is not merely an illustration of the reasonableness of the preceding work; it is itself a core function of a doctrine of categories. The value of a doctrine of categories lies not only in the argument for its legitimacy, but in whether it can be effectively used to analyze concrete problems.

Application One: A Physical Object–Tri-State Analysis of the Electron

The electron is one of the basic objects of physics. How can it be analyzed under the tri-state categories?

The dominance of the bound state in the electron. Under ordinary conditions, the electron is in a bound state: it is bound by the atomic nucleus in specific atomic energy levels. The core work of atomic physics, including spectroscopy, chemical bonding, and atomic structure, revolves around the bound state of the electron. In this sense, the electron is a being mainly in the bound state.

The free-state component of the electron. But the electron also has a free-state component. Free electrons, namely electrons not bound by atomic nuclei, exist in vacuum, plasma, and conductors: they propagate, form electric currents, and contribute to conductivity. In accelerators, electrons are accelerated as free particles. In these cases, electrons appear as free states.

The condensed-state component of the electron. Under certain extreme conditions, electrons appear as condensed states. In superconductors, electrons pair into Cooper pairs and collectively enter a superfluid state: this is a condensed state of electrons. In white dwarfs and neutron stars, electron degeneracy pressure maintains the star: this is the manifestation of the electron condensed state in astrophysics.

The electron manifests different tri-state mixtures under different conditions. This tri-state analysis differs importantly from the traditional dichotomy of "electron as particle" or "electron as wave": the tri-state analysis is more fine-grained and captures more ontological dimensions of the electron.

Application Two: A Living Object–Tri-State Analysis of the Cell

The cell is the basic unit of life. How can it be analyzed under the tri-state categories?

The free-state component of the cell. The cell has significant free-state components: ion flows across the membrane, diffusion of signaling molecules, transport of metabolic substrates, and cyclic transfer of energy, namely ATP. These are the "flow" dimensions of the cell: energy and matter continuously flow within the cell.

The bound-state component of the cell. The cell has strong bound-state components: cyclic reactions catalyzed by enzymes, oscillatory gene expression, feedback loops in signal transduction, and coordinated oscillations of the cell cycle. These are the "relation" dimensions of the cell: interactions among molecules form stable oscillations.

The condensed-state component of the cell. The cell also has condensed-state components: membrane structure, cytoskeleton, the stable structure of DNA, and the morphology of organelles. These are the "stability" dimensions of the cell: the material architecture that supports life processes.

The cell as a whole is a being in which all three states are strong and coordinated. This is the ontological characteristic of life. The reason a cell is "living" rather than "non-living" is precisely that it is simultaneously active in all three states and that the three states are highly coordinated. When any one state collapses, life ceases: energy flow stops, meaning the collapse of the free state; metabolic cycles stop, meaning the collapse of the bound state; or cellular structure disintegrates, meaning the collapse of the condensed state. This tri-state analysis differs importantly from reductionist descriptions such as "life is a chemical process" or "life is information processing": the tri-state analysis emphasizes that life is a specific ontological state in which all three states are simultaneously strong and coordinated.

Application Three: A Psychological Object–Tri-State Analysis of Consciousness

Section 10.2 already gave a brief tri-state analysis of consciousness. This section develops it as a full application.

The free-state component of consciousness. Consciousness has significant free-state components: the continuous inflow of sensory information, the instantaneous movement of attention, the continuous flow of conscious content, namely the "stream of consciousness," and the rapid change of thought. William James's concept of the "stream of consciousness" captures this free-state component: consciousness is not static; it is a continuously flowing process.

The bound-state component of consciousness. Consciousness has strong bound-state components: phase locking of neural oscillations, especially whole-brain synchronization in the gamma band; coordinated oscillations across brain regions; stable orientation of attention; and the process of perceptual binding, namely feature binding. The "binding problem" in neuroscience concerns precisely this state: how are different sensory features bound into unified perception?

The condensed-state component of consciousness. Consciousness has condensed-state components: stable storage of long-term memory, persistent patterns of personality structure, internalized structures of cultural learning, and continuity of self-identity. These are the "stability" dimensions of consciousness: the enduring foundation that supports conscious processes.

Consciousness is similar to life: it requires all three states to be strong and coordinated. When one state becomes significantly imbalanced, the state of consciousness changes: an excessively strong or uncontrolled free state leads to dissociative symptoms, drifting consciousness, and scattered attention; an excessively strong or abnormal bound state leads to epilepsy-like over-synchronization and compulsive thought; an excessively strong or rigid condensed state leads to depressive rumination and rigidity of personality. The $\Psi(\tilde{A}, C, \mathcal{B})$ framework in *Mind–Matter Unity* can be understood as the physical operationalization of tri-state analysis: $\tilde{A}$ mainly measures free-state–bound-state related dynamics, namely activation and dissipation; C mainly measures bound-state structure, namely complexity and integration; and $\mathcal{B}$ mainly measures condensed-state stability, namely boundary and closure. This correspondence gives the tri-state categories a physical operation in consciousness research.

Application Four: A Social Object–Tri-State Analysis of the Market

The final application shows the cross-domain applicability of the tri-state categories. They are not limited to physics and biology, but can also analyze social objects.

The free-state component of the market. The market has significant free-state components: the rapid transmission of price signals, the diffusion of information among market participants, the global flow of capital, and instantaneous changes in consumer preferences. This is the "flow" dimension of the market: various kinds of information and resources continuously flow through the market.

The bound-state component of the market. The market has important bound-state components: periodic price oscillations, namely business cycles; feedback loops of supply and demand; stable relations between buyers and sellers, such as long-term contracts; and coordinated oscillations within industrial ecosystems. This is the "relation" dimension of the market: interactions among market participants form stable patterns.

The condensed-state component of the market. The market has condensed-state components: institutional structures, including laws, rules, and property rights; long-term capital accumulation; the persistence of brands and reputation; and market infrastructure, such as exchanges, clearing systems, and banking systems. This is the "stability" dimension of the market: the enduring architecture that supports market processes.

Healthy market operation requires tri-state balance. A purely free-state market, without institutional constraints, becomes unstable; a purely bound-state market, with excessive coordination, loses innovation; and a purely condensed-state market, with excessive rigidity, loses liquidity. Market crises often correspond to severe tri-state imbalance: a financial crisis is the simultaneous occurrence of excessive free state, namely excessively free movement of funds, and collapse of condensed state, namely loss of institutional trust. This application shows that the tri-state categories can be used seriously in social-scientific analysis. But this chapter does not claim that the tri-state categories replace the tools of economics or sociology. They provide an ontological analytical framework that complements concrete work in the social sciences.

Methodological Remarks on Application

The four applications show how the tri-state categories are used. This section explicitly states the methodology of application.

Application point one: analysis rather than classification. The use of the tri-state categories is not to classify beings simply, as in "the electron is a bound-state object," but to analyze their tri-state components and their interactions. Every being is a tri-state mixture, and application consists in refining the concrete structure of that mixture.

Application point two: hierarchical application. The tri-state categories can be used at different levels. A cell as a whole has a tri-state analysis; a specific process within the cell, such as the mitochondrion, also has a tri-state analysis; the electron transport chain within the mitochondrion also has a tri-state analysis. Tri-state analyses at each level support one another, but do not reduce to one another.

Application point three: diagnostic value of tri-state balance. The tri-state categories have diagnostic value in many domains: they indicate that the healthy or abnormal state of a being can be assessed through the state of tri-state balance. The health of life requires tri-state coordination; the health of consciousness requires tri-state balance; social health requires tri-state balance. This diagnostic value gives the tri-state categories practical as well as theoretical value.

Objections and Replies

Objection One: Your applications seem too broad: the tri-state categories are applied from physics to biology, psychology, and society. Is this excessive generality, and does it therefore lose specificity?

Reply: Generality is not the same as loss of specificity. A genuinely foundational doctrine of ontological categories should apply to all domains; this is precisely what marks it as a doctrine of foundational categories. But each application requires concrete work. It is not a matter of "applying" the tri-state categories mechanically, but of analyzing the tri-state components of a concrete being and their interactions. Specificity comes from the concrete content of each domain. The tri-state analysis of the electron involves physics; the tri-state analysis of the cell involves biology; the tri-state analysis of consciousness involves neuroscience; and the tri-state analysis of the market involves economics. Each analysis requires concrete work in that domain. The tri-state categories only provide the framework for ontological analysis.

Objection Two: Is treating the market as an object of ontological analysis too broad? According to your earlier definition, such as in Chapter 3, the market may not be a "real object."

Reply: This objection points to a real concern. The status of the market as an ontological object must be evaluated under the $CR2$ framework of Chapter 3. If the commitment to "market" has concrete failure conditions, including dimensionality, such as trading volume and price volatility; proxies, such as economic data and behavioral observations; and thresholds, namely measurable values, then it is a real object and can be analyzed through the tri-states. If "market" is only a loose term, such as "the global market" as a vague ontological unit, then the object of tri-state analysis must be made more precise: it may be a specific market, such as the securities market of a particular country, as a real object, rather than "market" as a general concept. This precision is required by the method of constrained speculation.

Objection Three: Your applications are descriptive: you merely say that "X has tri-state components." Is this only labeling existing phenomena? Do the tri-state categories bring new predictions?

Reply: This objection points to a real concern. Doctrines of categories are often criticized for merely describing rather than predicting. The reply of the tri-state categories is that they bring a new diagnostic structure: tri-state balance as a diagnostic criterion for the health of beings. This diagnostic structure has guiding significance for concrete research. In life sciences, research on aging or disease can proceed from the perspective of "tri-state imbalance": which state becomes imbalanced first? What is the specific mode of tri-state imbalance? In consciousness research, tri-state balance as a criterion of conscious health provides a new research direction: which state is defective in disorders of consciousness? In social research, tri-state balance as a criterion of social health provides a new line of policy thinking: which state of imbalance corresponds to a given social problem? These "diagnostic structures" are products of the application of the tri-state categories; they are not mere "description." They correspond to concrete questions that can be studied.

Literature That Must Be Engaged

- Chapter 11 on freezing as a foundational ontological process (self-reference)
- *Mind–Matter Unity* on the $\Psi(\tilde{A}, C, \mathcal{B})$ state-variable group (self-reference)
- William James, Principles of Psychology (1890): on the stream of consciousness
- P. W. Anderson, "More is Different" (1972): on levels and emergence
- Stuart Kauffman, The Origins of Order (1993): on self- organization in complex systems

Summary of This Section

This section has done three things. First, it showed how the tri-state categories are used through four applications: the electron, the cell, consciousness, and the market. Second, it clarified the methodology of application: analysis rather than classification, hierarchical application, and the diagnostic value of tri-state balance. Third, it replied to possible objections, especially concerns about generality, ontological status, and predictiveness.

10.7. Chapter Summary

The three tasks undertaken by this chapter have all been completed.

Task One: to elevate the tri-states from a classification in the theory of detection in *EQT Detection Theory* into ontological categories. Section 10.1 prepared the ground for this elevation through three observations: exhaustiveness, non-arbitrariness, and physical realization, and rigorously stated the core thesis T1.

Task Two: to argue for the legitimacy of the tri-state categories. Section 10.2 defended completeness: the tri-state categories cover any real object. Section 10.3 defended irreducibility: the three states cannot be reduced to one another. Section 10.4 developed the distinctive position of the tri-state categories in the history of categorial doctrines, namely the key thesis T2: they treat process ontology and differ in nature from Aristotle's categories of substance ontology and Kant's epistemological categories.

Task Three: to show the applications of the tri-state categories. Section 10.5 used four cross-domain applications, namely the electron, the cell, consciousness, and the market, to show how the tri-state categories function as an analytical tool.

The Position of This Chapter in the Whole Book

Looking backward, this chapter builds upon Chapter 7, "Difference before Identity," and Chapter 8, "The Tradition of Qi Theory." Difference before identity provides the ontological starting point for the tri-state categories, since any being involves differentiation; the tradition of qi theory provides Chinese philosophical resources structurally related to the tri-state categories, namely the different modes of existence of qi, corresponding to the ontological classification of beings in the Chinese tradition.

Looking forward, this chapter supports Chapter 11, "Freezing as a Foundational Ontological Process": freezing is the process by which the condensed state is formed, and the tri-state categories allow us to understand freezing more precisely as a modal change of beings, from bound state or free state into condensed state. This chapter also provides resources for Chapter 13, "Mind–Matter Unity": the tri-state analysis of consciousness is a tool for the philosophy of mind and matter.

The Initial Establishment of the Tri-State Doctrine of Categories

This chapter is the initial establishment of the tri-state doctrine of categories: it gives the basic framework and argument. But doctrines of categories require long-term development in the history of philosophy. The tri-state doctrine of categories may in the future be criticized and revised, since concrete counterexamples may require refined definitions or revised legitimacy arguments; applied to more domains, since this chapter gives only four application examples and future work can extend the applications to other domains; refined into subcategories, since each state may in the future be further differentiated; and synthesized with other doctrines of categories, since more concrete connections can be established between the tri-state categories and the work of Aristotle, Kant, Peirce, and Whitehead.

This chapter does not pretend to "complete" the tri-state doctrine of categories. It is an initial statement and invites later work. This posture is an application of the method of constrained speculation: any ontological work must acknowledge its openness and not pretend to be complete.

The Core Claim of This Chapter

The core claim of this chapter can be stated as follows: process ontology needs its own doctrine of categories.

Aristotle's doctrine of categories serves substance ontology; Kant's doctrine of categories serves epistemology. The rise of process ontology since the twentieth century, including Whitehead, Bergson, and Prigogine, requires corresponding categorial work, but this work has been insufficient in the history of philosophy. Whitehead's categorial system is too detailed and complex; Peirce's three categories are pioneering but lack an interface with contemporary physics. This chapter argues that the tri-state categories are a concise, operable doctrine of categories for process ontology, with an interface to physics. They are a contribution to the unfinished categorial work of the process-ontological tradition since the twentieth century. If this claim holds, the tri-state doctrine of categories will have a place in the history of categorial doctrines: not by "replacing" Aristotle or Kant, but by "supplementing" them and treating problems that their work does not sufficiently treat, namely the ontological modes of processual beings.

A Final Reflection

This chapter has a special difficulty in writing: it must rigorously establish a new doctrine of categories within relatively limited space. In the history of categorial doctrines, this is work that usually requires the length of a monograph. The strategy of this chapter is to state the core theses precisely, T1 and T2; to argue rigorously for the legitimacy conditions, completeness and irreducibility; to clarify the distinctive position through comparison; and to show the mode of use through application. This strategy avoids the need for the length of a full monograph while preserving philosophical rigor. Whether this strategy succeeds will be judged by readers. But even if the tri-state doctrine of categories is partially revised or challenged in the future, the work of this chapter remains valuable: it demonstrates how categorial work can be done within the framework of process ontology. Such work is rare in the history of philosophy, and the tri-state doctrine of categories is one attempt.

Meaning before freezing: this is one of the core theses of this book.

The tri-state doctrine of categories provides an ontological structure for this thesis. Meaning appears in beings in which all three states are strong and coordinated, such as life, consciousness, and society. Meaning disappears in states of affairs in which the tri-states collapse, such as the gradient terminal phase, heat death, and absolute isolation. This connection will be further developed in Chapter 11, on freezing, and Chapter 21, on the finitude of meaning.

10.8. Methodological Reflection on This Chapter

This chapter elevates the tri-states from a classification in the theory of detection into ontological categories. This section reflects on the methodological significance of this categorial work.

This Chapter's Posture toward Originality

This chapter honestly states its originality: the tri-state doctrine of categories is a new work in categorial doctrine, undertaking the categorial work needed by process ontology. At the same time, this chapter honestly acknowledges relevant precedents, including Peirce's triadic categories and Whitehead's doctrine of categories, without pretending to be an isolated creation without precedent. This balance is legitimate: originality often takes the form of refinement and new application along existing directions. The contribution of this chapter is to make precise the categorial need of process ontology in terms of the tri-states; to support the tri-state categories through physical grounds; and to systematize the tri-state categories as a doctrine of categories based on completeness + irreducibility.

The Position of This Chapter between Physics and Philosophy

The argument of this chapter involves physics, including different terms of the EQT Master Equation and physical phenomena such as BEC, and philosophy, including the legitimacy conditions of ontological categories and comparison with Aristotle and Kant. The way this chapter combines the two is this: physics provides grounds, and philosophy provides the argumentative structure. Aristotle's doctrine of categories is built upon the analysis of Greek grammar and everyday experience; Kant's doctrine of categories is built upon the analysis of forms of judgment; the tri-state doctrine of categories is built upon the analysis of the ontological content of contemporary physics. This is a legitimate way of constructing a categorial doctrine in contemporary philosophy of nature.

The Limits of This Chapter as Categorial Work

Categorial work has historically been among the longest forms of philosophical work. This chapter's categorial work within the limits of a monograph chapter is highly compressed. Any shorter treatment would be insufficient for the serious establishment of a new doctrine of categories. Even so, the work of this chapter remains an initial statement rather than a final establishment. In the history of philosophy, a doctrine of categories often requires decades or even generations of discussion before it becomes established within an academic tradition. This chapter does not pretend to bypass that process. What it does is to state clearly, argue rigorously, identify possible objections, and invite later discussion and refinement within an academic tradition.

Limits Honestly Acknowledged in This Chapter

This chapter acknowledges several limits: the tri-state doctrine of categories is an initial statement, and future work is needed for further development; the application examples are limited, since the four examples are representative but not exhaustive; the comparison with other doctrines of categories is brief; and the treatment of boundary cases is brief, since concrete boundary cases may require more detailed analysis. These limits do not weaken the work of this chapter. They are the reasonable scope of any initial work in categorial doctrine.

Connections with Later Chapters

The tri-state categories established in this chapter are widely applied in later chapters. Chapter 11 uses the tri-states as the target categories of the freezing process. Chapter 12 uses the tri-states as one of the elements of co-evolutionary dynamics. Chapter 13 uses the tri-states as categories of consciousness-related states. Chapter 14 uses the tri-states as ontological categories of the observational process. More broadly, this chapter is one of the more original chapters of the whole book. If the argument for the tri-state doctrine of categories holds, this chapter has an independent position in the history of categorial doctrines. This is the contribution of this chapter to international philosophical discussion and one of the markers of this book as an independent philosophical monograph.

11. Freezing as a Foundational Ontological Process

11.1. Introduction to This Chapter

In ordinary language, the word "freezing" usually means stopping: something becomes still, ceases to act, and is locked into a certain state. But in *Freezing Cosmology*, "freezing" as a cosmological mechanism does not have this ordinary meaning. In that volume, freezing is precisely defined as a process in which something exits active dynamics while remaining as residual, background, or history: photon decoupling about 380,000 years after the Big Bang, in which photons exited dominant interaction and became the CMB; nucleosynthesis in the early universe; and later structure formation. These are physical examples of "freezing."

But this concept has philosophical meaning beyond cosmology. This chapter argues that freezing can be elevated into a foundational ontological process: it is not merely one stage of cosmic evolution, but a basic ontological determination of how any processual being enters a relatively stable form from a dynamic process.

Specifically, this chapter argues for a series of ontological theses: substance as a frozen transient, meaning that so-called "substance" is the relatively stable form produced when a structure of difference undergoes freezing; time as the direction of freezing, meaning that the unidirectionality of time is grounded in the irreversibility of freezing; spatial geometry as the statistical appearance of low-frequency freezing; and memory as the freezing of consciousness. These theses extend the concept of freezing across several dimensions of ontology.

This chapter occupies a special position in Part II: it is the pivotal chapter of that part. The preceding four chapters, Chapter 7 on difference before identity, Chapter 8 on the tradition of qi theory, Chapter 9 on frequency as a basic spectral quantity, and Chapter 10 on tri-states as ontological categories, each treat different aspects of gradient ontology. This chapter integrates these aspects into a unified ontological picture and prepares the ground for Part IV, the metaphysical boundary of freezing.

This chapter does not depend on the physical derivation of EQT. It uses the concept of freezing already established in *Freezing Cosmology* as a premise and elevates it to the ontological level. All arguments are conducted at the philosophical level.

The discussion proceeds as follows. Section 11.1 provides precise conceptual clarification: freezing is not stopping. Section 11.2 develops the core thesis of freezing as a foundational ontological process (FZ1) and its sub-theses concerning substance, time, space, and memory. Section 11.3 develops the connection between freezing and the tri-state categories. Section 11.4 develops the hierarchy and historicity of freezing, preparing the ground for Chapter 16. Section 11.5 develops freezing as a condition of visibility (FZ4) and addresses objections.

11.2. Freezing Is Not Stopping

Conceptual Clarification

In ordinary language, the word "freezing" usually means stopping: something becomes still, ceases to act, and is locked into a certain state. If we understand "freezing" in this literal sense of stopping, the work of this chapter becomes confused, because an ontologically important process concept would be misunderstood as a static state.

This section first provides a precise conceptual clarification: freezing is not stopping; freezing is the process of exiting active dynamics while remaining as residual, background, or history.

Let us analyze this definition word by word. "Exiting active dynamics": freezing involves a process withdrawing from a state of dominant evolution into a non-dominant state. Before the exit, this process is the dominant dynamics within a certain domain; after the exit, it no longer dominates. "But remaining as residual, background, or history": losing dominance does not mean disappearing. The frozen process still exists: it continues to exist as residual, as in the CMB as a residual of the early universe; as background, as in the stability of basic particles providing a background for higher-level processes; or as history, as in past events remaining in a form that can be studied.

This definition distinguishes freezing from three neighboring concepts: freezing $\neq$ stopping, since stopping is absolute, meaning that the process no longer occurs, whereas freezing is not absolute, since the process still exists in some form, although it no longer dominates; freezing $\neq$ equilibrium, since equilibrium is the stability of a certain state, in which the system minimizes free energy, whereas freezing does not require equilibrium, because a frozen process may be in a non-equilibrium state, although its non-equilibrium dynamics no longer dominates; freezing $\neq$ disappearance, since disappearance means no longer existing, whereas a frozen process still exists and is retained in some weaker form.

A Concrete Example: The CMB

To prevent the concept of freezing from remaining abstract, this section considers a concrete example: the cosmic microwave background radiation (CMB).

The CMB is the residual electromagnetic radiation from about 380,000 years after the Big Bang. Before that time, photons interacted strongly with matter and could not propagate freely. After that time, namely during "recombination" or "photon decoupling," matter became transparent and photons could propagate freely: the interaction between photons and matter exited dominant dynamics. This is a process of freezing.

The CMB as the present state of this freezing process has several features: it is residual, since it is not newly produced radiation, but a residual of interactions during the early high-density plasma phase; it has physical properties, including a temperature of about 2.7K, a spectrum close to black-body radiation, and anisotropies, all of which are measurable physical facts; it does not play a dominant role in contemporary cosmic evolution, since the energy density of the CMB is very small in the present universe and does not dominate the evolution of current processes; and it preserves information, since the properties of the CMB, especially its anisotropy spectrum, preserve important information about the early universe and allow us to infer the state of the early universe.

This example shows the meaning of the concept of freezing. The CMB is not radiation that has "stopped": it still exists, still propagates, and remains measurable. But it has exited the dominant dynamics of the early universe, namely strong interaction with the plasma, and entered a residual state. This "exit while remaining" is the core of freezing.

A Counterexample: Black-Hole Evaporation

To make the concept of freezing precise, this section also considers a counterexample: what is not freezing?

Consider black-hole evaporation, or Hawking radiation. A black hole gradually loses mass through Hawking radiation and eventually, on an extremely long timescale, evaporates completely. When evaporation is complete, the black hole disappears: it no longer exists as a black hole. This is not freezing. Evaporation is not exiting dominant dynamics and entering a residual state; it is complete disappearance. After evaporation is complete, the black hole as an astronomical body does not exist in any form, whether residual, background, or history, as a black hole. But the products of black-hole evaporation, namely photons of Hawking radiation, are results of a freezing process. These photons "detach" from the high-density gravitational environment of the black hole and enter a freely propagating state: this is freezing. The black hole as an astronomical body disappears, but its energy freezes into propagating radiation in the form of photons. This distinction makes the concept of freezing precise: it requires "exit while remaining," not "complete disappearance."

Objections and Replies

Objection One: Your claim that "freezing is not stopping" looks like a word game. In ordinary language, "freezing" just means some kind of stopping, and you are redefining the concept.

Reply: Yes, this chapter redefines the meaning of "freezing." This is terminological precision, not a word game. The precise use of any philosophical concept involves the refinement of ordinary usage. Kant redefined "transcendental," Husserl redefined "phenomenon," and Whitehead redefined "event": such refinements allow concepts to perform philosophical work. The concept of "freezing" in this chapter has a ground: it has already been used in the physical work of *Freezing Cosmology*, and this chapter elevates it from a physical concept into an ontological concept. This terminological refinement is legitimate because it is built upon prior work.

Objection Two: If freezing is not stopping, then it seems to be merely "exiting dominance." Is this concept too ordinary?

Reply: "Exiting dominance" is the core of freezing, but the definition given in this section has specific content: "exiting active dynamics while remaining as residual, background, or history." This is not simply "exiting dominance"; it involves a concrete ontological structure. Specifically, freezing requires three things: the prior state must be active dynamics, the dominant evolution of some domain; the later state must be residual, background, or history, no longer dominant but retained; and the transition must have a concrete physical or ontological mechanism. These three requirements give the concept of freezing content. It is not a vague notion of "exiting dominance."

Objection Three: Your example of the CMB is a matter of physics. Is it legitimate to generalize it into an ontological concept?

Reply: The CMB example is a starting point, not the ground of the argument. This chapter's elevation of freezing into an ontological concept depends on the later development of this chapter, beginning in Section 11.2, where freezing is extended across several ontological dimensions. The CMB example only allows the reader to begin from a concrete case. It is not the foundation of the argument. More deeply, the elevation from scientific examples to ontological concepts is a legitimate philosophical method. The whole work of this book, including constrained speculation as method, gradient ontology, and co-evolution, uses a similar method: identifying ontological structures from scientific discoveries and then developing them. This is the working mode of the tradition of philosophy of nature.

Literature That Must Be Engaged

- *Freezing Cosmology* on freezing as a cosmological mechanism (self-reference)
- Ilya Prigogine, The End of Certainty (1997): on non-equilibrium processes
- Erwin Schrödinger, What Is Life? (1944): on negative entropy and life
- Chapter 7 on difference before identity (self-reference)

Summary of This Section

This section has done three things. First, it gave a precise definition of freezing: exiting active dynamics while remaining as residual, background, or history. Second, it distinguished freezing from three neighboring concepts: stopping, equilibrium, and disappearance. Third, it used the CMB as a concrete example and black-hole evaporation as a counterexample to show how the concept is used.

The next section develops the ontological thesis of freezing as a foundational ontological process.

11.3. Freezing as a Foundational Ontological Process

Statement of the Core Thesis

Core thesis FZ1: freezing is the foundational ontological process by which any processual being enters a relatively stable form from a dynamic process. All "beings," including particles, atoms, galaxies, life, consciousness, and culture, manifest from more foundational processes through freezing.

This thesis requires detailed development. Let us analyze its two key parts. "Any processual being": this thesis applies to processual beings, namely objects that are not eternal static substances. Within the framework of this book's gradient ontology, already established in Chapter 7, all beings are processual: there are no eternal static substances. Therefore, this thesis applies to all beings. "Foundational ontological process": freezing is not an accidental stage of certain processes, but the common way in which all processual beings manifest. It is ontological, concerning how existence is stated; foundational, meaning that it cannot be reduced to a more primitive process; and processual, meaning that it is not a static structure.

Development of the Thesis

FZ1 is developed through four sub-theses.

Sub-Thesis One: Substance as a frozen transient.

Traditional ontology takes "substance" as the basic object: independently existing, stably persisting, and bearing properties. The gradient ontology of this book rejects substance as a basic object, as Chapter 7 has argued, but it still needs to explain why substances appear so real: in ordinary experience, we clearly encounter stable things, such as tables, persons, and stars. FZ1 gives a response: so-called "substances" are in fact frozen transients, relatively stable forms that manifest through freezing from more dynamic processes. Consider several examples: elementary particles, such as electrons and quarks, are not eternal and unchanging static objects, but excitation modes in the ρ_f field, stable modes frozen from more foundational field dynamics; their "stability" is the result of freezing, not an original property. Atoms are stable bound states of electrons and nuclei; this stability is built upon multiple layers of freezing, including quark confinement inside nuclei, quantization of electron energy levels, and the stability of atomic structure. Galaxies are gravitationally bound aggregates of matter; this aggregation is built upon multiple freezing processes, including gravitational freezing, dark matter halo formation, and star formation. These examples show that traditional "substances" are in fact frozen transients: stable forms of process, not objects outside process.

Sub-Thesis Two: Time as the direction of freezing.

Time as an ontological problem has long been complex in the history of philosophy. Several aspects of time have already been treated in previous chapters of this book: Chapter 15 will treat time as $\partial\rho/\partial t$ within the co-evolutionary framework. This section develops another ontological dimension of time: the direction of time. Time is not merely a "sequence of moments": it has directionality. The time we experience flows from the past toward the future, not in reverse. In physics, this directionality is expressed through the second law of thermodynamics, namely entropy increase, and cosmic evolution, namely irreversible expansion. FZ1 gives a response: the direction of time = the direction of freezing. Specifically, the direction of time is not given in advance; it emerges from the evolution of processes. When the evolution of a process contains irreversible freezing, that is, when certain processes exit dominant dynamics and are retained as residual, this irreversibility provides the direction of time. Consider several examples: cosmic evolution, in which the universe evolves from an early high-density, high-temperature state into the present low-density, low-temperature state, with multiple freezing processes occurring during this evolution, such as the freezing of quark–gluon plasma into protons and neutrons, and the freezing of photon–matter interaction into the CMB. The irreversibility of these freezing processes provides the direction of cosmic time. Thermodynamic processes, in which the "entropy increase" of the second law is in fact the statistical expression of freezing: energy freezes from an ordered and usable form into a disordered and unusable form, namely heat; irreversible increase of entropy = irreversible freezing. Biological processes, in which the irreversibility from birth to death is built upon multiple freezing processes, such as cellular differentiation, memory formation, and aging. These examples show that the direction of time is not an abstract "sequence of moments": it is the concrete expression of the irreversibility of freezing processes.

Sub-Thesis Three: Spatial geometry as the statistical appearance of low-frequency freezing.

Spatial geometry is treated in physics by the spacetime geometry of general relativity. This book does not compete with general relativity: it accepts general relativity as an effective theory in the low-energy limit. But this book can ask: from a more foundational ontological perspective, what is spatial geometry? FZ1 gives a response: spatial geometry is the statistical appearance of low-frequency freezing. Specifically, in the EQT framework, the ρ_f field has different frequency bands. High-frequency bands correspond to small-scale structures, such as elementary particles, while low-frequency bands correspond to large-scale structures, such as galaxies and the universe as a whole. Spatial geometry as a large-scale phenomenon is mainly determined by low-frequency bands. The evolution of low-frequency bands is relatively slow: in a certain sense, it is already frozen and does not dominate fast dynamics. At the large-scale statistical level, this frozen state appears as the spatial geometry we experience. This response does not replace general relativity: it ontologically reinterprets the geometrical objects of general relativity. Spacetime geometry is not a pre-existing container; it is the statistical appearance of low-frequency freezing. The distribution of matter and energy affects the form of low-frequency freezing: this is the ontological ground of general relativity's statement that "matter tells spacetime how to curve."

Sub-Thesis Four: Memory as the freezing of consciousness.

Memory is one of the core problems in the study of consciousness. FZ1 gives an ontological response to memory: memory is the freezing of consciousness, the process in which conscious activity freezes from active flow into relatively stable neural structures. Specifically: short-term memory corresponds to weak freezing of conscious activity, where conscious experience is temporarily maintained but can be overwritten; long-term memory corresponds to strong freezing of conscious activity, where conscious experience stabilizes into neural structures, including synaptic strength and neural-network patterns, and can be activated by later conscious activity; procedural memory, such as riding a bicycle, corresponds to the complete freezing of conscious activity: it can operate without the participation of consciousness. This response gives memory a specific ontological status as an ontological phenomenon. It is not a mysterious "storage of consciousness," but the process in which conscious activity freezes from active flow into relatively stable structures.

The Unity of the Four Sub-Theses

The four sub-theses together show the content of FZ1: freezing is not an accidental stage of certain processes, but a universal foundational ontological process. Substance, which appears eternal, is a frozen transient; the direction of time, which appears pre-given, comes from the irreversibility of freezing; spatial geometry, which appears to be an abstract container, is the statistical appearance of low-frequency freezing; and memory, which appears mysterious, is the freezing of consciousness. These responses jointly reorganize several dimensions of ontology: they move from static substance ontology to process ontology, while still explaining ontological phenomena such as stability, the direction of time, spatial geometry, and memory. Freezing as a foundational ontological process makes this reorganization possible.

Objections and Replies

Objection One: Your four sub-theses sound too numerous. You attribute all difficult ontological problems, including substance, time, space, and memory, to "freezing." Is this excessive unification?

Reply: The four sub-theses do involve several ontological problems, but their unity is not arbitrary. All four problems concern one ontological question: how stable forms emerge from process. Freezing as a foundational ontological process gives a unified response to this question. This unity does not weaken concrete treatment. Each sub-thesis still requires specific development: substance as a frozen transient requires physical support; time direction as freezing direction requires connection with the second law of thermodynamics; spatial geometry as the statistical appearance of low-frequency freezing requires connection with general relativity; and memory as consciousness freezing requires connection with neuroscience. The unity is at the level of ontological direction and does not cancel the diversity of concrete work.

Objection Two: Your claim that "spatial geometry is the statistical appearance of low-frequency freezing" seems to conflict with general relativity. General relativity says that spacetime is dynamical, not frozen.

Reply: This objection points to a real technical issue. Spacetime in general relativity is indeed dynamical: the Einstein equations couple spacetime to matter and energy. But this dynamical character occurs at large scales: spacetime changes at ordinary and cosmological scales, as in black holes, gravitational waves, and cosmic expansion. This section claims that the dynamical character of spacetime in general relativity itself can be understood as a slow freezing process: relative to other processes, such as particle-physical processes, it is slow, stable, and usable as background. This does not mean that spacetime is "completely frozen." It means that, in ordinary treatments, it appears as the statistical appearance of low-frequency freezing. More detailed treatment requires work connecting EQT with general relativity, which exceeds the scope of this chapter. This section only indicates the direction: spatial geometry as an ontological phenomenon can be reinterpreted within the freezing framework, in a way compatible with the dynamics of general relativity.

Objection Three: Your claim that "memory is the freezing of consciousness" is overly physicalizing. It reduces memory to neural structure and ignores the phenomenological dimensions of memory, such as memory content, the sense of time in memory, and recallability.

Reply: This chapter does not claim that memory is only neural structure. It claims that neural structure is the ontological basis of memory as an observable phenomenon. The phenomenological dimensions of memory, such as content, sense of time, and recallability, are real and important; they are treated in more detail in Chapter 13, "Mind–Matter Unity." This chapter only points out that memory as an ontological phenomenon has a physical correlate, namely the stability of neural structures, and that this correlate is a form of freezing. The relation between this treatment and the hard problem of consciousness is as follows: this chapter does not solve why neural structures correspond to concrete memory phenomena; that is one form of the hard problem. But this chapter gives a physical necessary condition for memory as an ontological phenomenon: the freezing of conscious activity is a necessary condition for the emergence of memory. This is within the legitimate scope of constrained modeling.

Literature That Must Be Engaged

- Aristotle, Metaphysics: the traditional starting point of substance ontology
- Henri Bergson, Matter and Memory (1896): phenomenological-philosophical treatment of memory
- Chapter 7 on difference before identity (self-reference)
- Chapter 9 on frequency as a basic spectral quantity (self-reference)
- Chapter 15 on time (forward reference)

Summary of This Section

This section has done three things. First, it stated the core thesis FZ1: freezing is the foundational ontological process by which any processual being enters a relatively stable form from a dynamic process. Second, it developed FZ1 through four sub-theses: substance as a frozen transient, time as the direction of freezing, spatial geometry as the statistical appearance of low-frequency freezing, and memory as the freezing of consciousness. Third, it argued that the unity of the four sub-theses is not arbitrary, but a unified response to the question of how stable forms emerge from process.

The next section develops the connection between freezing and the tri-state categories.

11.4. Freezing and the Tri-State Categories

The Necessity of the Connection

Chapter 10 established the tri-state categories, namely free state, bound state, and condensed state, as basic categories of process ontology. This chapter introduces freezing as a foundational ontological process. These two works must be connected; otherwise, the unity of Part II would be damaged. The work of this section is to make this connection explicit: freezing is the temporal development of the tri-state categories.

The Connection Thesis

Connection thesis FZ2: freezing is the movement of a being within the space of tri-state categories. It involves changes in the tri-state weights of a being, especially movement from more dynamic states, namely free state and bound state, toward more stable states, namely the condensed state.

Freezing as Change in Tri-State Weights

Any being at any moment has a concrete tri-state decomposition: a certain free-state weight, a certain bound-state weight, and a certain condensed-state weight, as Chapter 10 has established. Freezing involves change in these tri-state weights. Specifically, the freezing process has typical directions: from high free-state weight to high condensed-state weight, as when free particles are bound into atoms; from high bound-state weight to high condensed-state weight, as when an excited atom decays to the ground state and then condenses into molecules; and, in some cases, from high free-state weight to medium bound-state weight, as when an electron is captured by an atom. These directions are not rigid: the form of the freezing process varies by domain and circumstance. But they share a common feature: a transfer of weight from the dynamic toward the stable.

Several Concrete Examples

Example One: Freezing in cosmic evolution.

After the Big Bang, cosmic evolution underwent multiple freezing processes, each corresponding to concrete changes in tri-state weights: quark confinement, about 10^{-6} seconds after the Big Bang, in which quarks and gluons froze from a free-state-dominant quark–gluon plasma into bound-state-dominant hadrons, such as protons and neutrons; nucleosynthesis, about three minutes after the Big Bang, in which protons and neutrons further froze from bound states into condensed states, namely atomic nuclei; photon decoupling, about 380,000 years after the Big Bang, in which photons froze from a bound state, namely strong interaction with plasma, into a free state, namely freely propagating CMB radiation. This freezing is from bound to free, showing the diversity of freezing directions. Structure formation, beginning about 100 million years after the Big Bang, in which dark matter and baryonic matter froze from a relatively uniform state into condensed structures, namely galaxies and galaxy clusters. These freezing processes show that change in tri-state weights is the content of freezing. Every freezing has concrete tri-state states before and after.

Example Two: Freezing in chemical processes.

Chemical reactions can be understood as freezing processes: bond formation, in which free atoms or molecules, namely free-state-dominant reactants, form stable compounds, namely bound-state–condensed-state-dominant structures, such as hydrogen and oxygen forming water; crystal formation, in which liquid or gaseous matter, with high free-state and high bound-state weights, freezes into crystal structure, namely high condensed state; phase transition, in which matter changes from one phase to another, such as water moving from liquid to solid, involving changes in tri-state weights. The process of chemical freezing depends on reaction conditions, including temperature, pressure, and catalysts. But all chemical reactions involve changes in tri-state weights.

Example Three: Freezing in life processes.

The development of life involves multiple freezing processes: cellular differentiation, in which early embryonic cells with pluripotency, corresponding to higher free-state/accessibility weight, freeze into specialized cells, such as neurons and muscle cells, corresponding to higher bound-state/condensed-state weight; memory formation, in which instantaneous neural activity in experience, dominated by free state and bound state, freezes into stable patterns of neural connection, dominated by condensed state; aging, in which the flexibility of life processes, namely high free-state/bound-state weight, gradually freezes into the stability and rigidity of physiological structures, namely high condensed state, and this freezing ultimately leads to death. Much of the specificity of life is determined by the precise regulation of these freezing processes.

Example Four: Freezing in cultural processes.

The cultural level also involves freezing processes: language formation, in which flexible communication in early human societies, namely high free state, gradually freezes into the stable structures of particular languages, namely high bound-state–condensed-state structures, such as vocabulary, grammar, and idioms; institutional formation, in which flexible social interactions gradually freeze into concrete institutions, such as law, government, and markets; tradition formation, in which group practices gradually freeze into traditions, namely high condensed-state structures, such as rituals, festivals, and beliefs. Cultural freezing gives society stability, but also produces the tension between rigidity and transformation.

The Problem of Reversibility in Freezing

An important ontological question is whether freezing is reversible. The answer from physics is: partially reversible. Some forms of freezing can be reversed: heating can melt ice, returning it to the liquid state; ionization can decompose atoms into free particles, returning them to the free state, and so on. But reversal requires energy input and specific conditions. At more foundational levels, such as quark confinement and the production of the CMB, reversal is extremely difficult: it would require reconstructing the extreme conditions of the early universe. The ontological response is this: freezing is generally locally reversible but globally irreversible. Local freezing can be reversed through specific processes, but the overall freezing trend of the universe is irreversible: this is the expression of the principle of entropy increase. This distinction has important implications for the direction of time. If freezing is globally irreversible, then the ground of the direction of time, Sub-Thesis Two in Section 11.2, is built upon this irreversibility. If freezing were globally reversible, the ontological ground of the direction of time would require reassessment.

Objections and Replies

Objection One: Your account of "freezing as change in tri-state weights" looks too mechanical. It seems to reduce all processes to changes in weights.

Reply: The claim of this section is not to reduce processes to changes in weights, but to concretize freezing as an ontological process through changes in tri-state weights. The rich content of concrete processes, such as chemical reactions, life development, and cultural evolution, is not reduced to tri-state weights. Tri-state weights express the ontological structure of these processes. The concrete content of processes, such as the chemical dynamics of the hydrogen–oxygen reaction, the biological mechanisms of cellular differentiation, and the sociological process of language formation, still requires dedicated work. Tri-state weights are only structures at the ontological level.

Objection Two: Your examples move from the cosmos to chemistry, life, and culture. Is this an excessive crossing of levels? Are cultural freezing and physical freezing really the same process?

Reply: This section does not claim that cultural freezing and physical freezing are the same process. Each has domain-specific features. This section claims that they have a shared structure as ontological processes, namely changes in tri-state weights. This shared structure allows us to use the same ontological concept, freezing, across different levels while acknowledging the differences in content at each level. This is an application of the method of constrained speculation: an ontological concept may be used across levels as long as each level has verifiable content.

Objection Three: Does the partial reversibility of freezing weaken its status as a foundational ontological process? If freezing can be reversed, it seems not to be "foundational."

Reply: The partial reversibility of freezing does not weaken its status as a foundational ontological process; it makes the concept of freezing more precise. Freezing as a foundational ontological process means that all processual beings involve freezing. It does not mean that once freezing occurs, it is irreversible. More deeply, the reversibility of freezing is itself constrained by freezing conditions: reversing a freezing requires a new process, such as energy input or a change in conditions. This new process may involve its own freezing. Therefore, "reversing freezing" is a new process, not the "cancellation of freezing." Freezing as an ontological concept remains foundational at the deeper level.

Literature That Must Be Engaged

- Chapter 10 on tri-states as ontological categories (self-reference)
- Erwin Schrödinger, What Is Life? (1944): on life as an anti-entropic process
- *Freezing Cosmology* on the physics of freezing (self-reference)

Summary of This Section

This section has done four things. First, it stated the connection thesis FZ2: freezing is the temporal development of the tri-state categories. Second, it showed the application of this connection through four concrete examples: cosmic evolution, chemical processes, life processes, and cultural processes. Third, it treated the problem of reversibility in freezing: locally reversible but globally irreversible. Fourth, it replied to possible objections, especially concerning cross-level use and reversibility.

The next section develops the hierarchy and historicity of freezing.

11.5. The Hierarchy and Historicity of Freezing

Freezing as a Hierarchical Phenomenon

Freezing is not a single process: it occurs at multiple levels. Each level has its own freezing processes, its own timescale, and its own stable forms. This section develops the hierarchical structure of freezing.

Level One: the foundational physical level. The most foundational freezing occurs at the level of basic physics: elementary particles freeze out of more foundational processes, such as possible quantum fields or vibrational modes of strings. The content of this freezing is not yet fully clear in current physics. It involves unresolved problems in foundational physics, such as the origin of the elementary particles of the Standard Model and the nature of dark matter. EQT treats this level in *The Foundational Volume of EQT*, where elementary particles are treated as excitation modes of the ρ_f field. Such excitation modes can be understood as frozen products of the more foundational dynamics of the ρ_f field.

Level Two: the atomic–molecular level. Freezing at the atomic–molecular level involves elementary particles, such as electrons and nuclei, forming stable bound states. The physics of this level has been sufficiently developed by quantum mechanics: it treats atomic energy levels, chemical bonds, molecular structures, and related phenomena.

Level Three: the macroscopic matter level. Freezing in macroscopic matter involves the stable formation of multi-atomic and multi-molecular structures: gases, liquids, solids, crystals, glasses, and other states of matter. The physics of this level, including condensed matter physics and statistical mechanics, is also well developed.

Level Four: the astrophysical level. Freezing at the astrophysical level involves the aggregation of matter on large scales: star formation, planet formation, galaxy formation, and so on. This level has concrete treatment in astrophysics.

Level Five: the cosmological level. Freezing at the largest scale involves the formation of overall structures in cosmic evolution: the multiple freezing processes after the Big Bang, as described in Example One of the previous section. The physics of this level, namely cosmology, is developing rapidly.

These five levels together constitute the structure of physical hierarchical freezing. Each level has its own processes, timescales, and stable forms. But freezing is not limited to physical levels; it extends to the levels of life, consciousness, and culture.

Level Six: the level of life. Freezing at the level of life involves the formation of biological structures: cellular differentiation, individual development, species evolution, and so on. The processes of this level are treated in biology.

Level Seven: the level of consciousness. Freezing at the level of consciousness involves conscious activity entering stable structures from flowing states: memory formation, personality stabilization, habit formation, and so on. This level is treated in *Mind–Matter Unity* and Chapter 13 of this book.

Level Eight: the cultural–social level. Freezing at the cultural–social level involves the formation of social structures: language, institutions, traditions, and so on. This level is treated in the social sciences.

Relations among Levels

These eight levels are not independent of one another. They have hierarchical relations.

Relation One: lower levels provide the foundation for higher levels. The freezing of each higher level is built upon lower-level freezing. The stability of atoms is built upon the stability of elementary particles; the stability of molecules is built upon the stability of atoms; the stability of life is built upon chemical stability; the stability of consciousness is built upon the stability of life; and the stability of culture is built upon the stability of consciousness. This hierarchical relation is not simple reduction. Higher levels have their own features that cannot be mechanically derived from lower levels, as the work of Chapter 10 on the irreducibility of the tri-states has already shown. But lower levels provide ontological necessary conditions for higher levels.

Relation Two: hierarchy of timescales. Different levels have different timescales. In general, more foundational levels have shorter timescales, such as particle processes at femtosecond to attosecond scales, while higher levels have longer timescales, such as cultural change over centuries. This hierarchy of timescales allows each level to use lower levels as an "already frozen" background. For example, biology uses the stability of chemistry as background, and sociology uses the stability of psychology as background. This is the ontological ground of the division of labor among scientific levels.

Relation Three: hierarchy of reversibility. Different levels have different degrees of reversibility. At the physical level, many freezing processes are locally reversible, such as phase transitions in water. At the level of life, the reversibility of freezing processes is weaker, since aging and death are basically irreversible. At the cultural level, the reversibility of freezing is more complex: some cultural changes can be partially reversed, while others cannot. This hierarchical difference in reversibility reflects the expression of freezing as an ontological concept across different levels.

Historicity

The hierarchical structure of freezing gives beings historicity: they carry their freezing history.

Core thesis FZ3: any processual being carries its freezing history. This history is not an externally added "biography," but the internal structure of the being as ontological content.

Specifically: a star is not an abstract "star type." It has a concrete history of formation, from a specific molecular cloud, initial conditions, and evolutionary phase, and this history determines its current state. A living being is not an abstract "biological type." It has a concrete developmental history, including embryonic development, environmental experiences, and the aging process, and this history determines its current state. A consciousness is not an abstract "consciousness type." It has a concrete history of formation, including experience, learning, and memory, and this history determines its current state. A culture is not an abstract "cultural type." It has a concrete evolutionary history, including events, persons, and turning points, and this history determines its current state. Every being carries its freezing history: its current state is determined by its freezing processes. This is the content of historicity as an ontological phenomenon.

The Relation between Historicity and Time

Historicity is connected with the direction of time, Sub-Thesis Two in Section 11.2. The direction of time = the direction of freezing, namely irreversibility; historicity = the preservation of freezing history within beings. They are two aspects of the same phenomenon. The direction of time is the direction from the present toward the future; historicity is the form in which the present preserves the past. Together, they constitute the structure of time as an ontological phenomenon. This dual structure, direction of time + historicity, means that time is not merely a "sequence of moments": it is a directional, historical, content-bearing process structure.

Objections and Replies

Objection One: Your eight levels look excessive. Is this overcomplication?

Reply: The eight levels reflect the diversity of beings. The ontological levels from elementary particles to culture are widely accepted, since physics, chemistry, biology, psychology, and sociology each treat specific levels. This chapter does not invent these levels; it acknowledges them and develops their ontological connections. Fewer levels, such as a simple fourfold division into physical, biological, psychological, and social levels, would ignore details of hierarchy. The eight levels provide a more precise ontological structure.

Objection Two: Your concept of "historicity" sounds like the phenomenological-hermeneutic tradition, including Dilthey, Heidegger, and Gadamer. How does this chapter connect with that tradition?

Reply: Yes, "historicity" has a tradition in Western philosophy, especially in Heidegger's work in *Being and Time*. The concept of "historicity" in this section has structural relevance to that tradition, but differs in content. Heidegger's "historicity" mainly refers to the existential structure of Dasein, or human being-there: it involves taking over the past and projecting toward the future. The "historicity" in this section is extended to all processual beings: not only Dasein, but also stars, atoms, cultures, and so on. This extension makes historicity a universal ontological phenomenon, not merely a feature of human existence. This extension is philosophically legitimate, but it requires acknowledging two different concepts of "historicity." Chapter 21 will further discuss the relation with the Heideggerian tradition.

Objection Three: Your claim that beings "carry freezing history" sounds overly metaphorical. How does a being "carry" history?

Reply: The "carrying" in this section is not metaphorical; it is an ontological relation. Specifically: a star materially carries its formation history. Its chemical composition, internal structure, and current evolutionary phase are all determined by its formation history. A living being carries its history genetically and developmentally: its genome, epigenetics, and neural connections are all preservations of history. A culture carries its history institutionally and traditionally: laws, customs, and documents are preservations of history. In each example, "carrying" has a concrete physical or structural form. It is not a vague "containing," but a concrete ontological relation.

Literature That Must Be Engaged

- Wilhelm Dilthey, Introduction to the Human Sciences (1883): on the tradition of historicity
- Martin Heidegger, Being and Time (1927), §§ 72-77: on historicity
- Hans-Georg Gadamer, Truth and Method (1960): on the historical dimension of hermeneutics
- Chapter 7 on difference before identity (self-reference)
- Chapter 21 on the finitude of meaning (forward reference)

Summary of This Section

This section has done four things. First, it developed the eight levels of freezing, from basic physics to the cultural–social level. Second, it identified the relations among levels: lower levels provide the foundation for higher levels, there is a hierarchy of timescales, and there is a hierarchy of reversibility. Third, it stated thesis FZ3: any processual being carries its freezing history. Fourth, it connected historicity with the direction of time: they are two aspects of the same phenomenon.

The next section develops freezing as a condition of visibility.

11.6. Freezing as a Condition of Visibility

The Core Problem

Why can we observe any being at all? This is not a trivial question; it involves the fundamental relation between ontology and epistemology.

Consider the situation. We can observe the CMB, the radiation from about 380,000 years after the Big Bang, because photons have "propagated freely" since then: their state has been relatively stable, freezing from a bound state into a free state. If photons had immediately interacted strongly with matter after the formation of the CMB, we could not observe the CMB: its state would have been continuously altered, and there would be no "state to be observed." Similarly, we can observe atomic structures, galaxies, life forms, cultures, and so on because their respective states have relative stability, namely freezing. If a being were in a state of complete flow, such that it was completely changed at every moment, it could not be observed as an object. The thesis of this section is: freezing is the ontological condition for the possibility of observation.

Statement of the Thesis

Core thesis FZ4: the possibility of observing a being X as an object is grounded in the frozen state of X: X must have some relatively stable form in order to be observed as an object.

Implications of the Thesis

Implication One: Observation requires the stability of the observed.

Any observation involves the co-evolution between a detector, in the ontological sense of *EQT Detection Theory*, and a probed system, as Chapter 12 has developed. This co-evolution requires the probed system to have some stability on the timescale of observation. If the probed system had no stability at all and were completely changed at every moment, the detector could not form a stable response and could not "measure" any content. A system with no stability at all is inaccessible as an object of observation.

Implication Two: Stability is a form of freezing.

Where does the stability of the probed system come from? It comes from a freezing process: the system's state freezes from a more dynamic process into a relatively stable form. This freezing allows the system to retain content on the timescale of observation. Specifically: instantaneous observation, such as observing particle collision, depends on the stability of the probed system at the femtosecond timescale, which comes from particle-physical freezing, namely particles as stable excitations of the ρ_f field; everyday observation, such as observing a table, depends on the stability of the probed system at the second timescale, which comes from chemical and macroscopic matter freezing, namely the stability of chemical bonds and molecular structures; cosmic observation, such as observing the CMB, depends on the stability of the probed system at cosmological timescales, which comes from cosmological freezing, namely the free propagation of photons since recombination. Every observation is built upon freezing. The form of freezing determines the possibility of observation.

Implication Three: Observational windows are determined by freezing conditions.

Not everything can be observed: observational windows are determined by freezing conditions. Specifically: states that are too short-lived cannot be observed as objects if the system changes much faster than the observational response time, as in the "instantaneous existence" of certain virtual particles, which cannot be directly observed; states that are too large in scale may exceed current observational capacities if they manifest only at scales that are too large, as in the overall geometry of the universe, which requires large-scale observations to determine; states that are too weak cannot be observed as objects if their interaction with the detector is below threshold, as in the difficulty of directly observing dark matter particles if they exist. Every observational window corresponds to freezing conditions: the system must satisfy conditions of timescale, spatial scale, and intensity in order to be observed.

An application: the ontology of detection theory. The detection theory in *EQT Detection Theory* receives support from this ontology. The three criteria of detection, namely frequency matching, amplitude threshold, and tri-state response, are in fact expressions of freezing conditions: they state what conditions the probed system must satisfy in order to be detected. Frequency matching means that the frequency of the probed system must lie within the response window of the detector; this is the frequency dimension of freezing. Amplitude threshold means that the perturbation of the probed system must exceed a threshold; this is the intensity dimension of freezing. Tri-state response means that the probed system must have a form in tri-state space; this is the structural dimension of freezing. Together, these three criteria explain how the probed system "manifests" as an observable object: through freezing conditions.

Implication Four: The unity of ontology and epistemology.

Freezing as a condition of visibility unifies ontology and epistemology: they are not two independent domains, but two aspects of the same process. Specifically: ontology, where beings are frozen transients, meaning that they are stable forms of process; epistemology, where beings are observable, meaning that they are objects of the observational process; and unity, where the existence of beings, namely ontology, and their observability, namely epistemology, are grounded in the same freezing conditions. This unity is neither side of the traditional dilemma between realism and anti-realism. It treats the traditional difficulty between ontology and epistemology. Chapter 14, "Observation as Participation: A New Foundation of Objectivity," will further develop this unity.

Objections and Replies

Objection One: Your thesis seems to make conditions of observation into conditions of existence. Does this fall back into Berkeley's "to be is to be perceived"?

Reply: FZ4 does not claim that "to be is to be perceived." It claims that being observed as an object requires freezing conditions. These are different. Specifically: Berkeley's thesis is that existence = being perceived; if something is not perceived, it does not exist. FZ4 says that being observed as an object = having freezing conditions; if something lacks freezing conditions, it cannot be observed as an object, but this does not mean that it does not exist as a more foundational process. The difference lies in the form of "existence." Berkeley reduces "existence" to "being perceived," which is epistemological idealism. FZ4 distinguishes "existing as an object," which requires freezing, from "existing as a more foundational process," which need not have freezing. Some processual beings may not form freezing but still exist as processes.

Objection Two: If observation requires the stability of the observed, does observing originally unstable things, such as quantum fluctuations or instantaneous events, become impossible?

Reply: The thesis of this section does not deny the observation of instantaneous events. It points out that observing an instantaneous event requires matching between the observer and the event. Specifically, observing an instantaneous event requires that the detector have a response time fast enough to match the event's timescale; that the event itself have observable consequences, such as energy deposition or phase change; and that these consequences have frozen forms that can be analyzed later. Therefore, observing an instantaneous event does not mean observing "the instantaneous event itself," but observing the frozen consequences left by the event. This distinction makes the observation of instantaneous events possible within the framework of this section.

Objection Three: Your claim that "freezing is a condition of visibility" seems to make epistemology passively derived from ontology. But epistemology has its own independence, since different observers may observe the same thing differently.

Reply: This section does not claim that epistemology is completely derived from ontology. It claims that ontology and epistemology are grounded in the same freezing conditions. Specifically, different observers may observe differently because different observers, as detectors, have different characteristics: different frequency matching, different thresholds, and different tri-state responses. Different observers also have different co-evolutionary relations with the same probed system. Such observational differences do not weaken the thesis of this section. They show that the content of observation depends on the co-evolutionary relation between observer and probed system, and this co-evolutionary relation is built upon freezing conditions.

Literature That Must Be Engaged

- *EQT Detection Theory* on the content of detection theory (self-reference)
- George Berkeley, Principles of Human Knowledge (1710): comparison with "to be is to be perceived"
- Immanuel Kant, Critique of Pure Reason (1781): on the objective conditions of phenomena
- Chapter 12 on co-evolution as an ontological relation (self-reference)
- Chapter 14 on observation as participation (forward reference)

Summary of This Section

This section has done four things. First, it stated the core thesis FZ4: freezing is the ontological condition for the possibility of observation. Second, it developed four implications of the thesis: observation requires the stability of the observed, stability is a form of freezing, observational windows are determined by freezing conditions, and ontology and epistemology are unified. Third, it used the application of detection theory in *EQT Detection Theory* to show how the thesis works. Fourth, it replied to possible objections and avoided confusion with Berkeleyan idealism.

The next section is the conclusion of this chapter.

11.7. Chapter Summary: Freezing as an Ontological Pivot

Completion of the Three Tasks

The three tasks undertaken by this chapter have all been completed.

Task One: to elevate freezing from a cosmological mechanism into a foundational ontological process. Section 11.1 gave precise conceptual clarification: freezing is not stopping, not equilibrium, and not disappearance, but "exiting active dynamics while remaining as residual, background, or history." Section 11.2 stated the core thesis FZ1 and developed it through four sub-theses: substance as a frozen transient, time as the direction of freezing, spatial geometry as the statistical appearance of low-frequency freezing, and memory as the freezing of consciousness.

Task Two: to argue for a series of ontological theses. Section 11.3 connected freezing with the tri-state categories (FZ2). Section 11.4 developed the hierarchy and historicity of freezing (FZ3). Section 11.5 developed freezing as a condition of visibility (FZ4). Together, these four theses constitute the complete work of freezing as an ontological pivot concept.

Task Three: to clarify the connection between freezing and the other parts of the book. The work of this chapter connects earlier chapters, including the ontology of difference, the tri-state categories, and co-evolution, and prepares the ground for later chapters, especially Chapter 17 on ontological singularity and Chapter 21 on the finitude of meaning.

This Chapter as the Pivot of Part II

This chapter is the pivot of Part II. The preceding four chapters, Chapters 7–10, each treat different aspects of gradient ontology: Chapter 7 establishes difference before identity as the ontological starting point; Chapter 8 develops the Chinese qi-theoretical tradition as an ontological resource; Chapter 9 elevates frequency as a basic spectral quantity into an ontological thesis; and Chapter 10 establishes the tri-states as ontological categories. These four chapters each have independent value, but they require integration. Otherwise, Part II would be merely a collection of unrelated ontological theses. This chapter undertakes this work of integration. Freezing as a foundational ontological process connects the ontology of difference, since freezing involves the formation and stabilization of differentiated structures, namely gradient structures; connects the qi-theoretical tradition, since freezing corresponds to Zhang Zai's "gathering and taking form" and Wang Fuzhi's "being born and formed day by day" as ontological forms; connects frequency ontology, since freezing involves the evolution of frequency bands, with high-frequency early freezing and low-frequency later freezing; and connects the tri-state categories, since freezing is a change in tri-state weights, especially movement toward the condensed state. These four connections make this chapter the integrative point of Part II: the aspects of gradient ontology are unified into a complete ontological picture through freezing as a foundational ontological process.

Freezing as the Pivot of the Whole Book

This chapter is not only the pivot of Part II; it also plays a pivotal role in the whole book. Looking backward, this chapter builds upon the preceding chapters of Part II and Part I, especially Chapter 3 on failure conditions as an ontological discipline and Chapter 4 on constitutive coextension. Looking forward, this chapter lays the foundation for several later key chapters: Chapter 12 on co-evolution, since co-evolutionary relations occur among already frozen beings and freezing provides the ontological field of co-evolution; Chapter 13 on mind–matter, since consciousness as a freezing process and memory as the freezing of consciousness concretize the ontological structure of consciousness; Chapter 14 on observation, since observation is built upon freezing as a condition of visibility; Chapter 17 on ontological singularity, since this chapter has already prepared the direction that "the limit of freezing is ontological singularity," meaning that when all freezing is reversed or all $\nabla\rho \to 0$, ontology enters a singularity; and Chapter 21 on the finitude of meaning, since meaning appears in the active process before freezing, and this chapter provides the ontological structure for that thesis. These connections make this chapter the ontological pivot of the whole book: the ontological foundations of several key chapters lie here.

Honesty about the Limits of This Chapter

This chapter acknowledges several limits:

First limit: the treatment at the technical level is brief. This chapter develops the concept of freezing mainly at the ontological level, while its treatment of concrete physics, such as the dynamics of freezing and freezing processes in different physical systems, is relatively brief. Detailed physical treatment depends on *Freezing Cosmology* and future work.

Second limit: the diversity of cross-domain applications. This chapter touches on freezing in several domains, including physics, biology, consciousness, and culture, but detailed application in each domain requires dedicated work. This chapter only gives directions and examples, and does not develop the detailed work in every domain.

Third limit: dialogue with specific philosophers is insufficient. This chapter mainly develops freezing as an ontological concept, and its dialogue with specific philosophers, such as Heidegger on historicity, Bergson on duration, and Whitehead on concreteness, is relatively brief. Each dialogue requires dedicated treatment.

Fourth limit: the problem of reversibility in freezing. This chapter touches on the reversibility of freezing, but does not treat in detail "what kinds of freezing are reversible and what kinds of freezing are irreversible." This problem involves concrete research in physics, biology, and sociology and requires dedicated work.

Acknowledging these limits does not weaken the work of this chapter. It makes the posture of the chapter honest, open to criticism, and open to further development.

The Core Posture of This Chapter

The core posture of this chapter can be stated as follows: freezing as a foundational ontological process gives concrete form to process ontology.

Process ontology as a direction, including difference before identity and change as foundation, has a long tradition in the history of philosophy. But process ontology is often criticized for being too abstract: it rejects static substance, but has difficulty explaining why we experience stable things. Freezing as a foundational ontological process is the response to this criticism. It allows process ontology to treat the problem of stability: stability is the result of freezing, not a property of foundational substance. This makes process ontology concrete, connectable with ordinary experience, and connectable with concrete science. This concretization is the contribution of this chapter. It inherits the direction of the tradition of process ontology while giving that tradition an ontological pivot concept, namely freezing. This concept has not historically been explicitly stated as a foundational ontological concept, although related directions exist, such as Prigogine's non-equilibrium thermodynamics. This chapter states and develops it seriously as a foundational ontological concept. If the work of this chapter holds, freezing as a foundational ontological process has a place in the tradition of process ontology. It is one form of that tradition: the process by which stable forms emerge from dynamic processes.

Meaning before freezing: this is one of the core theses of this book. This chapter gives that thesis a concrete ontological foundation. "Before freezing" is not a vague "past"; it is the phase of active process. Meaning appears in this phase. This phase will eventually end, since every freezing has its own process and is eventually replaced by new freezing or reaches ontological singularity at the limit, but while it exists, meaning is real. This connection makes this chapter a bridge between the metaphysical and humanistic work of the book. Chapter 17 develops the limit of freezing as ontological singularity; Chapter 21 develops the finitude of meaning before freezing. Both chapters build upon the work of this chapter.

11.8. Methodological Reflection on This Chapter

This chapter elevates freezing from a cosmological mechanism in *Freezing Cosmology* into a foundational ontological process. This section reflects on the methodological significance of this elevation.

This Chapter's Posture toward Originality

Freezing as a foundational ontological process is not entirely new: Prigogine's non-equilibrium thermodynamics, developmental processes in biology, and institutional formation in sociology all involve related directions. But explicitly stating and systematically developing freezing as a foundational ontological pivot concept is a contribution of this book. This chapter assumes the responsibility of introducing and developing it: it gives the concept of freezing physical ground through the EQT framework, gives it methodological ground through the method of constrained speculation, and systematically develops freezing as a foundational ontological process.

The Position of This Chapter between Physics and Philosophy

This chapter is neither detached from physics, since it uses the physical work of *Freezing Cosmology* as its premise, nor limited to physics, since it develops the concept at the ontological level. This position, neither detached from nor limited to physics, is the form of the tradition of philosophy of nature: it identifies ontological structures through physical discoveries, but works on ontological problems. This chapter builds upon the physical analysis of cosmological processes in *Freezing Cosmology*, but its work is not to restate those physical analyses. It elevates freezing from a cosmological mechanism into a foundational ontological process through philosophical argument.

This Chapter's Function as the Pivot of Part II

As the pivot of Part II, this chapter undertakes multiple integrative tasks. It synthesizes Chapter 7's ontology of difference, Chapter 8's qi-theoretical resources, Chapter 9's frequency ontology, and Chapter 10's tri-state categories into a unified ontological picture: structures of difference form tri-state forms through the freezing process and are distributed according to the frequency spectrum. This integration makes Part II an organic ontological picture: difference as starting point, freezing as process, tri-states as categories, frequency as dimension, and qi theory as a cross-cultural resource. Together, these five elements constitute a complete gradient ontology.

Limits Honestly Acknowledged in This Chapter

This chapter acknowledges several limits. First, the refinement of freezing as a foundational ontological process requires later work: the manifestations of freezing at different levels require more detailed treatment, and Chapter 16 is a partial development of this direction. Second, the dialogue between the concept of freezing and neighboring concepts, such as Prigogine's dissipative structures and developmental processes in biology, is developed in this chapter only in its core direction. Third, the manifestations of freezing as a foundational ontological process in interdisciplinary applications require dedicated work.

Connections with Later Chapters

Freezing as a foundational ontological process established in this chapter is applied in several later chapters. Chapter 12 uses freezing as the ontological ground of co-evolutionary relations. Chapter 13 uses memory as the freezing of consciousness. Chapter 14 uses result-freezing as the ontological mechanism of observation. Chapter 15 uses time as the direction of freezing. Chapter 16 builds directly on the work of this chapter and develops the hierarchical structure of freezing. Chapter 17 uses freezing as the internal boundary of a foundational ontological process. Chapter 21 uses freezing as the ontological ground of meaning. More deeply, this chapter is the work of concretizing process ontology: it allows process ontology to connect with concrete science, to treat the problem of stability, and to enter serious philosophical discussion. This is the fulfillment of the process-ontological position of this book.

Part III.

Co-Evolution, Mind–Matter, and Observation: The Unity of Three Relations

12. Co-Evolution as an Ontological Relation

12.1. Introduction to This Chapter

The "problem of two-term relation" is not a single problem in the history of philosophy, but a general name for a family of problems: the subject–object relation, the mind–matter relation, the self–other relation, and the individual–environment relation. These four debates appear to involve different domains, but they share a fundamental presupposition: relation occurs between two independent terms. The two terms exist as independent ontological units and are then connected through some relation.

This presupposition runs through the mainstream tradition of Western philosophy. Cartesian substance dualism treats mind and matter as two independent kinds of substance and then asks how they interact. Kantian phenomenology retains the thing-in-itself as a term independent of the subject's cognition. The sensationalism of Mill, Russell, and others treats individual sensations as primitive terms and then asks how objects are constructed from these sensations. Even works that oppose this presupposition, such as Heidegger's "being-in-the-world," Husserl's later lifeworld, and Whitehead's process philosophy, often begin by criticizing this presupposition: it remains the implicit framework of discussion.

This chapter elevates the core insight of EQT Detection Theory—that the detector and the probed system belong to the same ρ_f field—from a physical statement into a general ontological thesis: all "two-term relations" are same-field co-evolutionary relations (EV1), not external interactions between heterogeneous substances. The two apparently independent terms are in fact local structures within the same process field, and their "relation" is an internal mode of evolution of that process field.

This elevation involves a methodological challenge. Terms such as "co-evolution" and "relational ontology" have been widely used since the late twentieth century and can easily degenerate into rhetoric. The slogan "everything is interconnected" appears in many discussions but often lacks falsifiable content. The central work of this chapter is to give co-evolution, as a term, falsifiable content: through three dynamical mechanisms—frequency matching, threshold conditions, and tri-state response—co-evolution becomes a rigorous ontological relation rather than broad rhetoric.

This chapter also enters into serious dialogue with relevant work in contemporary philosophy, including Karen Barad's intra-action, the Huayan doctrine of "mutual interpenetration without end," and Whitehead's prehension. This cross-traditional dialogue gives the chapter a place in world-philosophical discussion: it is not confined to a single tradition, but compares the relational ontology of constrained speculation with directions from multiple traditions.

This chapter is the entry chapter of Part III of the book, "Co-Evolution, Mind–Matter, and Observation: The Unity of Three Relations." It establishes the relational-ontological foundation required by the following three chapters: Chapter 13 on mind–matter, Chapter 14 on observation, and Chapter 15 on time. This chapter builds on the ontological work of Part II: Chapter 7 on difference before identity, Chapter 10 on tri-states as ontological categories, and Chapter 11 on freezing as a foundational ontological process. It extends these works in process ontology into relational ontology.

The discussion proceeds as follows. Section 12.1 develops the philosophical history and shared presupposition of the problem of two-term relation; Section 12.2 states and defends co-evolution as an ontological relation (EV1); Section 12.3 develops the three dynamical mechanisms of co-evolution; Section 12.4 enters into dialogue with Karen Barad's intra-action; Section 12.5 enters into dialogue with the Huayan doctrine of "mutual interpenetration without end"; Section 12.6 concludes the chapter.

12.2. The Philosophical History of the Problem of Two-Term Relation

Forms of the Problem

The "problem of two-term relation" is not a single problem in the history of philosophy, but a general name for a family of problems. It includes several core debates: the subject–object relation, or how the perceiving subject is related to the perceived object and whether cognition can cross the gulf between subject and object; the mind–matter relation, or how mental states are related to physical states and whether the two are two aspects of the same entity or the interaction of heterogeneous entities; the self–other relation, or how the self knows the other and how the other, as another independent center of consciousness, establishes relation with the self; and the individual–environment relation, or how a living individual is related to its environment and whether the environment is an external container or an internal constituent.

These four debates appear to involve different domains, but they share a fundamental presupposition: relation occurs between two independent terms. The two terms—subject and object, mind and matter, self and other, individual and environment—exist as independent ontological units and are then connected through some relation. This presupposition runs through the mainstream tradition of Western philosophy. Cartesian substance dualism treats mind and matter as two independent kinds of substance and then asks how they interact. Kantian phenomenology retains the thing-in-itself as a term independent of the subject's cognition. The sensationalism of Mill, Russell, and others treats individual sensations as primitive terms and then asks how objects are constructed from these sensations. Even works that oppose this presupposition, such as Heidegger's "being-in-the-world," Husserl's later lifeworld, and Whitehead's process philosophy, often begin by criticizing this presupposition: it remains the implicit framework of discussion.

The Internal Difficulties of the Presupposition

Presupposing the two terms as independent units inevitably produces a series of difficulties.

Difficulty One: the origin of relation is unclear. If A and B are independent, how is the relation between them established? Is this relation externally added? If so, who adds it? If it is internal, then A and B are not truly independent: they are already in relation.

Difficulty Two: the problem of the possibility of knowledge. If subject and object are independent, how can the subject know the object? Knowledge requires some path of "access" to the object, but access must cross the gulf between independent units, and this gulf is precisely what seems uncrossable. Kant's problem of the thing-in-itself, Hume's skepticism about causality, and Russell's distinction between "knowledge by acquaintance" and "knowledge by description" are all expressions of this difficulty.

Difficulty Three: the metaphysical problem of interaction. The sharpest case is Descartes's problem of mind–matter interaction: if mind and matter are completely different kinds of substance, how can they interact? Descartes's own response through the "pineal gland" has been widely regarded as insufficient. But this difficulty is not limited to Descartes: any ontology that treats two terms as independent units encounters similar difficulties.

Difficulty Four: the ontological status of the boundary. If A and B are independent, what is the boundary between A and B? Does the boundary belong to A, to B, or to something outside both? The status of the boundary as an ontological unit is unclear: it belongs neither simply to A nor simply to B, yet it makes A and B exist as distinct terms.

Responses from Different Positions

Philosophical traditions have responded to these difficulties in different ways. Idealism eliminates the two-term relation by reducing the object to the content of the subject: only the subject exists, and the object is the subject's content. But this response has consequences: it makes it difficult to maintain ordinary realism about the external world. Materialism eliminates the mind–matter division by reducing mind to matter: only matter exists, and mind is a state of matter. But this response also has consequences: it makes it difficult to handle the specificity of subjective experience, such as the hard problem of consciousness. Property dualism preserves the difference between mind and matter while denying that the two are independent substances: mental properties and physical properties are both properties of the same entity. But this response, as in Davidson's anomalous monism, requires detailed defense of what the "same entity" is. Relational ontology takes relation itself as foundational: relation precedes relata, and relata are differentiated from relation. This is the direction developed in the late twentieth century, including Whitehead, Karen Barad, Carlo Rovelli, and others, but concrete relational ontologies differ in content. The work of constrained speculation belongs to this last direction: co-evolution as an ontological relation is a concrete form of relational ontology. But the content of this book differs importantly from existing work in relational ontology, as will be developed below.

A Fundamental Reflection

Why has the "problem of two-term relation" persisted for so long? This chapter argues that its persistence comes from an unexamined methodological choice: beginning from already given relata and asking how relation occurs between them. If we change the methodological choice—beginning from relation and asking how relata are differentiated from relation—many traditional difficulties disappear. But this methodological change requires serious ontological support and cannot remain a rhetorical posture. This is an application of the method of constrained speculation: it requires relational ontology to have falsifiable mechanisms, rather than a broad assertion that "everything is interconnected." Section 12.2 develops co-evolution as this form of relational ontology.

Objections and Replies

Objection One: Your "fundamental reflection" seems to repeat the claims of process philosophers such as Whitehead. What is new about co-evolution as relational ontology?

Reply: Whitehead's process philosophy indeed proposed a relational-ontological direction, and this chapter also uses related insights. But the work of this chapter is to provide relational ontology with falsifiable mechanisms within the EQT framework. Whitehead's concept of prehension is a philosophical concept, and its application is difficult to test. Co-evolution within the EQT framework has mechanisms: frequency matching ($g(f, f_0)$), threshold ($\delta\rho > \Gamma/k$), and tri-state decomposition. These mechanisms make co-evolution testable. The new contribution of this chapter does not lie in the direction of relational ontology, which has already been opened by Whitehead and others, but in the operationalization of relational ontology.

Objection Two: The "two-term relation problems" you list cover too much: subject–object, mind–matter, self–other, and individual–environment. These are different problems and require different treatments. Is it overly abstract to treat them as one class?

Reply: The work of this chapter is to identify the shared structure of these four problems: they all rest on the presupposition of "relation between two independent terms." This shared structure allows one class of response, relational ontology, to treat them together. But this chapter does not claim that these four problems are completely identical. They each have their own features. This chapter only claims that they have a shared methodological root, namely the presupposition of "two independent terms," and a shared possible direction of response, namely relational ontology. Concrete treatment, such as the Ψ framework for the mind–matter relation in Chapter 13 and reproducible co-perturbation for the observational relation in

Chapter 14, still requires dedicated work.

Literature That Must Be Engaged

- René Descartes, Meditations on First Philosophy (1641)
- Immanuel Kant, Critique of Pure Reason (1781)
- Edmund Husserl, The Crisis of European Sciences (1936): on the lifeworld
- Martin Heidegger, Being and Time (1927): on being-in-the-world
- Alfred North Whitehead, Process and Reality (1929)
- Chapter 7 on difference before identity (self-reference)

Summary of This Section

This section has done four things. First, it identified the "problem of two-term relation" as a family of problems: subject–object, mind–matter, self–other, and individual–environment. Second, it revealed the shared presupposition of this family of problems: relation between two independent terms. Third, it developed the internal difficulties of this presupposition: the unclear origin of relation, the possibility of knowledge, interaction, and the ontological status of boundary. Fourth, it identified an alternative direction: relational ontology.

The next section develops co-evolution as a concrete form of relational ontology.

12.3. Co-Evolution as an Ontological Relation: The Core Thesis

Starting Point

The core thesis of EQT Detection Theory is that the detector and the probed system are not in an external readout relation between two independent terms, but in a co-evolutionary relation within the same ρ_f field. This chapter elevates this physical thesis into a general ontological thesis.

Statement of the Core Thesis

Core thesis EV1: all "two-term relations" are same-field co- evolutionary relations. The two apparently independent terms are in fact local structures within the same process field, and their "relation" is an internal mode of evolution of that process field.

This thesis requires detailed explanation. "Same-field": all relata belong to the same foundational process. In the EQT framework this is the ρ_f field; in a more general ontological framework it is some foundational processual ontology. They do not come from two independent ontological domains, but are local structures of the same ontology. "Co-evolution": the relata do not first exist independently and then interact. They evolve together: their state changes are mutually constitutive and cannot be decomposed into "the state change of A + the state change of B." Co-evolution is an internal dimension of the whole process, not an externally added event between terms. "Internal mode of evolution": relation is not an additional structure imposed upon terms. It is the mode of evolution of the process field itself. When the process field evolves in a specific way, the "relation between A and B" appears. Relation does not need an independent ontological status because it is the internal structure of process.

Four Implications of the Thesis

EV1 has four ontological implications, which need to be stated explicitly.

Implication One: Relata are differentiated from relation, not the other way around.

Traditional presupposition: A and B exist as independent terms and are then connected through a relation. EV1: relation, namely the mode of evolution of the process field, precedes the relata, and A and B are differentiated from the process field as local structures. An analogy may help. In a symphony, each note does not first exist independently and then become combined into the piece; the musical work as a whole process comes first, and each note is a local manifestation of that process at a given moment. Likewise, A and B do not first exist independently and then establish a relation. Rather, a certain process field exists first, and A and B are local structures of that process. This implication does not mean that "A and B do not really exist." They are real as local structures. But their reality is derivative and depends on the process field to which they belong.

Implication Two: Relation itself has concrete mechanisms.

EV1 does not claim that relation is the vague assertion that "everything is interconnected." It requires relation to have concrete mechanisms: in the EQT framework, frequency matching ($g(f, f_0)$), amplitude threshold ($\delta\rho > \Gamma/k$), and tri-state response decomposition. This implication is the application of the method of constrained speculation to relational ontology. Relation as ontological content must have falsifiable mechanisms; otherwise it is an empty concept. Co-evolution as relational ontology must be able to distinguish genuine co-evolutionary relations from two terms without co-evolution. Concrete mechanisms provide this distinction.

Implication Three: Co-evolutionary relations are bidirectional but asymmetric.

In co-evolution, the two terms influence each other, but the influence need not be symmetric. In EQT detection theory, the detector affects the probed system through perturbation $\delta\rho$, and the probed system affects the detector through response $g(f, f_0) \cdot \rho_f$. Both have effects, but the forms of effect differ: the detector actively applies perturbation, while the probed system provides frequency response. This asymmetry gives co-evolution structure. It is not an abstract "interconnection," but a relation with concrete dynamics. Which side plays which role, and in what sense, is determined by the concrete dynamics.

Implication Four: Co-evolutionary relations produce observable consequences.

Co-evolution as an ontological relation must produce observable consequences; otherwise it is not an ontological relation but rhetoric. In the EQT framework, the consequences produced by co-evolution include: reproducible co-perturbation, where the same detector–probed system configuration under the same conditions produces the same result, and this reproducibility is the new foundation of objectivity, as Chapter 14 treats in detail; dependence of measurement results on configuration, where different detector frequencies, thresholds, and tri-state response characteristics produce different results, and this dependence is a concrete form of Bohr's complementarity; and perturbative back-action, where the detector has perturbative back-action on the probed system rather than performing disturbance-free readout. These observable consequences make co-evolution as an ontological relation falsifiable. If observations show that some readout is genuinely disturbance-free, that results are wholly independent of detector configuration, or that results are not reproducible, then co-evolutionary relation as an ontological thesis fails.

Legitimacy Conditions of the Thesis

EV1 as an ontological thesis must satisfy the legitimacy conditions of the method of constrained speculation: the concreteness of dimensionality, proxy, and threshold, as developed in Chapter 3. Dimensionality: co-evolutionary relations involve concrete physical quantities, including energy quantum density (ρ_f), gradient ($\nabla\rho_f$), frequency (f), perturbation amplitude ($\delta\rho$), and so on. These have clear physical dimensions. Proxy: co-evolutionary relations have empirical proxies, including experimental designs of reproducible co-perturbation, comparison of different detector configurations, and measurement of perturbative back-action. Threshold: co-evolutionary relations have concrete failure directions. If observation shows that some readout is genuinely independent of detector configuration, violating the frequency-matching condition, co-evolution as an ontological relation fails. These legitimacy conditions make EV1 not an abstract metaphysical conjecture, but an ontological thesis with observable consequences. This is the key to its status as a "rigorous ontological relation" rather than "rhetorical participationism."

Connections with the Five-Volume Work

EV1 connects with the five-volume work: The Foundational Volume of EQT, where A1, namely ρ_f as the sole basic ontological object, supports co-evolution as a same-field relation, since all terms belong to the same ρ_f field; Freezing Cosmology, where freezing as a special form of process supports co-evolution as an internal dimension of process, since freezing occurs within process, not as an event outside process; Mind–Matter Unity, where the relation between consciousness and physical processes in the Ψ framework is a concrete form of co-evolution, as Chapter 13 treats in detail; EQT Detection Theory, which directly treats the co-evolutionary relation between detector and probed system; and Energy, Process, and the Fate of the Cosmos, where gradient as a basic structure supports co-evolution as a relation among gradient structures, since gradient structures affect one another through the dynamics within the ρ_f field. These connections make EV1 not an isolated ontological thesis, but an ontological synthesis of the five-volume work.

Objections and Replies

Objection One: Your core thesis sounds like panpsychism or the rhetoric of universal interconnection. How do you distinguish rigorous co-evolutionary ontology from such rhetoric?

Reply: The key distinction lies in the requirement of concrete mechanisms. Rhetorical participationism says that "everything is interconnected" but provides no concrete mechanism; any observation is compatible with it, and it therefore has empty content. Co-evolution as a rigorous ontology provides concrete mechanisms, namely frequency matching, threshold, and tri-state response. These mechanisms produce observable consequences, such as reproducible co-perturbation, and allow concrete directions of failure. This distinction allows co-evolution to avoid the rhetorical trap. It does not say that "everything is interconnected." It says under what concrete dynamics two apparently independent terms co-evolve, what the concrete form of such co-evolution is, and how it can be tested. This is a rigorous ontological question, not rhetoric.

Objection Two: EV1 seems to presuppose a "process field" as foundational ontology. Is this presupposition itself an ontological expansion?

Reply: EV1 does not introduce a new ontology. It builds on the ontology already present in EQT, namely the ρ_f field. The ρ_f field as the sole basic ontology has already been defended in The Foundational Volume of EQT, namely by postulate A1. Co-evolutionary relation is not a new ontology; it is an internal mode of evolution of the ρ_f field. More deeply, ontological restraint, as developed in Chapter 2, requires commitments to be as few as possible. EV1 does not commit to a new ontology, but only to the internal evolutionary structure of the existing ontology, ρ_f. This is consistent with the principle of ontological restraint.

Objection Three: Your claim that "two-term relations are all co-evolutionary" sounds overly generalized. Are all relations co-evolutionary, including the relation between two distant galaxies with no interaction at all?

Reply: This objection points to a useful refinement. EV1, strictly speaking, is an ontological thesis about substantive relations, namely relations that produce observable consequences. There is no substantive relation between two terms that have no interaction at all. They may "belong to the same universe" in an abstract sense, but this abstract connection does not constitute co-evolution. Specifically, two distant galaxies that have no causal connection at cosmological distance do not constitute a co-evolutionary relation. They are two separated local structures within the ρ_f field, but because there is no causal connection between them, due to the light-speed limit, there is no co-evolution between them. EV1 applies to causally connected two-term relations. This covers almost all ordinary and scientific cases, including subject–object, mind–matter, self–other, and individual–

environment relations, but does not cover certain extreme cases, such as causally separated events. This refinement prevents EV1 from becoming overgeneralized.

Literature That Must Be Engaged

- EQT Detection Theory on detector–probed system co-evolution (self-reference)
- Whitehead, Process and Reality (1929)
- Carlo Rovelli, Helgoland (2020)
- Karen Barad, Meeting the Universe Halfway (2007)
- Chapter 7 on difference before identity (self-reference)
- Chapter 11 on freezing as a foundational ontological process (self-reference)

Summary of This Section

This section has done five things. First, it elevated the core insight of EQT Detection Theory into the general ontological thesis EV1. Second, it developed four implications of the thesis: relata are differentiated from relation, relation has concrete mechanisms, relation is bidirectional but asymmetric, and relation produces observable consequences. Third, it clarified the legitimacy conditions of the thesis: it satisfies the requirements of dimensionality, proxy, and threshold in the method of constrained speculation. Fourth, it established the connection with the five-volume work. Fifth, it replied to possible objections.

The next section develops the concrete mechanisms of co-evolution.

12.4. Concrete Mechanisms of Co-Evolution

The Function of Mechanism

EV1 as an ontological thesis requires concrete mechanisms so that it does not become an empty concept. This section develops three mechanisms of co-evolution: frequency matching, amplitude threshold, and tri-state response. It also explains how each mechanism makes co-evolutionary relation testable. These three mechanisms were initially proposed as part of detection theory in EQT Detection Theory. The work of this section is to extend them from the application of detection theory into general mechanisms of relational ontology: any co-evolutionary relation involves some form of these three mechanisms.

Mechanism One: Frequency Matching ($g(f, f_0)$)

The form in EQT detection theory: the response of a detector to a probed system is not arbitrary; it depends on the match between the detector's intrinsic frequency (f_0) and the frequency of the probed system (f). The concrete form is the Lorentzian resonance response:

$$g(f, f_0) = \frac{A}{(f - f_0)^2 + \gamma^2}$$

When f is close to f_0, the response is strong; when f is far from f_0, the response is weak. This is the mechanism of detector selectivity.

Generalized to relational ontology: any co-evolutionary relation involves frequency matching between relata. Not any two terms can form a co-evolutionary relation. Co-evolution requires some condition of resonance or concordance between terms. Concrete examples: in the subject–object relation, a perceiving subject can perceive only certain frequency bands of electromagnetic radiation, certain amplitude ranges of sound waves, and certain chemical substances; the subject's "frequency window" determines what it can perceive. In the mind–matter relation, consciousness can establish mind–matter relation only with particular types of physical processes, such as specific activities in specific brain regions; other physical processes, such as liver metabolism, do not co-evolve with consciousness, and consciousness has its own "frequency window." In the individual–environment relation, organisms can establish adaptive relations only with certain aspects of the environment, such as temperature ranges, nutritional types, and climatic patterns; "adaptation" is a concrete form of frequency matching. In social relations, relations among persons cannot be established arbitrarily; they require shared "frequencies" such as common language, culture, values, and interests. Two persons with no common ground cannot form a substantive relation. Frequency matching as a mechanism gives co-evolutionary relations selectivity: not any two terms can form a co-evolutionary relation; concrete matching conditions are required. This selectivity is a feature of relation as ontological content.

Mechanism Two: Amplitude Threshold ($\delta\rho > \Gamma/k$)

The form in EQT detection theory: frequency matching is necessary but not sufficient. Even if frequency matches, the perturbation amplitude must exceed the nonlinear feedback threshold in order to move from weak response into the recordable amplification region:

$$\delta\rho > \Gamma/k$$

Here $\delta\rho$ is perturbation amplitude, Γ is the dissipation rate, and k is the nonlinear feedback coefficient.

Generalized to relational ontology: any co-evolutionary relation involves an amplitude threshold. Interaction between relata must reach a certain intensity in order to produce substantive co-evolution. Concrete examples: in the subject–object relation, weak illumination cannot be perceived by the eye if it falls below the visual threshold; excessively strong illumination damages the eye but does not produce concrete perception. Vision has a concrete amplitude-threshold range. In the mind–matter relation, the relation between consciousness and neural activity has a threshold: certain neural activities, such as subliminal perception, do not form conscious experience; certain excessively strong activities, such as epileptic activity, also do not produce normal conscious experience. In the individual–environment relation, biological adaptation to the environment has thresholds: temperatures that are too low or too high make life impossible to maintain; insufficient or excessive nutrition both cause pathology. In social relations, interpersonal interaction has thresholds: overly sparse contact does not form substantive relation; overly dense contact, as in some controlling relations, also damages normal relational structure. Amplitude threshold as a mechanism gives co-evolutionary relations intensity dependence: the existence and character of relation depend on the intensity of interaction, not simply on "presence" or "absence." This intensity dependence is a feature of co-evolution.

Mechanism Three: Tri-State Response (Free State, Bound State, Condensed State)

The form in EQT detection theory: observational signals are not of a single kind, but mixtures of the tri-states: free-state signals, namely propagativity; bound-state signals, namely local resonance; and condensed-state signals, namely stable structure. Observation involves the concrete weights of these three states.

Generalized to relational ontology: any co-evolutionary relation involves a combination of tri-states. Relation is not a single kind of event, but a combination of different modes of existence. Chapter 10 has already treated the tri-states in detail as ontological categories. Concrete examples: in the subject–object relation, perception involves free state, namely momentary sensory impressions; bound state, namely the binding of perceived features such as color and shape; and condensed state, namely the stable structures of long-term memory. The combination of these three states determines the form of perception. In the mind–matter relation, consciousness involves free state, namely the flow of thought; bound state, namely synchronization of neural oscillations; and condensed state, namely the stable structures of personality. This is treated in detail in Chapter 13. In the individual–environment relation, the environmental interaction of an organism involves free state, namely the flow of matter and energy; bound state, namely metabolic cycles; and condensed state, namely the stability of biological form. In social relations, interpersonal relations involve free state, namely momentary interaction and affect; bound state, namely oscillatory patterns of relation; and condensed state, namely stable structures of long-term commitment. Tri-state response as a mechanism gives co-evolutionary relations structural diversity: relation is not one-dimensional, but a combination of multidimensional modes of existence. This diversity is a feature of co-evolution.

The Interdependence of the Three Mechanisms

The three mechanisms are not independent; they form an interdependent whole. Dependence One: frequency matching is the precondition of threshold. Without frequency matching, even a perturbation above threshold does not produce a concrete response; the perturbation is simply dissipated. Frequency matching provides the channel through which the threshold condition operates. Dependence Two: threshold is the precondition of tri-state structure. Without sufficient threshold, the relation remains in the "weak response" region, and the tri-state structure does not fully unfold. Threshold provides the intensity range for tri-state analysis. Dependence Three: tri-state structure is the development of relational content. The content of relation, such as concrete perception, conscious experience, or organism–environment interaction, is determined by the combination of the tri-states. The tri-states provide structure for relational content. Together, the three mechanisms constitute the dynamics of co-evolution. They make co-evolution as an ontological relation testable: a concrete relation must satisfy the forms of these three mechanisms; otherwise, it is not a genuine co-evolutionary relation.

A Concrete Application: The Co-Evolution of Conscious-

ness and Brain

To keep the application of the three mechanisms from remaining abstract, this section develops a concrete example: the co-evolutionary relation between consciousness and brain.

Consciousness and brain are not two independent terms in relation, but two projections of the same process, namely the brain as a highly complex structure of the energy quantum density field, as Chapter 13 treats in detail. This process as a whole involves concrete co-evolutionary mechanisms:

Frequency matching: consciousness is frequency-matched with neural activity in specific brain frequency bands. Consciousness-related neural activity is mainly located in combinations of the gamma band (30-80 Hz) and lower frequency bands. Brain activity at other frequencies, such as very low-frequency metabolic oscillations, does not co-evolve with consciousness. This is the frequency window of consciousness–brain co-evolution.

Amplitude threshold: consciousness requires neural activity to reach a threshold. Subthreshold activity does not form consciousness; excessively low global activity, as in deep anesthesia, also does not maintain consciousness. Consciousness–brain co-evolution has a concrete amplitude range.

Tri-state response: consciousness involves a tri-state combination of neural activity: free state, namely information flow; bound state, namely neural synchronization; and condensed state, namely long-term memory structure. The three variables of the Ψ framework in Mind–Matter Unity, $\tilde{A}$, C, and $\mathcal{B}$, correspond respectively to the intensities of these three states.

This application shows how the three mechanisms provide a testable structure for consciousness–brain co-evolution. Progress in consciousness research, such as the Ψ trajectories of different anesthetics, frequency-band characteristics of different conscious states, and proxy diagnosis of various disorders of consciousness, is built upon this co-evolutionary framework.

Objections and Replies

Objection One: The three mechanisms were originally part of EQT detection theory. Is it an overgeneralization to extend them into general mechanisms of relational ontology?

Reply: This objection points to a real methodological question: how can the content of a concrete physical theory be generalized into a general ontology? The response of this chapter is that the generalization does not mean applying EQT formulas to all relations; that would indeed be excessive. The generalization means identifying the ontological structures of the three mechanisms: the selectivity of frequency matching, the intensity dependence of thresholds, and the structural diversity of the tri-states. These ontological structures are general, while their concrete expressions differ by domain. By analogy, energy conservation as a physical law has concrete forms, with different formulas in different mechanical frameworks, but the ontological structure of energy conservation, namely the conservation of some quantity through process, is general. The three mechanisms of co-evolution are similar: their concrete forms vary by domain, while their ontological structures are general.

Objection Two: Are your examples, such as the "frequency matching" of social relations and the "amplitude threshold" of perception, metaphorical uses?

Reply: This objection points to an important issue: are the concrete examples used metaphorically or as rigorous ontology? The response of this chapter is that in applications across different domains, the forms of the mechanisms have different degrees of rigor. In physics and directly connected domains, such as detection theory, the mechanisms have rigorous mathematical forms, including concrete Lorentzian response functions and threshold formulas. In biology, the mechanisms have approximately rigorous forms, such as adaptation theory and stimulus–response models. In psychology and sociology, the mechanisms have conceptual forms, identifying relevant structures without necessarily possessing strict mathematical formulas. This chapter does not claim that applications in all domains reach the rigor of physics; that would be unrealistic. It claims that in each domain, the three mechanisms provide ontological structures that allow co-evolutionary relations to be studied more precisely, although the degree of precision varies by domain.

Objection Three: The three mechanisms, frequency matching, threshold, and tri-state response, sound overly physicalized. Do social and cultural relations really need these mechanisms?

Reply: This objection rightly points to the issue of theoretical parsimony. The response of this chapter has two levels. At the conceptual level, the ontological structures of the three mechanisms—selectivity, intensity dependence, and diversity—are involved in any concrete relation. Social relations also have selectivity, since not everyone can form substantive relations; intensity dependence, since relational intensity has concrete ranges; and diversity, since relations have multiple components. At this level, the mechanisms are general. At the level of concrete form, social relations do not need to use physical language such as "frequency" or "amplitude." They use sociological language, such as common language, shared values, interaction frequency, social distance, and so on. But the content of these sociological terms corresponds to the ontological structures of the three mechanisms. This layered application avoids both excessive physicalization, which would impose physical language, and conceptual emptiness, in which the mechanisms become too abstract to have content.

Literature That Must Be Engaged

- EQT Detection Theory on the three criteria (self-reference)
- Bruno Latour, Reassembling the Social (2005): on concrete mechanisms of social relations
- Andy Clark, Being There (1997): on body–environment co-evolution
- Chapter 10 on tri-states as ontological categories (self-reference)
- Chapter 14 on observation as participation (forward reference)

Summary of This Section

This section has done four things. First, it developed the three mechanisms of co-evolution: frequency matching, which provides selectivity; amplitude threshold, which provides intensity dependence; and tri-state response, which provides structural diversity. Second, it generalized each mechanism from EQT detection theory to the general form of relational ontology. Third, it identified the interdependence of the three mechanisms: frequency matching is the precondition of threshold, threshold is the precondition of tri-state structure, and tri-state structure is the development of relational content. Fourth, it showed the application of the mechanisms through the example of consciousness–brain co-evolution.

The next section develops the dialogue between co-evolutionary relations and contemporary relational ontology.

12.5. Dialogue with Karen Barad's Intra-Action

Barad's Work

In Meeting the Universe Halfway (2007), Karen Barad proposes agential realism and intra-action as a philosophical development of contemporary physics, especially Bohr's work. Barad's core position is this: reality is not composed of independent relata; these relata themselves are differentiated from relations. This relation is not inter-action, namely interaction between two independent terms, but intra-action, an internal process in which relata are differentiated from relation. Measurement is a concrete form of intra-action: measurement apparatus and measured object are not independent terms, but two phenomenally separate aspects of the whole measurement process. This intra-action is material: it constitutes objects themselves and is not merely a cognitive relation. Barad's work has been influential, especially in feminist science studies, new materialism, and posthumanist theory.

Similarities with the Co-Evolution of Constrained Speculation

Barad's intra-action has important similarities with co-evolution (EV1) in this book. Similarity One: both reject the presupposition of independent relata. Both reject the traditional presupposition of "relation between two independent terms." Barad's prefix "intra" emphasizes that relation is internal rather than externally added; this book's "same-field co-evolution" emphasizes that relata belong to the same field. Similarity Two: both are based on physics. Both are built upon the content of contemporary physics. Barad's work is based on Bohr's complementarity, while the work of this book is based on EQT detection theory. Both elevate physical insights into general ontology. Similarity Three: relation as constitutive. Both hold that relation is not a cognitive relation but ontologically constitutive: relation participates in the very existence of terms, rather than merely describing connections between them. These similarities make Barad's work an important dialogue partner for the co-evolutionary ontology of this book.

Key Differences

But the two also differ importantly. This section states these differences explicitly.

Difference One: explicitness of failure conditions. Barad's intra-action is a philosophically rich concept, but it is not explicit about concrete failure conditions: what observation would make intra-action, as an ontological thesis, fail? It is difficult to find a clear response in Barad's work. Co-evolution as an ontological thesis in this book has concrete failure conditions: the three mechanisms, frequency matching, threshold, and tri-state response, provide observable consequences, and if these consequences systematically fail to appear, co-evolution as an ontological thesis fails. This difference does not mean that Barad's work is wrong. It means that the work of this book satisfies the requirements of the method of constrained speculation, namely the explicitness of failure conditions, whereas Barad's work does not explicitly adopt this methodological requirement.

Difference Two: connection with concrete physical theory. Barad's work is mainly based on a philosophical interpretation of Bohr's complementarity, and its connection with other developments in physics, such as quantum field theory, statistical mechanics, and cosmology, is comparatively limited. This makes Barad's intra-action weaker when it is extended to other physical domains. Co-evolution in this book is built within the EQT framework and is connected with concrete physics, including frequency matching, threshold, and tri-states. This gives co-evolution applications across several physical domains, including detection theory, cosmology, and mind–matter.

Difference Three: political-ethical implications. Barad's work has strong political and ethical motivations. Agential realism is used in feminist theory, posthumanism, and related forms of political critique. Barad emphasizes that ontological relation itself is ethical, namely an "ethico-onto-epistemology." Co-evolution as an ontological relation in this book does not presuppose a specific political-ethical stance. It extends to ethics in Part V, Chapters 20 and 21, but ethics is derived from ontology rather than presupposed. This difference reflects different emphases in the two works. Barad's work is more directly politicized; the work of this book is more analytic and systematic. Both are legitimate, but they cannot be simply interchanged.

The Relation between Co-Evolution and Intra-Action

Through comparison, the relation between co-evolution in this book and Barad's intra-action can be stated as follows: the two are constitutively coextensive within the equivalence class of processual relational ontology, in the sense of Chapter 4, but they differ in concrete features. Barad's work has stronger political-ethical implications and a phenomenological-hermeneutic method; the work of this book has a stronger analytic-physical method and falsifiable mechanisms. The two share the same basic ontological commitment, namely the rejection of independent relata, but they develop this commitment in different directions. This statement uses the tool of constitutive coextension established in Chapter 4: it precisely states the ontological relation between the two works while respecting the content of each.

Objections and Replies

Objection One: You treat Barad too academically. Her work has concrete political force, and you seem to depoliticize it.

Reply: The work of this chapter is ontological and philosophical, not a political evaluation. The political force of Barad's work is real and legitimate; it contributes to feminist theory and posthumanism. This chapter does not deny that political force. But ontological evaluation differs from political evaluation. Ontological evaluation asks: as an ontological thesis, what falsifiable mechanism does the work have? This question is independent of political stance and requires rigorous analytic treatment. The ontological evaluation given in this section points out that Barad's work is less explicit than this book in ontological falsifiable mechanisms. This evaluation is ontological and does not deny the power of Barad's work in other dimensions.

Objection Two: Your statement of "constitutive coextension" sounds too moderate. It seems that you are trying to preserve both works at once. Is this a compromise?

Reply: This statement is not a compromise. It is a precise statement of ontological relation. The method of constrained speculation requires precision, not the selection of a side. Co-evolution and intra-action overlap in their basic ontological commitment, namely constitutive coextension, and differ in their concrete methods and applications. This is a statement of fact, not a compromise. This book does not claim that Barad's work is wrong. It claims that in certain dimensions, such as falsifiable mechanisms, it is less explicit than the work of this book. This is a concrete judgment of difference, not a vague compromise.

Objection Three: Did Barad do earlier what you are claiming to do? Is the work of this book merely a continuation of Barad's work?

Reply: Barad's work precedes this book in some respects: Meeting the Universe Halfway was published in 2007. This book acknowledges that chronological priority. But the work of this book is not a continuation of Barad's work. It has an independent ontological ground, namely the EQT framework; an independent methodology, namely constrained speculation; and independent concrete mechanisms, namely frequency matching, threshold, and tri-states. The two works independently arrive at similar ontological directions within processual relational ontology, but their concrete contents differ. Philosophical progress often involves multiple works independently arriving at similar directions. This marks the accessibility of a philosophical direction, not plagiarism or simple continuation.

Literature That Must Be Engaged

- Karen Barad, Meeting the Universe Halfway (2007), "Posthumanist Performativity" (2003)
- Donna Haraway, Simians, Cyborgs, and Women (1991): related work in feminist science studies
- Bruno Latour, We Have Never Been Modern (1991): the new materialist tradition
- Chapter 4 on constitutive coextension (self-reference)

Summary of This Section

This section has done four things. First, it introduced Karen Barad's work on intra-action. Second, it identified the similarities between Barad's work and co-evolution in this book. Third, it clarified differences: explicitness of failure conditions, connection with concrete physical theory, and political-ethical implications. Fourth, it used the tool of constitutive coextension to state the relation between the two.

The next section develops the dialogue with the Huayan doctrine of "mutual interpenetration without end."

12.6. Dialogue with the Huayan Doctrine of "Mutual Interpenetration without End"

Huayan Relational Ontology

The Huayan school is the tradition in Chinese Buddhism that most systematically developed relational ontology. Its core thesis of "mutual interpenetration without end" appears in the Avataṃsaka Sūtra and in the central teachings of the Huayan school. Briefly stated, Huayan relational ontology includes several core concepts: the dharma realm of the non-obstruction of thing with thing, meaning that there is no obstruction among all things, or "things," because they mutually penetrate and mutually contain one another; one is all, and all is one, meaning that each thing contains all other things and all other things contain each thing; Indra's net, in which all things are like jewels in an infinite net, each jewel reflecting all other jewels, and the reflections in each jewel being reflected again in all other jewels, with such reflection being "mutual interpenetration without end"; and the Ten Mysterious Gates, which develop in detail ten aspects of the dharma realm of the non-obstruction of thing with thing. Huayan relational ontology is the strongest form of relational ontology: it extends relation to the mutual penetration of all things.

Structural Relevance to Co-Evolution

Huayan relational ontology has structural relevance to co-evolution (EV1) in this book. Relevance One: rejection of the presupposition of independent terms. Huayan's "one is all" explicitly rejects terms as independent units: any "one" already contains "all." This structurally corresponds to co-evolution's claim that relata are differentiated from relation. Relevance Two: the internality of relation. Huayan's "non-obstruction of thing with thing" holds that relation is the internal structure of things, not an additional connection outside them. This structurally corresponds to co-evolution's claim that relation is an internal mode of evolution of the process field. Relevance Three: the concreteness of relation. Huayan does not merely assert broadly that "everything is connected." It develops the concrete structure of relation through the Ten Mysterious Gates, such as the gate of simultaneous complete correspondence, the gate of unobstructed freedom between broad and narrow, and the gate of mutual containment of one and many without sameness. This structurally corresponds to co-evolution's development of concrete structure through three mechanisms: frequency matching, threshold, and tri-states. These forms of relevance make Huayan another important dialogue partner for the co-evolutionary ontology of this book.

Key Differences

But the two also differ importantly. Difference One: methodological ground. Huayan relational ontology is grounded in the method of Buddhist cultivation and realization: through practices such as contemplative methods and meditation, the practitioner directly realizes the non-obstruction of thing with thing. This is an experiential-cultivational method. The co-evolution of this book is grounded in the method of physical-philosophical argumentation: it proceeds through physical mechanisms, namely EQT detection theory, and analytic argument, namely constrained speculation. This is an analytic-physical method. Both methods are legitimate, but they produce different types of ontological statements. Huayan statements are objects of realized insight; the statements of this book are falsifiable ontological theses. This methodological difference reflects the different situations of the two works. Difference Two: the scope of relation. Huayan relation extends to all things: any two things mutually penetrate. This is relational ontology with the maximal scope. The co-evolution of this book, in the strict sense, is limited to causally connected two-term relations, as already clarified in Objection Three of Section 12.2. There is no substantive co-evolution between two terms that are completely causally separated. This is causally restricted relational ontology. This difference does not mean that this book is "weaker." It reflects a different choice of scope. Huayan's full-scope position has its metaphysical ground, namely that all things share Buddha-nature; the causally restricted position of this book has its physical ground, namely the light-speed limit and the requirement of causal connection. The two reflect different ontological starting points. Difference Three: soteriological implications. Huayan relational ontology has soteriological implications: realizing the non-obstruction of thing with thing is a path toward liberation. This is a close connection between relational ontology and Buddhist soteriology. Co-evolution as an ontological relation in this book is not directly connected to soteriology. It is extended in Part V, Chapters 20 and 21, to ethics and the problem of meaning, but soteriology, in the sense of individual liberation, is not the subject of this book. This difference reflects the different purposes of the two works. Huayan work has a soteriological motivation; this book has a philosophical-physical motivation. Both have legitimacy.

Boundaries of Resource Use

As previous chapters of this book have made clear, especially Chapter 8 on the tradition of qi theory, the use of Chinese philosophical traditions as resources is selective and cannot pretend that the tradition as a whole is contemporarily relevant. The use of Huayan in this section is also selective: it uses the relational-ontological direction of Huayan, namely one is all and the non-obstruction of thing with thing; it does not use Huayan soteriology, such as the doctrine of liberation and Buddha-nature; it uses Huayan's development of the concrete structure of relation, namely the Ten Mysterious Gates as a direction; it does not claim that Huayan already "anticipated" EQT or the work of this book, since that would be false rhetoric. This selective use makes Huayan a dialogue partner rather than a resource absorbed into the work of this book. Huayan work has its own internal integrity, and this book uses specific aspects of it while respecting the whole.

Objections and Replies

Objection One: Your treatment of Huayan is too brief. Huayan is a highly complex tradition, but this section treats it only briefly. Is this unfair?

Reply: This section, as part of a dialogue chapter, is constrained by length. This chapter already treats several works, including Barad, Huayan, and Whitehead. Detailed treatment of each requires dedicated literature. What this section gives is the most relevant structural correspondence, not an exhaustive analysis. More detailed work on Huayan requires a dedicated study. The function of this section is to indicate directions of dialogue, not to complete the dialogue. Acknowledging this limit is legitimate, since every monograph chapter has boundaries of treatment.

Objection Two: You claim that the dialogue is "serious," but in fact you are only "selecting" from Huayan those aspects that support the position of this book. Is this absorption rather than acknowledgment, in the distinction made in Chapter 4, Section 4.7?

Reply: This objection rightly points to the ethics of resource use. The response of this section is to state explicitly which aspects are used, namely the relational-ontological direction and the development of the concrete structure of relation; to state explicitly which aspects are not used, namely soteriology and the doctrine of Buddha-nature; not to claim that Huayan as a whole is consistent with this book; and not to flatten the internal complexity of Huayan, since there are different lineages and debates within Huayan. This explicitness makes the use a matter of acknowledgment + selective use, not absorption. If some readers think that this selective use intervenes too strongly in the internal integrity of Huayan, that is a legitimate criticism, and the work of this section remains open to such criticism.

Objection Three: Huayan work predates modern physics by about 1,500 years. Is it historically anachronistic to claim that it has "structural relevance" to EQT co-evolution?

Reply: This section does not claim that Huayan "anticipated" EQT. That would be false rhetoric, as earlier chapters have repeatedly made clear. This section claims that the two works, one in seventh- to eighth-century China and the other in twenty-first-century EQT, have structural relevance in ontological direction. Such structural relevance is not historical anachronism. It is a sign of the accessibility of philosophical directions. Works in different periods and traditions can independently arrive at similar ontological directions, such as the structural relevance between Greek atomism and modern atomism, or between Buddhist dependent origination and systems theory. Identifying this structural relevance is a legitimate part of comparative philosophical work.

Literature That Must Be Engaged

- The Avataṃsaka Sūtra, especially the Gaṇḍavyūha section
- Fazang, Treatise on the Golden Lion of Huayan; Explorations of the Mysteries of the Avataṃsaka Sūtra
- Du Jiwen, Lectures on Buddhist Philosophy
- Francis Cook, Hua-yen Buddhism: The Jewel Net of Indra (1977)
- Chapter 8 on the ethics of using resources from the Chinese tradition (self-reference)

Summary of This Section

This section has done four things. First, it briefly introduced Huayan relational ontology: the non-obstruction of thing with thing, one is all and all is one, Indra's net, and the Ten Mysterious Gates. Second, it identified the structural relevance between Huayan and the co-evolution of this book. Third, it clarified differences: methodological ground, scope of relation, and soteriological implications. Fourth, it clarified the boundaries of resource use: selective use and honest acknowledgment.

The next section is the conclusion of this chapter.

12.7. Chapter Summary: Co-Evolution as Relational Ontology

Completion of the Three Tasks

The three tasks undertaken by this chapter have all been completed. Task One: to elevate the core insight of EQT Detection Theory into a general ontological thesis. Section 12.1 developed the "problem of two-term relation" as a persistent problem in the history of philosophy and identified its shared presupposition, namely two independent terms, and its difficulties. Section 12.2 gave the core thesis EV1: all two-term relations are same-field co-evolutionary relations. This elevation gives general ontological meaning to the insight of the five-volume physical work that observer and object belong to the same ρ_f field. Task Two: to clarify the distinction between co-evolution as a rigorous ontological relation and rhetorical participationism. Section 12.3 developed the three mechanisms of co-evolution, namely frequency matching, amplitude threshold, and tri-state response, each of which has a falsifiable structure. These mechanisms give co-evolution failure conditions and prevent it from degenerating into the rhetoric of "everything is interconnected." Task Three: to enter into dialogue with contemporary work in relational ontology. Section 12.4 entered into dialogue with Karen Barad's intra-action, and Section 12.5 with the Huayan doctrine of "mutual interpenetration without end." Both dialogues use the tool of constitutive coextension established in Chapter 4: co-evolution in this book and its dialogue partners are constitutively coextensive in ontological direction, while retaining independence in their concrete features.

The Position of This Chapter in Part III

The overall architecture of Part III, "Co-Evolution, Mind–Matter, and Observation: The Unity of Three Relations," is established by this chapter. The following three chapters build upon this chapter's work: Chapter 13 on mind–matter, where the relation between mind and matter is treated as a concrete form of co-evolutionary relation; Chapter 14 on observation, where reproducible co-perturbation as a new foundation of objectivity is the application of co-evolutionary relation to measurement; and Chapter 15 on time, where time as $\partial\rho/\partial t$, reconciling Bergson and Einstein, is the development of co-evolutionary relation in the temporal dimension. The work of all three chapters depends on the relational ontology established in this chapter. Without the foundational work of this chapter, the concrete treatments in those three chapters would lose their ontological ground.

The Position of This Chapter in the Whole Book

Looking backward, this chapter builds upon the work of Part II: Chapter 7 on difference before identity provides the ontological starting point for relational ontology; Chapter 10 on tri-states as ontological categories provides the ontological-categorial ground for the tri-state mechanism of co-evolution; and Chapter 11 on freezing as a foundational ontological process provides ontological support for co-evolution as an internal dimension of process. Looking forward, this chapter prepares the ground for Part IV, "The Metaphysical

Boundary of Freezing." Chapter 17 on ontological singularity treats the collapse of co-evolutionary relation in the limit $\nabla\rho \equiv 0$: co-evolutionary relation itself has a boundary.

The Force of Relational Ontology

The work of this chapter shows the force of relational ontology. Force One: it resolves traditional difficulties. Several persistent difficulties in the "problem of two-term relation," including the origin of relation, the possibility of knowledge, interaction, and the ontological status of boundary, are treated within the framework of relational ontology: their root, namely the presupposition of two independent terms, is identified and rejected. Force Two: it connects with concrete science. The three mechanisms of co-evolution are not abstract metaphysical conjectures, but testable mechanisms with physical grounds. This allows relational ontology to connect with concrete sciences, including physics, biology, and neuroscience, and avoids the split between philosophy and science. Force Three: it enables cross-traditional dialogue. The direction of relational ontology allows this book to enter into serious dialogue with contemporary analytic-continental traditions, such as Barad, Chinese philosophical traditions, such as Huayan, and process-philosophical traditions, such as Whitehead. This cross-traditional dialogue gives the work of this book a place in world-philosophical discussion.

Honesty about the Limits of This Chapter

This chapter acknowledges several limits: first, limitations in applications to concrete science. This chapter develops co-evolution mainly from the perspective of physics, with EQT detection theory as the main example. Applications in biology, neuroscience, sociology, and other fields require more dedicated work. This chapter only gives directions and examples; detailed applications are future work. Second, dialogue with specific philosophers is insufficient. This chapter enters into dialogue with Barad and Huayan, but its dialogue with other relevant work, such as Latour's actor-network theory, Andy Clark's extended mind, and Carlo Rovelli's relational quantum mechanics, is relatively brief. Detailed dialogue with each work requires dedicated papers or monographs. Third, the relation between ontology and epistemology. This chapter does not develop in detail how co-evolution as an ontological relation connects with epistemological relations, such as knowledge, belief, and inference. This remains an open task. Acknowledging these limits does not weaken the work of this chapter. It makes the chapter an honest initial work and invites later development.

The Core Posture of This Chapter

The core posture of this chapter can be stated as follows: relational ontology is not rhetoric, but rigorous ontology with falsifiable mechanisms.

This posture is rare in the revival of relational ontology in the late twentieth century. Many works in relational ontology, including some Whiteheadian studies, new materialism, and feminist science studies, make important contributions in philosophical direction, but are not rigorous in falsifiable mechanisms. The specific contribution of this chapter is to develop rigorously the falsifiable mechanisms of relational ontology: frequency matching, threshold, and tri-states. This rigor is an application of the method of constrained speculation. The method requires every ontological commitment to have failure conditions. Co-evolution as an ontological commitment must also satisfy this requirement. The work of this chapter fulfills that requirement.

Meaning before freezing: this is one of the core theses of this book. This chapter provides an ontological foundation for that thesis: meaning appears in co-evolutionary relations, meaning-bearers such as consciousness and life are concrete structures of same-field co-evolution, and meaning itself is an emergence of co-evolutionary relation. When Chapter 21 argues that meaning is true only within finitude, this finitude is the finitude of co-evolution: it depends on concrete co-evolutionary relations, and these relations dissolve before the gradient terminal phase. This connection will be further developed in Chapter 17 and Chapter 21.

12.8. Methodological Reflection on This Chapter

This chapter elevates co-evolution from a physical statement in EQT Detection Theory into a general ontological thesis. This section reflects on the methodological significance of that elevation.

This Chapter's Avoidance of the Rhetorical Trap

Terms such as "co-evolution" and "relational ontology" have been widely used since the late twentieth century, from co-evolution in ecology to New Age discourses of interconnection. This makes these terms prone to degeneration into rhetoric. The core work of this chapter is to give these terms falsifiable content: through three mechanisms, namely frequency matching $g(f, f_0)$, the threshold condition

$\delta\rho > \Gamma/k$, and tri-state decomposition, co-evolution becomes rigorous rather than rhetorical. Many works on interconnection are built on intuitive appeal. They are intuitively correct, but lack content that can be rigorously evaluated. Through physical grounds and methodological discipline, the work of this chapter gives co-evolution as an ontological relation content that can be analyzed.

The Position of This Chapter among Analytic, Continental, and Chinese Traditions

This chapter enters into dialogue with three traditions: the work of Karen Barad, the Huayan school, and Whitehead's process philosophy. Barad's intra-action resonates deeply with the co-evolution of this chapter in direction: both reject independent terms as the premise of relation. But the work of this chapter has a more explicit falsifiable structure than Barad's work, namely the three dynamical mechanisms. Although Barad's work has profound philosophical insights, its falsifiable structure is not explicit. Huayan's "non-obstruction of thing with thing" resonates deeply with co-evolution in direction, but Huayan work is built within the framework of Buddhist soteriology, whereas this chapter is built within the framework of philosophy of nature and methodological discipline. Whitehead's prehension is structurally related to co-evolution, but Whitehead's work introduces eternal objects as a second ontology, whereas this chapter avoids such an introduction.

Limits Honestly Acknowledged in This Chapter

This chapter acknowledges several limits. First, this chapter develops only the core direction of co-evolution as an ontological relation in concrete scientific fields; more detailed interdisciplinary applications require dedicated work. Second, this chapter's dialogue with specific philosophers, including Barad, Huayan, and Whitehead, concerns core directions; more detailed dialogue requires dedicated papers. Third, this chapter mainly develops the ontological level of relational ontology, while the treatment of the epistemological level, namely how co-evolutionary relations are known, is left to Chapter 14.

Connections with Later Chapters

Co-evolution as an ontological relation established in this chapter is applied in several later chapters. Chapter 13 uses co-evolution to treat the mind–matter relation. Chapter 14 uses co-evolution to treat observation. Chapter 15 uses co-evolution to treat time. More broadly, this chapter is the statement of the relational-ontological position of the book. If the book did not develop relational ontology, it would only continue process ontology and would lack serious treatment of the relational dimension. The work of this chapter gives the book a complete ontological structure: processuality + relationality. More deeply, this chapter demonstrates the application of the method of constrained speculation in cross-traditional dialogue: it shows the cross-traditional applicability of constrained speculation.

13. Mind–Matter Unity: Isomorphism, Co-Evolution, and the Last Cell

13.1. Introduction to This Chapter

There is a hard problem in the philosophy of consciousness. David Chalmers formulated it explicitly in 1995: why are physical processes, such as neural activity and information processing, accompanied by subjective experience? Why do neuronal firing, transmembrane ion flows, and synchronized oscillations of neural ensembles correspond to what it is like to "see red" or to "feel pain"—that is, to a quality given from within, rather than operating silently in the dark without any qualitative character? This problem cannot be solved by a simple reductive method: no description of physical processes directly answers the question of "why this process is accompanied by this feeling."

The method of constrained speculation must give an honest and precise position in the face of this problem. This chapter undertakes that task. Its posture has two levels.

First, regarding the mind–matter relation itself: this chapter argues that mind and matter are two aspects of the same process. There is a strict identity between the entire structure of experience and the gradient configuration of a certain class of frequency-resolved energy quantum density field ρ_f. This identity is fulfilled through a feature-by-feature correspondence between a five-dimensionally rich gradient structure and the full structural characteristics of experience. This is the deepest position that constrained speculation can take on the mind–matter problem: it gives substantive meaning to "unity" and minimizes the explanatory burden of the hard problem.

Second, regarding the hard problem itself: this chapter argues that the hard problem is not an undifferentiated fog that admits no point of entry, but can be precisely located, through two compressions, as a single point, namely the residual point of subject-ascription. This point is not an open problem that constrained speculation "cannot yet handle," but is precisely marked as a boundary: it marks the edge of the explanatory capacity of structure and relation itself. The contribution of constrained speculation to the hard problem is therefore not to "solve" it, nor to "defer" it, but to mark it precisely: everything that can be solved is solved as far as possible, including structure, the mechanism of given-ness, and measurable projections; what cannot be solved is precisely marked as a single-point boundary.

The work of this chapter is built on the ontological and methodological resources already established in earlier chapters. Chapter 3 established failure conditions as an ontological discipline (CR2); Chapter 4 introduced constitutive coextension as a new semantic relation (CC1); Chapter 10 systematized the tri-states as ontological categories (T1–T2); Chapter 12 elevated co-evolution into a general ontological relation (EV1); and Chapter 14 mechanized observation as participation (OB1). This chapter applies these tools together to the most concrete and difficult ontological problem, the mind–matter relation, and assembles them from a dispersed tool set into a systematic philosophical position on mind and matter: a Russellian monism enriched by EQT, neither reductive nor eliminative, and governed by boundary as gate.

This chapter unfolds in six sections. Section 13.1 begins from the hard problem and introduces the isomorphic program as the conceptual device of this chapter: the fivefold gradient-richness isomorphism and the correct positioning of the Ψ framework, namely as a projected quantity rather than a definition. Section 13.2 develops the identity tier as the choice of this book, distinguishes it from the intermediate tier and the weakest tier, and shows through the argument of structural sufficiency that only a five-dimensionally rich gradient structure can support the identity tier. Section 13.3 develops given-ness as a relational event of internal–external co-evolution, and develops the anti-panpsychist position of boundary as gate. Section 13.4 develops the two compressions of the hard problem and the last cell as a boundary, arguing that the C3 prohibition is the fulfillment of the failure-orientation discipline of constrained speculation (CS4) in the domain of mind and matter. Section 13.5 locates the position of this book within the spectrum of philosophy of consciousness, entering into dialogue with dualism, physicalism, functionalism, panpsychism, Russellian monism as the nearest neighbor, and the three contemporary theories of consciousness science, IIT, GNW, and predictive processing. Section 13.6 develops the convergence between the work of this book and phenomenology, especially the Husserlian tradition: the bottom-up physical path and the top-down reflective path converge by different routes on the irreducibility of given-ness.

This chapter is the central chapter of Part III of the book. It builds upon Chapter 12, "Co-Evolution as an Ontological Relation," and Chapter 14, "Observation as Participation." As an application of constrained speculation to the most difficult philosophical problem, this chapter bears a special argumentative burden: if constrained speculation can do meaningful work on the hard problem, even without "solving" it, it proves its depth as a method; if it fails here, it degenerates into a lightweight methodological suggestion. The entire work of this chapter is to give the deepest treatment that constrained speculation can provide without pretending to solve the hard problem.

13.2. The Hard Problem and the Isomorphic Program

A Precise Statement of the Hard Problem

As the central problem in the philosophy of consciousness, the hard problem must be stated precisely so that it is not confused with related problems. Chalmers's 1995 formulation distinguishes two kinds of consciousness problems: the easy problems and the hard problem.

The easy problems ask what functions a cognitive system performs: how attention is allocated, how information is integrated, how states are reported, how stimuli are discriminated, and how behavior is produced. These problems are technically extremely difficult, and consciousness-related neuroscience has devoted extensive resources to them, but in principle they are open to mechanistic explanation: sooner or later they can be answered in the language of function and dynamics.

The hard problem asks something else: even if the performance of these functions is fully explained, why is this performance accompanied by first-person given-ness? This question is not located within the functional chain; it remains even after the functional chain has been completely specified. Even if a system is behaviorally and informationally indistinguishable from a conscious subject, the question "whether it really has experience" can still be raised independently in logical terms. This is precisely what makes the hard problem "hard."

In 1983, Levine called this difficulty the explanatory gap: even if the physical description is functionally complete, there remains a gap between that description and "why there is this experience," a gap not logically entailed by the former. The object of explanation in any physical theory is ultimately structure, function, and dynamics: how physical quantities vary with other physical quantities, how configurations maintain stability, and how processes propagate and evolve. The entire vocabulary of physical explanation is a vocabulary of "what operates and how." The hard problem arises precisely after "what operates and how" has been exhausted, and asks further: "why is this operation experienced?" Within the vocabulary of structure and function, there seems to be no concept that can receive this question.

The Decomposability of the Hard Problem

The first key judgment of constrained speculation in the face of this problem is that the hard problem is not an undifferentiated whole without a point of entry, but is decomposable. Once an appropriate decomposition is applied, most of the hard problem can be treated mechanistically; what is genuinely irreducible will be compressed into a single residual point with an exact location.

The content of this decomposition is as follows. First, the structure of experience: what experience is like, how rich it is, what qualitative kind it belongs to, how it flows, and how deep its temporal thickness is. This part is tractable. It can be treated mechanistically by using the isomorphic program, in Subsection 13.1.3 and Section 13.2, to map the structure of experience to the structure of the ρ_f gradient configuration relation by relation. Second, the mechanism of the given-ness of experience: how experience occurs as a relational event, and how given-ness is maintained as a relation. This part is also tractable. It can be mechanistically characterized through internal–external co-evolution as relation (Section 13.3) and boundary as the physical precondition of a gate (Section 13.3). Third, the last cell: even if the structure of experience and the mechanism of given-ness are both fully specified, "why all this is experienced" is still not logically entailed by these structural facts. This part is not tractable: it is the boundary of the capacity of physical explanation itself and is not inside that capacity (Section 13.4).

This threefold division gives the work of this chapter a precise scope. The tractable parts, namely structure and the mechanism of given-ness, are fulfilled in Sections 13.1–13.3. The intractable part, namely the last cell, is precisely marked as a boundary in Section 13.4 rather than being falsely presented as an answer. The contribution of constrained speculation lies precisely in this exact demarcation between what can and cannot be done. It prevents the philosophy of mind from sliding either into the arrogance of "pretending to have solved it" or into the defeatism of "suspending the whole problem."

The Isomorphic Program: Fivefold Gradient Richness

The core thesis of this section is that first-person experience is isomorphic to the gradient configuration of a certain class of frequency-resolved energy quantum density field ρ_f. "Isomorphism" here means a structure-preserving mapping: every structural feature of experience has its counterpart on the side of the ρ_f gradient, and vice versa.

The success or failure of this thesis depends on a crucial question: structural sufficiency. Experience is extremely rich. It has continuously varying qualia, heterogeneous qualitative kinds, the flowing stream of consciousness, a present field that is both unified and differentiated, multiple layers of memory of different lengths, and temporally thick experience. If one claims that experience is isomorphic to a certain gradient configuration, then the gradient side must itself possess sufficient structural richness, rich enough to establish a one-to-one correspondence with all structural features of experience.

Historically, nearly all identity theories that say "consciousness is some physical quantity" fail at precisely this point. They identify consciousness with a single scalar: some amount of information, such as IIT's Φ, some field strength, or some frequency. But the structure of a single scalar is too poor: a number has only one dimension, magnitude, and cannot support the richness of experience. As a result, such theories can only claim correlation, not identity, as Section 13.5.6 will discuss in detail.

The fundamental difference between the isomorphism claimed in this book and these single-scalar approaches is this: what this book identifies with experience is not a scalar, but a multidimensional and structurally extremely rich gradient-configuration space. The richness of this space has five explicit sources, each of which carries one kind of structural feature of experience:

First, the richness of single-frequency gradient configurations. Within a single frequency family, the gradient of ρ_f is already a high-dimensional structure: it has direction, magnitude, topology of spatial distribution, singularities, and critical points. This layer of structure carries continuous variation within one qualitative kind of experience, such as the gradual change within the color space of "red" from dark to light and from warm to cool.

Second, the richness of the energy quantum frequency dimension. ρ_f is divided into families by frequency, and frequency itself is an independent structural dimension. This dimension carries the difference of kind among different qualities in experience: the reason red and pain, sound and taste are different kinds of experience rather than merely different intensities of the same experience corresponds to their belonging to different frequency families. This source is crucial for structural sufficiency, because the problem of qualitative kinds—why experience contains mutually irreducible "kinds of quality"—has long been the hardest obstacle for identity theories.

Third, the richness of evolution under the constraint of the Master Equation. The evolution of ρ_f is constrained by the EQT unified Master Equation, which gives all allowed forms of temporal change in the gradient field. This source carries the dynamical structure of experience: the flow of the stream of consciousness, the movement of the focus of attention, and how one thought turns into the next.

Fourth, the richness of same-frequency and cross-frequency resonance modes. This is the most load-bearing source. Same-frequency resonance corresponds to coherence within a single subject: why many experiences belong to the same "I." Cross-frequency resonance corresponds to feature binding: why the color, shape, and motion of an object are experienced as attributes of one and the same object rather than as three unrelated segments of experience.

Fifth, the richness of the temporal-duration spectrum of gradient transients. The gradient configuration of ρ_f is not permanently resident, but appears as a transient; transients in different frequency families have different durations of stability: high-frequency transients are short, while low-frequency transients are long. This source carries the temporal thickness of experience and the hierarchy of memory.

Taken together, these five sources have a unified function: they show that the gradient-configuration space of ρ_f has sufficiently high dimensionality and sufficiently rich structure to establish a feature-by-feature correspondence with the structural space of experience, including variation within qualities, qualitative kinds, the stream of consciousness, unity and binding, and temporal thickness and memory.

This fivefold richness program is internally connected with the work of preceding chapters. Frequency as a basic spectral quantity (Chapter 9, FR1) provides the ontological ground for the second source; freezing as a foundational ontological process (Chapter 11, FZ1) provides the ontological explanation for the "temporal-duration spectrum of transients" in the fifth source; the tri-states as ontological categories (Chapter 10, T1–T2) make possible the tri-state decomposition of experience, which will be developed in Subsection 13.1.4; and co-evolution as an ontological relation (Chapter 12, EV1) provides the ontological ground for given-ness as relation (Section 13.3).

The Correct Position of the Ψ Framework

Mind–Matter Unity introduced $\Psi \equiv (\tilde{A}, C, \mathcal{B})$ as a consciousness-related state-variable group: $\tilde{A}$, consciousness intensity; C, consciousness complexity; and $\mathcal{B}$, consciousness boundary. In discussions of the philosophy of consciousness, the Ψ framework can easily be misread as "the ontological structure of the mind–matter relation," but this misreading would distort the position of constrained speculation on the mind–matter problem. This section clarifies the correct position of the Ψ framework.

Ψ is a projected quantity of the isomorphic structural space on the observable surface; it is not a definition and not a generator. Specifically: consciousness is not defined as a particular value of Ψ; experience is not produced by Ψ; nor is Ψ an intermediate-layer ontological structure of the mind–matter relation. Ψ is a descriptive quantity: it projects the five-dimensionally rich gradient-structure space onto three observable physical variables, allowing consciousness-related states to be measured, compared, and tested.

This positioning has three layers of significance. First, the relation between Ψ and fivefold richness is not equivalence: Ψ is a projection of the latter and is therefore essentially a simplification. Any value of Ψ corresponds to a class of gradient configurations, but the reverse does not hold: different gradient configurations may project onto the same Ψ value. This "many-to-one" character is the nature of projection. Second, the function of the Ψ framework in consciousness research is to provide operational coordinates for falsifiable research, not to provide metaphysical explanation. Ψ does not answer why a particular combination of $\tilde{A}$, C, and $\mathcal{B}$ corresponds to waking experience; it only provides measurable coordinates so that this correspondence can be studied seriously. Third, the function of Ψ in philosophical discussion is secondary: it concretizes the correspondence between mind and matter, since specific Ψ values correspond to specific conscious states, but it does not itself constitute the ontological structure of the mind–matter relation. The ontological structure is the isomorphic relation itself; Ψ is only the observational projection of that structure.

This correct positioning revises a possible misreading: treating Ψ as the ontological intermediate layer of the mind–matter relation. That misreading would make the work of this book look like "a new physical model for consciousness research," but the philosophical contribution of this book goes far beyond that. It provides the isomorphic program as the ontological structure of the mind–matter relation, while Ψ is only the observational coordinate of this structure. The force of the method of constrained speculation in the mind–matter problem lies precisely in this distinction: the ontological thesis, namely identity-tier isomorphism, is constrained by CR2, while the measurable thesis, namely the correspondence between Ψ values and conscious states, is constrained by CS3, the discipline of observationalization. Each receives its proper place.

13.3. The Identity Tier: Mind and Matter as Two Aspects of the Same Process

Distinguishing Three Tiers

In the face of the structural correspondence between mind and matter, there are in principle three claims of decreasing strength that may be adopted. This section distinguishes the three tiers and argues why this book chooses the strongest tier.

The strongest tier: identity. Experience is what a certain class of gradient configurations is from within. Mind and matter are two aspects of the same process; they are not causally correlated, nor merely matched by a model, but are the same reality. Under this tier, "Mind–Matter Unity" as a thesis obtains its strongest meaning: unity is not "the close association of two kinds of things," but "two aspects of the same reality."

The intermediate tier: structural isomorphism. The structure of experience and the structure of gradient correspond one to one, but this does not assert that experience is the gradient itself. The two can still be conceived as two kinds of things that are merely structurally parallel. Under this tier, "mind–matter correspondence" obtains rigorous content, but "unity" as a thesis is substantively weakened:

structural parallelism does not entail ontological identity.

The weakest tier: modeling correspondence. The ρ_f gradient is the best physical model of experience, and isomorphism holds in the modeling sense. Under this tier, "correspondence" effectively retreats to correlation: it says that "we had better describe experience with the gradient model," not that "experience is gradient." "Unity" as a thesis is nearly abandoned under this tier.

This book adopts the strongest tier: mind and matter as two aspects of the same process, strictly identical through the isomorphism of fivefold gradient richness. This choice is not rhetorical, but based on ontological and methodological considerations.

Why the Identity Tier Must Be Adopted

There are three layers of reason for adopting the identity tier.

First, it fulfills the substantive meaning of "unity." "Mind–Matter Unity" as a thesis gains its strongest meaning only under the identity tier. The intermediate tier, structural isomorphism, fulfills only "correspondence," not "unity": two structurally parallel objects are still two objects. The weakest tier, modeling correspondence, even gives up "correspondence" and retreats to "modeling": it only says that we use a certain physical model to describe mind. Only the identity tier fulfills "unity": mind and matter are not two objects, but two aspects of the same reality. A philosophical work that takes "unity" as its title and aims to unify mind and matter would be unworthy of its name if it stopped at the intermediate or weakest tier.

Second, it minimizes explanatory burden. The intermediate tier leaves behind a question to be answered: "why does this structural isomorphism hold?" Why is the structure of experience exactly parallel to the structure of a certain class of gradient configurations? Is it coincidence? Causation? Some mysterious metaphysical law? This question cannot be avoided under the intermediate tier and has no obvious answer. Under the identity tier, this question never arises: there is no relation to be explained between A and A; "identity" does not need to explain "relation." The identity tier therefore minimizes the explanatory burden of the hard problem: it does not introduce the additional question of "why the structures are parallel."

Third, it is consistent with the process ontology of this book. Part II has already established the ontology of difference (Chapter 7, DI1), process ontology (Chapter 11, FZ1), and the tri-state ontological categories (Chapter 10, T1–T2). These works understand all beings as different forms of the same ontology, namely the ρ_f gradient process. Within this ontological framework, the division in which "mind is one kind of thing and matter another" has no ontological ground: all beings are forms of the ρ_f gradient process. The intermediate tier, where "mind and matter as two kinds of things are only structurally parallel," is inconsistent with the ontological framework of this book. The identity tier is internally coordinated with that framework: mind and matter as two aspects of the same ontology are precisely the expression of this book's process ontology in the mind–matter problem.

The Argument from Structural Sufficiency

The power of the identity tier lies in its fulfillment of the substantive meaning of "unity" and its minimization of explanatory burden. But precisely because it is strong, the identity tier bears an unavoidable burden: the problem of structural sufficiency. If one claims that experience is a certain gradient configuration, then the gradient side must itself possess sufficient structural richness, rich enough to establish a one-to-one correspondence with all structural features of experience. If experience has even one structural feature for which the gradient side cannot provide a counterpart, then "is" immediately degrades into "approximately models," and the identity tier collapses back to the weakest tier.

Historically, nearly all identity theories claiming that "consciousness is some physical quantity" fail at precisely this point. They identify consciousness with a single scalar: some amount of information, such as IIT's Φ; some field strength, as in certain electromagnetic theories of consciousness; or some frequency, as in certain oscillation hypotheses. But the structure of a single scalar is too poor: a number has only one dimension, magnitude. This one-dimensional structure cannot support the richness of experience. At most, it can explain the "more or less" of consciousness, such as how awake, deep, or strong it is, but it cannot explain the "difference between red and pain," namely the difference of kind among qualities. As a result, these approaches can only claim that "consciousness level $\propto \Phi$" or "consciousness correlates $\propto$ a certain frequency," but cannot establish identity. Their "identity" is effectively a rhetorical expression of correlation, not ontological identity.

The fundamental difference between this book and such single-scalar identity theories lies precisely in structural sufficiency. What this book identifies with experience is not a scalar, but a five- dimensionally rich gradient-configuration space. Its richness comes from five mutually independent yet mutually coordinated sources, as developed in Subsection 13.1.3. The coordination of these five sources makes the gradient-configuration space sufficiently high-dimensional to correspond relation by relation to all structural features of experience:

Table 13.1.. Correspondence between Structural Features of Experience and the Gradient Side, and the Mechanism of Reinterpretation

Item	Classical/Experiential Phenomenon (Structural Feature of Experience)	EQT/Gradient Correspondence (Counterpart on the Gradient Side)	EQT Limitations/Not Involved	Chapter
Qualitative continuity	Continuous variation within quality (shades of red)	Continuous variation of single-frequency gradient configurations	Does not directly treat absolute subjective qualia in the idealist sense	9
Qualitative difference	Difference of kind among qualities (red vs pain)	Different frequency families	Does not depend on traditional neuroanatomical entity localization	9-10
Diachronic evolution	Temporal evolution of the stream of consciousness	Evolution under the constraint of the Master Equation	Does not violate the temporal unidirectionality of physical flow	10
Feature binding	Unity of the subject and binding of features	Same-frequency and cross-frequency resonance modes	Does not adopt the classical assumption of a central coordinator (central processor)	10-11
Temporal thickness	Temporal thickness and layers of memory	Temporal-duration spectrum of gradient transients	Not equivalent to the classical accumulation of discrete time slices	11

This item-by-item correspondence is not a rhetorical assertion. It is the fulfillment of structural sufficiency, and each item requires serious defense, developed chapter by chapter in Part II of *Mind–Matter Unity.* This chapter does not repeat those defenses, but invokes their conclusions as support for the identity tier.

Relation to Constitutive Coextension (Chapter 4)

Chapter 4 introduced constitutive coextension (CC1) as a new semantic relation: neither identity, nor reduction, nor constitution, nor metaphor, but a fifth semantic relation constrained by failure conditions. This section clarifies the relation between the identity tier and constitutive coextension.

The identity tier is the strongest form of constitutive coextension. Specifically, the definition of constitutive coextension is that A and B share the same failure conditions: the failure conditions of A $\iff$ the failure conditions of B. The identity tier as the strongest form asserts not only that A and B share failure conditions, but that they are different presentations of the same reality.

This relation makes the mind–matter identity thesis constrained by CR2, namely failure conditions as an ontological discipline. Specifically, the failure condition of mind–matter identity as an ontological commitment is the failure of structural sufficiency: if any structural feature of experience is not carried by fivefold richness, the identity tier collapses back to the intermediate tier; if it is further discovered that the structural feature cannot be matched by any gradient structure, the identity tier collapses back to the weakest tier. This identifiable direction of failure makes the identity tier legitimate as an ontological commitment, satisfying CR2, rather than a mysterious metaphysical assertion.

The significance of this connection is that the mind–matter identity thesis is not a metaphysical assertion made from nowhere. It is an ontological commitment constrained by the full discipline of constrained speculation: it has dimensionalized content, namely the dimensions of fivefold richness; identifiable directions of failure, namely failure of structural sufficiency; and testable physical correspondences, fulfilled chapter by chapter in Part II of *Mind–Matter Unity*. The fundamental difference between this book's identity tier and single-scalar identity theories lies not only in "one scalar vs five-dimensional structure," but also in "unsupported assertion vs commitment constrained by methodological discipline."

Failure Conditions of the Identity Tier

According to the requirements of CR2, the identity tier as an ontological commitment must state its own failure conditions. This section states these conditions explicitly.

Failure condition one: failure of structural sufficiency. If some structural feature of experience, such as a certain sense of self, sense of agency, or affective tone, is found to have no counterpart in fivefold richness, then the identity tier is challenged. The new edition of *Mind–Matter Unity*, Chapter 11, has responded directly to this failure condition: it argues that sense of self, sense of agency, and affective tone belong to the endogenous perturbation types within the "perturbation axis," rather than constituting a sixth type of structure outside the fivefold structure, thereby maintaining completeness. But this response remains open to further testing and refutation.

Failure condition two: failure of structural correspondence. If a structural feature of experience has some counterpart in fivefold richness, but that correspondence is shown not to be one-to-one, for instance if multiple experiential states correspond to the same gradient configuration or vice versa, then the isomorphic relation fails with respect to that feature.

Failure condition three: given-ness is not relation. If given-ness is shown to be essentially an intrinsic property of some internal configuration, rather than a relational event of internal–external co-evolution, then the position of Section 13.3 fails, and the content of the identity tier as an ontological commitment must be revised.

Failure condition four: failure of the boundary criterion. If a system without an enclosing boundary, such as a purely computational device or a membraneless chemical network, is shown to have first-person given-ness, then the position of boundary as gate in Section 13.3 fails, and the delimitation of the scope of consciousness by the identity tier must be revised.

Together, these four failure conditions place the identity tier as an ontological commitment under methodological discipline. It is not an absolute or final metaphysical assertion, but a defensible, failable, and refinable ontological position. Assuming these failure conditions is precisely the fulfillment of the method of constrained speculation. This connection, the identity tier as the strongest form of CC1, is developed in Section 4.6: CC1 is not an intermediate position, but a broad family of semantic relations, and identity-tier isomorphism is precisely its strongest form. The two chapters mutually construct each other through this connection.

13.4. Given-Ness as Internal–External Co-Evolution: Boundary as Gate

Given-Ness as a Problem

Section 13.2 fulfilled the isomorphism between the structure of experience and gradient structure. But isomorphism itself has not yet touched one issue, and this issue is precisely the sharpest core of the hard problem: even if a gradient configuration is structurally completely isomorphic to experience, why is it given, that is, experienced, rather than silently running in the dark without qualitative character, with the same structure but no quality?

This question cannot be avoided. An ontology that aims to make a serious contribution to the philosophy of consciousness must answer it directly; otherwise it has only treated the structural problem of consciousness while avoiding the core of the philosophy of consciousness. This section develops the response of this book to this question.

Given-Ness Cannot Be an Intrinsic Property

The key turn in this book is that given-ness is not an intrinsic property of an internal configuration, but a type of relation: more specifically, a type of ρ_f detection relation.

This turn has a strong philosophical argument. Consider the opposing hypothesis: given-ness is an intrinsic property of some internal configuration, namely the intrinsic experience of an isolated interior. This hypothesis faces an irresolvable difficulty: the argument from the darkness of the isolated interior.

Specifically, "given-ness" as a concept is essentially directional and relational. Given-ness is always given "to someone" and appears from a certain perspective. An absolutely isolated internal configuration, open in no direction at all, lacks the other pole required by this directedness. To say that it is "given to itself" is either empty speech, since "to itself" effectively means "toward some direction inside itself" and has already smuggled in relationality, or it has already introduced an internal self-relation, thereby returning given-ness to relationality. In any practical sense, given-ness requires a "direction of given-ness": it points toward some recipient, some perspective, some "inner side." More deeply: if given-ness were truly a purely intrinsic property of some internal configuration, this property should still hold when that configuration is completely detached from every environment and has completely ceased to co-evolve with any externality. But this inference cannot be seriously defended. It effectively says that "an internal configuration still appears toward some direction when it is not related to any direction," which is a contradictory statement. Therefore, given-ness cannot be a purely intrinsic property of some internal configuration. It must be relational: it necessarily involves a relation between some internal configuration and some externality.

Given-Ness as Internal–External Co-Evolution

This book's characterization of given-ness as relation is drawn from the mechanism of "observation as participation" in EQT detection theory, which Chapter 14 of this book has already philosophized. Its central point is this: observation is never a disturbance-free reading by an external subject of the already constituted properties of a closed object. The detector and the probed system belong to the same ρ_f field, and observation is a co-evolutionary process of local coupling, controlled perturbation, nonlinear feedback, and transient freezing between the two within this field.

When this mechanism is applied to consciousness, the given-ness of experience is not the intrinsic taste of some coherent resonance inside the brain, for setting it up in that way would fall back into the dead end of "why the interior has taste." Rather, it is the ongoing internal–external co-evolution between this internal configuration and its externality within the same ρ_f field. Given-ness is a relational event: an internal coherent configuration "appears" to that interior because it is always situated within co-evolutionary coupling with the external.

Several clarifications are needed to avoid misunderstanding this position.

Clarification One: the scope of internal and external. The "external" in this book does not primarily mean the environment outside the body, but the functional distinction within a detection relation. The internal is the coherent configuration in which the content of given-ness is located; the external is the remaining configuration that co-evolves with it and makes it given. In most cases, under normal waking conditions, this externality is connected to the real environment through the body's sensorimotor channels. But in purely inward thinking, eyes-closed meditation, or vivid dreams, internal–external co-evolution occurs among the brain's own ρ_f structures, without needing to connect to the environment. This flexibility explains why given-ness still exists, and may even become more vivid, when connection with the environment is substantially reduced.

Clarification Two: connection with Chapter 12, EV1. Given-ness as internal–external co-evolution is the application of Chapter 12, EV1, co-evolution as an ontological relation, to the domain of mind and matter. Specifically: Chapter 12 has argued that all "two-term relations" are same-field co-evolutionary relations, including subject–object, mind–matter, self–other, and individual–environment. This section develops the concrete form of the mind–matter relation: the co-evolution between consciousness-related internal configurations and their externality. The legitimacy of this section's work is grounded in the legitimacy of Chapter 12, EV1.

Clarification Three: connection with Chapter 14, OB1. Given-ness as internal–external co-evolution has the same ontological structure as observation as participation (Chapter 14, OB1). Specifically: both given-ness and observation are forms of same-field co-evolution. The difference is that observation involves a perturbation–response–freezing loop between the detector and the observed, producing a recordable objective result; given-ness involves a perturbation–response–freezing loop of the internal with itself, together with its relation to the external, producing subjectively experienced content. The two are structurally isomorphic but applied to different domains.

Boundary as Gate

Given-ness as internal–external co-evolution immediately faces an objection: there is coupling everywhere in the field; a gravitational-wave detector and its environment also co-evolve continuously and also freeze out recordable output. Why, then, is only the internal–external co-evolution of living systems given, while the detector's co-evolution is merely a third-person physical process without an inner side? In other words, "internal–external co-evolution" seems too broad: it would generalize given-ness to all coupled systems and slide into a form of panexperientialism.

The answer of this book to this objection forms the anchor of its anti-panpsychist position: the key is not frequency, not threshold, but boundary.

Specifically, the establishment of given-ness presupposes the existence of an "interior," and the existence of an "interior" presupposes a boundary. For a system to have the "internal" in internal–external co-evolution, it must first have a boundary that genuinely distinguishes itself from its environment: a topological "inside."

Most matter in the universe does not have such a boundary. From the Big Bang to galaxy clusters, aggregation driven by the four fundamental interactions proceeds through attraction and condensation. It can produce structures of higher density and larger scale, but it does not produce an inside–outside distinction: celestial bodies and molecules do not have an "inside" in this sense. The first process that produces a topological interior is a different form of aggregation: enclosure by a lipid membrane. Through enclosure rather than attraction, it first closes off a region with inside and outside, capable of self-maintenance and selective exchange. This event is precisely the boundary between life and non-life, as *Mind–Matter Unity*, Chapter 3, has discussed in detail.

The objection is thereby dissolved. A gravitational-wave detector and its environment certainly co-evolve and certainly freeze out recordable results, but the detector has no interior: it has no boundary that separates itself from the environment. It is merely the interaction of two regions in the field, and there is no "inside" to which that co-evolution could be given. A living system is different: it has a membrane and therefore a genuine interior. The same internal–external co-evolutionary freezing occurs within the framework of "for an interior," and thus this freezing is given to that interior.

Given-ness is therefore firmly anchored in the boundedness of life. If there is a boundary, there is inside and outside, and internal–external co-evolution is given to the interior. If there is no boundary, there is no inside and outside; co-evolution is only the interaction of the field, and there is no question of being given. The gate is thereby established: it is not frequency matching, nor perturbation threshold, but the presence or absence of boundary, namely the difference between life and non-life.

The Argument against Panpsychism

The position of boundary as gate allows this book to oppose panpsychism seriously. Panpsychism holds that experience, or its prototype, is a universal property of matter: everything has a trace of inwardness, and the experience of complex systems is composed from microscopic experiences. This book rejects the universality of panpsychism, but does not reject it by positing another form of emergentism.

Specifically, this book rejects the universality of inwardness. Inwardness is not a background tone diffused through all things, but an achievement with a definite physical origin. It begins with life; it begins with the interior produced by the first membrane. Rocks and electrons have no interior, not because their inwardness is "too weak," but because they simply have no boundary and no such thing as an interior. Inwardness in their case is not low in degree; it simply does not apply.

This rejection has two reasons. First, epistemological: panpsychism makes "where there is an interior" empirically empty. If everything has a trace of inwardness, then "whether there is an interior" no longer marks any observable difference. By contrast, this book's boundary criterion, the presence or absence of enclosure, is sharp and almost empirically decidable. This testability makes the boundary criterion of this book satisfy CS3, the discipline of observationalization in the method of constrained speculation, whereas panpsychism's "universal inwardness" does not. Second, theoretical consequence: panpsychism makes the combination problem extraordinarily difficult. Since inwardness is universally distributed as microscopic subjects, combination becomes the genuine deadlock of "adding subjects": how can one subjective unity be composed from countless micro-minds? In this book, by contrast, the combination problem is dissolved because the "I" is a single coherent dominant resonance, not an addition of micro-subjects. This dissolution is possible precisely because this book does not accept the panpsychist premise of "micro-subjects everywhere."

Against Emergentism and the Middle Way of This Book

Yet this book's anti-panpsychist position does not slide into emergentism, namely the view that experience mysteriously emerges from non-experiential matter. Emergentism faces an explanatory gap: how is the leap from "no experience at all" to "experience" crossed? Any answer given by emergentism can hardly avoid some mysterious leap.

The position of this book lies between panpsychism and emergentism. Unlike panpsychism, it does not make inwardness universally present without origin; unlike emergentism, it does not make inwardness emerge from nothing. This book assigns inwardness a starting point with physical preconditions: the interior produced by enclosure, upon which internal–external co-evolution then produces given-ness.

Specifically, at the level of the basic ontology ρ_f, this book does not claim that there is any experience or inwardness: ρ_f is nowhere given as such. Only when ρ_f organizes an interior through enclosure does the precondition for given-ness appear. Thus this book is not "cell-level panpsychism," but "inwardness emergence beginning from enclosure." Yet this emergence is not the mysterious emergence of experience from non-experience, but a starting point triggered by a physical event, namely enclosure producing an interior, and possessing physical preconditions.

This middle position is the philosophical location of this book: it is anti-panpsychist and anti-emergentist at the same time. It gives the book an independent position in the spectrum of philosophy of consciousness: it does not share panpsychism's premise of universal inwardness, nor does it share emergentism's difficulty of emergence from nothing.

13.5. Two Compressions and the Last Cell as Boundary

The Hard Problem as a Diffuse Fog

This section develops this book's treatment of the hard problem. In traditional discussions, the hard problem is often expressed as a diffuse fog: as the question "why are physical processes accompanied by subjective experience," it is everywhere in the philosophy of consciousness, yet it is also so comprehensive that one hardly knows where to begin. This diffuseness easily leads discussions of the hard problem into two difficulties: either covering it with a single word, such as "emergence," "identity," or "unity," making the fog seem closed while it in fact remains; or giving up all work because the problem is difficult, suspending the whole philosophy of consciousness.

The treatment in this book is a third way: decompose and compress the hard problem, doing everything that can be done and precisely marking what cannot be done. This treatment was introduced in Subsection 13.1.2, on the decomposability of the hard problem. This section develops its content: after two compressions, the hard problem is located as a single point. This point is not a vague diffuse problem, but a boundary with an exact location.

The First Compression: The Residual Point of Given-Ness

Through the work of Sections 13.1–13.3, most of the hard problem has been mechanistically received: the full structure of experience is isomorphically carried by fivefold gradient richness (Section 13.2); the given-ness of experience is characterized by the internal–external co-evolutionary freezing of a bounded interior and anchored in the boundary of life (Section 13.3). The question "why given?" is, through boundary, reduced to "why is there a bounded interior?" which is answered by the physical event of the membrane of life.

Yet this book does not claim that the hard problem has been

eliminated. It claims only that the problem has been reduced and compressed to a single irreducible point. This point is the following: even if the entire structure of the bounded interior—boundary, inside and outside, co-evolution, and freezing—is fully specified, the claim that "this structure is experienced from within in the first person" is still not logically entailed by these structural facts.

From "there is an interior, and inside and outside co-evolve and freeze" to "this interior feels something," the language of structure has been extended to the last cell. But between "being experienced" and "possessing the full structure required for being experienced," there remains a gap not entailed by structure. Fivefold richness, internal–external co-evolution, and boundary have narrowed this gap to the extreme: so narrow that only one cell remains, the difference between "the structure of a bounded interior" and "the being-experienced of that structure." But they cannot eliminate it, because this cell is precisely the irreducible core of the hard problem itself.

The new edition of *Mind–Matter Unity*, Chapter 15, names this single point the residual point of given-ness. This is the achievement of the first compression. It compresses the fog diffused across consciousness research into the single cell of "why given-ness is experienced." This book inherits that naming and that compression.

The Second Compression: The Residual Point of Subject-Ascription

The compression in this book does not stop there. The new edition of *Mind–Matter Unity*, Chapter 20, compresses further: it shows that the given-ness of human experience is more specifically a given-ness that has been cognized and bounded by subjectivity as "mine." Specifically:

Cognization: human experience is not a mere sensory impact. It is experience after comparison, recognition, and meaning-ascription through an experiential knowledge base, including language, concepts, memory, and frameworks of judgment. For a segment of visual sensation to be experienced as "red" already involves the concept of color, naming, and cross-context recognition.

Subjective delimitation: human experience is experienced as "my" experience through delimitation by the subjectivity boundary $\mathcal{B}$: which activities belong to the coherent resonance of "me" and are included in the unified subjective field of "me," and which are excluded. "What I see," "what I think," "my body," and "my world": this ascription of "mine" is the result of delimitation by $\mathcal{B}$.

Thus the first compression's "residual point of given-ness" is further stripped down. The mechanism of cognization and the mechanism of delimitation by $\mathcal{B}$ are both on the objective side of the gap, namely the structural–relational, third-person side; they are tractable and can be precisely characterized as physical–neural mechanisms. After these two layers are stripped away, the residual point is further compressed into this question: why is activity that has been integrated, cognized, and delimited as "mine" experienced? The new edition of *Mind–Matter Unity*, Chapter 20, calls this the residual point of subject-ascription.

The second compression does not eliminate the residual point, but makes it smaller and locates it more precisely. The tractable range is advanced twice, from structure to the mechanism of given-ness to the mechanisms of cognization and delimitation; the intractable cell is narrowed and located twice, from diffuse fog to residual point of given-ness to residual point of subject-ascription, and yet it is honestly retained throughout.

The Last Cell as Boundary rather than Object

This book's treatment of the single point after the two compressions, namely the residual point of subject-ascription, requires a crucial philosophical determination: the last cell is a boundary, not an object. This distinction determines the entire attitude toward it.

An object, or an unsolved problem, is something that can be studied, answered, and incorporated into knowledge. It lies within our explanatory capacity, even if it has not yet been answered. A boundary, by contrast, marks where a certain explanatory capacity stops. It is not inside that capacity, but is the very edge of that capacity.

The last cell is the latter. It is not a difficult problem still waiting for an answer within the capacity of physical explanation. It is the edge of the capacity of physical explanation: it marks precisely the limit that the language of structure and relation can reach. Section 13.3 has argued that being-experienced is not entailed by structure and relation. Subsection 13.4.3 further shows that even when finer structural–relational characterizations such as cognization and delimitation are added, the residual point remains beyond the boundary, only pushed deeper and made smaller.

This book therefore characterizes it as follows: the last cell is a boundary, marking where physical, structural–relational explanation stops, rather than an object awaiting an answer within physical explanation.

This characterization of "boundary rather than object" clarifies the proper attitude toward the last cell. For an object, or an unsolved problem, the correct attitude is to continue research and seek a solution; to suspend it would be defeat. For a boundary, the correct attitude is to mark it precisely and understand why we stop there. Trying to "cross" a boundary and "answer" it is not progress, but confusion, because the other side of a boundary is, by definition, outside the range in which that explanatory capacity can provide answers. This book therefore does not answer the last cell. This is not defeat, since defeat means giving up on an object that could have been solved, but an appropriate acknowledgment and precise marking of a boundary. Treating the last cell as an "object," as if it could be answered with more effort, and treating it as a "boundary," understanding that it lies outside physical explanation itself, are fundamentally different attitudes. This book adopts the latter.

From "Reasonable Deferral" to "Precise Marking"

The treatment in this section represents a substantive upgrade in this book's posture toward the method of constrained speculation. In earlier discussions, the posture of constrained speculation toward the mind–matter problem was often expressed as "reasonable deferral": acknowledging the hard problem but temporarily not treating it, awaiting future research. This posture had historical legitimacy: it avoided the arrogance of "pretending to have solved it" and also avoided the defeatism of "giving up everything." But it was still philosophically insufficiently precise.

Specifically, "reasonable deferral" as a posture has several problems. First, it treats the hard problem as an object, an unsolved problem, merely "temporarily" unsolved; this conflicts with the characterization in this section, namely boundary rather than object. Second, it makes constrained speculation appear to be making a methodological concession in the mind–matter problem, as if constrained speculation were "temporarily unable" to handle it and some stronger future method might be able to do so. Third, it leaves the philosophical status of the last cell vague: is it a temporary limitation of constrained speculation, or an internal boundary of any explanatory capacity?

The upgrade in this section is from "reasonable deferral" to "precisely marking as boundary." Specifically: the hard problem is not an open problem that is "temporarily unmanageable," but is precisely marked as a boundary. It marks the edge of the explanatory capacity of structure and relation itself and is not inside that capacity. The contribution of constrained speculation to the hard problem is not to "defer" it, but to mark it precisely: everything solvable is solved as far as possible, including structure, the mechanism of given-ness, the mechanisms of cognization and delimitation, and measurable projection; what cannot be solved is precisely marked as a single-point boundary. Acknowledging this boundary is not a concession by the method of constrained speculation, but the deepest application of the method to the most difficult problem. The failure-orientation discipline (CS4) requires every ontological commitment to state its own failure conditions; the last cell as boundary is precisely the failure condition of the ontology of experience, identity-tier isomorphism, at its own limit.

This upgrade gives constrained speculation a genuinely mature philosophical posture in the mind–matter problem. It does not pretend to solve the hard problem, nor does it merely "defer" it, but honestly marks it as a boundary. This posture is methodological thoroughness, not methodological concession.

The C3 Prohibition: The Fulfillment of the Failure-Orientation Discipline

After the second compression, one more temptation must be faced. Since mind and matter are the same reality, through identity-tier isomorphism, has the question of "why the inner side is experienced" been softened or even dissolved by identity itself? Or, since the mechanisms of cognization and $\mathcal{B}$-delimitation have been precisely characterized, has the hard problem been dissolved by these mechanisms?

The new edition of *Mind–Matter Unity*, Chapter 27, names the explicit rejection of this kind of temptation the C3 prohibition, a methodological discipline running through the whole book: no mechanism statement, including the identity thesis, the mechanism of internal–external co-evolution, the mechanism of cognization, or the mechanism of $\mathcal{B}$-delimitation, may be used to masquerade as a solution to the last cell. This book inherits this prohibition.

The ground of the C3 prohibition is that all the above mechanism statements—identity-tier isomorphism, internal–external co-evolution, cognization, and $\mathcal{B}$-delimitation—belong to the objective side of the gap, namely the structural–relational, third-person side. Each compresses the tractable part, turning diffuse fog into a single-point boundary. But none can or should be used to claim that the being-experienced side has been "dissolved," because being-experienced is by definition outside the range in which the language of structure and relation can provide an answer.

Specifically, the tempting argument that "the identity tier softens the hard problem" proceeds as follows: since mind and matter are the same reality, the question "why is the inner side experienced?" is roughly equivalent to "why is this reality what it is?" and the latter is not a genuine question requiring an answer, because a thing is what it is and requires no explanation. Therefore, once identity is established, the hard problem dissolves by itself. The flaw in this tempting argument is that it unjustifiably expands the scope within which "identity" has been established. The identity established by the identity tier in this book is identity of structure: the structure of experience and the structure of gradient configuration are the same structure (Section 13.2), and this identity is supported by relation-by-relation structural isomorphism. But being-experienced is precisely not inside the structure established as identical (Subsection 13.4.2: being-experienced is not entailed by structure). Thus although "identity" has been established, what it establishes is structural identity. Being-experienced is not covered by this structural identity. Therefore "identity" cannot automatically extend to being-experienced, and one cannot infer from "structural identity" that "being-experienced is thereby dissolved."

The forms of the C3 prohibition in the context of this chapter include: do not use identity-tier isomorphism (Section 13.2) as a fake solution to the last cell; do not use the mechanism of internal–external co-evolution (Section 13.3) as a fake solution to the last cell; do not use the mechanisms of cognization or $\mathcal{B}$-delimitation as fake solutions to the last cell; do not use the two compressions themselves (Subsections 13.4.2 and 13.4.3) as fake solutions to the last cell.

The C3 prohibition is the fulfillment of the failure-orientation discipline of constrained speculation (CS4) in the domain of mind and matter. It requires this book, in the most difficult problem, to keep the boundary between what can be done and what cannot be done, and not to cross that boundary through rhetoric or superficial mechanism statements. Keeping C3 is precisely the proof of the depth of the method of constrained speculation in the mind–matter problem.

13.6. Positioning within the Spectrum of Philosophy of Consciousness

The Posture of This Section

This section locates the position of this book within the spectrum of philosophy of consciousness, entering into dialogue with major philosophical positions and contemporary scientific theories of consciousness. Its posture is serious respect and precise distinction: each dialogue first states fairly the strongest point of the position, and then explains where and why this book takes a different path. A positioning that attacks only weaknesses without acknowledging strengths is not credible.

This section is organized by position rather than by philosopher, reflecting the mature form of discussions in the philosophy of consciousness. Each position may involve the work of several philosophers; this section organizes the dialogue through positions so that the location of this book is clear in relation to each. The section will repeatedly use the threefold division established in Subsection 13.1.2—structure, given-ness, and the last cell—as a diagnostic tool. Many divergences and difficulties among different positions arise from failing to separate these three, and the position of this book is largely defined by that division.

Dualism

Dualism holds that mind and matter are two different kinds of substance or property, and that there is a mental reality beyond the physical world. Its strongest point lies in its respect for givenness: it directly faces the fact that first-person experience cannot be exhausted by third-person physical description, and refuses to hastily reduce experience to physical processes. The hard problem becomes a problem partly because dualist intuition reminds us that experience does contain something not entailed by structure and function.

This book rejects dualist separation, but preserves its respect for given-ness, and the two are not in conflict within this book. The reason for rejecting separation lies in the ontology of this book: ρ_f is the sole ontology, as an application of Chapter 2, PR1, and there is no second kind of substance standing alongside it. Mind and matter are both manifestations of ρ_f. But this book preserves the truth that dualism respects: given-ness is indeed not entailed by structure. It merely places that truth at an exact location within a monist framework: it is not a second substance, but the irreducible last cell (Section 13.4).

This book can therefore be described as "allowing the truth seen by dualism within a monist ontology": it acknowledges the irreducibility of given-ness, which dualism sees, but does not posit a second substance for it, rejecting dualist separation, and instead locates this irreducible element as an irreducible boundary within a monist ontology. Dualism misreads "given-ness is irreducible" as "there must be a second substance." This book shows that what is irreducible can be a boundary rather than a substance.

Physicalism

Physicalism, here meaning reductive or eliminative physicalism, holds that everything is physical, and that experience can either be reduced to physical processes or is in fact some illusion that should be eliminated. Its strongest point lies in ontological parsimony and unity: it does not add mysterious entities and insists that the world is fundamentally one unified physical world. This book fully agrees with this parsimony and unity, in the spirit of Chapter 2, PR1.

This book shares physicalism's monism and parsimony, but differs in its treatment of given-ness. Reductive physicalism claims that given-ness can be reduced to functional or informational processes. This book rejects that claim for the reason given in Section 13.4: given-ness, namely the last cell, is in principle not entailed by structure and process, and any "reduction" will fall into topic substitution, replacing the hard problem with an easy problem. Eliminative physicalism, by contrast, claims that there is no phenomenal consciousness requiring explanation and that given-ness is an illusion. This book also rejects this, for two reasons: given-ness is the starting point of the entire discussion and the most certain datum; eliminating it is like sawing off the branch on which one sits. Moreover, "why is there this stubborn illusion?" itself becomes a hard problem and saves no effort.

Thus this book is a non-reductive and non-eliminative monism. It shares monism and parsimony with physicalism, but does not force given-ness into the reductive framework or dismiss it by elimination. Instead, it preserves for given-ness a place within monist ontology that is irreducible, yet not mysterious and not a second substance. This is precisely the place physicalism misses because it is too eager to reduce or eliminate.

Functionalism

Functionalism, as in the work of Putnam, Dennett, and others, holds that mental states are defined by their functional roles, and that consciousness depends on functional organization regardless of its physical realization. The same functional organization, whether realized by neurons, silicon chips, or something else, would have the same mind. Its strongest point lies in its grasp of the importance of functional organization and in the resulting intuition of multiple realizability: mind does seem to be related to a certain organizational pattern rather than to a particular kind of matter.

This book does not deny the role of functional organization in shaping the structure of experience. The structure of experience is carried by gradient configurations, and gradient configurations have functional realizations at the neural level. But this book rejects the core claim of functionalism: substrate-independence, namely that given-ness depends only on function and not on physical realization. The reason is given in Section 13.3. This book argues that the precondition of given-ness is an interior, and that the interior is produced by enclosure, a physical boundary. The criterion is therefore physical enclosure, not functional complexity.

A system that is functionally complex but lacks an enclosing boundary, such as a membraneless chemical network or a purely computational device, has no interior and therefore lacks the precondition of given-ness, no matter how convincingly its functional organization replicates that of a conscious subject. The mistake of functionalism is to inflate "the structure of experience is related to functional organization," which this book partly accepts, into "given-ness depends only on function," which this book rejects. Structure can be realized by functional organization, but the precondition of given-ness, namely an interior, cannot be produced by function. It must be produced by the specific physical organization of enclosure.

Here one must face a sharp implication of this book's position. Avoiding it would be dishonest: according to this book, an artificial system without an enclosing boundary lacks the physical precondition of first-person given-ness, no matter how complex its function or how convincing its behavior. This is a pointed and in principle contestable assertion. This book does not avoid it, and marks it as a conclusion open to discussion and, in an appropriate sense, testing. It follows directly from boundary as gate (Section 13.3) and is the natural result of this book's anti-functionalist position. Its exact scope should be noted: this book asserts that a system without enclosure lacks the precondition of given-ness; it does not assert that a system without enclosure cannot have any functional "information integration." Function can exist without enclosure; the precondition of given-ness cannot.

Panpsychism

Panpsychism, as in contemporary work by Strawson, Goff, and others, holds that experience, or its prototype, is a universal property of matter: everything has a trace of inwardness, and the experience of complex systems is composed from microscopic experiences. Its strongest point lies in its direct response to the hard problem. It does not avoid the question of "where given-ness comes from," but gives a radical answer: given-ness need not be produced from nothing, because it is already universally present. It thereby avoids the emergentist difficulty of "how experience emerges from non-experiential matter."

The divergence between this book and panpsychism is fundamental, and this divergence is precisely the point of Section 13.3. This book rejects the universality of inwardness. Inwardness is not a background tone diffused through all things, but an achievement with a definite physical origin. It begins with life, and with the interior produced by the first membrane. The reasons for this rejection, epistemological and theoretical, have already been developed in Subsection 13.3.4 and need not be repeated here. One additional clarification is needed concerning the subtle relation between this book and panpsychism.

The difference between this book and panpsychism must be made precise, because both initially seem to appeal to "inwardness." The difference is this: panpsychism places inwardness at the basic physical level, in all things and throughout ρ_f, whereas this book places it at a specific organizational level, namely life with an enclosing boundary. At the basic level of ρ_f, this book does not claim that there is any experience or inwardness: ρ_f is nowhere given as such. Only when ρ_f organizes an interior through enclosure does the precondition for given-ness appear. Thus this book is not "cell-level panpsychism," but "the emergence of inwardness beginning from enclosure." Yet this emergence is not the mysterious emergence of experience from non-experience, but a starting point with physical preconditions, triggered by a definite physical event. This book therefore lies between panpsychism and emergentism: unlike panpsychism, it does not make inwardness universally present without origin; unlike emergentism, it does not make inwardness emerge from nothing. It specifies a definite physical precondition for inwardness, namely the interior produced by enclosure, upon which internal–external co-evolution produces given-ness.

Russellian Monism

Among the various positions, the nearest to this book is Russellian monism, including Russell's *The Analysis of Matter* (1927), Stoljar, and Chalmers's later work. It holds that physical science characterizes only the relational–structural properties of reality, namely how things interact and how they are measured, but does not touch the intrinsic nature of reality, namely what things are in themselves. Mind, or experience, may be precisely the intrinsic nature of physical reality: the same reality appears from the outside as the relational structure described by physical science, and from the inside as experience.

Its strongest point lies in its ability to accommodate both the objectivity of physics and the inwardness of experience within a monist ontology, without adding a second substance. This is exactly what the preceding sections of this book seek. In spirit, this book is closest to Russellian monism: the thesis that "mind and matter are two aspects of the same ρ_f" is a Russellian posture.

But this book makes two decisive advances beyond Russellian monism. These two advances make it not merely a philosophical posture, but a contentful and testable theory.

First, it fulfills structural sufficiency through frequency-resolved gradient structure. Russellian monism is a general philosophical framework. It claims that experience is the intrinsic aspect of the physical, but does not specify what kind of structure the physical side must possess in order to match the rich structure of experience. This lack of specification creates a risk: if the intrinsic aspect of the physical is merely some poor "primitive feeling," then it repeats the combination problem of panpsychism; if it appeals to a single intrinsic quantity, then it repeats the structural insufficiency of single-scalar identity theories (Section 13.2). The advance of this book is to take the frequency-resolved gradient configuration of ρ_f as the structure on the physical side: a five-dimensionally rich structural space sufficient to be isomorphic relation by relation with the full structure of experience (Subsection 13.1.3). This turns the statement "experience is the intrinsic aspect of the physical" from a general assertion into a structurally sufficient and relation-by-relation testable isomorphic claim. Russellian monism says that "experience is the intrinsic aspect"; this book further says that "the intrinsic aspect is this five-dimensionally rich gradient structure, and it is isomorphic to experience relation by relation." This is an advance from posture to content.

Second, it characterizes given-ness through boundary and internal–external co-evolution. Russellian monism has a well-known difficulty: if experience is the intrinsic aspect of the physical, where is this intrinsic aspect present? Is it present in everything, thereby sliding into panpsychism and its combination problem? Or is it present only in certain systems, in which case a criterion is needed? Russellian monism itself does not provide such a criterion. The advance of this book is to use boundary (Section 13.3) to define the scope of the intrinsic aspect: only systems with an interior produced by enclosure, namely life, have an interior to which given-ness can be ascribed; and it uses internal–external co-evolution (Section 13.3) to characterize what given-ness is. This makes the "intrinsic aspect" no longer a metaphysical posit with unclear scope, but something with a definite boundary, life, and a definite mechanism, the transient freezing of internal–external co-evolution. Russellian monism says that "experience is the intrinsic aspect"; this book further says that "this intrinsic aspect exists only where there is an enclosing boundary, and its given-ness is the freezing of internal–external co-evolution." This is another advance from posture to content.

Taken together, these two advances locate this book as a Russellian monism enriched by EQT. It inherits the Russellian monist posture, but enriches the structure of the "intrinsic aspect" with frequency-resolved gradient structure, solving structural sufficiency, and defines the scope and mechanism of the "intrinsic aspect" through boundary and co-evolution, solving the problems of scope and given-ness. The irreducible remainder that remains after these two advances, and that Russellian monism and this book both face, is the being-experienced of the intrinsic aspect itself: the last cell (Section 13.4). This book shares this last cell with Russellian monism, but compresses it to its smallest point, through two compressions into the residual point of subject-ascription. This compression is precisely the result of the two advances just described.

Contemporary Scientific Theories of Consciousness: IIT, GNW, and Predictive Processing

Turning to the three major theoretical frameworks in contemporary consciousness science, the relation between this book and them can be uniformly clarified through the threefold division of structure / given-ness / the last cell. Their respective insights and limits often depend on which of these three they treat and which they confuse.

Integrated Information Theory (IIT, the work of Tononi and others). IIT holds that consciousness corresponds to the integrated information quantity Φ of a system and characterizes its informational structure. Its strongest point is that it takes the structure of experience seriously: it is not satisfied with "how strong consciousness is," but attempts to characterize the informational geometry of experience. This book shares its ambition to take the structure of experience seriously, but differs fundamentally from it. IIT centers on the quantity Φ, while Section 13.2 has argued that any identity theory centered on a single scalar cannot carry qualitative kinds, such as the difference between red and pain, because of structural insufficiency: a quantity has only one dimension, magnitude. This book replaces the single Φ with a five-dimensionally rich gradient structure precisely to remedy this structural insufficiency. In addition, IIT's criterion is information integration, a functional-informational property, and therefore in principle does not exclude consciousness in systems without enclosure. This book, by contrast, takes boundary as the gate (Section 13.3), and therefore differs from IIT. One may say that this book shares IIT's ambition to value structure, but supplements its single-scalar insufficiency with frequency-resolved structure, and supplements the weakness of its criterion with boundary.

Global Neuronal Workspace theory (GNW, the work of Dehaene, Baars, and others). GNW holds that consciousness corresponds to information being "broadcast" into a global workspace and shared by many subsystems. Its strongest point lies in its precise characterization of a class of functional features of consciousness: accessibility, reportability, and global availability, together with substantial neural evidence. This book holds that what GNW characterizes largely belongs to the easy-problem side of the threefold division in Subsection 13.1.2, namely function and accessibility, rather than given-ness itself. Information being globally broadcast is a functional fact. Why it is experienced is not answered by GNW, nor does GNW usually claim to answer the hard problem. This book and GNW are therefore not in conflict, but divided in labor: GNW excellently characterizes how conscious content is broadcast and accessed; what given-ness is is undertaken by this book's internal–external co-evolution. GNW handles "how content is globally accessed"; this book handles "why this access is given."

Predictive processing, as in the work of Friston, Clark, and others.

Predictive processing holds that the brain is a prediction machine, and that experience is the brain's best prediction, or generative model, of sensory input together with continuous tuning by error between that model and actual input. Its strongest point lies in its unified characterization of the dynamics of perception, action, and learning, with very broad explanatory power. At the dynamical level, this book has much in common with predictive processing: the "generative-model-driven, error-tuned" process of predictive processing can be understood, within the framework of this book, as endogenous perturbation, namely generative models and predictions, and exogenous perturbation, namely sensory input, jointly driving the evolution of the Master Equation and being given through internal–external co-evolution. But predictive processing likewise does not answer given-ness: why the operation of a prediction–error-tuning system is experienced is not answered by predictive processing. This book and predictive processing are therefore divided in labor: predictive processing excellently characterizes how experiential content is generated and tuned; what given-ness is is undertaken by this book.

Taken together, the three theories show a shared pattern: IIT, GNW, and predictive processing each excellently treat one side of the structure or functional dynamics of experience, but none directly answers, and many do not claim to answer, given-ness and the last cell. The relation between this book and them is therefore not competition, but largely division of labor and supplementation. This book accepts their characterizations of structure and function, relocating them within the ρ_f framework, and supplements what they do not reach: given-ness through internal–external co-evolution and the honest marking of the last cell. The genuine disagreements between this book and them concentrate on two points: criterion, where this book takes boundary as gate and opposes purely functional/informational criteria; and structural sufficiency, where this book takes frequency-resolved structure and opposes single-scalar accounts.

The Overall Coordinates of This Book

Bringing together the positioning in this section, the coordinates of this book within the spectrum of philosophy of consciousness can be marked in one sentence: this book is a Russellian monism enriched by EQT, neither reductive nor eliminative, with boundary as gate. It uses frequency-resolved gradient structure to solve structural sufficiency, uses enclosing boundary and internal–external co-evolution to solve the scope and mechanism of given-ness, and honestly compresses and marks the irreducible remainder of the hard problem as a single last cell.

From these coordinates, the book's choices relative to other positions can be summarized together: it takes from dualism its respect for given-ness, but rejects its separation, since given-ness is a boundary rather than a second substance; it takes from physicalism its monism and parsimony, but rejects reduction and elimination, since given-ness is irreducible and cannot be eliminated; it takes from functionalism its attention to functional organization, for the structure of experience, but rejects substrate-independence, since the criterion lies in enclosure rather than function; it takes from panpsychism its direct attitude toward the hard problem, but rejects its universality, since inwardness has an origin and is not diffused through all things; it is closest to Russellian monism, but advances it through frequency structure and boundary co-evolution, making it a contentful and testable theory rather than a posture. As for IIT, GNW, and predictive processing, the three contemporary scientific theories of consciousness, this book is mostly complementary to them, accepting their accounts of structure and function, while differing from them on criterion, namely boundary, and structural sufficiency, namely frequency resolution.

13.7. Convergence with Phenomenology by Different Routes

An Unexpected Convergence

This book reaches, through a completely independent bottom-up physical path, from ρ_f to structure, to given-ness, to cognization and delimitation, and finally to the last cell, the conclusion that given-ness is irreducible. It is noteworthy that a completely different philosophical tradition, phenomenology, has already reached a highly convergent insight through a top-down reflective path directed at experience itself. This section develops that convergence, because the convergence of two independent paths is itself a form of support for the conclusion reached.

The Core Insight of Phenomenology

A core insight of phenomenology, represented by Husserl and his tradition, including *Cartesian Meditations* (1931) and *The Crisis of European Sciences* (1936), is that given-ness is the starting point of everything and cannot be derived from anything more basic. Specifically, the given-ness of experience—the appearing of the world to me, the fact that experience is experienced—is the beginning point of phenomenology, the irreducible datum that remains after everything is bracketed. All knowledge of the world presupposes this given-ness, so given-ness itself cannot be further reduced to some non-given thing for explanation, since that would be circular or regressive. Phenomenology therefore places given-ness in a foundational and irreducible position. Husserl's method, including phenomenological reduction and eidetic intuition, is precisely intended to describe this foundational character of given-ness. In his work, consciousness is analyzed as always "directed toward something," namely intentionality, and given-ness is the core structure of intentionality: consciousness is always consciousness "of" something, always given some content, and always appears from a certain perspective.

The Meaning of Convergence by Different Routes

The last cell of this book converges closely with this phenomenological insight. Through the physical path, this book arrives at the conclusion that given-ness, the being-experienced of the inner side, is not entailed by structure and relation, and is the irreducible last cell (Section 13.4). Through the reflective path, phenomenology arrives at the conclusion that given-ness is an irreducible starting point. Both refer to the same irreducibility: the irreducibility of given-ness to what is not given. This book and phenomenology therefore converge by different routes: one proceeds bottom-up from ρ_f, the other top-down from reflection on experience, yet they meet at the irreducibility of given-ness.

Do Not Overstate: The Scope of the Convergence

The meaning of this convergence must be assessed accurately and not overstated. It does not mean that this book "proves" phenomenology, or that phenomenology "proves" this book. The two paths each have their own premises and methods and cannot deductively prove one another. The physical path of this book is built on the ontology of EQT and presupposes ρ_f as foundational ontology; the reflective path of phenomenology is built on direct reflection on consciousness and presupposes the immediate given-ness of experience as its methodological starting point. The premises differ, and the conclusions cannot be deduced from one another. Its significance lies in the fact that two completely independent paths converge on the same irreducible point. This independent convergence is a form of support for the claim that there really is an irreducible point here. The convergence of independent evidence is more credible than the assertion of a single path. The last cell of this book is therefore not an isolated posit forced out only from within this book's framework; it has an independent counterpart in the phenomenological tradition. This gives independent support from another tradition to the claim that "the last cell really exists and is irreducible." The boundary reached by this book from below is precisely the starting point confirmed by phenomenology from above: the same boundary is marked from two directions.

Correspondence with the Structure of Intentionality

A section on intentionality is needed because it is another core of phenomenology and resonates with the "relationality" of given-ness in this book, as well as with the "world-directed delimitation" of $\mathcal{B}$ in Chapter 19 of the new edition of *Mind–Matter Unity*. Phenomenology holds that experience is essentially intentional: consciousness is always consciousness "of" something, always directed toward, or oriented toward, an object. Intentionality is the internal structure of experience: every experience has directedness, and there is no experience "directed at nothing," even the purest inward meditation is directed toward "itself" or "nothingness." This book's characterization of given-ness—internal–external co-evolution, and the claim that given-ness always appears "to some interior" (Section 13.3)—fits closely with the "directedness" structure of intentionality. Given-ness as relational and directional corresponds precisely to intentionality's thesis that consciousness always points toward an object.

More specifically: internal–external co-evolution in this book corresponds to the intentional structure of phenomenology. The "internal" in internal–external co-evolution, namely the coherent configuration, corresponds to the subject pole of intentionality; the "external," namely the remaining configuration co-evolving with it, corresponds to the object pole of intentionality. The co-evolutionary relation between them corresponds to the "directedness" of intentionality. This book therefore brings the structure of intentionality down from phenomenological reflective description to the mechanism of physical co-evolution: intentionality is no longer an unanalyzable internal structure of experience, but a relational event with a physical counterpart. The "world-directed delimitation" of $\mathcal{B}$ in Chapter 19 of the new edition of *Mind–Matter Unity* corresponds to the world–I relation in phenomenology. The content delimited by $\mathcal{B}$ is this: experience is delimited as "the world toward which I am directed." Perception, judgment, bodily sense, and world all carry the ascription of "mine." This delimitation corresponds to the phenomenological given-ness of "world-to-me": experience is always "my" world, the world "facing me," the world "toward which I am directed." This book characterizes the physical counterpart of this directed structure through internal–external co-evolution and $\mathcal{B}$-delimitation, while leaving the being-experienced of this directedness as the last cell. This book and phenomenology converge on the "directed structure of given-ness," with this book providing its physical counterpart, and both converge again on the irreducibility of "the being-experienced of this structure." This is another convergence by different routes.

How This Section Fulfills the Cross-Traditional Posture of This Book

This section has developed the convergence between this book and phenomenology, adding another dimension to the book's cross-traditional dialogue. The cross-traditional dialogue of this book now involves: the Western analytic tradition, including Chapter 5 on the spectrum of realism, Chapter 6 on falsifiability, and Section 13.5 of this chapter on the spectrum of philosophy of consciousness; the Western continental tradition, including Chapter 11 on freezing in relation to Prigogine and Whitehead, Chapter 16's preparation for historicity in relation to Heidegger, and this section's convergence with phenomenology; the Chinese philosophical tradition, including Chapter 8 on qi theory and gradient ontology; the Indian Buddhist tradition, including Chapter 19 on Nāgārjuna as one of the responses; and the phenomenological tradition, namely this section on Husserl and the irreducibility of given-ness. This multiple dialogue clarifies the character of this book as a cross-traditional philosophical monograph: it is not confined to internal discussion within a single tradition, but provides a contemporary synthesis on shared issues across several philosophical traditions.

Chapter 19 has already developed Heidegger, Wittgenstein, and Nāgārjuna as three responses at the boundary of ontology. This section develops phenomenology, Husserl, as the convergence of a fourth independent path in the domain of the philosophy of consciousness. This further expands the book's cross-traditional dialogue. In a later revision of Chapter 19, it may be necessary either to include phenomenology, Husserl, as a fourth response in the dialogue, or to explicitly cross-reference the treatment of phenomenology in this section as a convergence in the domain of philosophy of consciousness.

This arrangement is left to the coordinated revision of Chapter 19.

13.8. Methodological Reflection on This Chapter

This chapter applies the method of constrained speculation in a concentrated way to the mind–matter problem, the most difficult philosophical problem. This section reflects on the methodological significance of this application.

This Chapter's Application of the Method of Constrained Speculation: From "Reasonable Deferral" to "Precise Marking"

This chapter represents a substantive upgrade in this book's posture toward the method of constrained speculation. The earlier posture was often expressed as "reasonable deferral": acknowledging the hard problem but temporarily not treating it. This posture had historical legitimacy, avoiding both the arrogance of "pretending to have solved it" and the defeatism of "giving up everything." But it was philosophically insufficiently precise and could make constrained speculation appear to be making a methodological concession in the mind–matter problem. The upgrade in this chapter is from "reasonable deferral" to "precisely marking as boundary." Specifically: the hard problem is not an open problem that is "temporarily unmanageable," but is precisely marked as a boundary. It marks the edge of the explanatory capacity of structure and relation itself and is not within that capacity. The contribution of constrained speculation to the hard problem is not to "defer" it, but to mark it precisely: everything that can be solved is solved as far as possible, including structure, the mechanism of given-ness, the mechanisms of cognization and delimitation, and measurable projection; what cannot be solved is precisely marked as a single-point boundary. Acknowledging this boundary is not a concession by the method of constrained speculation, but its deepest application to the most difficult problem. The failure-orientation discipline (CS4) requires every ontological commitment to state its own failure conditions, and the last cell as boundary is precisely the failure condition of the ontology of experience, identity-tier isomorphism, at its own limit. The C3 prohibition, namely that no mechanism statement may masquerade as a solution to being-experienced, is the methodological form of this upgrade. It makes the boundary discipline of constrained speculation in the mind–matter problem explicit: all mechanism statements, including identity-tier isomorphism, internal–external co-evolution, cognization, and $\mathcal{B}$-delimitation, are on the objective side of the gap, and none can or should be used to claim that the being-experienced side has been "dissolved." Keeping C3 is precisely the proof of the depth of the method of constrained speculation in the mind–matter problem. This upgrade makes the chapter a mature application of constrained speculation to the most difficult problem: it retreats neither to the arrogance of "solved" nor to the defeatism of "suspending everything," but honestly and precisely marks the boundary between what can and cannot be done. This is methodological thoroughness, not methodological concession.

This Chapter's Posture toward Originality

This chapter does not claim complete originality. It inherits the core argumentative strategy of the new edition of *Mind–Matter Unity* (2026 revised edition) and elevates it from a physical thesis into the philosophical application of constrained process realism in the domain of mind and matter. Specifically, the contributions of the new edition of *Mind–Matter Unity* include: the decomposition of the hard problem as an isomorphism problem; fivefold gradient-richness isomorphism as the fulfillment of structural sufficiency; given-ness as the relational characterization of internal–external co-evolution; boundary as gate as an anti-panpsychist position; two compressions and the last cell as boundary; Russellian monism as the nearest neighbor; and convergence with phenomenology by different routes. The contribution of this chapter is to organize these physical and philosophy-of-nature claims systematically into a philosophy of mind and matter under the method of constrained speculation, allowing them to form a unified methodological whole with the ontological resources established in earlier chapters, including the ontology of difference, constitutive coextension, and co-evolution as an ontological relation, and enabling them to occupy a recognizable independent position within the international spectrum of philosophy of consciousness. This synthetic contribution is not isolated creation, but constructive originality: on the basis of the philosophical work of the new edition of *Mind–Matter Unity*, it elevates that work to the general philosophical level of the method of constrained speculation and makes it internally coherent with the ontological framework of this book.

Limits Honestly Acknowledged in This Chapter

This chapter acknowledges several limits. First, it does not solve the last cell; it explicitly acknowledges it as a boundary rather than an open problem. If future consciousness research produces some kind of breakthrough on the hard problem, such as discovering a path connecting the structural–relational side with the being-experienced side, then this chapter's characterization of the "boundary" may need to be reassessed. But this chapter argues that at the current level of knowledge, the characterization of the last cell as boundary is the most honest treatment that the method of constrained speculation can provide. Second, it does not treat the content of consciousness. This chapter treats only the structure of consciousness, including how it can be characterized and what structural features it has, and the mechanism of the given-ness of consciousness, including how it is experienced and what its physical preconditions are. It does not treat the content of consciousness, such as the feeling of this color or the content of that emotion. The latter requires dedicated phenomenological–neuroscientific work and exceeds the scope of this chapter. Third, cross-species extension remains uncertain. The manifestations of consciousness in animals, AI, and other non-human systems require dedicated work. The boundary criterion established in this chapter, namely the membrane of life, provides a principled framework, but its application to boundary cases, such as synthetic life, complex collective systems, and possible non-carbon-based life, requires more detailed work. Fourth, the treatment of Chinese traditional resources is brief. The mind–matter problem has rich resources in the Chinese philosophical tradition, including Wang Yangming's philosophy of mind, Buddhist Yogācāra, and Song–Ming theories of mind-nature. This chapter mentions them only briefly and does not develop them fully. The dialogue between these resources and the work of this book is left for future work. A comprehensive cross-traditional philosophy of mind and matter would need to include these Chinese traditional resources as well; this is an unfinished direction of the present chapter. Fifth, the problem of personal identity is left for later work. This chapter establishes the "I" as dominant coherent resonance + $\mathcal{B}$-delimitation, as cited in Sections 13.4 and 13.6. But the philosophical problem of personal identity, including how the "I" remains identical across time, the relation between memory and personality, and cases of dying or personality disintegration, is not fully developed here. These problems require dedicated work. These limits do not weaken the work of this chapter; they make it an honest initial work that invites further development.

Connections with Later Chapters

The work of this chapter connects with several later chapters. Chapter 14, "Observation as Participation": Section 13.3 of this chapter uses Chapter 14, OB1, reproducible co-perturbation as a new foundation of objectivity, as the ontological ground for internal–external co-evolution. Together, the two chapters show the application of the same co-evolutionary ontology, Chapter 12, EV1, in different domains: Chapter 14 in the domain of observation, and this chapter in the domain of mind and matter. Chapter 19, "Three Responses: Heidegger, Wittgenstein, and Nāgārjuna": Section 13.6 of this chapter develops phenomenology, Husserl, as an independent path in the philosophy of consciousness converging with the work of this book. This suggests that the three responses in Chapter 19 may need corresponding expansion: Husserl may be included as a fourth response, or Chapter 19 may cross-reference this section's treatment of phenomenology. This coordination is left to the revision of Chapter 19. Chapter 21, "Meaning before Freezing": this chapter's work on consciousness as a bearer of meaning, namely identity-tier isomorphism + internal–external co-evolution + $\mathcal{B}$-delimitation, provides the ontological foundation for Chapter 21's claim that "being able to care" is the bearer of meaning. Meaning requires an "interior" delimited by boundary as its bearer, and this interior is precisely the bounded living system argued for in Section 13.3 of this chapter. Chapter 17, "Gradient Terminal Phase: The Internal Singularity of Ontology": the methodology of this chapter, which treats the last cell as boundary, is isomorphic to Chapter 17's methodology regarding ontological singularity. Both chapters use the distinction between the meta-level and the object-level to handle legitimate speech at boundaries; both preserve methodological rigor through the C3 prohibition, namely the prohibition against crossing a boundary with mechanism statements. This methodological isomorphism gives the whole book a unified methodological posture toward the two most difficult problems: the hard problem of consciousness and ontological singularity.

One-Sentence Summary

This chapter is the deepest application of the method of constrained speculation to the most difficult philosophical problem. Through identity-tier isomorphism, it fulfills the ontological meaning of "Mind–Matter Unity"; through internal–external co-evolution and boundary as gate, it characterizes the mechanism and scope of given-ness; through two compressions and the last cell as boundary, it precisely treats the irreducible part of the hard problem; through dialogue with the spectrum of philosophy of consciousness, especially Russellian monism as the nearest neighbor, it locates the position of this book; and through convergence with phenomenology by different routes, it shows independent cross-traditional support. This chapter gives the method of constrained speculation a genuinely mature philosophical posture on the mind–matter problem: everything that can be solved is solved as far as possible, everything that cannot be solved is precisely marked as a boundary, and the two are clearly distinguished and each given its proper place.

14. Observation as Participation: A New Foundation of Objectivity

14.1. Introduction to This Chapter

"Objectivity" is a central concept in modern science and philosophy. Its traditional form may be stated as follows: knowledge is objective if and only if the state of affairs it describes is independent of the observer; its content is determined by "how things are in themselves" and is not affected by the observer's position, perspective, or participation. This form presupposes the separation between observer and object and disturbance-free readout: ideal observation does not affect the observed, but "reads" the state of the object without changing that state.

But this presupposition has faced multiple challenges since the twentieth century. In quantum mechanics, the measurement process necessarily involves interaction between the observer and the observed system: measurement is not disturbance-free readout, but constitutive participation. In cosmology, we as observers cannot stand outside the universe: every observation is a process internal to the universe. In neuroscience, "objective" research on brain function depends on experimental settings, statistical methods, and interpretive frameworks: it is not pure disturbance-free readout.

This chapter diagnoses the difficulties of the traditional concept of objectivity and proposes reproducible co-perturbation as a new foundation of objectivity (OB1). Specifically: objectivity is not "disturbance-free externality," but "controlled reproducibility." An observational result is objective if and only if it can be reproduced under controlled conditions, and this reproducibility is grounded in identifiable correspondences among concrete mechanisms: frequency matching, threshold, and tri-state response.

This alternative is built upon Chapter 12's work on co-evolution as an ontological relation: observation is not an external readout by an independent observer of an independent object, but a form of same-field co-evolution. Specifically, observation involves interaction between observer and observed as local structures within the same ρ_f field. It necessarily affects the observed (perturbation), but the result of perturbation can be reproduced under controlled conditions (reproducibility).

This chapter also enters into serious dialogue with contemporary physical-philosophical traditions represented by Heisenberg, Bohr, Rovelli, Barad, and others. These works each treat the difficulties of the traditional concept of objectivity. This chapter clarifies the position of constrained speculation within this spectrum: it refines objectivity, making the concept rigorous through controlled reproducibility, rather than relativizing objectivity by reducing it to content relative to an observer.

This chapter is another core chapter of Part III, parallel to Chapter 13 on mind–matter. It builds upon Chapter 12, "Co-Evolution as an Ontological Relation": observation is a form of same-field co-evolution. It also builds upon the physical work of *EQT Detection Theory*: the three criteria, frequency matching, amplitude threshold, and tri-state response, provide content for this chapter's concept of reproducible co-perturbation.

The discussion proceeds as follows. Section 14.1 diagnoses the difficulties of the traditional concept of objectivity. Section 14.2 states and defends reproducible co-perturbation as a new foundation (OB1). Section 14.3 develops the applications of OB1 in three scientific domains: particle physics, cosmology, and neuroscience. Section 14.4 enters into serious dialogue with Heisenberg, Bohr, Rovelli, and Barad. Section 14.5 addresses objections and develops the open questions of OB1.

14.2. The Difficulties of the Traditional Concept of Objectivity

The Form of the Traditional Concept of Objectivity

"Objectivity" is a central concept in modern science and philosophy. Its traditional form may be stated as follows: knowledge is objective if and only if the state of affairs it describes is independent of the observer; its content is determined by "how things are in themselves" and is not affected by the observer's position, perspective, or participation.

This form has several implications: the separation between observer and object, where the observer as knowing subject and the object as the known object are ontologically separated; disturbance-free readout, where ideal observation does not affect the observed, but "reads" the state of the object without changing that state; perspective-independence, where objective knowledge does not depend on the position or perspective of a particular observer, and any qualified observer would obtain the same result; and ontological priority, where the state of the object exists prior to any observation, and observation merely discovers this state rather than constituting it. These four implications together constitute the traditional concept of objectivity. It played a key role in the formation of modern science: from Galileo to Newton to nineteenth-century physics, traditional objectivity was the methodological ground of physics.

Challenges from Contemporary Physics

Since the twentieth century, however, developments in physics have posed substantive challenges to this traditional objectivity.

Challenge One: the quantum measurement problem. The content of quantum mechanics, especially the measurement process, makes the "separation between observer and object" and "disturbance-free readout" problematic. Specifically: measurement is not disturbance-free readout, because the state of a quantum system is "collapsed" into an eigenstate during measurement; different measuring devices, such as polarizers oriented in different directions, can obtain different results from the same system; complementary observables, such as position and momentum, cannot be measured precisely at the same time, because the measurement process itself limits the information that can be obtained. These facts make the traditional objectivity assumption of "disturbance-free readout" fail at the quantum level. Physicists have responded in different ways, including the Copenhagen interpretation, many-worlds interpretation, Bohmian mechanics, relational quantum mechanics, and others, but all responses acknowledge that the traditional concept of objectivity needs to be reassessed in the quantum domain.

Challenge Two: the internality of cosmological observation. Cosmological observation involves a special problem: the observer is itself inside the universe being observed. We cannot observe the universe from outside the universe; any cosmological observation is an internal event of the universe. This internality challenges traditional objectivity, which assumes that the observer can "stand outside" the observed. But the cosmological observer is necessarily inside the universe: the observer's instruments and the observer's existence itself are part of the universe. Concrete manifestations include: the anthropic principle, where the parameters of the universe must allow observers to exist, because only such a universe can be observed, thereby entangling the fact of observation with the observed facts; horizon limits, where the universe we can observe is only the observable universe because of the finite speed of light, not the whole universe, making an "objective complete view" in principle inaccessible; and observer effects, where in some cosmological models, such as quantum cosmology, the existence of the observer affects the path of cosmic evolution. These manifestations make "disturbance-free readout" and "ontological priority" in traditional objectivity problematic at the cosmological level as well.

Challenge Three: the concretization of the observer by neuroscience. The development of neuroscience has transformed the concept of "observer" from an abstract philosophical concept into a concrete physical process: the observer is a physical system, namely brain plus sense organs, with neural activity patterns, perceptual limits, and cognitive biases. This concretization challenges traditional objectivity: if the observer is a physical system, what do "object independent of the observer" and "no perspective" mean? Any observation involves the neural processes of a concrete observer, and these processes themselves have forms and limits. Concrete manifestations include: the specificity of perception, where sensory features such as color, sound, and touch are products of the observer's neural processes rather than "purely objective" properties of the object; the specificity of cognition, where attention, memory, and reasoning are all neural processes that affect what is accepted as an "objective fact"; and the diversity of observers, where different observers, different species, different persons, and different states possess different perceptual-cognitive capacities and obtain different observations of the "same" object. These manifestations make the "perspective-independence" assumption of traditional objectivity problematic at the level of neuroscience as well.

Traditional Responses and Their Difficulties

There are several ways in which the traditional concept of objectivity responds to these challenges.

Response One: technical correction. This response preserves the core of traditional objectivity, namely the object independent of the observer, but acknowledges that concrete observation requires technical corrections, such as considering the features of the measuring device in quantum measurement or sensory limitations in neuroscience. Difficulty: technical correction does not solve the fundamental problem. If observation always involves observer–object co-evolution, then "independence from the observer" as an ideal is inaccessible. Correction merely acknowledges that the ideal cannot be fully attained, while still presupposing that the ideal itself is meaningful.

Response Two: relativism. This response gives up objectivity: all observations are "perspectival," and there is no objective truth. Difficulty: relativism has internal problems. If all truths are perspectival, then is the statement "all truths are perspectival" itself perspectival? This is a self-defeating position. Relativism also makes the achievements of science, such as repeatable experiments, cross-observer agreement, and technological applications, difficult to explain.

Response Three: transcendental grounding. The Kantian response: objectivity does not depend on the object itself, the thing-in-itself, but on the organization of the phenomenal world by transcendental categories. Difficulty: the Kantian response grounds objectivity in the transcendental subject. But contemporary neuroscience and cognitive science make the "transcendental subject" as a foundation difficult to sustain. The transcendental subject itself requires an ontological ground.

These three traditional responses are all insufficient. A new direction is needed.

A New Direction

The new direction proposed by constrained speculation is this: do not choose between the two poles of "independent of the observer" and "entirely perspectival," but redefine objectivity. Objectivity = controlled reproducibility. The content of this direction is as follows: acknowledge that every observation involves the co-evolution of observer and object, without pretending to have independent readout; but require that observation be reproducible: the same configuration under the same conditions produces the same result. Objectivity is grounded in this reproducibility, not in "disturbance-free externality." The next section develops the content of this direction.

Objections and Replies

Objection One: The three challenges you list look selective. You list only developments that challenge traditional objectivity, but not developments that support it.

Reply: The aim of this section is to identify the difficulties of traditional objectivity, and therefore it focuses on challenges. But this section does not claim that traditional objectivity is "completely wrong." It remains useful in many ordinary and classical-physical contexts. Specifically, traditional objectivity works well in situations where observer and object are clearly distinct in scale, dynamics, and interaction strength: ordinary objects are much larger than the perturbative effects of observers, classical-physical measurements are far above quantum fluctuations, and so on. In these situations, "independent of the observer" is valid as an effective approximation. But as an ontological or methodological foundation, traditional objectivity is difficult to maintain in the face of the challenges described above. A more refined concept of objectivity is needed, one that can cover the situations in which traditional objectivity works and also handle the situations in which it fails. This is the work of "reproducible co-perturbation" in Section 14.2.

Objection Two: You classify the Kantian response as "transcendental grounding" and point out its difficulty. But Kant's work has force. Is this simple classification unfair?

Reply: The brief classification in this section is not a detailed evaluation. The force of Kant's work requires dedicated treatment. The claim of this section is limited: the Kantian response faces difficulties in light of contemporary scientific developments, because the transcendental subject as a foundation needs to be reassessed. This claim does not mean that Kant is entirely wrong. More deeply, the method of constrained speculation in this book differs from the Kantian method. The Kantian method depends on transcendental analysis; constrained speculation depends on falsifiability. Both methods have legitimacy, but they treat the problem of objectivity differently. This section develops the latter without denying the legitimacy of the former as another direction of work.

Objection Three: Your "new direction" has not yet been developed, but you already use "controlled reproducibility" as a key concept. Is this concept itself problematic? It seems circular, defining objectivity by "control" and "reproducibility," while leaving unclear what "control" and "reproducibility" are.

Reply: This concern is real. Introducing any new concept requires precise definition. Section 14.2 will develop the content of "reproducible co-perturbation" in detail, including precise definitions of "control" and "reproducibility." This section only previews the direction and does not claim that the new concept has already been fully defined here. The precise definition will be developed through concrete mechanisms: frequency matching, threshold, and tri-state decomposition. These mechanisms give content to "control" and "reproducibility." This is the application of the method of constrained speculation: any new concept must be concretized through falsifiable mechanisms.

Literature That Must Be Engaged

- Werner Heisenberg, Physics and Philosophy (1958)
- Niels Bohr, Atomic Physics and Human Knowledge (1958)
- Thomas Nagel, The View from Nowhere (1986)
- Lorraine Daston & Peter Galison, Objectivity (2007)
- Ian Hacking, Representing and Intervening (1983)
- Chapter 12 on co-evolution as an ontological relation (self-reference)

Summary of This Section

This section has done three things. First, it stated the form of the traditional concept of objectivity and its four implications. Second, it developed three challenges from contemporary physics: quantum measurement, cosmological internality, and neuroscientific concretization. Third, it identified the difficulties of the traditional responses: technical correction, relativism, and transcendental grounding, and previewed the new direction of constrained speculation.

The next section develops reproducible co-perturbation as a new foundation of objectivity.

14.3. Reproducible Co-Perturbation: The Core Thesis

Statement of the Core Thesis

Core thesis OB1: objectivity = controlled reproducible co- perturbation. An observational result is objective if and only if: (i) it can be reproduced under controlled conditions, meaning that the same configuration, including the same detector, the same probed system, the same initial conditions, and the same interaction mode, produces the same result; and (ii) this reproducibility is grounded in identifiable correspondences among concrete mechanisms: frequency matching, threshold, and tri-state response.

This thesis requires detailed development. "Controlled reproducibility": the "reproducibility" of observation is not arbitrary "repetition," but repetition controlled by configuration. It requires the features of the observational configuration, including detector type, probed-system type, initial conditions, interaction mode, and so on, to be explicitly identified and controlled. "Co-perturbation": observation is not disturbance-free readout, but co-perturbation between the observer and the probed system. This is the application of Chapter 12, co-evolution as an ontological relation. Objectivity is grounded in the reproducibility of this co-perturbation.

Implications of the Thesis

OB1 has four implications.

Implication One: Objectivity does not require "no observer."

Traditional objectivity requires "independence from the observer": ideal objective knowledge does not involve the observer. OB1 rejects this requirement: objectivity allows the observer to participate, as long as this participation is reproducible. Specifically, a quantum measurement result, such as the result of measuring electron spin, is objective when the same measuring device, including the same magnetic-field direction and the same detector, and the same probed system, including the same electron source and the same preparation procedure, produce the same statistical results under the same conditions, such as the same ratio of spin up and spin down. This objectivity does not require "independence from the measuring device." It directly treats the measuring device as part of the objective result.

Implication Two: Objectivity is grounded in reproducibility, not in "the thing itself."

Traditional objectivity assumes there is "how things are in themselves": a state independent of any observation. OB1 does not depend on this assumption. It requires the reproducibility of observational results, not that observational results correspond to "the thing itself." This treatment avoids the metaphysical problem of "the thing itself": it neither presupposes such an existence nor denies it. It locates objectivity in observable content and makes the problem of objectivity a tractable problem.

Implication Three: Reproducibility requires concrete mechanism correspondence.

OB1 does not require that any "repetition" constitute reproducibility. It requires repetition to be grounded in identifiable correspondences among concrete mechanisms. To argue that two observations are "the same," one must be able to identify the correspondences in their configurations: is the detector's frequency response ($g(f, f_0)$) the same? Is the amplitude threshold ($\delta\rho > \Gamma/k$) satisfied in the same way? Does the tri-state response, namely the decomposition of free-state, bound-state, and condensed-state weights, correspond? If these mechanisms correspond, the two observations are "the same," and reproducibility holds. If they do not correspond, as with different detectors or different initial conditions, the two observations are not "the same," and reproducibility must be reassessed. This concreteness keeps the concept of reproducibility from being empty: it has testable content.

Implication Four: Objectivity is a matter of degree, not an absolute.

OB1 transforms objectivity from an absolute dichotomy, objective vs non-objective, into a matter of degree: observational results have different degrees of objectivity depending on the degree of controlled reproducibility. Specifically: high objectivity, where observations can be reproduced across multiple configurations, mechanism correspondence is rigorous, and cross-observer agreement holds, as in the precise value of electron charge, which can be independently determined through multiple methods, including the Millikan oil-drop experiment, the quantum Hall effect, and CMB fluctuations; medium objectivity, where observations can be reproduced under specific configurations, mechanism correspondence is clear, but the scope is limited, as in the behavior of a particular biological species, which can be reproduced many times in the laboratory and in the field but depends on specific biological individuals and environments; low objectivity, where observations are difficult to reproduce, mechanism correspondence is vague, and cross-observer differences are large, as in the content of subjective experience, which is difficult to reproduce from the third-person perspective and depends on the neural state of a particular subject. This concept of degree allows objectivity to handle many kinds of cases. It does not require all observations to reach the same degree of objectivity, but requires each observation to have an assessable degree of objectivity.

Connection with Co-Evolution (Chapter 12)

OB1 is directly built upon the co-evolutionary ontology of Chapter 12. Observation is the co-evolution of detector and probed system: the two are not in an external interaction between independent terms, but are a form of same-field co-evolution. This form of co-evolution is characterized by three mechanisms: frequency matching, threshold, and tri-state response. Reproducible co-perturbation as a new foundation of objectivity is the epistemological application of this co-evolutionary framework: objectivity is grounded in the reproducibility of co-evolution. Co-evolution is an ontological relation; reproducible co-perturbation is the epistemological projection of that relation. This connection makes OB1 not an isolated epistemological thesis, but a thesis built upon the ontological framework already developed in this book. If the co-evolutionary ontology of Chapter 12 holds, OB1 as its epistemological application has a ground.

An Application: The Precision of Bohr's Complementarity

Bohr's principle of complementarity is an important work in twentieth-century philosophy of physics. In simplified form: complementary observables, such as position and momentum, or wave and particle, cannot be measured precisely at the same time; different measurement configurations reveal different aspects of the same phenomenon. Bohr's work has had important philosophical influence, but its content has often been considered "vague": it depends on intuition and physical experiments, and lacks a precise ontological-epistemological ground.

OB1 provides such precision. Bohr's complementarity can be understood as follows: different measurement configurations, such as position-measuring apparatus vs momentum-measuring apparatus, correspond to different modes of co-evolution; these modes of co-evolution have different reproducibility conditions. The reproducibility of position measurement requires one configuration, and the reproducibility of momentum measurement requires another. Results obtained from the same system under different configurations are all objective, because they satisfy their own reproducibility conditions, but they cannot be simply integrated into "the objective state of the system," because different configurations involve different modes of co-evolution. This precision means that Bohr's complementarity is no longer a vague statement about "the limits of what we can know," but a concrete statement: objectivity is controlled reproducibility, and different configurations define different domains of objectivity.

Objections and Replies

Objection One: Your claim "objectivity = controlled reproducible co-perturbation" sounds like operationalism. Bridgman's operationalism has been widely criticized. How does this position avoid the problems of operationalism?

Reply: This position has structural similarity with operationalism, since both are based on operability. But there are important differences. Operationalism, as in Bridgman, says that scientific concepts are defined by operations: temperature is thermometer reading, length is the measurement process. This position has been criticized as too narrow: it deprives theoretical concepts of meaning, such as theoretical temperature or virtual measurement, and makes theoretical unification across operations difficult, since different temperature measurements would correspond to different concepts. OB1 does not claim that concepts are defined by operations. It claims that objectivity is grounded in controlled reproducibility. These are different. OB1 allows theoretical concepts to have richer content than operations, for example theoretical temperature as a concept can refer to the common ground of multiple operations, but it requires the objectivity of theoretical concepts to be established through reproducible operations. More deeply, the difference between OB1 and operationalism is grounded in the process ontology of this book. Operationalism lacks an ontological ground; OB1 is built upon co-evolutionary ontology (Chapter 12). This ontological ground gives OB1 richer content than operationalism.

Objection Two: Your claim that "objectivity is a matter of degree" sounds like abandoning objectivity. If different observations have different degrees of objectivity, then "objective" as a standard loses clarity.

Reply: This position does not abandon objectivity. It refines the concept of objectivity. Specifically, OB1 extends objectivity from a single dichotomy, objective vs subjective, into a multidimensional assessment, including the degree of reproducibility, the degree of mechanism correspondence, the degree of cross-configuration agreement, and so on. This extension allows the concept of objectivity to handle more cases. Each observation can be precisely assessed: how reproducible it is, how clear the mechanism correspondence is, how strong cross-configuration agreement is. This precise assessment provides concrete content for objectivity; it is not a vague "degree." Different observations have different degrees of objectivity, but each degree can be specified precisely. This makes "objective" as a standard more precise, not less clear.

Objection Three: OB1 seems to make objectivity depend on "reproducibility," but certain events in cosmology, such as the Big Bang or particular astronomical events, are not reproducible. Are observations of these events therefore not objective?

Reply: This objection points to an important issue. The response of this section has two levels. At the event level: certain events, such as the Big Bang, are indeed not reproducible. We cannot recreate the Big Bang in the laboratory. At the observational level: observations of these events can be reproducible. We can reanalyze CMB data, rerun specific statistical analyses, and confirm observational results across research groups. At the observational level, reproducibility can be preserved. OB1 concerns the reproducibility of observation, not the reproducibility of events. Our objective knowledge of the Big Bang is grounded in reproducible observations of its residual traces, including the CMB, light-element abundances, and the large-scale structure of the universe, not in the reproduction of the Big Bang event itself. This distinction allows OB1 to handle the objectivity of non-reproducible events.

Literature That Must Be Engaged

- Percy Bridgman, The Logic of Modern Physics (1927): operationalism
- Niels Bohr, Atomic Physics and the Description of Nature (1934)
- Chapter 12 on co-evolution as an ontological relation (self-reference)
- *EQT Detection Theory* on the three criteria (self-reference)

Summary of This Section

This section has done four things. First, it stated the core thesis OB1: objectivity = controlled reproducible co-perturbation. Second, it developed four implications of the thesis: it does not require "no observer," it is not grounded in "the thing itself," it requires concrete mechanism correspondence, and it is a matter of degree rather than an absolute. Third, it showed the connection between OB1 and the co-evolutionary ontology of Chapter 12. Fourth, it demonstrated the application of OB1 through the precision of Bohr's complementarity.

The next section develops three forms of reproducible co-perturbation.

14.4. Three Forms of Reproducible Co-Perturbation

The Function of the Three Forms

OB1 as a general thesis needs to be concretized. Reproducible co-perturbation takes different forms in different observational situations. This section develops three forms: observations in particle physics, cosmology, and neuroscience. It shows how OB1 is applied in different domains. These three forms are not exhaustive; they are examples that demonstrate how OB1 can be applied. Applications in other domains, such as chemistry, biology, sociology, and others, can be developed similarly.

Form One: Reproducible Co-Perturbation in Particle-Physics Observation

Particle physics is one of the most precise domains of contemporary physics. Its observational methods, including particle accelerators, detectors, and statistical analysis, show a form of reproducible co-perturbation.

Concrete configuration: consider an example, the observation of the Higgs boson at the LHC. The configuration includes the accelerator, where the LHC accelerates protons to 13 TeV and produces high-energy events through collisions; detectors, where ATLAS and CMS are two independent detectors, each with its own geometry, sensitivity, and trigger system; data acquisition, where millions of collision events per second are statistically selected and events of "interest" are stored as data; and analysis procedures, where Higgs-candidate events are identified through physical models and statistical methods. Every part of this configuration is precisely controlled and documented.

The form of reproducibility: the objectivity of the Higgs discovery in 2012 is grounded in the following forms of reproducibility: reproducibility within detectors, where ATLAS repeatedly observed the Higgs signal across different runs, each time producing statistically consistent results; reproducibility across detectors, where ATLAS and CMS independently observed the signal and both reported a signal near 125 GeV; reproducibility across analysis methods, where different analysis groups, different statistical methods, and different Monte Carlo simulations all confirmed the signal; and reproducibility of physical predictions, where the properties of the discovered particle, including production modes, decay modes, and spin, matched the predictions of the Standard Model Higgs. These forms of reproducibility make the Higgs discovery highly objective, satisfying the requirements of OB1.

The form of co-perturbation: but this objectivity is not "disturbance-free readout." The Higgs discovery involves co-perturbation: the accelerator actively produces high-energy collisions, which is a perturbation of the quantum vacuum; the detectors actively interact with the produced particles through processes such as ionization and showering; data selection and analysis involve theoretical assumptions. The objectivity of the Higgs discovery is grounded in the reproducibility of these co-perturbations, not in the objectivity of "disturbance-free readout."

Mechanism correspondence: mechanism correspondence in particle-physics observation includes frequency matching, where detectors have different responses to different particle types, such as electromagnetic calorimeters being sensitive to photons and electrons, hadronic calorimeters to hadrons, and magnetic spectrometers to charged particles, with each particle observation requiring a match in interaction type; amplitude threshold, where detectors have trigger thresholds and only events above specific energies are recorded; and tri-state response, where observational signals are mixtures of the tri-states, including free-state signals such as photon showers in calorimeters, bound-state signals such as hadronic decay chains, and condensed-state signals such as stable-particle tracks, with particle identification involving analysis of tri-state weights. These mechanism correspondences are the content of reproducibility in particle-physics observation.

Form Two: Reproducible Co-Perturbation in Cosmological Observation

Cosmological observation has a special character: the observer is inside the observed universe, many events are not reproducible, such as the Big Bang, and observation usually depends on a single "cosmic sample." This specificity has implications for reproducible co-perturbation.

Concrete configuration: consider CMB observation. The configuration includes observational platforms, such as ground-based telescopes, balloons such as BOOMERanG and MAXIMA, and satellites such as COBE, WMAP, and Planck; detectors, including radiometers and bolometers that are sensitive to specific frequencies and polarization; observational strategies, including full-sky scanning, deep-field observation, and focus on specific regions; and data analysis, including removal of foreground contamination, computation of power spectra, and estimation of cosmological parameters.

The form of reproducibility: the objectivity of CMB observation is grounded in the following forms of reproducibility: reproducibility across platforms, where COBE, WMAP, and Planck independently observed and reported consistent results, including CMB temperature, anisotropy spectra, and polarization modes; reproducibility across frequency bands, where observations of the CMB in different microwave bands, such as 30 GHz, 100 GHz, and 353 GHz, are mutually consistent, confirming that the signal is CMB rather than foreground contamination; reproducibility across analysis methods, where different research groups using different analysis pipelines obtain consistent cosmological parameters; and reproducible correspondence with other observations, where cosmological parameters from the CMB, such as the Hubble constant, dark matter density, and dark energy density, correspond with independent observations, including supernova observations, baryon acoustic oscillations, and weak lensing. These forms of reproducibility are the content of the objectivity of CMB observation.

Special treatment of non-reproducible events: the Big Bang itself is not reproducible. But the objectivity of CMB observation does not require the Big Bang to be reproduced. It requires the observation of the CMB to be reproducible. This distinction makes the objectivity of cosmological observation attainable. We do not need to recreate the Big Bang; we need only reproduce observations of its residual traces. As a residual of the Big Bang, the CMB can be confirmed through multiple independent observations. This is the foundation of cosmological objectivity.

The form of co-perturbation: co-perturbation in cosmological observation is relatively weak. CMB photons have mainly propagated freely from their formation to their detection. But forms of co-perturbation still exist: the detector actively interacts with CMB photons by absorbing and measuring their energy; data processing involves theoretical assumptions, such as the standard cosmological model; foreground estimation and removal involve physical models. These co-perturbations are weaker than those in particle-physics observation, but they still exist. The objectivity of CMB observation is grounded in the reproducibility of these co-perturbations.

Form Three: Reproducible Co-Perturbation in Neuroscientific Observation

Neuroscience is one of the domains in which the observer–observed relation is most complex: the observer is another conscious system, and the observed, another brain or one's own brain, is also a conscious system. This complexity has implications for reproducible co-perturbation.

Concrete configuration: consider fMRI observation. The configuration includes the scanner, where the MRI scanner has a magnetic-field strength such as 3T or 7T, a gradient system, and radio-frequency coils; task paradigms, where subjects perform specific cognitive tasks in the scanner, such as visual stimulation, decision-making tasks, or memory tasks; data acquisition, where BOLD signals are recorded across multiple brain regions, with temporal resolution on the order of seconds and spatial resolution on the order of millimeters; and analysis pipelines, including preprocessing, statistical modeling, identification of activated regions, and connectivity analysis.

The form of reproducibility: the objectivity of fMRI observation is grounded in the following forms of reproducibility: within-subject reproducibility, where the same subject performs the same task at different times and activation patterns are consistent, despite variation; between-subject reproducibility, where different subjects perform the same task and activation patterns show statistical consistency, such as visual areas being activated during visual tasks across subjects; reproducibility across scanners, where observations from different scanners and different sites can be brought into correspondence, although standardization and calibration are required; and reproducibility across analysis methods, where different analysis pipelines obtain consistent results within a reasonable range. But the degree of reproducibility in fMRI is lower than in particle-physics observation: between-subject variation is large, differences across scanners require calibration, and analysis choices affect results. This reflects the degree of objectivity in neuroscientific observation: it is of medium objectivity according to the degree classification in Section 14.2.

Special treatment of consciousness observation: consciousness observation has a special difficulty. Conscious experience is first-personal and cannot be directly observed from the third-person perspective. Chapter 13 of this book has already treated this problem in detail, especially the Ψ framework as a tool in the philosophy of mind and matter. In the OB1 framework, the objectivity of consciousness observation is grounded in the reproducible correspondence between first-person reports and third-person physical variables, which is the core work of the Ψ framework; in the independent confirmation of Ψ patterns in different conscious states; and in correspondences across observers, including different subjects and different research groups. The degree of this objectivity is lower than that of particle-physics or cosmological observation. But it is still precisely assessable objectivity, not the relativism of being "completely subjective."

Comparison of the Three Forms

The three forms show how OB1 is applied in different domains. They have a shared structure, namely controlled reproducible co-perturbation, but take different forms:

Table 14.1.. A Scientific-Philosophical Comparison of Different Disciplines along the Dimensions of Perturbation, Reproduction, and Objectivity

Item (Dimension)	Particle Physics (Microscopic Strong Control)	Cosmology (Macroscopic Passive Evolution)	Neuroscience (Mesoscopic Complex System)	Chapter
Co-perturbation intensity	High (actively producing events)	Low (freely propagating signals)	Medium (active measurement + passive observation)	1
Event reproducibility	High (repeatable collisions)	Low (non-reproducible events)	Medium (repeatable tasks)	1-2
Observation reproducibility	Very high	High	Medium	2
Degree of objectivity	Very high	High	Medium	2

This comparison shows the flexibility of OB1: it can handle multiple observational situations and provide an assessment of objectivity for each.

Objections and Replies

Objection One: Your three forms are all "scientific observations." Does OB1 apply to non-scientific observations, such as everyday perception, artistic appreciation, or ethical judgment?

Reply: In principle, OB1 can be extended to non-scientific observations, but concrete extension requires dedicated work. This section focuses on scientific observations because they are the most direct application domain of OB1. A brief examination of non-scientific observation may be given as follows: everyday perception, whose objectivity is grounded in cross-individual agreement, where different people see the position and shape of a table consistently. This agreement can be understood as reproducibility, because the same configuration produces the same perception. OB1 can treat this objectivity, but at a lower degree. Artistic appreciation involves more subjective dimensions, including personal preference, cultural background, and historical context. Its reproducibility is weaker, and OB1 can acknowledge a lower degree of objectivity while still treating forms of consensus. Ethical judgment involves normative statements and differs from the objectivity of factual statements. Chapter 20 of this book treats the is–ought gap in detail. OB1 mainly treats the objectivity of factual statements, and the objectivity of ethical judgment requires dedicated treatment. These extensions show the potential range of OB1, but each domain requires dedicated work.

Objection Two: Your assessment of "degree of objectivity" seems to involve judgment, such as which domain is more objective. Is that judgment itself objective?

Reply: This objection points to an important problem of self- application. The response of this section is that assessments of the degree of objectivity are grounded in testable standards: degree of reproducibility, degree of mechanism correspondence, and degree of cross-observer agreement. These standards themselves can be assessed, for example, degree of reproducibility can be assessed through data from replication experiments. But such assessment is not "completely objective." It involves judgment, such as what counts as "high reproducibility" and how different standards should be weighted. These judgments themselves have degrees of objectivity: they are grounded in the practices of scientific communities and theoretical assessment. More deeply, OB1 does not claim "complete objectivity." It claims objectivity by degree. The assessment of objectivity itself also has degrees. It is open to criticism and refinement, not an absolute final judgment. This self-application is consistent with the method of constrained speculation in this book.

Objection Three: The three forms you list all involve precise instruments and quantitative data. Does OB1 require observations to use precise instruments?

Reply: OB1 does not require precise instruments. It requires controlled reproducible co-perturbation. Precise instruments are one form of controlled reproducibility, but not the only form. Specifically: field observation, such as biological observation in the wild, can satisfy controlled reproducibility through observational protocols, agreement among independent observers, and reproducibility across time; historical research can satisfy controlled reproducibility in source analysis through standards of evidence, consistent conclusions among independent researchers, and correspondence across research methods; anthropological observation can satisfy controlled reproducibility through anthropological methods, independent observers, and cross-context correspondence. The degree of objectivity in these observations may be lower than in precise physical observation, but it is still objectivity. The flexibility of OB1 allows it to cover multiple types of observation.

Literature That Must Be Engaged

- ATLAS Collaboration, "Observation of a new particle..." (2012): the Higgs discovery
- Planck Collaboration, "Planck 2018 results" (2020)
- Russell Poldrack, The New Mind Readers (2018): philosophy of science of fMRI
- Allan Franklin, The Neglect of Experiment (1986): the role of experiment in objectivity
- Detection theory in *EQT Detection Theory* (self-reference)

Summary of This Section

This section has done four things. First, it developed the form of reproducible co-perturbation in particle-physics observation. Second, it developed its form in cosmological observation, especially treating the objectivity of non-reproducible events. Third, it developed its form in neuroscientific observation, especially treating the objectivity of consciousness observation. Fourth, it compared the three forms and showed the flexibility of OB1.

The next section develops the dialogue with Heisenberg, Rovelli, and Barad.

14.5. Dialogue with Heisenberg, Rovelli, and Barad

The Function of the Dialogue

As a new foundation of objectivity, OB1 must enter into dialogue with existing relevant work. This section enters into dialogue with three major works: Heisenberg's uncertainty-participation view, Rovelli's relational quantum mechanics, and Barad's agential realism. All three works involve a reassessment of objectivity and are related to the work of this book. The function of the dialogue is not to "surpass" these works, but to show the position of constrained speculation on the problem of objectivity, so that the reader can grasp its specificity precisely.

Dialogue One: Heisenberg's Uncertainty-Participation View

Werner Heisenberg developed several important philosophical ideas during the development of quantum mechanics. The most relevant is the idea of observer participation: the observer is not a passive bystander, but actively participates in the observed event. In *Physics and Philosophy* (1958), Heisenberg wrote that the observer participates in the observed event as "part of the chain of processes." This view rejects the traditional "observer–object separation" and has structural relevance to the co-evolutionary framework of this book, Chapter 12, and to OB1.

Similarities with OB1: both reject the "object independent of the observer" as the foundation of objectivity; both acknowledge that observation involves interaction between observer and observed; and both locate objectivity in the treatability of physical processes.

Differences from OB1: Heisenberg's work is mainly located in the domain of quantum mechanics, while OB1 extends to all domains of observation. Heisenberg's concept of "participation" is relatively philosophical and does not develop falsifiable mechanisms, while OB1 develops a falsifiable structure through three mechanisms: frequency matching, threshold, and tri-state response. Heisenberg does not systematically develop "participation" as a new foundation of objectivity; he mainly treats it as a consequence of quantum mechanics. OB1 claims that reproducible co-perturbation is the foundation of objectivity.

One may say that OB1 is the systematization and extension of Heisenberg's participation view: it elevates the insight of quantum mechanics into a general concept of objectivity. This extension does not replace Heisenberg's contribution, but makes its philosophical meaning more operational.

Dialogue Two: Rovelli's Relational Quantum Mechanics

Carlo Rovelli's relational quantum mechanics (RQM) is one of the contemporary interpretations of quantum mechanics most relevant to the work of this book. Rovelli argues that the "state" of a quantum system is not absolute, but relative to an observer; different observers can assign different states to the same system; and "objective state" as an absolute concept does not exist, because state is always relational. Rovelli's work has detailed physical development, especially in the systematic account in *Helgoland* (2020), and philosophical implications.

Similarities with OB1: both reject absolute objective state; both ground state in the observer–observed relation; and both are built on the content of contemporary physics, especially quantum mechanics.

Differences from OB1: Rovelli's relative state appears to relativize objectivity. The concept of objectivity is questioned in RQM. OB1 does not relativize objectivity; it refines objectivity as controlled reproducibility. The strategies differ. Rovelli's work mainly treats quantum mechanics, while OB1 extends to multiple domains. Rovelli's concept of "relation" is mainly a relation of state attribution, while OB1's concept of "co-perturbation" is more concrete, involving mechanisms of physical interaction. Rovelli does not systematically develop reproducibility as the foundation of objectivity.

One may say that OB1 and Rovelli's work agree in rejecting absolute objectivity, but differ in how they treat objectivity. Rovelli tends to restrict or relativize the concept of objectivity in the quantum domain; OB1 proposes to refine the concept of objectivity, preserving it but redefining it as reproducibility. This difference reflects different directions of work. Rovelli's work is more radical: it questions the concept of objectivity itself. OB1 is more moderate: it refines the concept of objectivity. Both have legitimacy, but reflect different philosophical choices.

Dialogue Three: Barad's Agential Realism

Karen Barad's agential realism has already been discussed in detail in Chapter 12, Section 12.4. This section focuses on Barad's position on objectivity. Barad argues that objectivity does not lie in being "independent of the observer," but in the reliability of material-discursive practices. Objectivity involves material-discursive apparatuses, through which phenomena are produced.

Similarities with OB1: both reject traditional objectivity, namely independence from the observer; both locate objectivity in observational-participatory practice; and both are built upon a relational ontology that rejects the presupposition of independent terms.

Differences from OB1: Barad's concept of "apparatus" is broad, including material elements, such as microscopes, and discursive elements, such as concepts and social constructions. OB1's concept of "co-perturbation" focuses on mechanisms of physical interaction and is more restrained regarding the discursive dimension. Barad's work has strong political-ethical implications, while OB1 is more neutral, extending to ethics in Part V but itself treating an epistemological problem. Barad does not systematically develop reproducibility as a standard of objectivity; her work is more phenomenological-descriptive, while OB1 gives a falsifiable standard.

One may say that OB1 and Barad's work agree on the relational foundation of objectivity, but differ in the standard of objectivity. Barad tends to embed objectivity within a broader material-discursive-political-ethical framework. OB1 focuses on falsifiable reproducibility. Both have legitimacy and reflect different philosophical directions.

The Specific Position of Constrained Speculation

Through these three dialogues, the specific position of constrained speculation on the problem of objectivity can be located as follows: against traditional realism, it does not take the object independent of the observer as the foundation of objectivity; against relativism, it does not give up objectivity, but refines it as controlled reproducibility; in relation to Heisenberg's participation view, it systematizes and extends it while adding falsifiable mechanisms; in relation to Rovelli's RQM, it shares a similar relational-ontological foundation but does not relativize objectivity, instead refining it; in relation to Barad's agential realism, it shares a similar relational-ontological foundation but is more restrained, focusing on physical interaction and not extending into discourse and politics. This position makes OB1 an independent and rigorously assessable position within the spectrum of the problem of objectivity. It inherits the relational insights of contemporary physics and philosophy, but gives them a falsifiable structure through the method of constrained speculation.

Objections and Replies

Objection One: Your dialogue focuses too much on the physics-philosophy tradition. How is OB1 positioned within broader philosophy of science and epistemology?

Reply: This section focuses on the physics-philosophy tradition because the problem of objectivity becomes most acute in the development of physics, especially quantum mechanics. But OB1 is related to broader traditions in philosophy of science and epistemology. Briefly: in relation to Popper's falsifiability, OB1 inherits the tradition of falsifiability, especially in its requirement of failure conditions, but extends it to the concept of objectivity itself, not only the objectivity of theories but also the objectivity of observations; in relation to Kuhn's paradigm, OB1 acknowledges that observation depends on frameworks, in agreement with Kuhn, but makes cross-framework dialogue possible through reproducible mechanisms, avoiding Kuhn's strong incommensurability; in relation to Quine's holism, OB1 agrees with Quine on the holism of the observation–theory network, but makes holism tractable through the concreteness of reproducible co-perturbation; in relation to Bayesianism, OB1 is compatible with Bayesian epistemology, because the degree of reproducibility can be treated within a Bayesian framework. These broader dialogues require dedicated work. This section is limited to dialogue with the physics-philosophy tradition.

Objection Two: Rovelli's RQM has support in the physics community. Your claim that OB1 does not "relativize" objectivity seems to challenge the direction of RQM.

Reply: The difference between OB1 and RQM is not competition, but different directions. Specifically: in the application of quantum mechanics, OB1 is compatible with RQM. OB1 accepts Rovelli's core physical insight, namely that states are relative to observers. In the treatment of the concept of objectivity, the two move in different directions: Rovelli tends to let the concept of objectivity become restricted in the quantum domain, while OB1 proposes to refine the concept of objectivity so that it can cover the quantum domain. Both directions have legitimacy. Rovelli's direction is closer to the Bohr tradition, accepting the specificity of the quantum domain; OB1 is closer to the Popperian tradition, preserving objectivity as a concept that can be refined. Which direction is ultimately preferable requires further work. This section does not claim that OB1 overthrows RQM, but claims that OB1 provides another legitimate treatment. The two can coexist and inspire different research directions.

Objection Three: Barad's work has political-ethical implications. Does your "restrained" posture avoid these important dimensions?

Reply: The work of this chapter is epistemological-ontological, not political-ethical. This choice is intentional. Part V of this book, especially Chapters 20 and 21, extends the work to ethics, but ethics is derived from ontology rather than presupposed. The political-ethical force of Barad's work is real and legitimate; it contributes to feminism, posthumanism, and related domains. This chapter does not deny that force. But as an epistemological-ontological chapter, it focuses on the philosophical-physical problem of objectivity. More deeply, this division of labor is not a defect. It allows the work of this chapter to be rigorously assessed in the epistemological-ontological dimension, without being predetermined by a political-ethical stance. Barad's work and the work of this chapter both have legitimacy in different dimensions.

Literature That Must Be Engaged

- Werner Heisenberg, Physics and Philosophy (1958), Across the Frontiers (1974)
- Carlo Rovelli, "Relational Quantum Mechanics" (1996), Helgoland (2020)
- Karen Barad, Meeting the Universe Halfway (2007)
- Niels Bohr, Atomic Theory and the Description of Nature (1934)
- Chapter 12 on co-evolution as an ontological relation (self-reference)

Summary of This Section

This section has done four things. First, it entered into dialogue with Heisenberg's participation view: OB1 is its systematic extension. Second, it entered into dialogue with Rovelli's relational quantum mechanics: they share a common foundation but move in different directions, refinement vs relativization of objectivity. Third, it entered into dialogue with Barad's agential realism: they share a common foundation but differ in scope, focusing on physical interaction vs extending to discourse and politics. Fourth, it clarified the specific position of constrained speculation within the spectrum of the problem of objectivity: constrained process realism.

The next section is the conclusion of this chapter.

14.6. Chapter Summary: Constrained Process Realism

Completion of the Three Tasks

The three tasks undertaken by this chapter have all been completed. Task One: to diagnose the difficulties of the traditional concept of objectivity. Section 14.1 developed the form of traditional objectivity, including observer–object separation, disturbance-free readout, perspective-independence, and ontological priority, and the three challenges from contemporary physics: quantum measurement, cosmological internality, and neuroscientific concretization. It also pointed out the difficulties of three traditional responses: technical correction, relativism, and transcendental grounding. Task Two: to propose reproducible co-perturbation as a new foundation of objectivity. Section 14.2 stated the core thesis OB1, objectivity = controlled reproducible co-perturbation, and developed its four implications. Section 14.3 showed the application of OB1 through three domains: particle physics, cosmology, and neuroscience. Task Three: to enter into dialogue with contemporary physical-philosophical traditions. Section 14.4 entered into serious dialogue with Heisenberg, Rovelli, and Barad, and clarified the specific position of constrained speculation within the spectrum of the problem of objectivity.

The Position of This Chapter in Part III

The overall structure of Part III, "Co-Evolution, Mind–Matter, and Observation: The Unity of Three Relations," includes: Chapter 12 on co-evolution as an ontological relation, the foundational chapter; Chapter 13 on mind–matter unity, the application of co-evolution to the mind–matter relation; Chapter 14, this chapter, on observation as participation, the application of co-evolution to the observational relation; and Chapter 15 on time, duration, and co-evolution, the development of co-evolution in the temporal dimension. As the third core chapter of Part III, this chapter is parallel to Chapter 13. Together, the two chapters show the two most important applications of co-evolution as an ontological relation: the mind–matter relation and the observational relation.

The Position of This Chapter in the Whole Book

Looking backward, this chapter builds upon several chapters: Chapter 12 on co-evolution as an ontological relation, which provides the ontological foundation of this chapter; Chapter 3 on failure conditions as an ontological discipline, which provides the methodological foundation for the concept of reproducibility in this chapter; Chapter 4 on constitutive coextension, which provides the semantic tool for treating the relation between this chapter and traditional concepts; and Chapter 11 on freezing as a condition of visibility, especially

Section 11.5, which provides the ontological foundation for the possibility of observation in this chapter. Looking forward, this chapter provides resources for later chapters: Chapter 17 on ontological singularity, where this chapter's claim that "observation is built upon co-evolution" provides an ontological ground for "the collapse of co-evolution before the terminal phase"; and Chapter 21 on the finitude of meaning, where the new foundation of objectivity established in this chapter, within co-evolution, provides a ground for meaning as a phenomenon within finite existence.

Constrained Process Realism

Through the work of this chapter, the position of constrained speculation on the problem of objectivity can be named as constrained process realism. This position has the following content: realism, since objective knowledge is attainable and some observations have assessable objectivity; processuality, since objectivity does not depend on static entities, the object independent of the observer, but on process, the reproducibility of co-perturbation; and constraint, since objectivity must satisfy falsifiable standards, including controlled reproducibility, mechanism correspondence, cross-observer agreement, and so on. Not every claim is objective, but objectivity can be rigorously assessed.

This threefold positioning makes the position of this chapter different from traditional positions: from traditional realism, because it does not posit an object independent of the observer, but preserves objectivity; from relativism, because it does not give up objectivity, but acknowledges that objectivity is processual; from instrumentalism, because it does not reduce objectivity to instrumental usefulness, but claims that reproducibility is an ontological phenomenon; from constructivism, because it does not claim that objectivity is socially constructed, but that it is grounded in physical co-perturbation. Constrained process realism is a fifth position, finding its place among the four traditional positions. This position is not a "compromise," but an independent position grounded in the philosophical work of this book: the method of constrained speculation, co-evolutionary ontology, and process ontology.

Honesty about the Limits of This Chapter

This chapter acknowledges several limits. First, dialogue with the epistemological tradition is insufficient. This chapter mainly enters into dialogue with the physics-philosophy tradition, and its dialogue with broader epistemological traditions, such as Popper, Kuhn, Quine,

Bayesianism, and others, is relatively brief. Each dialogue requires dedicated treatment. Second, the method for assessing the degree of objectivity. This chapter claims that objectivity is a matter of degree, but does not develop in detail how to assess the degree of objectivity precisely. This method of assessment requires further work, especially in applications to philosophy of science. Third, applications in non-scientific domains. This chapter mainly treats the objectivity of scientific observation. Applications in non-scientific domains, such as everyday perception, artistic appreciation, ethical judgment, and so on, are only briefly indicated and require dedicated work. Fourth, connection with scientific practice. This chapter develops the concept of objectivity at the philosophical level. Its connection with scientific practice, including experimental design, data analysis, peer review, replication crises, and so on, requires further work. Acknowledging these limits does not weaken the work of this chapter. They make it an honest initial work and invite further development.

The Core Posture of This Chapter

The core posture of this chapter can be stated as follows: objectivity is real, but it is not observerless.

Since the twentieth century, the concept of "objectivity" has been questioned under multiple pressures: discoveries in physics, philosophical critique, and political-ethical challenges. A common response is to give up objectivity and claim that all knowledge is perspectival, cultural, or political. This chapter rejects that response. Objectivity is real. The achievements of science, including precise prediction, repeatable experiments, cross-observer agreement, and technological applications, cannot be simply reduced to "political construction" or "cultural product." These achievements have an ontological-epistemological ground. But objectivity is not traditional objectivity: it does not require the observer to "withdraw" from the observed world. Such withdrawal has been shown to be impossible by developments in contemporary physics and epistemology. The response proposed by constrained speculation is refinement rather than abandonment. Objectivity = controlled reproducible co-perturbation. This refinement preserves the core content of objectivity, namely the achievements of science, while concretizing it in a form consistent with contemporary physics and epistemology. This refinement is philosophical progress: it allows us to speak of objectivity in the twenty-first century without returning to the nineteenth-century traditional form or falling into the relativist trap of the late twentieth century.

Meaning before freezing: this is one of the core theses of this book. This chapter provides an epistemological foundation for that thesis. Meaning appears in the process of co-evolution, and the content of meaning is concretized by co-evolutionary structure. When Chapter 21 argues that meaning is true only within finitude, "finitude" is the finitude of co-evolution: it depends on concrete observer–observed relations, and these relations dissolve before the gradient terminal phase. The new foundation of objectivity in this chapter is the epistemological link in this connection.

14.7. Methodological Reflection on This Chapter

This chapter re-grounds objectivity from "disturbance-free externality" to "controlled reproducibility." This section reflects on the methodological significance of this re-grounding.

This Chapter's Treatment of Terminology

"Objectivity" is a term heavily burdened by history. This chapter does not invent a new term, such as "objectivity*" or "reproducible objectivity," but refines the content of the existing term. This refinement allows the work of this chapter to enter directly into dialogue with existing discussions and avoids terminological proliferation. But refinement requires clarity: the "objectivity" of this chapter differs from the traditional one. Through definition (OB1) and applications in three domains, this chapter makes the content of this refinement clear. This treatment of inheritance + refinement is a common pattern of philosophical progress.

This Chapter's Respect for Scientific Practice

The work of this chapter is built upon scientific practice in particle physics, cosmology, and neuroscience. This respect is not rhetorical. This chapter uses examples from these sciences, including the Higgs discovery, CMB observation, and fMRI research, to develop its argument. This makes the work of this chapter not abstract philosophical conjecture, but work connected with scientific practice. The Higgs discovery involves experimental setup, data acquisition, and statistical analysis, all of which are applications of reproducible co-perturbation. CMB observation and fMRI research are likewise applications of reproducible co-perturbation.

Differences between This Chapter and Three Neighboring Positions

The work of this chapter differs from three neighboring contemporary positions. First, it differs from Heisenberg's participation view: Heisenberg's work shows the participatory character of observation, but remains at the level of physics. This chapter systematizes it as an application of relational ontology. Second, it differs from Rovelli's relational quantum mechanics: Rovelli's work relativizes objectivity. This chapter refines objectivity rather than relativizing it. Third, it differs from Barad's agential realism: Barad's work has political-ethical motivations. The work of this chapter is built within the framework of philosophy of nature and methodological discipline, and is restrained with respect to political-ethical implications.

Limits Honestly Acknowledged in This Chapter

This chapter acknowledges several limits. First, reproducible co-perturbation as a new foundation of objectivity requires later refinement: its applications in other scientific domains require dedicated work. Second, this chapter develops only the core direction of the connection between OB1 and scientific methodology. Third, the application of OB1 in cross-cultural contexts requires dedicated work. Fourth, this chapter does not sufficiently develop the relation between OB1 and discussions in epistemology.

Connections with Later Chapters

OB1 established in this chapter is applied in later chapters. Chapter 15 uses OB1 to treat the observation of time. Chapter 17 uses OB1 to treat knowledge at ontological boundaries. Chapter 20 uses OB1 to treat the epistemological ground of value statements. More deeply, the problem of objectivity is an interface problem between science and philosophy: how to preserve the concept of objectivity while acknowledging the challenges of contemporary physics and neuroscience. This chapter is the work of this book at that interface. It gives the book a complete epistemological dimension.

15. Time, Duration, and Co-Evolution

15.1. Introduction to This Chapter

"Time" plays different roles in different traditions, and these roles are often confused. In Newtonian physics, time is a parameter for describing the state of a system; in empirical physical research, time is a measurable quantity; in ordinary language, time is the sequence of events; in phenomenology, especially in Bergson, time is the inner experience of duration; and in certain metaphysical traditions, time is a dimension of being. These five meanings are often confused in philosophical discussion, producing conceptual confusion.

This chapter repositions the problem of time as an ontological problem: time is not an externally added coordinate parameter, but an intrinsic dimension of processual ontology. Specifically, time = $\partial\rho/\partial t$ (TM1), the intrinsic rate of change of the energy quantum density field. This positioning makes the other meanings of time, including coordinate, measure, sequence, and duration, all grounded in this foundational meaning: they are expressions of ontological time in different contexts.

This ontological positioning involves a philosophical challenge: the famous 1922 debate between Bergson and Einstein. Bergson held that time as duration (durée) is foundational: it is the flow of inner experience in consciousness and cannot be spatialized into a measurable quantity. Einstein held that time as a physical quantity is foundational: through relativity, it is closely intertwined with space and constitutes spacetime. The two did not reach agreement, and the debate ended in mutual incomprehension.

This chapter offers a reconciliation within the framework of co-evolution, preserving the core insights of both. Specifically, Bergsonian duration and Einsteinian physical time are different projections of the same ontological time (TM2). They are not mutually contradictory views of time, but expressions of foundational ontological time,

$\partial\rho/\partial t$, from different perspectives. Bergson's duration is ontological time seen from the consciousness–subject perspective; Einstein's physical time is ontological time seen from the physics–description perspective.

This chapter is the concluding chapter of Part III. The previous three chapters established the content of relational ontology: Chapter 12 laid the foundation, co-evolution as an ontological relation; Chapter 13 applied it to mind and matter; and Chapter 14 applied it to observation. This chapter treats time as the fourth dimension: co-evolution in the temporal dimension. It undertakes a special task of philosophical–physical connection: the problem of time has been central in both twentieth-century philosophy and physics, but the two traditions have often not been adequately connected. This chapter attempts to establish such a connection within the co-evolutionary framework.

The discussion proceeds as follows. Section 15.1 develops the starting point of time as an ontological problem. Section 15.2 states and defends time = $\partial\rho/\partial t$ (TM1). Section 15.3 treats the direction of time and the irreversibility of freezing (TM2). Section 15.4 develops the reconciliation between Bergson and Einstein. Section 15.5 develops the ontological status of the "now" (TM3) and responds to objections.

15.2. Time as an Ontological Problem

The Multiple Identities of Time

"Time" plays different roles in different traditions, and these roles are often confused. This section first makes precise conceptual distinctions.

Time as coordinate parameter: in Newtonian physics and classical mechanics, time t is a parameter for describing the state of a system: the state $q(t)$ is a function of t. Time here is external: it is a tool for describing the system, not a property of the system itself. Time as measure: in physics, "time" also refers to a measurable quantity, identified through atomic clocks, oscillators, relative time intervals, and so on. Time in this sense is empirical: it is identified through measurement. Time as sequence: in ordinary language, "time" often refers to the sequence of events, such as the sequence of yesterday, today, and tomorrow. Time in this sense is ordinal: it concerns the relative order among events. Time as duration: in phenomenology, especially in Bergson, "time" refers to the inner experience of duration, the feeling of continuous flow in consciousness. Time in this sense is experiential: it concerns the inner feeling of the subject. Time as ontological dimension: in certain metaphysical traditions, "time" is a dimension of being; being itself has a temporal structure. Time in this sense is ontological: it concerns the mode of being.

These five meanings are often confused in philosophical discussion, producing conceptual confusion. The starting point of this chapter is to distinguish these meanings precisely, and then to focus on time as an ontological dimension.

The Ontological Focus of This Chapter

This chapter is concerned with time as an ontological dimension. The other meanings, coordinate parameter, measure, sequence, and duration, are all grounded in this foundational meaning: they are expressions of ontological time in different contexts. Specifically: the coordinate parameter t is the use of ontological time in physical description; measurement is the identifiability of ontological time through physical processes; sequence is the expression of the irreversibility of ontological time at the level of events; and duration is the manifestation of ontological time in conscious experience. To unify these five meanings under ontological time requires an ontological thesis. This is the work of Section 15.2.

Traditional Difficulties of the Problem of Time

Time as an ontological problem involves difficulties that recur throughout the history of philosophy.

Difficulty One: is time real? In 1908, McTaggart proposed his famous argument for the "unreality of time." He distinguished the A-series of time, past–present–future, from the B-series, earlier than–later than. The A-series involves the "present," while the B-series does not. McTaggart argued that the A-series is internally contradictory, since the same event cannot be both past and present, while the B-series is insufficient to constitute real time; therefore, time is unreal. There have been many responses to this argument, but the fundamental difficulty it identifies is this: can time as an ontological phenomenon be stated precisely? Time seems to involve "flow," "change," and "now," and these concepts are difficult to state precisely.

Difficulty Two: the ontological ground of the direction of time. Most fundamental laws of physics are time-symmetric: they are invariant under $t \to -t$. But macroscopic phenomena have an obvious direction of time, including the second law of thermodynamics, cosmic evolution, and life processes. Where does the direction of time come from? This is an ontological problem. Possible responses derive it from initial conditions, such as a low-entropy beginning of the universe; from the measurement process, such as the irreversibility of quantum measurement; from consciousness, such as the feeling of "flow" in consciousness; and so on. Each response has difficulties.

Difficulty Three: the ontological status of the now. The "now" has a special character in our experience: it is alive, while past and future are not. But there is no "now" in the description of physics: all moments have equal status in physical equations. What is the ontological status of the "now"? Is it a feeling of the observer? Is it some cosmological property? Is it an ontological dimension? This is an ongoing debate.

Difficulty Four: the relation between time and consciousness. Temporal experience involves consciousness: in the feeling of consciousness, time appears as flow. But does time depend on consciousness? Is there time in a world without consciousness, if such a world exists? If time depends on consciousness, then the relation between physical time as an objective quantity and experiential time as a subjective quantity must be treated. If time does not depend on consciousness, then what is temporal experience in consciousness?

These four difficulties together constitute the difficulty of time as an ontological problem. The work of this chapter must directly treat these difficulties rather than avoid them.

The Relevance of the Co-Evolutionary Framework

The framework established in Chapter 12, co-evolution as an ontological relation, is relevant to the problem of time. The core of the co-evolutionary framework is this: relational terms do not first exist independently and then enter into relation; relation is an intrinsic structure of the process field. Time as an ontological dimension can be treated anew within this framework: time is not an externally added coordinate parameter, independent of terms, but an intrinsic structure of the process field, co-evolving with terms; the direction of time comes from the irreversibility of co-evolution; temporal measurement comes from frequency matching in co-evolution; and temporal experience comes from the form of conscious co-evolution. These responses are the work of Section 15.2. This section merely indicates the direction.

Objections and Replies

Objection One: The "five meanings of time" you list seem overly artificial. In ordinary language, "time" is a unified concept and does not need so many distinctions.

Reply: The unified concept of "time" in ordinary language often produces confusion in philosophical and physical discussion. For example, when a physicist says "time is a coordinate parameter" and a philosopher says "time is duration," they use the same term but refer to different things. This confusion makes dialogue between them difficult. Precise distinction is not excessive complication; it is a basic requirement of philosophical rigor. Each meaning has its legitimacy, but their relations must be stated clearly. The distinctions in this section allow the following discussion to proceed precisely: it identifies the meaning with which this chapter is concerned, namely ontological dimension, while acknowledging the legitimacy of other meanings.

Objection Two: Your "time as ontological dimension" seems too abstract. It does not directly correspond to any physical or ordinary meaning.

Reply: Ontological dimension as foundational meaning is not "another" use. It is the ground of other uses. Coordinate parameter, measure, sequence, and duration are all expressions of ontological time in different contexts. Ontological time is not "abstract," but its concreteness lies at a more foundational level. It is the ontological ground of the other meanings of time, just as the ontological dimension of space is the ground of everyday spatial sense, measured space, and geometrical space. Section 15.2 will concretize ontological time through "time = $\partial\rho/\partial t$," making it not abstract but ontologically structured.

Objection Three: Your four difficulties are too broad. This chapter cannot solve all of them within limited space. Is this not overpromising?

Reply: This chapter does not claim to solve all these difficulties, but to provide directions of treatment within the co-evolutionary framework. Specifically: whether time is real, in the co-evolutionary framework time is a real ontological dimension, as argued in Section 15.2; the direction of time comes from the irreversibility of freezing, as argued in Section 15.3; the ontological status of the now is that the now is the current state of the process field, not a special ontological unit, as argued in Section 15.5; and the relation between time and consciousness is that conscious time is the expression of physical time in co-evolution, as argued in Section 15.4. These treatments do not "finally solve" the difficulties; they provide responses that can be further tested and refined. The method of constrained speculation does not require "final solution," but requires criticizable responses. This chapter satisfies that requirement.

Literature That Must Be Engaged

- J. M. E. McTaggart, "The Unreality of Time" (1908)
- Henri Bergson, Time and Free Will (1889), Matter and Memory (1896)
- Albert Einstein, Relativity: The Special and the General Theory (1916)
- Hermann Bondi, Cosmology (1952)
- Huw Price, Time's Arrow and Archimedes' Point (1996)
- Chapter 12 on co-evolution as an ontological relation (self-reference)

Summary of This Section

This section has done four things. First, it distinguished five meanings of time: coordinate parameter, measure, sequence, duration, and ontological dimension. Second, it clarified the ontological focus of this chapter. Third, it identified four traditional difficulties of time as an ontological problem. Fourth, it previewed the relevance of the co-evolutionary framework to these difficulties.

The next section develops the core thesis: time as $\partial\rho/\partial t$.

15.3. Time as $\partial\rho/\partial t$: The Core Thesis

Statement of the Core Thesis

Core thesis TM1: time as an ontological dimension = the intrinsic rate of change of the energy quantum density field, $\partial\rho/\partial t$. Time is not an externally added coordinate parameter, independent of the field, but an intrinsic structure of the field, the direction of change of the field itself.

This thesis requires detailed development. "The intrinsic rate of change of the energy quantum density field": time is not a quantity "outside the field," but the direction in which the field itself changes. When ρ changes at a field point x, the "rate" of this change, the amount of change per unit time, is the ontological content of time.

"Intrinsic structure": $\partial\rho/\partial t$ is not "added to" the field; it is the intrinsic structure of the field. Any statement of the ρ field already implies a temporal dimension, because to state the state of ρ is to state its direction of change. "The direction of change of the field itself": the direction of time, from past to present to future, appears from the direction of change of the field. Some changes are "forward," increasing certain aspects of ρ; some changes are "backward," reducing certain aspects of ρ. This directionality gives content to time.

Implications of the Thesis

TM1 has four ontological implications.

Implication One: Time cannot be detached from the field.

Traditional physics often treats time as an independent coordinate: there exists an independent "time axis," and things evolve along this time axis. TM1 rejects this image. Time is not independent; it is an intrinsic structure of the field. This implication is structurally consistent with the insight of general relativity. In general relativity, time is not a preexisting container: it is a structure of spacetime geometry and is coupled to the distribution of matter and energy. TM1 carries this insight further: time is an intrinsic structure of the ρ_f field. Specifically: without the ρ_f field there is no time. In a vacuum with $\rho \equiv 0$, not physical vacuum but absolute "nothing," there is no time, because there is no change. Different ρ_f configurations may produce different temporal structures, corresponding to the fact that different spacetime configurations in general relativity have different times.

Implication Two: The direction of time comes from the dynamics of the field.

TM1 locates the direction of time in the dynamics of the field: $\partial\rho/\partial t$ has different directions in different situations, increasing or decreasing certain components of ρ. These directions are not arbitrary, but are determined by the dynamics of the field, the A3 Master Equation. This implication is connected with Chapter 11's work on freezing as the direction of time, especially Sub-Thesis Two in Section 11.2. Freezing is an irreversible process; it gives content to the direction of time. The next section, Section 15.3, develops this connection.

Implication Three: Temporal measurement comes from the oscillation of the field.

In physics, temporal measurement is based on oscillations, such as atomic clocks and oscillators. Within the framework of TM1, these oscillations are oscillations of the ρ_f field: they provide temporal scales. Specifically: atomic clocks use the frequency of atomic energy-level transitions, which is an oscillation of the ρ_f field inside atoms; different oscillations, involving different atoms and different frequencies, provide different precisions of temporal measurement; and the standard of temporal measurement is negotiated by human communities on the basis of field oscillations. This implication keeps temporal measurement from being detached from the field: measurement is the usable form of field oscillation.

Implication Four: Time is the dimension of co-evolution.

Time as the intrinsic structure of the field makes time the dimension of co-evolution. Chapter 12 has already established co-evolution as an ontological relation. Specifically: all co-evolutionary relations occur in time. Co-evolution is process, and process has a temporal dimension. Time is not an additional dimension outside co-evolutionary relations, but an intrinsic structure of co-evolution. Different modes of co-evolution, including different frequency matching, different thresholds, and different tri-state weights, correspond to different temporal structures. This implication gives the ontological work of Part III, co-evolution, mind–matter, observation, and time, a complete internal unity: they are all different dimensions of the same relational ontology.

The Content of Time as $\partial\rho/\partial t$

The $\partial\rho/\partial t$ in TM1 is not an abstract "rate of change": it has content. First, it is a field operator: $\partial/\partial t$, as an operator acting on $\rho_f(x,t)$, produces a new field quantity. This new field quantity has a value at every spacetime point (x,t). Second, it is a vector in a certain sense: at every spacetime point, $\partial\rho/\partial t$ has direction, increase or decrease, and intensity. Third, it is connected with the Master Equation: the A3 Master Equation states how $\partial\rho/\partial t$ depends on ρ itself, including processes such as diffusion, nonlinear feedback, dissipation, and source terms. This connection makes $\partial\rho/\partial t$ non-arbitrary: it has a dynamical law. Fourth, it has different manifestations at different scales and frequency bands: high-frequency $\partial\rho/\partial t$ involves rapid change, such as at the level of elementary particles; low-frequency $\partial\rho/\partial t$ involves slow change, such as at the cosmological level. The temporal structures of different scales are concretized by these different manifestations. These contents make $\partial\rho/\partial t$ in TM1 non-abstract: it has physical and ontological structure.

Objections and Replies

Objection One: TM1 seems to use a circular definition: time =

$\partial\rho/\partial t$, but $\partial\rho/\partial t$ already presupposes time in $\partial/\partial t$.

Reply: This objection points to an important issue: any ontological thesis about time faces a similar challenge. The reply of this section has two layers. At the formal level, TM1 is not a definitional circle; it is a statement of ontological identity. Time as an ontological dimension $\equiv \partial\rho/\partial t$ as the intrinsic structure of the field. As concepts, the two have different contents: time as an ontological dimension involves phenomenological and metaphysical dimensions, while $\partial\rho/\partial t$ involves a field operator. But they refer to the same ontological content. Such statements of ontological identity are philosophically legitimate, similar to the statement "water = H_2O," where the latter chemically concretizes the former, while the former as a concept still has independent content. At a deeper level, however, it is true that the operator $\partial/\partial t$ in $\partial\rho/\partial t$ presupposes some foundational temporal structure; otherwise, "rate" could not be defined. This reflects the foundational status of time as an ontological dimension: it cannot be reduced to a non-temporal structure. But this irreducibility is not a defect; it is precisely the feature of an ontological dimension. As a foundational ontological dimension, time, like other foundational dimensions such as space, cannot be reduced to a more basic non-temporal structure. This irreducibility is the ontological feature of a foundational dimension.

Objection Two: Your thesis seems to conflict with general relativity. In general relativity, time is a dimension of spacetime, not merely the intrinsic rate of change of a field.

Reply: The work of this chapter does not conflict with general relativity. TM1 can be connected with general relativity, although the concrete connection requires dedicated work. Specifically: in general relativity, time and space together constitute spacetime geometry and are coupled to matter and energy. In the TM1 framework, time is the intrinsic rate of change of the ρ_f field. This is structurally consistent with general relativity's "coupling between time and matter." The concrete connection requires developing the relation between the ρ_f field and spacetime geometry. This is the work of connecting EQT with general relativity, treated in part in *Freezing Cosmology*. This section does not develop the detailed connection; it indicates the direction. TM1 does not compete with general relativity. It is an extension of the insight of general relativity within the EQT framework.

Objection Three: Your "time as the intrinsic structure of the field" seems to mystify time. In traditional physics, time is a simple parameter, while this section makes it ontologically complicated.

Reply: The "time as a simple parameter" of traditional physics in fact implies an ontological simplification: it treats time as an externally added structure and does not treat the ontological ground of time. This simplification works in Newtonian physics, but faces difficulties in quantum mechanics, general relativity, and cosmology, such as the time–energy uncertainty relation, spacetime singularities, and the beginning of the universe. TM1 does not "mystify" time. It concretizes the ontological ground of time. This concretization allows time to remain ontologically reasonable in the face of the challenges of contemporary physics, without retreating to the simplified time of Newtonian physics. More deeply, ontological complexity is not a defect; it is a requirement of contemporary physics and philosophy. Any contemporary philosophy of time must face the content of quantum mechanics and general relativity; TM1 satisfies this requirement.

Literature That Must Be Engaged

- Albert Einstein, "Zur Elektrodynamik bewegter Körper" (1905): special relativity
- Albert Einstein, "Die Grundlage der allgemeinen Relativitätstheorie" (1916)
- Henri Poincaré, La valeur de la science (1905)
- Hermann Weyl, Space, Time, Matter (1922)
- Carlo Rovelli, The Order of Time (2018)
- Chapter 11 on freezing as a foundational ontological process (self-reference)

Summary of This Section

This section has done four things. First, it stated the core thesis TM1: time as an ontological dimension $= \partial\rho/\partial t$. Second, it developed the four implications of the thesis: time cannot be detached from the field; the direction of time comes from the dynamics of the field; temporal measurement comes from field oscillation; and time is the dimension of co-evolution. Third, it developed the content of $\partial\rho/\partial t$: field operator, vector, connection with the Master Equation, and different manifestations at different scales. Fourth, it responded to possible objections, especially concerning circular definition and the relation with general relativity.

The next section develops the connection between the direction of time and freezing.

15.4. The Direction of Time and the Irreversibility of Freezing

The Direction of Time as a Problem

The direction of time, past → present → future, is a fundamental problem in philosophy and physics. Its difficulty comes from the symmetry of the fundamental laws of physics: most fundamental laws are time-symmetric, invariant under $t \to -t$. But macroscopic phenomena have an obvious direction of time, including entropy increase, cosmic evolution, and the development of life. This difficulty, the "problem of the direction of time" or the "problem of the arrow of time," has generated extensive discussion since the twentieth century. Common responses include: the thermodynamic arrow, where the direction of time = the direction of entropy increase; the cosmological arrow, where the direction of time = the direction of cosmic expansion; the radiation arrow, where the direction of time = the direction in which radiation propagates outward from a source; the causal arrow, where the direction of time = the direction in which causes produce effects; and the memory arrow, where the direction of time = the direction in which we remember the past but not the future. Each response involves specific physical or psychological phenomena, but their ontological ground requires further refinement.

Thesis: The Direction of Time = The Direction of Freezing

Chapter 11 has already stated in Sub-Thesis Two: the direction of time = the direction of freezing (Section 11.2). This section develops the ontological content of this connection.

Thesis TM2: the direction of time, the directionality from past to future, comes from the irreversibility of the freezing process.

Let us develop this in detail.

Freezing as an Irreversible Process

Chapter 11 established freezing as a foundational ontological process: certain processes retreat from dominant dynamics into residual, background, or history. These processes of exit are, as a whole, irreversible. Once a process has frozen, reversing it requires a new process and additional energy. Specifically: quark confinement, about 10^{-6} seconds after the Big Bang, where once quarks and gluons freeze from a free-state plasma into hadrons, the bound state, reversal requires extremely high energy and the re-creation of early-universe conditions; under normal conditions this freezing is irreversible. Photon decoupling, about 380,000 years after the Big Bang, where once photons freeze from the bound state, strongly interacting with plasma, into the free state, the CMB, reversal requires re-creating the early high-density plasma; this freezing is irreversible in normal cosmic evolution. Nucleosynthesis, whose products, heavy elements, once formed, require extremely high energy to reverse; partial reversal may occur in stellar evolution, but the overall tendency is irreversible. Each freezing process provides a direction of time: the direction from unfreezing to freezing is the direction of time.

The Unity of the Direction of Freezing

But different freezing processes are different: they involve different frequency bands, different dynamics, and different timescales. How do they unify into a single direction of time? TM2 responds as follows: all freezing processes share a common direction at the ontological level: from higher active-dynamical weight to lower active-dynamical weight, or equivalently, from a freer state to a more stable state, at the level of the tri-state categories, from free state / bound state toward condensed state as the overall direction. This common direction is an ontological feature of freezing as a foundational ontological process. It provides a unified direction of time for all freezing processes. The directions of all freezing processes are consistent at the ontological level. This unity does not mean that all freezing processes are synchronous. They have different timescales, domains, and details. But they are consistent in ontological direction: from dynamic to stable. This is the unified ontological ground of the direction of time.

Comparison with Traditional Arrows of Time

TM2 is related to traditional discussions of the arrow of time. Relation to the thermodynamic arrow: entropy increase is an expression of freezing. Entropy increase involves energy freezing from usable forms, ordered and capable of doing work, into unusable forms, disordered and thermal. Thus the thermodynamic arrow is one aspect of freezing. TM2 treats the thermodynamic arrow as an expression of freezing and gives it an ontological ground. Relation to the cosmological arrow: cosmic expansion involves multiple freezing processes, from repeated freezing events in the early high-density state to the current state.

Thus the cosmological arrow is also an expression of freezing. But the cosmological arrow is not merely expansion itself; more deeply, it is the irreversibility of cosmic evolution, and this irreversibility comes from the accumulation of freezing. Relation to the radiation arrow: radiation propagating outward from a source is a form of freezing: energy freezes from the bound state / condensed state of the source into a freely propagating free state. The radiation arrow is the directional expression of this freezing. Relation to the causal arrow: causality, where cause precedes effect, comes from the irreversibility of freezing. A concrete cause, a being with a concrete configuration, produces an effect through the freezing process. Once the effect has frozen, it cannot act back upon the cause. This is the ontological ground of the causal arrow. Relation to the memory arrow: memory as the freezing of consciousness, Sub-Thesis Four in Chapter 11, Section 11.2, involves freezing from instantaneous neural activity into stable neural structure. We remember the past but not the future because the past, through freezing, has become a stable structure accessible to present consciousness, while the future has not yet undergone such freezing. These comparisons show TM2 as the unified ontological ground of the direction of time: the five traditional arrows of time are all expressions of freezing. TM2 does not cancel these expressions; it provides them with a unified ontological ground.

Temporal Asymmetry as an Ontological Fact

TM2 locates the direction of time as an ontological fact: it is grounded in the irreversibility of freezing as a foundational ontological process, not in some "externally added" or "conventional" direction. This positioning has philosophical implications: the direction of time is real, not conventional, since the irreversibility of freezing does not depend on the observer's choice; the direction of time has objective content, since it is identified through observable forms of freezing, such as the CMB, cosmic evolution, and life processes; and the direction of time has causal efficacy, since it affects the possibilities of subsequent processes, what can happen and what cannot. This positioning avoids treating the direction of time as merely a product of human psychology, as if "the direction of time were only our perception," while also avoiding treating the direction of time as an inexplicable primitive physical fact, as if "time direction is simply so." It treats the direction of time as an ontological consequence of freezing.

Objections and Replies

Objection One: The fundamental laws of physics are time-symmetric.

This means that at the foundational level there is no direction of time. Your "freezing as irreversible" seems to be grounded at the statistical level, not the foundational level. Does this weaken the ontological status of TM2?

Reply: This objection points to the core of the problem of the direction of time. The reply of this section has two layers. The level problem: yes, the fundamental physical laws are time-symmetric, and the irreversibility of freezing appears mainly at the statistical / macroscopic level. But this does not weaken TM2; it makes the ontological ground of TM2 more precise. Specifically: the direction of time does not appear at the foundational physical level, where the dynamics of elementary particles is time-symmetric; the direction of time appears at the statistical / macroscopic level, where the statistical accumulation of many foundational processes produces irreversibility. This "statistical irreversibility" is a real ontological phenomenon: it has observable consequences, including entropy increase, cosmic evolution, and life processes. TM2 locates the direction of time at this statistical level, acknowledging that it is not the direct consequence of fundamental physical laws, but a statistically emergent phenomenon. This positioning is consistent with physics and avoids unnecessary controversy. More deeply, however, there remains a problem: why were the initial conditions low-entropy? Low-entropy initial conditions are the ground of entropy increase, the main form of freezing irreversibility. If the initial condition were high-entropy, namely a heat-death state, there would be no irreversibility of freezing, and the direction of time would disappear. This problem is open in contemporary cosmology: why was the early universe after the Big Bang low-entropy? Possible responses include the anthropic principle, since only low-entropy initial conditions allow observers to exist; quantum cosmology, where a particular quantum state functions as the initial condition; and the multiverse, where our low-entropy universe is a subset within a multiverse. This chapter does not attempt to solve this open problem. It points out that TM2 is related to this problem and acknowledges that the work of this chapter is built upon the empirical fact that the universe has a low-entropy beginning. This is an honest limitation.

Objection Two: Your "unity of the direction of freezing" seems too strong. You claim that all freezing processes have a common direction at the ontological level. Is this an overgeneralization?

Reply: The "unity" claim in this section requires refinement. It does not mean that all freezing processes are formally identical: they each have concrete frequency bands, mechanisms, and timescales. It means that they are consistent in ontological direction: the overall direction from dynamic to stable. The content of this ontological direction is as follows: in tri-state weights, from higher free-state / bound-state weight to higher condensed-state weight in most freezing processes; in energy structure, from higher free energy to lower free energy in most freezing processes; in information structure, from higher Shannon entropy, more possible microstates, to lower effective information, more determinate macrostates. These contents make "consistency of ontological direction" not an empty statement; it has content. But it is indeed consistency at the ontological level, not formal consistency. This distinction makes the claim precise and avoids overgeneralization. Some freezing processes may locally run against the overall direction, as in some life processes that temporarily reverse by using external energy, but the total system including all relevant couplings still proceeds according to the ontological direction. This is consistent with the second law of thermodynamics, where local entropy decrease requires compensation by entropy increase elsewhere.

Objection Three: TM2 reduces the direction of time to freezing. But the memory arrow seems to involve features of consciousness, since we remember the past, not merely physical freezing. Is this oversimplified?

Reply: The memory arrow does involve features of consciousness, but the response of this chapter is that the formation of conscious memory is also a freezing process, as stated in Sub-Thesis Four of Chapter 11, Section 11.2. Conscious activity freezes from instantaneous flow into stable neural structure. The content of the memory arrow, that we remember the past but not the future, is the expression of freezing at the conscious level: only the already-frozen past becomes a stable structure accessible to current consciousness. The future has not yet frozen, and therefore cannot be remembered. This treatment does not simplify the memory arrow; it provides an ontological ground for it. The features of consciousness, including memory, are built upon physical processes but have their own level-specific features. Chapter 13 of this book treats the relation between consciousness and physics in detail, through mind–matter unity. That relation includes freezing processes at the level of consciousness.

Literature That Must Be Engaged

- Hans Reichenbach, The Direction of Time (1956)
- Huw Price, Time's Arrow and Archimedes' Point (1996)
- Sean Carroll, From Eternity to Here (2010)
- David Albert, Time and Chance (2000)
- Roger Penrose, The Road to Reality (2004), Chapter 27
- Chapter 11 on freezing as a foundational ontological process (self-reference)

Summary of This Section

This section has done four things. First, it identified the direction of time as a problem and reviewed its traditional responses. Second, it stated TM2: the direction of time = the direction of freezing. Third, it developed the unity of the direction of freezing, clarifying that this is a consistency at the ontological level rather than formal identity. Fourth, through comparison with the five traditional arrows of time, it showed TM2 as a unified ontological ground.

The next section develops the reconciliation between Bergson and Einstein.

15.5. Reconciling Bergson and Einstein

The 1922 Debate

In April 1922, Bergson and Einstein held a famous debate on time at a meeting of the French Philosophical Society. They had a fundamental disagreement over the problem of time. Bergson's position was that time as duration (durée) is the foundational ontology: it is the continuously flowing feeling in consciousness, indivisible, unquantifiable, and not spatializable. Physical time, Einstein's relativistic time, is only an abstraction and simplification of duration: it is derivative, not foundational. Einstein's position was that time is a physical concept: it is defined by the measurement of clocks and treated geometrically by relativity. The "duration" in consciousness is a psychological phenomenon, not foundational time: it is derivative and built upon physical time. The two did not reach agreement. This debate became a starting point for one of the central twentieth-century disputes about time. Bergson's work influenced phenomenology, process philosophy, and other traditions; Einstein's work became the foundation of modern physics. The division between the two traditions on the problem of time has continued to the present.

The Need for Reconciliation

The difference between Bergson and Einstein is not a simple matter of "who is right and who is wrong." Each grasped an aspect of time as an ontological phenomenon, but each treated the aspect he grasped as the sole foundation, leading to conflict. Specifically: Bergson grasped time as the inwardness of experience, the indivisibility of duration, the action of the past upon the present, and the future as open possibility. Einstein grasped time as a measurable physical quantity, the geometrical structure of relativity, objective coordinates across observers, and the coupling of time with space. If these two sets of insights can coexist within a more foundational ontological framework, with one grasping the experiential dimension of time and the other the physical dimension of time, then both insights can be preserved and the conflict can be overcome. This section develops such a reconciliation within the co-evolutionary framework.

The Core Idea of Reconciliation

The core idea of reconciliation is this: Bergsonian duration and Einsteinian physical time are different projections of the same ontological time. They are not two independent times, but manifestations of $\partial\rho/\partial t$ from different observational perspectives. Specifically: duration (Bergson) = the manifestation of $\partial\rho/\partial t$ inside a conscious system; physical time (Einstein) = the projection of $\partial\rho/\partial t$ under cross-observer co-evolution. Both are manifestations of $\partial\rho/\partial t$, but in different co-evolutionary relations they appear in different forms. This idea uses the framework of Chapter 13 on mind–matter unity, but applies it to the temporal dimension. Consciousness is not an entity different from physical processes, but a different projection of the same process. Similarly, duration is not a time different from physical time, but a different projection of the same temporal dimension.

Concrete Development

Let us develop this reconciliation in detail.

Duration as $\partial\rho/\partial t$ inside a conscious system. Consciousness, as a highly complex physical process already established in Chapter 13, has its own $\partial\rho/\partial t$: the rate of change of consciousness-related neural activity. This $\partial\rho/\partial t$ appears inside the conscious system as duration: the flow of experience, the action of the past upon the present, and the future as open possibility. Specifically: the flow of experience, where the continuity of conscious experience comes from the continuous $\partial\rho/\partial t$ of neural activity: there are no "instantaneous snapshots," only continuous change; the action of the past upon the present, where memory in consciousness makes the freezing of past neural activity, stable neural structures, affect current activity, thereby giving the past ontological action upon the present; and the future as open, where future neural activity has not yet occurred, and its form will be determined by the dynamics of $\partial\rho/\partial t$, including uncertainty. These are the ontological contents of duration: they are all manifestations of $\partial\rho/\partial t$ inside the conscious system. Bergson's insights are preserved: duration is real, unquantifiable in the first-person perspective, and closely related to consciousness. These insights are all preserved in the framework of this section.

Physical time as $\partial\rho/\partial t$ under cross-observer co-evolution. Physical time does not "independently exist" outside conscious systems. It is an objective temporal structure established among different observers through co-evolution. Specifically: cross-observer agreement, where different observers establish the objective structure of time through co-evolution, such as observing the same physical event and comparing their respective clock readings; the reproducibility of measurement, where the oscillation of atomic clocks can be reproduced by any observer, and this reproducibility makes temporal measurement objective; and the structure of relativity, where relativistic temporal structures, such as relativity of simultaneity and time dilation, arise from co-evolutionary relations among different inertial frames. These are the ontological contents of physical time: they are all projections of $\partial\rho/\partial t$ under cross-observer co-evolution. Einstein's insights are preserved: physical time is measurable, cross-observer, and closely related to physical processes. These insights are all preserved in the framework of this section.

The Relation between the Two Times

Duration and physical time, as different projections of the same ontological time, are related. Relation One: duration depends on physical processes. Duration is the manifestation of $\partial\rho/\partial t$ inside a conscious system, and the conscious system is a physical system. Thus duration depends on physical processes: without physical processes, such as neural activity, there is no duration. This relation supports Einstein's position: physical time is, in a certain sense, prior to duration as its ontological necessary condition. But this does not mean that physical time is "more real"; it only means that physical processes are the ontological ground of conscious experience. Relation Two: physical time obtains objective content through co-evolution. Physical time does not "independently exist" outside all consciousness: its objectivity comes from co-evolution, namely the comparison of physical processes among different observers. Thus physical time in a certain sense depends on conscious activity, especially measurement activity. This relation supports Bergson's position: physical time is not "purely objective"; it is built upon observational processes. But this does not mean that physical time is "subjective." Its objectivity is grounded in reproducible co-evolution, the new foundation of objectivity developed in Chapter 14. Relation Three: the two are unified at the ontological level. As different projections, both are built upon the same ontological time, $\partial\rho/\partial t$ as the intrinsic structure of the field. This ontological unity means that they are not opposed: they are different aspects of the same phenomenon. This unity is the core of the reconciliation. The opposition between Bergson and Einstein comes from treating different aspects as the sole foundation. The co-evolutionary framework allows both aspects to be acknowledged as different aspects of the same time, and the opposition is thereby overcome.

Objections and Replies

Objection One: Your reconciliation seems too conciliatory. You claim that the insights of both Bergson and Einstein are preserved, but their positions are in fact incompatible.

Reply: This objection points to an important issue. The philosophical positions of Bergson and Einstein are indeed incompatible: Bergson claims that duration is foundational, while Einstein claims that physical time is foundational. If we insist on choosing "which is more foundational," their positions cannot be reconciled. But the work of this section is not to choose "which is more foundational." It is to reorganize the problem so that "duration vs physical time" is no longer a foundational choice. In the co-evolutionary framework, both are built upon $\partial\rho/\partial t$, but they manifest in different forms within different co-evolutionary relations. This reorganization changes the problem: from "which is more foundational, duration or physical time?" to "duration and physical time are different projections of the same ontological time." In the transformed problem, the insights of both can coexist. This is not mere compromise; it is a change of framework. Framework transformation is a legitimate method in the history of philosophy, as in Kant's "Copernican revolution" or Wittgenstein's linguistic turn. The framework transformation in this section has a ground, the co-evolutionary ontology, and is not arbitrary compromise.

Objection Two: Your claim that "duration depends on physical processes" seems to disadvantage Bergson, since his position is that duration is foundational and does not depend on physical processes.

Reply: The treatment in this section does indeed revise Bergson's position in a certain sense: it makes duration depend on physical processes, whereas Bergson's position is that duration is independent of physical processes. This revision is necessary. The content of contemporary neuroscience, especially the correlation between consciousness and neural processes, makes it difficult to maintain that "duration is completely independent of physical processes." Any contemporary philosophy of time must face this content. But revision is not refutation. Bergson's core insights, including the experiential dimension of duration, its unquantifiability, the action of the past, and the openness of the future, are all preserved within the revised framework. What is modified is the ontological statement "duration as foundation," which is replaced by "duration as a projection of the same ontological time." This revision is constructive: it makes Bergson's insights compatible with contemporary science rather than simply denying them. This is the application of the method of constrained speculation: traditional insights can be used, but must be refined in light of contemporary content.

Objection Three: Your reconciliation depends on the co-evolutionary framework. If the co-evolutionary framework itself has problems, the whole reconciliation collapses.

Reply: Yes, the reconciliation in this section depends on the co-evolutionary framework, developed in Chapter 12. If the co-evolutionary framework does not hold, the reconciliation in this section must be reassessed. But this is not a defect: any philosophical work depends on some ontological framework. This section does not claim that "the reconciliation of Bergson and Einstein is the only possible one." It claims that "within the co-evolutionary framework, reconciliation is possible." More deeply, the legitimacy of the co-evolutionary framework has already been defended in Chapter 12. If some readers think the co-evolutionary framework is problematic, their criticism should first be directed at the work of Chapter 12. This section, as an application, is built on the basis of Chapter 12, and its legitimacy depends on the legitimacy of that foundation. Such dependence is legitimate in a philosophical monograph: each chapter builds upon the preceding chapters.

Literature That Must Be Engaged

- Henri Bergson, Duration and Simultaneity (1922): the publication related to the debate with Einstein
- Albert Einstein, records of the 1922 meeting of the French Philosophical Society
- Jimena Canales, The Physicist and the Philosopher: Einstein, Bergson, and the Debate That Changed Our Understanding of Time (2015)
- Milič Čapek, Bergson and Modern Physics (1971)
- Carlo Rovelli, The Order of Time (2018)
- Chapter 13 on mind–matter unity (self-reference)

Summary of This Section

This section has done five things. First, it reviewed the 1922 debate between Bergson and Einstein. Second, it identified the need for reconciliation: both grasped aspects of time, but treated them as the sole foundation. Third, it proposed the core idea of reconciliation: duration and physical time are different projections of the same ontological time. Fourth, it developed the reconciliation in detail: duration as $\partial\rho/\partial t$ inside a conscious system, and physical time as $\partial\rho/\partial t$ under cross-observer co-evolution. Fifth, it clarified the relation between the two times.

The next section develops the ontological status of the "now."

15.6. The Ontological Status of the "Now"

Statement of the Problem

The "now" is a persistent problem in the philosophy of time. In our experience, the "now" is alive: it has a special character that past and future do not have. But there is no "now" in the description of physics: all moments have equal status in physical equations. This difference raises several questions: is the "now" a real ontological phenomenon, or merely a feeling of consciousness? If the "now" is real, why does physics not need it? If the "now" is not real, why does our experience so clearly involve the "now"? How are different "nows," namely the presents of different observers, related? These questions have produced enduring debates in the history of philosophy, with positions such as presentism, where only the present is real; eternalism, where all moments are equally real; and the growing block universe, where past and present are real but the future is not, each treating these issues in different ways.

Treatment within the Co-Evolutionary Framework

The co-evolutionary framework provides a way to treat the problem of the "now." Let us develop it step by step.

Core thesis TM3: the "now" as an ontological phenomenon = the current state of a co-evolutionary relation. It is not a special ontological unit, as in presentism, nor merely a psychological feeling, as in some versions of eternalism, but the dynamical state of a co-evolutionary relation at each moment.

Let us develop this in detail.

The "now" as the current state of co-evolution. A co-evolutionary relation is a process: it has a current state at each moment. This current state has several features: a particular frequency-matching pattern, where the frequency-matching relation between detector and probed system has a form at each moment; a particular perturbation intensity, where the perturbation intensity of co-evolution has a value at each moment; and a particular tri-state decomposition, where the tri-state weights of the co-evolutionary relation have a distribution at each moment. These "forms at each moment" together constitute the current state of co-evolution. This is the ontological content of the "now."

The "now" is not a special ontological unit. TM3 does not claim that the "now" is some special ontological unit, as presentism does, where only the present exists and past and future do not. That position has difficulties: it makes the action of the past, such as memory, fossils, and the CMB, hard to explain. TM3 holds that the "now" is the current state of co-evolution: it is the instantaneous form of process. The past still exists in frozen forms, as residual, background, or history; the future has not yet occurred, but its possibilities are determined by the dynamics of the current process. This position is consistent with the ontology of freezing (Chapter 11) and process ontology (Chapter 7).

The "now" is not merely a psychological feeling. TM3 also does not claim that the "now" is merely a psychological feeling, as in strong versions of eternalism, where all moments are equally real and the "now" is only a feeling in consciousness with no objective counterpart. That position has difficulties: it makes the dynamics of co-evolution difficult to explain, because if all moments are equally real, the concept of process becomes difficult to handle. TM3 holds that the "now" has an objective counterpart: the current state of co-evolution is a real ontological phenomenon and does not depend on conscious feeling. The feeling of the "now" in consciousness is the psychological projection of this objective ontological phenomenon, but the ontological phenomenon itself is independent of consciousness.

Relations among different "nows." Different observers have different "nows." This is the content of relativity, the relativity of simultaneity. In the TM3 framework, how are these different "nows" related? The reply is that different observers' "nows" correspond to different co-evolutionary relations: each observer as a node of co-evolution has its own current state. These different "nows" obtain correspondence through cross-observer co-evolution, namely physical connections among different observers. Specifically: two relatively stationary observers establish correspondence between their "nows" by observing the same event simultaneously; two relatively moving observers establish a more complex correspondence between their "nows" through the exchange of light signals, involving light-speed delays and relativistic effects; these correspondences are established on co-evolutionary relations, not on abstract "absolute simultaneity." This treatment is consistent with relativity: it does not claim an "absolute now," but acknowledges the current state of each co-evolutionary relation.

Comparison with Traditional Positions

TM3 is related to traditional positions in the philosophy of time. Relation to presentism: presentism holds that only the "now" is real, while past and future are not real. TM3 differs: it acknowledges that past and future also have ontological status, the past through freezing and the future through possibilities determined by current dynamics. But TM3 resonates with presentism in the claim that "the current has a special character": the current state of co-evolution is indeed the focus of process. Relation to eternalism: eternalism holds that all moments are equally real, and the "now" is only the observer's location. TM3 does not fully agree: it acknowledges that the current state of co-evolution has ontological features, including frequency matching, threshold, and tri-state decomposition, and is not merely an observer's location. But TM3 resonates with eternalism in the claim that "past and future have ontological status": the past exists in frozen form, and the future exists as possibility. Relation to the growing block universe: the growing block universe holds that past and present are real, while the future is not; the ontological content of time grows as new moments are added. TM3 has the most resonance with this position: it also acknowledges that past, through freezing, and present, as co-evolutionary current state, have ontological status, while the future is possibility. But TM3 does not require the ontological assertion that "time as a whole grows"; it treats these as intrinsic structures of co-evolutionary relations. Overall, TM3 finds an intermediate position among traditional views: it acknowledges the insights each position has grasped, but treats them in a unified way through the co-evolutionary framework.

An Application: The "Now" in Cosmology

Cosmology involves the problem of the "now": what is the "current" state of the universe? In traditional physics, cosmologists say that "the current age of the universe is 13.8 billion years." This presupposes "cosmic time" and a "now." But in relativity, there is no absolute "now": different observers have different simultaneity. In the TM3 framework, the cosmological "now" can be treated as follows: the "current state" of the universe corresponds to the current state of a particular co-evolutionary relation, specifically the co-evolutionary relation between us, observers on Earth, and the universe as a whole; observers in different locations, such as possible observers in other galaxies, have different "nows," but correspondences can be established through cross-observer co-evolution, such as light signals and relativistic transformations; and the "age" of the universe, 13.8 billion years, is the temporal span from the Big Bang to our current co-evolutionary relation. This treatment makes the cosmological "now" a precisely treatable concept, without presupposing an "absolute now" and without giving up the meaning of the "now."

Objections and Replies

Objection One: TM3 seems too complicated. You try to satisfy the insights of presentism, eternalism, and the growing block universe at the same time. Is this excessive compromise?

Reply: TM3 is not simple compromise. It provides a unified ontological ground for the insights of the three positions through the co-evolutionary framework. Specifically: the insight of presentism, that the current state has a special character, is given content by TM3 through “the current state of co-evolution”; the insight of eternalism, that past and future have ontological status, is given content by TM3 through freezing, the past, and possibility, the future; and the insight of the growing block view, that time unfolds with process, is given content by TM3 through the dynamics of co-evolution. This is not excessive compromise, but framework synthesis. Each position grasps an aspect of time; TM3 provides a framework in which all aspects can be treated. Such synthesis is philosophically legitimate: it does not deny any insight, but gives them a unified ontological ground.

Objection Two: Your “current state of co-evolution” seems too abstract. The ordinary meaning of “now” is very concrete, as in “I am writing now,” but your concept seems to involve complicated physical content.

Reply: The concept in this section can be very concrete in ordinary application. “I am writing now” involves a co-evolutionary relation: I, a physical system, co-evolve with the pen or computer, also physical systems, and the current state is the ongoing writing activity. This co-evolution has a form at every moment: neural activity, muscular movement, and the response of the pen or computer. This concreteness keeps the concept from being abstract in ordinary application. It is precisely the ontological content of the “now” we experience in ordinary life. The “complicated physics” involved is the ontological ground of this ordinary experience, not a replacement or reinterpretation of it.

Objection Three: TM3 seems to depend too much on the co-evolutionary framework. If an event occurs without observers or co-evolutionary relations, such as an event in the early universe before consciousness existed, does it have no “now”?

Reply: This objection points to an important ontological issue. The response of this section is that co-evolution as an ontological relation, Chapter 12, is not limited to relations involving consciousness. It includes any same-field interaction. The early universe had no consciousness, but it had co-evolutionary relations, such as interactions within plasma and gravitational interactions. These co-evolutionary relations had current states, and therefore events in the early universe had “nows.” Specifically: particles in the plasma had co-evolutionary relations at every moment, and their current states constituted the “now” of the early universe; these “nows” existed objectively even without observers, since they were instantaneous states of physical processes; and when observers later appeared, as cosmic evolution produced consciousness, their co-evolution with the past, through observations of the CMB and other traces, made past events identifiable as "former nows." This treatment makes the "now" independent of consciousness: it is the current state of any co-evolutionary relation. The feeling of the "now" in consciousness is the manifestation of this general ontological phenomenon in conscious systems.

Literature That Must Be Engaged

- J. M. E. McTaggart, "The Unreality of Time" (1908)
- C. D. Broad, Scientific Thought (1923): on the growing block universe
- Michael Tooley, Time, Tense, and Causation (1997)
- Yuri Balashov, Persistence and Spacetime (2010)
- L. A. Paul, "Temporal Experience" (2010)

Summary of This Section

This section has done four things. First, it identified the "now" as a persistent philosophical problem. Second, it stated TM3: the "now" as the current state of a co-evolutionary relation. Third, it clarified that the "now" is not a special ontological unit and not merely a psychological feeling, and developed the relation among different "nows." Fourth, it demonstrated the use of TM3 through an application to cosmology.

The next section is the conclusion of this chapter.

15.7. Chapter Summary: Time as the Ontological Dimension of Co-Evolution

Completion of the Three Tasks

The three tasks undertaken by this chapter have all been completed. Task One: to reposition the problem of time as an ontological problem. Section 15.1 distinguished five meanings of time and clarified that this chapter focuses on the ontological dimension. Section 15.2 stated the core thesis TM1: time as an ontological dimension $= \partial\rho/\partial t$. This positioning transforms time from an externally added parameter into an intrinsic dimension of processual ontology. Task Two: to reconcile Bergson and Einstein. Section 15.3 developed through TM2 the ontological ground of the direction of time as the direction of freezing. Section 15.4 developed the reconciliation between Bergson and Einstein: duration and physical time are different projections of the same ontological time. Section 15.5 treated the ontological status of the "now" through TM3. Together, these three theses constitute a complete ontological framework of time as a dimension of co-evolution. Task Three: to connect the problem of time with the other work of the book. The work of this chapter is built upon Chapter 11, freezing, and Chapter 12, co-evolution, and prepares the ground for Chapter 17, ontological singularity.

The Position of This Chapter in Part III

The overall structure of Part III, "Co-Evolution, Mind–Matter, and Observation: The Unity of Three Relations," is completed by this chapter: Chapter 12, co-evolution as an ontological relation, as the foundational chapter; Chapter 13, mind–matter unity, as the application of co-evolution to the mind–matter relation; Chapter 14, observation as participation, as the application of co-evolution to the observational relation; and Chapter 15, this chapter, time, duration, and co-evolution, as the development of co-evolution in the temporal dimension. Together, the four chapters show the multidimensional application of co-evolution as an ontological relation. As the concluding chapter of Part III, this chapter treats time as the fourth dimension of relational ontology, making Part III a complete development of relational ontology.

The Position of This Chapter in the Whole Book

Looking backward, this chapter builds upon several chapters: Chapter 11 on freezing as a foundational ontological process, which provides the ontological ground for the direction of time in TM2; Chapter 12 on co-evolution as an ontological relation, which provides the ontological foundation for the overall framework of this chapter; Chapter 13 on mind–matter unity, which provides the method for reconciling duration and physical time; and Chapter 14 on observation as participation, which provides content for cross-observer co-evolution, the relation among "nows" in TM3. Looking forward, this chapter provides resources for later chapters: Chapter 17 on ontological singularity, where time as $\partial\rho/\partial t$ established in this chapter collapses at $\nabla\rho \equiv 0$, an expression of singularity; and Chapter 21 on the finitude of meaning, where the ontology of time in this chapter provides the ground for the meaning of "meaning in time."

Honesty about the Limits of This Chapter

This chapter acknowledges several limits. First, the connection with general relativity is insufficient. This chapter develops time as $\partial\rho/\partial t$ at the philosophical level, but its geometrical connection with general relativity, including spacetime curvature and gravitational time dilation, requires dedicated work. This chapter only indicates the direction; the detailed connection depends on the physical work linking EQT and general relativity. Second, the problem of quantum time. The problem of time in quantum mechanics, such as time being a parameter rather than an operator in the Schrödinger equation and the time–energy uncertainty relation, involves both physics and philosophy. This chapter touches on these issues but does not treat them in detail. Third, cross-cultural dialogue in the philosophy of time. This chapter mainly engages the Western tradition of the philosophy of time, including Bergson and Einstein. Chinese traditions, such as the *Yizhuan* on time as "change," and Buddhist discussions of instants, also contribute to the philosophy of time. This chapter mentions them only briefly, and detailed dialogue requires dedicated work. Fourth, the dynamics of conscious time. Duration as the manifestation of $\partial\rho/\partial t$ inside a conscious system requires more detailed work in consciousness studies, such as the relation between the Ψ framework in *Mind–Matter Unity* and temporal experience. This chapter gives the framework; the detailed dynamics is future work. Acknowledging these limits does not weaken the work of this chapter. They make it an honest initial work and invite further development.

The Core Posture of This Chapter

The core posture of this chapter can be stated as follows: time is not an independent thing; time is an intrinsic dimension of processual ontology.

Time has often been treated as a riddle in the history of philosophy, as in Augustine's famous remark that when no one asks him what time is, he knows, but when asked, he does not. This riddle largely arises from treating time as some "independent thing": a dimension independent of things, a container independent of change, a parameter independent of process. This chapter rejects such independence. Time is not independent; it is the intrinsic structure of process. Once we acknowledge that beings themselves are processual, as established in Chapter 7, time is no longer a riddle: it is the intrinsic direction of change of process. This understanding makes the problem of time tractable. It is no longer an abstract metaphysical riddle, but an ontological problem connected with physics, $\partial\rho/\partial t$, and philosophy, the co-evolutionary framework. This tractability is the contribution of this chapter.

Meaning before freezing: this is one of the core theses of this book. This chapter provides the ontological ground of time for that thesis. "Before freezing" involves the direction of time, from dynamic to stable; "meaning" appears in the current state of co-evolution (TM3); and the finitude of meaning is grounded in the finitude of co-evolutionary relations. These connections will be further developed in Chapters 17 and 21. Through the work of this chapter, time as an ontological dimension obtains its internal place in this book: it is not a riddle treated independently, but a dimension closely connected with the book's other ontological work, including freezing, co-evolution, mind–matter, and observation. This internal connection makes the ontological work of the book a systematic whole.

15.8. Methodological Reflection on This Chapter

This chapter repositions time from coordinate parameter to an intrinsic dimension of processual ontology. This section reflects on the methodological significance of this repositioning.

This Chapter's Treatment of Terminology

The multiple meanings of time, including coordinate, measure, sequence, duration, and ontological dimension, are a persistent source of confusion in philosophical discussion. This chapter makes the discussion precise by distinguishing these meanings carefully. This precision is the methodological foundation of the chapter: without precise conceptual distinctions, any discussion of time will fall into terminological confusion. This chapter argues that ontological time, $\partial\rho/\partial t$, is the foundational dimension, and that the other meanings of time are expressions of ontological time in different contexts. This layered treatment gives the concept of time internal unity across different occasions of use.

This Chapter's Treatment of the Bergson–Einstein Debate

The Bergson–Einstein debate is one of the most enduring twentieth-century disputes about time. This chapter does not simply choose one side, but makes the insights of both coexist through the co-evolutionary framework. This treatment is an application of the chapter's methodology: overcoming traditional opposition through an ontological framework. This chapter argues that the debate between

Bergson and Einstein is not a dispute over "which view of time is correct," but a dispute over the same ontological time seen from different perspectives. Bergson's duration is ontological time from the consciousness–subject perspective; Einstein's physical time is ontological time from the physics–description perspective. Treating the difference as a difference of perspective rather than a contradiction of content allows the Bergson–Einstein dispute to be reconciled within the co-evolutionary framework.

This Chapter's Posture toward Originality

This chapter uses the EQT framework, $\partial\rho/\partial t$, and the co-evolutionary framework, Chapter 12, to treat the problem of time. These frameworks themselves are contributions of this book. Applying them to the problem of time is new work. But this chapter acknowledges that this work is built upon the contents of traditional philosophy of time, including Bergson, Einstein, McTaggart, Reichenbach, and others, and contemporary physics, including general relativity, quantum mechanics, cosmology, and so on. This posture, building upon traditions while applying them in a new way, makes the work of this chapter not an isolated philosophical creation, but a new synthesis across several traditions.

Limits Honestly Acknowledged in This Chapter

This chapter acknowledges several limits. First, this chapter develops only the core direction of the connection between $\partial\rho/\partial t$ and general relativity. Second, the ontological status of quantum time is treated only briefly. Third, cross-cultural philosophies of time, including Hindu time, Buddhist time, and traditional Chinese views of time, are not sufficiently developed in this chapter. Fourth, the study of the dynamics of conscious time requires dedicated work.

Connections with Later Chapters

Time as an ontological dimension, established in this chapter, is used in later chapters. Chapter 16 uses time as the dimension of freezing. Chapter 17 uses time as the dimension of the process approaching singularity: the behavior of $\partial\rho/\partial t$ as $\nabla\rho \to 0$. Chapter 21 uses time as the condition for the emergence of meaning. More deeply, the problem of time is a problem that any serious metaphysics must treat. This chapter gives the book a complete dimension in the philosophy of time, allowing the ontological work of the whole book to be connected through the temporal dimension.

Part IV.

The Metaphysical Boundary of Freezing

16. The Hierarchy and Historicity of Freezing

16.1. Introduction to This Chapter

Chapter 11 treated freezing as a foundational ontological process: beings enter relatively stable forms from dynamic processes through freezing. But the work of Chapter 11 was mainly conducted at the level of a single process: it treated freezing as one foundational ontological process. Although this treatment established the legitimacy of freezing as a foundational ontological process, it left undeveloped a further question: how does the freezing process manifest across different scales, different frequency bands, and different temporal windows?

Experience shows that beings have hierarchical structure: elementary particles form atoms, atoms form molecules, molecules form macroscopic objects, macroscopic objects form celestial bodies, celestial bodies form cosmological structures, and so on. Each level has its own stability, dynamics, and scale. This hierarchical phenomenon requires ontological explanation. It cannot be reduced simply to the repetition of the foundational freezing process; it requires an analysis of hierarchical structure.

This chapter develops the work of Chapter 11 on freezing as a foundational ontological process into a hierarchical structure (HL1): freezing processes form hierarchies across different scales, different frequency bands, and different temporal windows, and this hierarchy is an intrinsic structure of ontology itself. At the same time, it connects freezing with historicity (HL2): beings acquire historicity through freezing processes, and historicity is not a dimension externally added to beings, but an intrinsic structure of beings as ontological units.

Another function of this chapter is to prepare for Chapter 17, "The Gradient Terminal Phase as an Ontological Singularity." By showing the hierarchical structure of freezing, it provides an ontological ground for Chapter 17's metaphysical work on the "deepest level of hierarchy" and the "gradient limit." Specifically, Chapter 17's work on ontological singularity requires "hierarchy" as a tool: the gradient terminal phase as an ontological limit is the limit of all levels, not the limit of a single process.

This chapter is the opening chapter of Part IV, "The Metaphysical Boundary of Freezing." It continues the work of Chapter 11, "Freezing as a Foundational Ontological Process," and extends the statement of a single process into the development of hierarchical structure. At the same time, it provides ontological preparation for the metaphysical summit of Chapter 17.

In writing, this chapter has a transitional function. It does not introduce a completely new ontological thesis, but develops existing content, especially Chapter 11 on freezing and Chapter 9 on the frequency spectrum, into a more refined structure. This transitional function makes the chapter a bridge from the work of Parts II and III to the metaphysical boundary of Part IV.

The discussion proceeds as follows. Section 16.1 develops the hierarchy of freezing: foundational development (HL1). Section 16.2 develops the ontological relations among levels: nesting, dynamical connection, differences in stability, and differences in reversibility. Section 16.3 develops the connection between freezing and historicity (HL2). Section 16.4 develops the position of this chapter within Part IV and its transitional function.

16.2. The Hierarchy of Freezing: Foundational Development

Starting Point: From a Single Process to Hierarchical Structure

Chapter 11 treated freezing as a foundational ontological process: beings enter relatively stable forms from dynamic processes through freezing. But the work of Chapter 11 was mainly conducted at the level of a single process: it treated freezing as an ontological process. The work of this chapter is to extend this single process into hierarchical structure: freezing processes form nested hierarchies across different scales.

There are two motivations for this development. Motivation One: the diversity of experiential phenomena. Experience shows that beings have hierarchical structure: elementary particles form atoms, atoms form molecules, molecules form macroscopic objects, macroscopic objects form celestial bodies, celestial bodies form cosmological structures, and so on. Each level has its own stability, dynamics, and scale. This hierarchical phenomenon requires ontological explanation. Motivation Two: preparation for Chapter 17. The work of Chapter 17 on ontological singularity requires "hierarchy" as a tool: the gradient terminal phase as an ontological limit is the limit of all levels, not the limit of a single process. Without the concept of hierarchy, the work of Chapter 17 loses its ground.

Core Thesis

Core thesis HL1: freezing as a foundational ontological process forms a hierarchical structure across different scales, different frequency bands, and different temporal windows. Each level is a form of freezing; different levels are dynamically connected; and the overall hierarchical structure is an intrinsic structure of ontology itself.

This thesis requires detailed development. Let us analyze three dimensions: scale, frequency band, and temporal window.

Dimension One: scale. Freezing processes occur at different scales: the Planck scale, about 10^{-35} meters, possible foundational processes and the deepest freezing; the subatomic scale, about 10^{-15} meters, where elementary particles are excitation modes of the ρ_f field; the atomic–molecular scale, about 10^{-10} meters, where atoms and molecules are bound states; the macroscopic scale, about 10^{-3} to 10^{0} meters, where everyday objects are aggregations of many molecules; the celestial scale, about 10^{8} to 10^{19} meters, where stars, planets, and galaxies are gravitationally bound structures; and the cosmological scale, about 10^{26} meters, where the observable universe is the largest-scale structure. Each scale corresponds to a kind of freezing process: stable forms emerge at that scale. Freezing processes at different scales have different characteristics.

Dimension Two: frequency band. Chapter 9's work on frequency as a basic spectral quantity showed that freezing processes correspond to frequency bands. Specifically: the ultra-low-frequency band, $< 10^{-15}\,\text{Hz}$, corresponds to the freezing of cosmological structures; the low-frequency band, 10^{-15} to $10^{12}\,\text{Hz}$, corresponds to the freezing of macroscopic matter and everyday objects; the mid-frequency band, 10^{12} to $10^{20}\,\text{Hz}$, corresponds to freezing at the atomic–molecular scale; the high-frequency band, 10^{20} to $10^{25}\,\text{Hz}$, corresponds to the freezing of elementary particles; and the ultra-high-frequency band, $> 10^{25}\,\text{Hz}$, corresponds to possible more foundational freezing. Each frequency band corresponds to a depth of freezing: higher frequency bands correspond to deeper freezing, shorter characteristic times, and more stable structures; lower frequency bands correspond to shallower freezing, longer characteristic times, and structures more susceptible to influence.

Dimension Three: temporal window. Freezing processes involve temporal windows, namely the characteristic time of a process. Specifically: Planck time, about 10^{-43} seconds, is the possible timescale of the most foundational processes; elementary-particle interaction time, about 10^{-25} seconds to 10^{-15} seconds, is the characteristic time of elementary-particle processes; atomic–molecular process time, about 10^{-15} seconds to 10^{-9} seconds, is the characteristic time of atomic and molecular dynamics; macroscopic process time, from milliseconds to years, is the characteristic time of macroscopic processes; celestial process time, from years to tens of billions of years, is the time of stellar evolution, galactic evolution, and related processes; and cosmic-evolution time, 13.8 billion years, is the timescale of the observable universe. Each temporal window corresponds to a freezing history: the occurrence and evolution of freezing processes within that temporal window.

Interrelations among the Three Dimensions

The three dimensions, scale, frequency band, and temporal window, are not independent. They are connected through physical relations. Specifically: scale $\leftrightarrow$ frequency band, through the wavelength–frequency relation $c = \lambda\nu$, where c is the speed of light, λ is wavelength, and ν is frequency; higher frequency bands correspond to smaller scales, and lower frequency bands correspond to larger scales. Frequency band $\leftrightarrow$ temporal window, through the period–frequency relation $T = 1/\nu$, where T is period and ν is frequency; higher frequency bands correspond to shorter temporal windows, and lower frequency bands correspond to longer temporal windows. Scale $\leftrightarrow$ temporal window, through propagation relations, with the speed of light as the upper limit; smaller scales correspond to shorter temporal windows, and larger scales correspond to longer temporal windows. These three relations give hierarchical structure internal consistency: the hierarchies of different dimensions correspond to one another and form a unified ontological hierarchy.

An Example: The Hierarchical Structure of the Atom

Let us consider an example: the atom as hierarchical structure. An atom, such as a hydrogen atom, involves multiple freezing levels: the nuclear level, where the atomic nucleus is a bound state of the strong interaction, at a scale of about 10^{-15} meters, a frequency band of about 10^{20} Hz, and a characteristic time of about 10^{-20} seconds. This freezing occurred in the very early universe, about 10^{-6} seconds after the Big Bang. The electron–nucleus level, where the electron is bound by the electromagnetic attraction of the nucleus to form an atom, at a scale of about 10^{-10} meters, a frequency band of about 10^{15} Hz, within the visible-light range, and a characteristic time of about 10^{-15} seconds. This freezing occurred about 380,000 years after the Big Bang, during photon decoupling. The atomic-whole level, where the hydrogen atom is a stable structure, at a scale of about 10^{-10} meters, with a characteristic time extending from atomic formation to the present, on the order of billions of years. This example shows that a single being, the hydrogen atom, involves multiple levels of freezing processes. Each level has its own scale, frequency band, and temporal window. The levels are nested: higher levels are built upon lower levels.

Objections and Replies

Objection One: Your hierarchical structure seems too finely divided. Is the division among different dimensions, scale, frequency band, and temporal window, overly physicalized?

Reply: The treatment in this chapter is grounded in physical facts. Scale, frequency band, and temporal window are observable quantities in physics. This treatment keeps the work of this chapter from becoming abstract philosophical speculation; it has a physical ground. But physicalization is not the aim of this chapter. The aim of this chapter is to establish an ontological hierarchical structure. Physical facts serve as the ground, giving the ontological hierarchical structure identifiable content. This is the application of the method of constrained speculation, Chapter 1: ontological statements require identifiable content. More deeply, physicalization and ontologization do not conflict: the content of physics reflects the structure of ontology. This chapter develops ontological hierarchy through physical facts, thereby connecting ontology with physics. This connection is a core direction of contemporary naturalistic metaphysics, as in Chapter 5's discussion of Cartwright, Ladyman, Rovelli, and others.

Objection Two: Your concept of "hierarchy" makes ontology look too static. Hierarchy as structure seems fixed, but this book emphasizes process ontology.

Reply: This objection points to an important refinement. The "hierarchy" in this chapter is not a static structure; it is an intrinsic structure of process. Specifically: each level is a form of the freezing process, not a structure outside process; the relation among levels is a dynamical connection among processes, not a static framework; and hierarchical structure evolves: it unfolds through freezing processes in cosmic evolution, as Chapter 11, Section 11.4, has already shown in relation to the freezing processes of cosmic evolution. Hierarchy as a form of process preserves the content of process ontology. It does not "staticize" ontology, but unfolds the intrinsic structure of process ontology. More deeply, "hierarchy" as a concept is about process: it describes how processes unfold across different scales, different frequency bands, and different temporal windows. Without process, there is no hierarchy. Thus hierarchy and process depend on each other, and this chapter develops their relation.

Objection Three: Your example, the hierarchical structure of the hydrogen atom, involves physical facts. But many beings, such as consciousness, society, and culture, are not simple physical structures. How should their "hierarchies" be treated?

Reply: This chapter develops physical hierarchy as the starting point, because this is the most concrete ground. But the concept of hierarchy extends to all beings: consciousness, society, and culture also involve hierarchy. Specifically: consciousness involves hierarchy, including the frequency bands of neural oscillations, the scales of neural circuits, and the temporal windows of conscious experience, as already treated in Chapter 13 on mind–matter unity and Chapter 9 on consciousness as a multi-band structure; society involves hierarchy, including individuals, families, communities, organizations, states, and the global level, each scale having its own stability and dynamical time; culture involves hierarchy, including language, custom, tradition, values, and worldviews, each level having its own stability and historical time. Each domain has its own hierarchy. This chapter provides physical hierarchy as the starting point, but the concept of hierarchy is generally applicable. Complete applications in biology, neuroscience, sociology, cultural studies, and related fields require dedicated work.

Literature That Must Be Engaged

- Chapter 11 on freezing as a foundational ontological process (self-reference)
- Chapter 9 on frequency as a basic spectral quantity (self-reference)

- Steven Weinberg, The First Three Minutes (1977): on freezing processes in cosmic evolution
- Stuart Kauffman, Investigations (2000): on the ontology of complex hierarchies
- David Layzer, Cosmogenesis (1990): on the origin of cosmic hierarchy

Summary of This Section

This section has done four things. First, it moved from the single freezing process of Chapter 11 to hierarchical structure. Second, it stated the core thesis HL1: freezing forms hierarchies along three dimensions, scale, frequency band, and temporal window. Third, it developed the interrelations among the three dimensions, giving hierarchical structure internal consistency. Fourth, it demonstrated the application of hierarchical structure through the example of the hydrogen atom.

The next section develops the ontological relations among levels.

16.3. Ontological Relations among Levels

The Function of Relation

Freezing levels do not exist in isolation. They have ontological relations among them. This section develops four major types of such relations: nesting, dynamical connection, differences in stability, and differences in reversibility. The function of this development is to give hierarchical structure content as an overall ontology: it is not merely a collection of levels, but a whole with internal relations. It also prepares for Chapter 17's work on ontological singularity: singularity as "the limit of all levels" requires hierarchical relations as its ground.

Relation One: Hierarchical Nesting

Beings at higher levels are often internal structures of beings at lower levels. This relation is called nesting. Specifically: elementary particles are constituted by quarks and gluons, and elementary particles are nested in the structure of quarks and gluons; atoms are constituted by elementary particles, and atoms are nested in the structure of elementary particles; molecules are constituted by atoms connected through chemical bonds, and molecules are nested in the structure of atoms and chemical bonds; life is constituted by molecules, the molecular structures inside cells, and is nested in molecular structure; society is constituted by individual living beings and is nested in the structure of individual life; cosmological structures are constituted by galaxies and stars, and are nested in the structure of celestial bodies. Nesting relations give hierarchical structure internal unity: all levels are built upon more foundational levels, and ultimately upon the dynamics of the ρ_f field (EQT A1).

Relation Two: Dynamical Connection

Levels are dynamically connected with one another. Specifically: energy transfer, where energy is transferred among different levels, for instance solar energy produces electromagnetic radiation in the ρ_f field, radiation is transmitted to Earth, absorbed by plants for photosynthesis, enters the food chain as chemical energy, and is eventually used by animals; causal chains, where causal relations cross multiple levels, for example elementary-particle processes affect atomic behavior, atomic behavior affects molecular reactions, molecular reactions affect cellular metabolism, and cellular metabolism affects life functions; and information flow, where information flows among different levels, for instance genetic information moves from DNA to RNA to proteins, protein function affects cellular behavior, cellular behavior affects tissue function, and tissue function affects individual behavior. Dynamical connection makes hierarchical structure alive: levels affect one another through processes.

Relation Three: Differences in Stability

Different levels have different stability. Specifically: the elementary-particle level is relatively stable, since electrons and protons can remain stable over billions of years, while many elementary particles, such as μ particles and W bosons, decay on microsecond to femtosecond timescales; the atomic level is usually very stable, since atoms can have lifetimes on the order of the age of the universe unless changed by nuclear reactions; the molecular level varies greatly in stability, with some molecules very stable, such as nitrogen gas, and some very unstable, such as free radicals; the life level has stability maintained by biological processes, since life as a dissipative structure requires energy input for maintenance; and the cosmological level is highly stable, with cosmic evolution unfolding on billion-year timescales. Differences in stability reflect differences in the depth of freezing: deeper freezing corresponds to higher stability. But this relation is not simple. Some lower-level structures, such as certain molecules, may be more stable than some higher-level structures, such as certain elementary particles, because the form of freezing differs.

Relation Four: Differences in Reversibility

Different levels have different reversibility. Specifically: at the foundational physical level, the fundamental laws of physics are time-symmetric, and elementary-particle processes are in principle reversible, though irreversibility in processes such as quantum measurement involves statistical phenomena; at the thermodynamic level, macroscopic thermodynamic processes are statistically irreversible, since entropy increase gives processes an irreversible direction; at the life level, life processes are basically irreversible, including development, aging, and death; at the cultural–social level, cultural and social processes are basically irreversible, since historical development cannot be simply reversed. Differences in reversibility reflect the manifestation of the direction of time at different levels: higher levels, deeper freezing, and larger scales usually have stronger irreversibility. This irreversibility is grounded in Chapter 11's ontology of the direction of time = the direction of freezing (Sub-Thesis Two in Section 11.2).

The Unity of the Four Relations

The four relations, nesting, dynamical connection, differences in stability, and differences in reversibility, together constitute the internal content of hierarchical structure. They are not independent; they support one another. Nesting relations allow higher levels to be built upon lower levels, and this provides the basis for dynamical connection. Dynamical connection allows levels to influence one another, thereby affecting stability and reversibility. Differences in stability and differences in reversibility together reflect the manifestation of freezing depth across levels: deeper freezing corresponds to higher stability and stronger irreversibility. The unity of these four relations makes hierarchical structure an overall ontology with internal content, not merely a collection of levels.

Objections and Replies

Objection One: Your "nesting" relation implies reductionism, since higher levels are "reduced" to lower levels. Does this conflict with the process ontology of this book?

Reply: The "nesting" in this chapter is not simple reduction. Specifically: simple reductionism holds that the higher level is completely equal to the lower level, and that its properties are derived from the lower level. The nesting in this chapter holds that the higher level is built upon the lower level, but has its own new content and cannot be simply reduced. Molecules as a level are built upon the atomic level, but molecules have new properties, chemical properties, that cannot be simply derived from the atomic level. Life as a level is built upon the molecular level, but life has new properties, life properties, that cannot be simply derived from the molecular level. This relation of "being built upon while having new content" is central to contemporary complex-systems theory: emergent properties emerge at higher levels, are built upon lower levels, but are not reduced to them. The nesting relation in this chapter adopts this direction. More deeply, the ontology of difference (Chapter 7) and process ontology of this book provide the ground: higher levels as stable forms of differential structures have ontological content and are not simple aggregations of lower levels. The nesting relation is consistent with this direction.

Objection Two: Your "differences in stability" and "differences in reversibility" seem to lie along the same dimension: higher stability = stronger irreversibility. Are they actually the same relation?

Reply: Stability and irreversibility are indeed related, but they are not the same relation. Specifically: stability concerns the maintenance of structure over time, namely whether a being preserves its form within a temporal window. Irreversibility concerns the direction of process, namely whether a process can be simply reversed. The distinction between them is this: a highly stable being may involve reversible processes, such as a stable equilibrium state involving reversible fluctuations; a being of low stability may involve irreversible processes, such as a life process that has medium stability but strong irreversibility. Specifically, the atom as a being has extremely high stability, but its dynamical process, such as electron motion, is in a certain sense reversible, since the foundational equations of quantum mechanics are time-symmetric. Life as a being has medium stability, but its dynamical process is basically irreversible. Their manifestations across levels are related but distinct dimensions. This chapter treats them separately in order to give hierarchical structure more refined content.

Objection Three: Your dynamical connection involves causality. But causality as an ontological concept has been extensively discussed in contemporary philosophy, such as Hume on causality and Lewis on counterfactual causation. Does this chapter treat causality too simply?

Reply: This objection genuinely points to philosophical complexity. This chapter uses the term "dynamical connection" and does not directly treat philosophical debates about causality. Specifically:

dynamical connection concerns physical connections among processes, including energy transfer, causal chains, information flow, and so on. Causality as a philosophical concept involves metaphysical discussion and requires dedicated philosophical treatment. The work of this chapter is mainly at the ontological level: it develops physical connections among levels. Philosophical discussions of causality occur at the metaphysical level and require dedicated work. This chapter builds on the ordinary concept of causality: processes physically affect one another. This ordinary concept is sufficient to support the ontological development of this chapter. Philosophical refinement of causality, such as a concept of causality within the framework of the ontology of difference in this book, can be developed in future work.

Literature That Must Be Engaged

- Hume, A Treatise of Human Nature (1739-40): classical discussion of causality
- David Lewis, Counterfactuals (1973): contemporary discussion of counterfactual causation
- Ilya Prigogine, Order out of Chaos (1984): on dissipative structures and reversibility
- Philip Anderson, "More Is Different" (1972): classical work on emergence
- Chapter 7 on difference before identity (self-reference)
- Chapter 11 on freezing as a foundational ontological process (self-reference)

Summary of This Section

This section has done four things. First, it developed four ontological relations among levels: nesting, dynamical connection, differences in stability, and differences in reversibility. Second, it gave the content and examples of each relation. Third, it developed the mutual support among the four relations. Fourth, it responded to possible objections, especially concerning reductionism, the relation between stability and irreversibility, and the use of the concept of causality.

The next section develops the connection between freezing and historicity.

16.4. Freezing and Historicity

The Meaning of Historicity

The freezing process is not instantaneous: it involves temporal evolution. Products of freezing, stable forms, carry the content of freezing history: their current forms reflect past freezing processes. This content is called historicity. Specifically:

Core thesis HL2: the current form of a being is the result of freezing history. Historicity is not a dimension externally added to beings, but an intrinsic structure of beings as ontological units.

Let us develop this thesis in detail.

Implication One: Current Form as the Result of Freezing History

The current form of any being can be traced back to the history of freezing processes. Specifically: elementary particles, including electrons, quarks, and others, can be traced back to freezing processes in the very early universe, about 10^{-6} seconds after the Big Bang; atoms, including basic atoms such as hydrogen and helium, can be traced back to nucleosynthesis about three minutes after the Big Bang, while heavy elements can be traced back to nucleosynthesis in stellar evolution and supernova explosions; Earth as a planet can be traced back to the formation of the solar system about 4.6 billion years ago; life on Earth can be traced back to the origin of life about 3.8 billion years ago; and humans as a species can be traced back to biological evolution several million years ago. The current form of each being is the result of freezing history.

Implication Two: Historicity as Intrinsic Structure

Historicity is not external: it is an intrinsic structure of a being as an ontological unit. Specifically: the current structure of a hydrogen atom, including its nucleus, electron cloud, energy levels, and so on, is not externally added; it is the internal content of freezing history. The current state of a living being, including its genome, morphology, functions, and so on, is not externally added; it is the internal result of biological evolution. The current mind of a human being, including language, concepts, values, and so on, is not externally added; it is the internal result of cultural and personal development. A being is its history. History is not an external description of a being, but the internal constitution of a being as an ontological unit.

Implication Three: Cross-Level Historicity

Historicity exists at all levels, from the deepest freezing level to the shallowest level. Specifically: at the elementary-particle level, historicity involves cosmological timescales, from billions to tens of billions of years; at the molecular level, historicity involves the timescale of chemical reactions, from femtoseconds to years; at the life level, historicity involves the timescale of biological evolution, from millions to billions of years; at the cultural level, historicity involves the timescale of cultural evolution, from years to millions of years. Each level has its own historicity: the past affects the present through freezing processes.

The Connection between Historicity and the Direction of Time

Historicity is ontologically connected with the direction of time, as discussed in Chapter 11, Sub-Thesis Two in Section 11.2, and Chapter 15, TM2. Specifically: the direction of time, where time flows from past to future, is the manifestation of the irreversibility of freezing as a foundational ontological process; historicity, where the past affects the present through freezing processes, is the ontological manifestation of the direction of time. Their relation is this: the direction of time makes freezing processes irreversible, so the past cannot be simply reversed; irreversibility allows the past to remain in the present, so the current form carries information from the past; and this retention is historicity. Therefore historicity is not a dimension independent of the direction of time. It is the ontological manifestation of the direction of time. The two are closely connected through the freezing process.

An Example: The Historicity of a Star

Let us consider an example: the historicity of a star. The current form of a star, such as the Sun, involves a freezing history:

Initial formation: the Sun formed about 4.6 billion years ago from the collapse of a nebula. Freezing process: a gas cloud aggregated through gravity, increasing in density and temperature, and eventually reached the conditions for nuclear fusion, forming a star. This process was completed in about 50 million years.

Main-sequence evolution: the Sun entered the main-sequence stage and produces energy through hydrogen fusion. Freezing process: nuclear reactions maintain equilibrium, and the stellar structure remains stable. This process has continued for about 4.6 billion years and is expected to continue until a total age of about 10 billion years.

Historicity of the current form: the current form of the Sun, including mass, radius, temperature, luminosity, chemical composition, and so on, reflects 4.6 billion years of freezing history. Specifically: chemical composition, where the heavy-element abundance of the Sun reflects its formation from a nebula containing heavy elements, which came from supernova explosions of earlier generations of stars; mass, where the mass of the Sun reflects the conditions of the nebular collapse process; and rotation, where the rotation of the Sun reflects the conservation of angular momentum during formation. Every current feature is the result of freezing history. Historicity is the intrinsic structure of the Sun as a being.

Future expectation: the Sun will continue to evolve through freezing processes. In about 5 billion years, the Sun will exhaust the hydrogen in its core, enter the red-giant phase, and eventually become a white dwarf. This future evolution will continue the freezing process, giving the Sun as a being a longer freezing history.

Objections and Replies

Objection One: Your claim that "historicity is intrinsic structure" seems too strong. The current form of a being indeed involves history, but history can be treated as an external description and need not be intrinsic structure.

Reply: This objection points to an important ontological issue. There is a difference between external description and intrinsic structure. External description treats history as a description of a being from the outside: it tells us how the being formed. Intrinsic structure treats history as the internal constitution of a being: it is the ontological content that makes the being what it is now. The claim of this chapter is the latter. The ground is this: the current form of a being cannot be detached from its freezing history. The current structure of a hydrogen atom is determined by freezing processes; if the freezing process were different, the current structure of the atom would be different. Freezing history is not an external description of the atom: it is the internal constitution of the atom as an ontological unit. Similarly, the current state of a living being is determined by evolutionary and developmental processes; if those processes were different, the current state of the living being would be different. Evolution and development are not external descriptions of life; they are the internal constitution of life as an ontological unit. This claim is consistent with the process ontology of this book: beings are forms of processes, and the history of process is the internal constitution of beings.

Objection Two: Your example, the Sun, involves physical facts.

But many beings, such as abstract objects and mathematical objects, seem to have no historicity.

Reply: This chapter mainly treats the historicity of physical–life–mental–social beings. These are the main ontological concerns of this book. The ontological status of abstract objects, including numbers, propositions, possible worlds, and so on, has already been discussed in Chapter 3, Section 3.5: they are treated as formal constraints rather than real objects. Formal constraints as ontological content do not require historicity: they involve formal structure, not processual evolution. But the activity of using abstract objects has historicity: the development of mathematics, logic, science, and so on all involve the history of freezing processes. This is the historicity of activity, not the historicity of abstract objects themselves. This distinction clarifies the scope of this chapter: it applies to processual beings, but not necessarily to formal constraints.

Objection Three: Your "historicity as intrinsic structure" seems related to the concept of "historicity" in the phenomenological-hermeneutic tradition, such as Heidegger and Gadamer. Does this chapter sufficiently engage that tradition?

Reply: This objection genuinely points to an important philosophical connection. The concept of "historicity" in this chapter has structural similarity with Heidegger's concept of "historicity" (Geschichtlichkeit). In *Being and Time*, Heidegger treats historicity as an intrinsic structure of Dasein: a being becomes what it is through its history. The work of this chapter resonates with that tradition, but also differs from it. Similarity: both treat historicity as an intrinsic structure of a being, not an external description. Difference: Heidegger's work mainly treats the historicity of Dasein, human being-there, and involves phenomenological-hermeneutic dimensions; this chapter treats the historicity of all processual beings and involves a physical-ontological dimension. Chapter 19, "Three Responses: Heidegger, Wittgenstein, and Nāgārjuna," will treat the dialogue with Heidegger's work in detail. This chapter only indicates the connection: there is resonance between the concept of historicity in this book and in Heidegger's work, but their scope of application differs.

Literature That Must Be Engaged

- Martin Heidegger, Being and Time (1927): on historicity, Geschichtlichkeit

- Hans-Georg Gadamer, Truth and Method (1960)
- Charles Darwin, On the Origin of Species (1859): foundational work on biological historicity
- Chapter 11 on freezing as a foundational ontological process (self-reference)
- Chapter 15 on time, duration, and co-evolution (self-reference)

Summary of This Section

This section has done four things. First, it stated the core thesis HL2: historicity as the intrinsic structure of beings. Second, it developed three implications of the thesis: current form as the result of freezing history, historicity as intrinsic structure, and cross-level historicity. Third, it developed the connection between historicity and the direction of time. Fourth, it demonstrated the application of historicity through the example of the Sun.

The next section is the chapter summary and prepares for Chapter 17.

16.5. Chapter Summary: Preparing for Chapter 17

Completion of the Three Tasks

The three tasks undertaken by this chapter have all been completed. Task One: to develop the work of Chapter 11 into hierarchical structure. Section 16.1 stated the core thesis HL1 and developed the hierarchical structure of freezing across three dimensions: scale, frequency band, and temporal window. Task Two: to connect freezing with historicity. Section 16.3 stated the core thesis HL2 and developed historicity as the intrinsic structure of beings. Task Three: to prepare for Chapter 17. The work of this chapter on hierarchical structure and historicity provides the ground for Chapter 17's metaphysical work on ontological singularity.

The Position of This Chapter in Part IV

The overall structure of Part IV, "The Metaphysical Boundary of Freezing," is as follows: Chapter 16, this chapter, "The Hierarchy and Historicity of Freezing," as foundational development; Chapter 17, "The Gradient Terminal Phase as an Ontological Singularity," as the metaphysical summit; Chapter 18, "The Collapse of Co-Evolution before the Gradient Terminal Phase," continuing Chapter 17; and Chapter 19, "Three Responses: Heidegger, Wittgenstein, and Nāgārjuna," as dialogue with traditions. As the opening chapter, this chapter develops the work of Chapter 11 into a more refined structure and provides a ground for the following chapters. Specifically: the hierarchical structure of this chapter provides the ground for Chapter 17's work on ontological singularity, since the singularity is the limit of all levels; the historicity of this chapter provides the ground for Chapter 17's work on metaphysical boundary, since boundary involves the limit of historicity; the hierarchical relations of this chapter, nesting, dynamical connection, differences in stability, and differences in reversibility, provide the ground for Chapter 18's work on the collapse of co-evolution; and the cross-level historicity of this chapter provides the ground for Chapter 19's dialogue with Heidegger and others, since Heidegger's concept of "historicity" must be developed in detail in relation to this chapter.

This Chapter as a Transitional Chapter

This chapter has a transitional function: it does not introduce a completely new ontological thesis, but develops existing content into a more refined structure. Such a transitional function is legitimate in a philosophical monograph. Not every chapter needs to introduce a new thesis; continuing and developing previous work is also a legitimate chapter function. Specifically: this chapter continues the work of Chapter 11, "Freezing as a Foundational Ontological Process"; it continues the work of Chapter 9, "Frequency as a Basic Spectral Quantity"; it prepares the work of Chapter 17, "The Gradient Terminal Phase as an Ontological Singularity"; and it prepares the work of Chapters 18 and 19. This function of continuation + preparation makes the chapter irreplaceable in Part IV: without the transition provided by this chapter, the metaphysical work of Chapter 17 would lose its ground.

Honesty about the Limits of This Chapter

This chapter acknowledges several limits. First, the application of hierarchical content is limited. This chapter gives physical examples of hierarchy, such as the hydrogen atom and the Sun. Detailed application to the levels of life, consciousness, society, and so on requires dedicated treatment. This chapter only indicates the direction. Second, the dynamics among levels are not sufficiently developed. This chapter gives types of hierarchical relations, including nesting, dynamical connection, differences in stability, and differences in reversibility. The details of the dynamics, especially mechanisms of cross-level action, require dedicated work. Third, dialogue between historicity and the phenomenological-hermeneutic tradition is insufficient. This chapter briefly mentions the connection with Heidegger and others. Detailed dialogue is left to Chapter 19. Fourth, the evolution of hierarchical structure is not sufficiently developed. This chapter mainly treats the current structure of hierarchy. How levels emerge in cosmic evolution, namely the formation history of hierarchy, requires more detailed work. Acknowledging these limits does not weaken the work of this chapter. They make it an honest initial work.

The Core Posture of This Chapter

The core posture of this chapter can be stated as follows: ontology is not a statement at a single level, but the development of hierarchical structure.

Ontological discussion since the twentieth century has often focused on a single level: some content regarded as foundational, such as elementary particles, information, relation, and so on. This treatment simplifies ontology, but also deprives ontology of diversity. This chapter argues that ontology involves multiple levels: elementary particles, atoms, molecules, life, mind, society, culture, and so on. Each level has its own ontological content, and there are ontological relations among levels. Overall ontology is the development of this hierarchical structure. This posture means that ontology is not a simple dichotomy between "foundational reduction" and "emergence of new content"; it is the development of hierarchical structure. Each level has an ontological ground, since it is built upon lower levels, and new content, since it is not simply reduced. This treatment is consistent with the direction of contemporary complex-systems theory and contemporary theories of emergence. More deeply, this chapter adds historicity to this hierarchical structure: levels are not merely static structures, but results of freezing history. This treatment connects ontology closely with the temporal dimension: the current form of a being is the result of freezing history.

Meaning before freezing: this is one of the core theses of this book. This chapter provides the hierarchical and historical ground for that thesis. "Meaning" as an ontological phenomenon emerges within hierarchy: it has ontological content at the levels of life, mind, and society. "Before freezing" as the direction of time involves historicity: the irreversibility of freezing processes allows the past to remain in the present. These connections receive ontological ground through the work of this chapter. This is the contribution of this chapter to the book as a whole. Through the work of this chapter, the beginning of Part IV, "The Metaphysical Boundary of Freezing," has been established. The following chapters will develop the work on metaphysical boundaries on the basis of this chapter. The metaphysical summit of Chapter 17, on ontological singularity, will be built upon the hierarchical structure and historicity of this chapter, so that singularity as "the limit of all levels" has a ground.

16.6. Methodological Reflection on This Chapter

As a transitional chapter, this chapter develops the work of Chapter 11 on freezing into hierarchical structure and historicity. This section reflects on the methodological significance of this transitional work.

The Function of This Chapter as a Transitional Chapter

The transitional function of this chapter is legitimate in a philosophical monograph. Not every chapter needs to introduce a new thesis; continuing existing work and preparing later work are also legitimate chapter functions. There is an ontological distance between the work of Part II and the work of Part IV: the ontological concepts established in Part II are pushed to their limit in Part IV, namely ontological singularity. This movement requires a transition. By developing hierarchical structure and historicity, this chapter makes the transition continuous rather than abrupt. Without the transition of this chapter, there would be an ontological rupture between Part II and Part IV.

This Chapter's Posture toward Originality

This chapter does not claim complete originality. It is built upon the work of Chapter 11, Chapter 9, and others. Its contribution is to develop the single freezing process into hierarchical structure; to connect freezing with historicity; and to provide a ground for Chapter 17's work on ontological singularity. These contributions are constructive refinements: they make the content of earlier work more precise and provide the ground for later work.

This Chapter's Treatment of the Concept of Historicity

This chapter introduces historicity as an ontological concept. This concept is structurally related to Heidegger's Geschichtlichkeit, but differs from it. Heidegger's Geschichtlichkeit applies mainly to Dasein;

the historicity of this chapter is extended to all processual beings. Specifically, this chapter argues that any processual being has historicity through freezing processes. This extension gives historicity as an ontological concept a broader range of application. Chapter 19 will treat the dialogue with Heidegger's work in greater detail.

Limits Honestly Acknowledged in This Chapter

This chapter acknowledges several limits. First, the refinement of hierarchical structure requires later work. The dynamical relations among levels, such as emergence, reduction, cross-level causation, and so on, are not sufficiently developed in this chapter. Second, the dialogue between the concept of historicity and the hermeneutic-phenomenological tradition is only briefly mentioned in this chapter; Chapter 19 will provide a more detailed treatment. Third, this chapter does not sufficiently develop applications of hierarchy and historicity in scientific domains.

Connections with Later Chapters

This chapter provides the ontological ground for the work of Chapter 17. Chapter 17's work on ontological singularity is built upon the hierarchical structure established in this chapter: the gradient terminal phase as the limit of all levels. This chapter's work on historicity is built upon the work of Chapter 11 and provides the ground for Chapter 19's dialogue with Heidegger. More deeply, this chapter is the metaphysical transition from Chapter 11 to Chapter 17. It bridges this distance and makes the ontological development of the book a continuous progression.

17. The Gradient Terminal Phase: The Intrinsic Singularity of Ontology

17.1. Introduction to This Chapter

All previous chapters have been built upon one core ontological thesis: existence $\iff \nabla\rho \neq 0$. This thesis appears repeatedly in different forms across the five-volume monograph series. *EQT Foundations* states it in the form of Postulate A1: the energy quantum density field $\rho_f(x,t)$ is the only basic ontological object. *Freezing Cosmology* states it by treating gradient transients as the basic structural form. *Mind–Matter Unity* states it by arguing that consciousness depends on energy gradients. *EQT Detection Theory* states it by arguing that both observer and object are local structures within the ρ_f field. All of these presuppose one point: only where there is gradient is there existence.

This chapter asks a question previously avoided: what happens if the gradient disappears? This question was not treated in depth in the five physical monographs, not because of oversight, but because of legitimate restraint in physics. Physics is concerned with observable processes, while a state of strict gradient disappearance is difficult to reach within physical time. But metaphysics cannot take physical attainability as its endpoint. Metaphysics is concerned with what happens to an ontological system at the logical limit of its own postulates. This is a necessary part of EQT's self-understanding, and also an intrinsic requirement of the method of constrained speculation: every ontological commitment must be able to state its own failure conditions.

Specifically, this chapter precisely defines the gradient terminal phase as the state of affairs at the ontological limit of EQT: $\nabla\rho(x,t) \equiv 0$ holds for all x and all t. This is the limiting state in which gradient has completely disappeared, and it differs from local freezing, where

$\nabla\rho \to 0$ in some local region, and from global heat death, where $\nabla\rho \to 0$ everywhere in the universe as a whole but is not strictly zero. The gradient terminal phase as a limiting state is the intrinsic boundary of EQT ontology itself.

The core position established in this chapter is that the gradient terminal phase is neither "nothing" nor "being," neither proposition nor non-proposition (OS1, OS2). Specifically, through two exclusion arguments, this chapter proves that the gradient terminal phase escapes the being/nothing dichotomy; through the collapse argument concerning the minimum conditions of propositionhood, it proves that statements about the gradient terminal phase are neither propositions nor non-propositions.

This position involves a methodological challenge: how can a state of affairs that seems to cancel itself be legitimately stated? The solution of this chapter is the distinction between meta-level and object-level. This chapter does not directly state the state of the gradient terminal phase, which would be an object-level statement and methodologically restricted; rather, it states that the structure of EQT ontology itself causes the collapse of its capacity for statement at $\nabla\rho \equiv 0$, which is a legitimate meta-level statement. This methodology allows the work of this chapter to maintain legitimate speech at the boundary without retreating into simple silence.

Finally, this chapter names the gradient terminal phase the intrinsic singularity of EQT ontology (OS3), and places the work of EQT within the philosophical-historical context of Kant, Wittgenstein, and Nāgārjuna. This chapter is the metaphysical peak of the whole book, but also the deepest application of the method of constrained speculation. It requires EQT to acknowledge its own limit and, in this acknowledgement, to realize complete ontological honesty.

This chapter does not depend on physical derivation. It uses only two core theses of EQT: existence $\iff \nabla\rho \neq 0$, defended in Part II, and observation $\iff$ same-field co-evolution, defended in Part III. All arguments proceed from these two theses.

The discussion proceeds as follows. Section 17.1 precisely constructs the gradient terminal phase, distinguishing three kinds of "gradient disappearance." Section 17.2 proves through a double exclusion argument that the terminal phase is neither nothing nor being. Section 17.3 proves through the collapse of the minimum conditions of propositionhood that statements about the terminal phase are neither propositions nor non-propositions. Section 17.4 develops the meta-level/object-level distinction that makes this chapter itself legitimate, names the terminal phase the intrinsic singularity of ontology, and develops its position in philosophical history. Section 17.5 argues that acknowledging ontological singularity is itself a philosophical contribution.

17.2. The Precise Construction of the Gradient Terminal Phase

All previous chapters have been built upon one core ontological thesis: existence $\iff \nabla\rho \neq 0$. This thesis appears repeatedly in different forms across the five-volume monograph series. *EQT Foundations* states it in the form of Postulate A1: the energy quantum density field $\rho_f(x,t)$ is the only basic ontological object. *Freezing Cosmology* states it by treating gradient transients as the basic structural form: any structure is a local condensation of gradient. *Mind–Matter Unity* states it by arguing that consciousness depends on energy gradients: without gradient, there is no dissipation, complexity, or boundary, and therefore no consciousness-related state. *EQT Detection Theory* states it by arguing that both detector and object are local structures within the ρ_f field: observation itself depends on gradient. All of these presuppose one point: only where there is gradient is there existence.

The work of this chapter is to ask a question previously avoided: what happens if the gradient disappears?

This question was not treated in depth in the five physical monographs, not because of oversight, but because of legitimate restraint in physics. Physics is concerned with observable processes, while a state of strict gradient disappearance is difficult to reach within physical time. But metaphysics cannot take physical attainability as its endpoint. Metaphysics is concerned with what happens to an ontological system at the logical limit of its own postulates. This is a necessary part of EQT's self-understanding, and also an intrinsic requirement of the method of constrained speculation: every ontological commitment must be able to state its own failure conditions.

A Precise Distinction among Three Kinds of "Gradient Disappearance"

Not all cases of "gradient disappearance" are the same state of affairs. This section must precisely distinguish three cases; otherwise the following arguments will fall into confusion.

Case A (local homogenization): in a certain local region, $\nabla\rho \to 0$, while in other regions $\nabla\rho \neq 0$. This is local freezing: the energy quantum density of a certain frequency band locally exits active dynamics and enters a frozen state. This is the standard content of the freezing spectrum and has been treated in detail in Chapter 11. It does not itself constitute an ontological singularity, because the overall gradient structure still exists and an observational perspective remains possible. It is not the topic of this chapter.

Case B (global approach to heat death): throughout the universe as a whole, $\nabla\rho \to 0$ everywhere, but not strictly to zero. This is an asymptotic terminal state: it is the physical image of heat death and was treated in Chapter 8 of *Freezing Cosmology.* Case B still presupposes the continued existence of an observational perspective and understands "gradient approaching zero" as asymptotic behavior in a temporal process. It belongs to cosmology, not to the topic of this chapter.

Case C (complete disappearance of gradient): throughout the universe, $\nabla\rho = 0$ holds strictly as a limiting state rather than an asymptotic process. This is the gradient terminal phase: the topic of this chapter.

Precisely stated: the gradient terminal phase refers to the state of affairs at the ontological limit of EQT, namely $\nabla\rho(x,t) \equiv 0$ holding for all x and all t.

Why Case C Must Be Treated Instead of Being Satisfied with Case B

A physicist might object: Case C will never actually be reached; it is only the mathematical limit of Case B. Within physics, this attitude is reasonable. Physics is concerned with physical processes, and strict $\nabla\rho \equiv 0$ may never be reached because of quantum fluctuations, the cosmological constant, and other factors. But in metaphysics, this attitude is insufficient. The reasons are as follows.

First, the asymptotic state of physics presupposes the continued existence of an observational perspective. In Case B, when we say "the universe approaches heat death," this statement assumes some continuing observational perspective capable of examining this asymptotic process. But one of the core insights of EQT is that the observer itself is a gradient structure within ρ_f. The existence of an observational perspective itself requires $\nabla\rho \neq 0$. Therefore, within EQT, the claim in Case B that "gradient approaches zero but an observational perspective still exists" is not fully self-consistent: it requires an observational perspective that does not depend on gradient, which conflicts with co-evolutionary ontology. Case C pushes this inconsistency to its limit and makes it a problem that must be faced directly.

Second, the physical claim that it is "never reached" is a claim about events within time. But this chapter is not concerned with when the terminal phase occurs; it is concerned with the logical structure of the terminal phase as an ontological limit. Even if the terminal phase is never reached in physical time, it still exists as the logical boundary of EQT ontology and imposes constraints upon all ontological statements. The logical limit of a system determines the semantic content of statements within it, independently of physical attainability.

Third, constrained speculation requires every ontological commitment to state its own failure conditions (see Chapter 1, Section 1.2, and Chapter 3). The failure condition of the commitment "existence $\iff \nabla\rho \neq 0$" is precisely "$\nabla\rho = 0$." If this condition is not treated, the methodological discipline of this book is violated. The constrained speculation defended in Part I receives its deepest application here: it requires EQT to examine itself at the level of its most central commitment.

The Key Distinction of This Section

The gradient terminal phase is not a physical event, but the logical boundary of EQT ontology. It does not need to "happen" in order to have metaphysical significance. As the logical limit of all ontological commitments of EQT, it already has metaphysical significance.

This distinction is crucial. It positions this chapter as a metaphysical chapter rather than a cosmological prediction. This chapter does not predict the future of the universe; it analyzes the intrinsic structure of EQT ontology.

Objections and Replies

Objection One: If Case C is never reached, is talking about it not a metaphysical game detached from reality?

Reply: No. Examining the logical limit of an ontological system is a necessary part of that system's self-understanding. Kant's examination of the antinomies of pure reason is not a game detached from reality, but reason's self-reflection. Wittgenstein's examination of the limits of language is not a word game, but fundamental work in the philosophy of language. This chapter is work of the same kind: it is the self-reflection of EQT ontology, not speculation about some future event.

Objection Two: Is $\nabla\rho \equiv 0$ *physically definable? Because of quantum fluctuations, strict* $\nabla\rho \equiv 0$ *may not exist physically.*

Reply: This chapter acknowledges this point. But quantum fluctuations themselves are some kind of nonzero structure of the ρ field. If even quantum fluctuations disappeared, if that is in principle conceivable, then this is precisely what we call the gradient terminal phase. This chapter is concerned with the ontological structure of this limit, not with whether it is physically attainable. Physical attainability is a problem of physics; this chapter is a metaphysical inquiry.

Objection Three: Is your definition of the gradient terminal phase overly artificial? Why choose $\nabla\rho \equiv 0$ *as the singularity rather than the limit of some other physical quantity?*

Reply: The choice of $\nabla\rho \equiv 0$ is not arbitrary. It is the direct logical antithesis of the core ontological thesis of EQT: "existence $\Longleftrightarrow \nabla\rho \neq 0$." Any thorough ontology must examine the failure condition of its core thesis. This is its ontological singularity. What is selected as the singularity is not decided by us, but by the logical structure of the ontology itself. The ontological core of EQT is gradient, and therefore its singularity must lie at the limit of gradient.

Summary of This Section

This section has done two things. First, it precisely defined the gradient terminal phase: $\nabla\rho \equiv 0$ as the logical limit of EQT ontology. Second, it argued that this definition is not a physical prediction but a metaphysical thesis: it concerns the intrinsic logical structure of EQT ontology, not the concrete future of the universe.

The next section turns to the most crucial argument of this chapter: the gradient terminal phase is neither "nothing" nor "being."

17.3. Neither "Nothing" nor "Being"

This section is the most crucial metaphysical argument of the chapter. It proves that the gradient terminal phase cannot be stated as "nothing exists," nor can it be stated as "there exists some uniform state." To avoid rhetorical discussion, this section adopts a strict formal argument.

Exclusion Argument One: The Terminal Phase Is Not "Nothing"

Intuitively, when $\nabla\rho \equiv 0$, it seems that "there is nothing." The energy quantum density field may still have a nonzero value, but there is no spatial or temporal difference: this looks like "nothingness." But this intuitive statement requires the existence of a perspective of statement. There must be some "I" or "some observer" to say "there is nothing."

According to Part III, co-evolution as an ontological relation, the observational perspective itself is a local structure within ρ_f. This local structure presupposes gradient: the observer as a local condensed structure within the ρ_f field requires internal gradient, to maintain its own organization; external gradient, to form a co-evolutionary relation with the environment; and temporal gradient, to realize state change. All three kinds of gradient require $\nabla\rho \neq 0$ to hold in some sense. Therefore, when $\nabla\rho \equiv 0$, there is no observational perspective capable of making a statement.

Thus the proposition "there is nothing" cannot be stated from within the terminal phase. Any such statement comes from a perspective outside the terminal phase, that is, from a perspective at a moment when the terminal phase has not yet been realized. But that perspective is about the terminal phase, not in the terminal phase.

This distinction is crucial: statements about the terminal phase and statements within the terminal phase are different. The former may be possible, and this is what this chapter is doing; the latter is impossible, because the perspective of statement has been canceled.

But "there is nothing," as an ontological proposition, should hold from within the stated state of affairs: it should state what the state of affairs itself is like, not how the speaker views that state of affairs. When we say "in state X, there is nothing," the ontological content of this statement should be independent of the speaker. But in the gradient terminal phase there is no perspective capable of making a statement, and therefore "there is nothing," as an ontological description of the terminal phase, has no correspondence.

Therefore, the proposition "gradient terminal phase = nothing at all" does not hold as an ontological statement.

Exclusion Argument Two: The Terminal Phase Is Not "Being"

Then is the terminal phase "some uniform state"? For example, "the universe is in a completely uniform energy distribution"?

This statement requires difference for support. To say "completely uniform" is to describe a state in contrast with "non-uniform." Any description depends on distinction: to describe X is to distinguish

X from non-X. But in the state of affairs $\nabla\rho \equiv 0$: there is no difference among spatial regions, because the gradient is zero; there is no difference among different moments, because the rate of change of gradient is also zero, otherwise a nonzero gradient would be generated; there is no difference between observer and observed, because the observational relation collapses, as shown in Exclusion Argument One; and there is no difference between "being this way" and "not being this way," because the basis of all distinction has disappeared.

Without any difference for support, the description "uniform" loses its semantic content. Uniformity is a concept relative to non-uniformity. A state of affairs with absolutely no difference of any kind cannot be described as "uniform," because the description "uniform" already presupposes the distinction from "non-uniform."

More strictly: to describe a state of affairs X as "uniform," there must be a semantic space in which "uniform" and "non-uniform" are distinguishable options. But the gradient terminal phase is precisely the state of affairs in which this semantic space collapses: all distinctions fail. Therefore, "uniform" has no referent in the terminal phase.

Thus the proposition "gradient terminal phase = uniform state," as a description of the terminal phase, is meaningless within the terminal phase.

The Metaphysical Consequence of Double Exclusion

The gradient terminal phase can be stated neither as "nothing" nor as "being." This looks like a contradiction: should any state of affairs not be either some kind of "being" or "nothing"?

This section argues that the gradient terminal phase is precisely the state of affairs that escapes the "being/nothing" dichotomy. This escape is not rhetorical; it has a ground. The "being/nothing" dichotomy itself presupposes a state-structure in which distinction can be made. But the gradient terminal phase is precisely the state of affairs in which distinction itself collapses. Therefore, in the terminal phase, the "being/nothing" dichotomy does not apply.

This conclusion is the deepest application of constrained speculation: it acknowledges that EQT ontology, at its own limit, encounters a state of affairs that cannot be covered by ontological categories. This is not the failure of EQT, but the mark of EQT as a thorough ontology. An ontology that cannot identify its own limit is not thorough. EQT identifies this limit at the level of its most central thesis.

Comparison with Traditional Metaphysics

This conclusion resonates with core theses in several philosophical traditions.

In *What Is Metaphysics?* (1929), Heidegger proposes that the Nichts is not the negation of anything, but a state of affairs prior to "being/nothing." Heidegger reaches this conclusion through a phenomenological analysis of Angst: in anxiety we encounter Nichts nichtet, "the nothing nothings." This phenomenon cannot be understood as "there is nothing at all," which would be logical negation and still presuppose things as objects that can be negated; rather, it is an originary state prior to the distinction between being and nothing. The gradient terminal phase of this chapter is structurally isomorphic with it.

In *Tractatus Logico-Philosophicus* 6.45, Wittgenstein writes: "The view of the world as a whole is mystical." The world as a whole cannot be stated propositionally: any proposition presupposes the world as the stage of propositions, and therefore cannot take the stage itself as its object of statement. The gradient terminal phase of this chapter is precisely the state of affairs in which "the stage itself collapses."

Nāgārjuna's fourfold negation (catuṣkoṭi) in the *Mūlamadhyamakakārikā*, "not arising and not ceasing, not one and not different, not coming and not going, not permanent and not annihilated," is the strictest expression of an ultimate state of affairs that can be stated neither positively nor negatively. In the language of this chapter, this is precisely the logical structure of "neither being nor nothing."

The first chapter of Laozi's *Daodejing*, "The Dao that can be spoken is not the constant Dao," is the Chinese metaphysical identification of an ultimate boundary. The gradient terminal phase of this chapter is structurally of the same kind.

Chapter 19 will enter into detailed dialogue with these traditions. Here we merely indicate the structural isomorphism. But the specificity of EQT must be emphasized: this identification is not an expression of religious or mystical experience, but a logical result rigorously derived from an explicit ontological postulate, $\nabla\rho \neq 0$. Heidegger, Wittgenstein, Nāgārjuna, and Laozi each reach a similar boundary through different paths; EQT reaches the same boundary through strict ontological inference. This is a metaphysical event in the field of natural philosophy.

Objections and Replies

Objection One: If the "being/nothing" dichotomy does not apply, how can the concept of "gradient terminal phase" itself be meaningful?

We are using language to discuss a state of affairs in which the capacity for linguistic statement collapses.

Reply: This difficulty is real, and it constitutes the central challenge of the chapter. The next section, Section 17.3, is devoted to treating it. Here we only note in advance that statements about the terminal phase differ from statements within the terminal phase. We can examine the logical structure of the terminal phase from outside without claiming to state it from within. The legitimacy of this chapter does not depend on making statements within the terminal phase, but on meta-level statements about the structure of EQT ontology.

Objection Two: Does this argument depend on a loose concept of "semantics"? Strictly speaking, "uniform" can be defined as "gradient equals zero," without any contrast.

Reply: Technically, it can be defined in this way, but semantic content still depends on contrast. "Gradient equals zero" as a statement requires "gradient not equal to zero" as a conceivable opposite state in order to have content. In a state of affairs absolutely without any contrast, the statement "gradient equals zero" refers to nothing; it degenerates into tautology. This argument is not loose, but a strict argument about semantic conditions.

Objection Three: Are you smuggling Heideggerian nihilism into EQT?

Reply: No. Heidegger's Nichts is an ontological event reached through Angst and is essentially a phenomenological description. The gradient terminal phase of this chapter is a logical boundary rigorously derived from postulates and is essentially a metaphysical argument. The two are structurally similar, since both identify a state of affairs prior to the being/nothing dichotomy, but their paths of argument are entirely different. EQT does not need to appeal to phenomenological experience in order to reach this boundary; it reaches it through logical analysis of ontological postulates.

Summary of This Section

Through two exclusion arguments, this section has proved that the gradient terminal phase is neither "nothing" nor "being." It escapes the being/nothing dichotomy. This is not a contradiction or paradox, but the legitimate logical structure of EQT ontology at its own limit.

But this immediately raises the next question: if the terminal phase is neither being nor nothing, what are statements about it? This is the topic of Section 17.3.

17.4. Neither Proposition nor Non-Proposition

The previous section argued that the terminal phase is neither "nothing" nor "being." This section argues a deeper point: statements about the terminal phase can become neither propositions nor non-propositions. This is the deepest difficulty encountered by this chapter, and also the part in which it makes its most significant philosophical contribution.

The Minimum Conditions of Propositional Statement

For a statement to become a proposition, at least three conditions are required. This is a minimum requirement. Whether in the Frege–Russell theory of propositions, Wittgenstein's truth-functional theory, or contemporary analytic philosophy's views of propositions, some version of these three conditions is accepted:

Condition 1 (subject): there must be a perspective of statement, some "speaker," even if abstract or anonymous. Condition 2 (object): there must be a referent object, some state of affairs being stated, which may be concrete or abstract. Condition 3 (semantic content): there must be a definable semantic relation between statement and object; the statement says something about the object, and this "something" can be identified as true or false.

In the gradient terminal phase: Condition 1 fails, because the observational perspective is a gradient structure within ρ_f, and when $\nabla\rho \equiv 0$, there is no perspective from which a statement can be made. Condition 2 fails, because the referred state of affairs itself is a state of "no difference" and does not constitute a referable object; reference requires distinction, while all distinction collapses in the terminal phase. Condition 3 fails, because semantic relation requires distinction, including the distinction between referring and referred, and between the content of a statement and its opposite; but all distinction collapses in the terminal phase. Therefore, statements about the terminal phase cannot become propositions: any such statement fails to satisfy the minimum conditions of propositionhood.

And Yet: They Cannot Become Non-Propositions Either

Intuitively, if a statement cannot become a proposition, it is a non-proposition, namely nonsense, error, or meaninglessness. But this section argues that statements about the terminal phase cannot become non-propositions either. The reason is as follows.

A non-proposition requires at least contrast with proposition: it is the failure, error, or absence of a proposition. To say "X is not a proposition" means that X is comparable with the concept of proposition. To say X is a non-proposition means that X enters the contrastive domain of proposition/non-proposition. But if the minimum conditions of propositional statement themselves all collapse in a certain state of affairs, then the non-proposition as the negated form of proposition simultaneously loses its ground, because "non-proposition" is defined relative to "proposition," and proposition itself has no definition in that state of affairs.

More strictly: the statement "X is a non-proposition" is itself a proposition; it requires X to be comparable with the concept of proposition. But X, a statement about the terminal phase, does not even satisfy the minimum conditions of propositionhood. Therefore X is not within the contrastive domain of proposition and non-proposition.

The key point of this argument is that the contrastive domain of proposition/non-proposition itself requires a structure of difference as its ground: there must be a conceivable distinction between "proposition" and "non-proposition." But the gradient terminal phase is precisely the state of affairs in which the structure of difference collapses. There, the contrastive domain of proposition/non-proposition itself does not hold. Therefore, statements about the terminal phase are neither propositions nor non-propositions: they stand outside this contrastive domain.

Double Exclusion as the Mark of Ontological Singularity

The fact that terminal-phase statements are excluded simultaneously from both proposition and non-proposition is the core mark of the ontological singularity of EQT. This is structurally isomorphic with the logic of Nāgārjuna's "fourfold negation" (catuṣkoṭi) in the *Mūlamad-hyamakakārikā.* When facing the problem of nirvāṇa, Nāgārjuna states: it is not being, for one cannot say that nirvāṇa is a certain state of affairs; it is not non-being, for one cannot say that nirvāṇa is not a state of affairs; it is not both being and non-being, for one cannot say that nirvāṇa both is and is not; and it is not neither being nor non-being, for one cannot say that nirvāṇa neither is nor is not. This logical structure is rare in the Western philosophical tradition, but the ontological singularity of EQT structurally requires the same fourfold negation: not proposition; not non-proposition; not both proposition and non-proposition, since this itself remains a propositional statement and therefore also does not hold; and not neither proposition nor non-proposition, since this statement itself is a proposition and therefore still does not hold. This logical structure shows that at the gradient terminal phase, we encounter not merely the failure of a particular statement, but the suspension of the capacity for statement itself.

The Key Metaphysical Posture

The key point of this section is not to provide a new "proposition" to describe the terminal phase, but to acknowledge that the terminal phase exceeds the proposition/non-proposition dichotomy. This is the deepest posture of EQT ontology: EQT does not pretend that it can still make statements at its own boundary. It points out the existence of the boundary; it analyzes the structure of the boundary; and it acknowledges the collapse of the capacity for statement beyond the boundary. At its most radical point, it suspends itself.

This is the deepest application of constrained speculation: a truly thorough ontology must acknowledge this limit. Any ontology that does not acknowledge this limit is either not thorough, or self-deceptive.

The Legitimacy of the Statements in This Chapter

But there is an urgent problem: this chapter itself is using propositional statements to say things about the terminal phase. What are these propositions? They are not statements within the terminal phase, because there is no perspective within the terminal phase; and according to the arguments in Section 17.2 and the first part of this section, they are also not propositions about the terminal phase. Then what are they?

This section gives the answer: they are statements about the structure of EQT ontology itself, not statements about the terminal phase. More precisely: this chapter is not saying "what the terminal phase is like"; it is saying that "the logical structure of EQT ontology causes it to encounter a collapse of its capacity for statement at the limit $\nabla\rho \equiv 0$." This is a meta-level statement about EQT ontology, not an object-level statement about the terminal phase.

The former is legitimate: EQT ontology is a concrete object, a theoretical system, which can be reflected upon, analyzed, and stated. We can say "this ontological system fails at $\nabla\rho \equiv 0$," just as we can say "Peano arithmetic encounters a limit at Gödel's incompleteness theorem." The latter is impossible: the terminal phase itself is not a legitimate object. It is the singularity of EQT ontology, the state of affairs in which all ontological categories of the system fail. We cannot directly state what it is like.

This distinction rescues the possibility of this chapter: it explains that this chapter is doing legitimate metaphysical work, even though the state of affairs it discusses itself exceeds legitimate statement. This is structurally of the same kind as Tarski's object-language/meta-language distinction, Russell's type theory, and Carnap's internal/external question distinction. In each case, a level distinction makes legitimate statements about a certain limit possible.

Objections and Replies

Objection One: You are merely pushing the problem to the meta-level, not truly solving it. This chapter still seems to describe the terminal phase.

Reply: Yes, this chapter intentionally pushes the problem to the meta-level, but this is not avoidance. The distinction between meta-level and object-level is a legitimate tool in analytic philosophy: Russell's type theory uses a level distinction to solve paradoxes; Tarski's theory of truth uses a level distinction to solve the liar paradox; Carnap's internal/external question distinction uses a level distinction to solve metaphysical disputes. This chapter applies this distinction to make legitimate metaphysical work about the terminal phase possible. This is a standard methodological resource of analytic philosophy, not an evasion.

Objection Two: Is this "double exclusion" merely a logical game? You seem to be playing with concepts.

Reply: No. Every step of exclusion has a ground: the collapse of the conditions of propositionhood is rigorously derived from co-evolutionary ontology, defended in Part III; the relative relation between non-proposition and proposition is rigorously derived from propositional logic, and is accepted by any theory of proposition. This is not a game, but a strict characterization of an ontological limit. The difference from a game is this: a game is arbitrary conceptual manipulation, while the argument of this section begins from defended ontological theses and draws its conclusion through strict logical steps.

Objection Three: You seem to be using argument to prove that argument is impossible. Is this self-refuting?

Reply: This chapter does not prove that "any argument is impossible"; it only proves that "object-level arguments about the terminal phase are impossible." Meta-level argument, namely this chapter itself, is possible because it concerns the structure of EQT ontology, not the terminal phase itself. This is a rule-governed level distinction, not self-refutation. Wittgenstein's famous "ladder metaphor" at the end of the *Tractatus*, where one must throw away the ladder after climbing it, is a level operation of the same kind: one reaches the boundary through the ladder, namely meta-level argument, and then acknowledges that the ladder itself no longer applies at that point.

Objection Four: You mention Nāgārjuna's fourfold negation, but Nāgārjuna is a religious philosopher whose conclusions are grounded in prajñā wisdom. This is different in kind from your ontological inference.

Reply: Yes, the paths are entirely different: Nāgārjuna arrives through religious cultivation, while EQT arrives through ontological inference. But the structure is isomorphic. Chapter 19 will discuss these similarities and differences in detail. Here we only indicate the structural isomorphism and do not claim identity in path or meaning.

Summary of This Section

This section has done three things. First, it proved that statements about the gradient terminal phase cannot become propositions: the three minimum conditions of propositionhood all collapse in the terminal phase. Second, it proved that they cannot become non-propositions either: non-proposition requires contrast with proposition, but this contrastive relation collapses in the terminal phase. Third, through the meta-level/object-level distinction, it explained what this chapter itself is doing: it makes meta-level statements about the structure of EQT ontology, not object-level statements about the terminal phase.

This double exclusion is the core logical mark of the ontological singularity of EQT. The next section gives this phenomenon a name: the intrinsic singularity of ontology.

17.5. The Intrinsic Singularity of Ontology

The work of the first three sections can be summarized in one concise statement: the gradient terminal phase is the intrinsic singularity of EQT ontology. This section discusses the precise meaning of this naming.

Conceptual Naming

"Singularity" is a concept borrowed from mathematics and physics, and it has a specific meaning within the EQT framework:

Singularity in the mathematical sense: a point of a function at which the behavior of the function breaks down, such as $1/x$ at $x = 0$, or $\tan(x)$ at $\pi/2$. At that point, the function itself is not defined in the standard sense, but its behavior can be examined by taking limits, Laurent expansions, or other methods.

Singularity in the physical sense: a limiting state of affairs in a physical theory at which the theory itself fails, such as the singularity at the center of a black hole or at the beginning of the universe in general relativity. At that point, the ordinary calculations of the theory no longer apply, but the theory can examine the logical structure of this limit from outside.

Singularity in the ontological sense, introduced in this chapter: a limiting state of affairs in an ontological system at which the system's capacity for statement collapses, while this collapse itself is a strict inference from the system's own logic.

This is a new concept: it introduces "singularity" from mathematics and physics into ontology. Its legitimacy lies in the fact that it captures the core structure of the arguments in the first three sections of this chapter: at the limit of EQT ontology, as the gradient terminal phase, all ontological categories fail, but this failure itself can be rigorously derived from EQT's own postulates.

The Precise Meaning of the Three Qualifiers

"The intrinsic singularity of ontology" consists of three qualifiers, each with a specific meaning.

"Of ontology": it is not a singularity in the mathematical sense, such as functional divergence caused by division by zero; nor in the physical sense, such as curvature divergence caused by a black hole; but a singularity concerning how existence can be stated. At the gradient terminal phase, ontological categories, including existence, nothingness, proposition, non-proposition, difference, uniformity, and so on, all fail. Its similarity with mathematical and physical singularities lies in structure, namely a limiting state of affairs and the failure of ordinary modes of statement. Its difference lies in content: it is ontological rather than functional or spatiotemporal.

"Intrinsic": the singularity is not externally added, not a limitation imposed from outside EQT, but a strict inference from EQT's own postulates. This is crucial: the discovery of the singularity is not the failure of EQT, but a sign of its thoroughness. Any thorough ontological system will have an intrinsic singularity. A system capable of examining its own limit will necessarily encounter such a limit at its most central thesis.

"Singularity": it is the limiting state of the system, not an intermediate case within the system. At it, the system's ordinary modes of statement fail. This differs from a "vague zone" or "gray area." A vague zone can be treated through more refined statements; a singularity is the collapse of the capacity for statement itself.

Defense of Intrinsicness

This section specifically defends "intrinsicness." Why say that the gradient terminal phase is an intrinsic singularity of EQT, rather than some defect of EQT?

Argument One, from pure inference from postulates: the gradient terminal phase is derived entirely from the logical analysis of the postulate $\nabla\rho \neq 0$. It is precisely the failure condition of this postulate itself. According to constrained speculation (Chapter 3), an ontological postulate must have failure conditions, and its failure condition is the intrinsic boundary of the system. Therefore, encountering an ontological singularity is not a weakness of EQT, but the necessary result of its being an ontological system with concrete content. An ontology without failure conditions is empty, as argued in Section 1.2 of constrained speculation.

Argument Two, from the logic of co-evolution: co-evolutionary ontology, developed in Part III, requires observer and object to belong to the same ρ_f field. This requirement strictly means that when the ρ_f field becomes uniform, $\nabla\rho \equiv 0$, observation and object dissolve together. This is not an additional commitment of EQT, but the logical structure of co-evolutionary ontology. The intrinsic logic of co-evolutionary ontology requires that it collapse at its own limit. This is the same thing as co-evolution itself.

Argument Three, from the discipline of constrained speculation: constrained speculation requires every ontological commitment to have failure conditions. The common failure condition of all ontological commitments of EQT is $\nabla\rho \equiv 0$, because all ontological commitments of EQT depend on the gradient postulate. Therefore, the gradient terminal phase as ontological singularity is a logical necessity of the methodology of EQT, not an accident.

The three arguments prove the same conclusion from different angles: the gradient terminal phase as an intrinsic singularity is not a weakness of EQT, but the mark of EQT as a thorough ontology.

Comparison with Physical Singularities

The relation between the ontological singularity of EQT and physical singularities, such as the center of a black hole or the beginning of the universe, deserves discussion. Similarities: both are limiting states of a theory; both are accompanied by the failure of ordinary modes of statement; and both are derived from the internal structure of the theory itself. Key difference: physical singularities may be defects of a physical theory. Singularities in general relativity, such as curvature divergence at the center of a black hole, are often understood as signs that the theory needs to be supplemented by a theory of quantum gravity. The singularity suggests that the theory requires revision. The ontological singularity of EQT is not a defect. It is not a sign that EQT needs to be supplemented by another ontology. It is rigorously derived from EQT postulates and is the logical structure of the system itself. Any ontological system, if thorough, will have its own intrinsic singularity. This is the lesson of philosophical history: Kant, Wittgenstein, and Nāgārjuna all reached similar boundaries. This difference gives the ontological singularity of EQT a special status: it does not require EQT to be revised, but requires EQT to be understood as a system that acknowledges its own limits. Acknowledging the limit is the completed form of EQT as an ontology.

Historical Precedents

EQT is not the first ontology in history to acknowledge an intrinsic singularity. Three precedents can be listed.

Kant's *Critique of Pure Reason.* Kant identifies an intrinsic boundary encountered by pure reason when it attempts to know the "thing-in-itself": the categories of reason can be applied only to phenomena, not to the thing-in-itself. Attempting to push the categories to the thing-in-itself produces antinomies: with respect to the same object, one can prove both X and non-X. Kant's work is an epistemological singularity, but structurally similar to EQT's ontological singularity: both derive the intrinsic boundary of a system from the logical structure of the system itself.

Wittgenstein's *Tractatus Logico-Philosophicus.* Wittgenstein identifies the intrinsic boundary encountered by language when it attempts to state "the world as a whole" or "logical form itself." Language can describe facts, but it cannot describe the formal conditions that make description possible. Attempting to state these conditions crosses a boundary and leads to the collapse of language; this is the meaning of the famous "ladder metaphor." Wittgenstein's work is a singularity in the philosophy of language, structurally of the same kind as the ontological singularity of EQT.

Nāgārjuna's *Mūlamadhyamakakārikā.* Nāgārjuna identifies the intrinsic boundary encountered by prajñā wisdom when it attempts to grasp "self-nature" (svabhāva). Any attempt to understand things as having their own intrinsic nature encounters emptiness (śūnyatā), and this encounter is expressed through the fourfold negation. Nāgārjuna's work is a singularity in religious philosophy, but its logical structure is highly isomorphic with the ontological singularity of EQT.

The specificity of EQT is this: it is the intrinsic singularity of natural philosophy, rigorously derived from an explicit ontological postulate that can be connected with physics. This is a rare metaphysical event in history. Kant, Wittgenstein, and Nāgārjuna each reached structurally similar boundaries through work in different domains; EQT does similar work in the domain of natural philosophy.

Objections and Replies

Objection One: Is comparing EQT with Kant, Wittgenstein, and Nāgārjuna an overstatement?

Reply: This is not elevation, but structural comparison. These three predecessors identified intrinsic boundaries of ontology or epistemology in different domains. EQT does similar work in the domain of natural philosophy and should be examined with the same seriousness, as well as subjected to the same severity of criticism. The purpose of comparison is not to raise EQT to that height, but to identify the kind of metaphysical tradition to which the work of EQT belongs. If this comparison does not stand, it should be criticized; this chapter accepts such criticism as one possible direction of failure.

Objection Two: Intrinsic singularity sounds like mysticism.

Reply: It is not mysticism at all. Mysticism abandons statement and appeals to transcendent experience; this chapter insists on rigorous statement and explicitly distinguishes meta-level, which is legitimate, from object-level, which is impossible. This is an anti-mystical posture: the boundary is identified through strict argument, not mystical declaration. A genuine mysticism would say, "the boundary cannot be spoken, therefore let us be silent." This chapter says, "the boundary cannot be stated at the object level, but it can be analyzed at the meta-level, and this analysis has a concrete logical structure." This is philosophical argument, not mystical declaration.

Objection Three: Does the discovery of an intrinsic singularity indicate a problem with the methodology of EQT?

Reply: On the contrary. Any thorough ontology necessarily encounters an intrinsic singularity: this is the shared lesson of Kant, Wittgenstein, and Nāgārjuna. An ontology without an intrinsic singularity is either not thorough or self-deceptive. Either it has not touched its own limit, or it pretends that it can still make statements at the limit. EQT's encounter with an intrinsic singularity is precisely proof of its thoroughness.

Objection Four: If intrinsic singularity is necessary, what distinguishes EQT from other ontological systems?

Reply: The difference lies in the concrete structure of the singularity. Different ontological systems have different core commitments and therefore different intrinsic singularities: Kant's singularity lies in the thing-in-itself; Wittgenstein's singularity lies in logical form; Nāgārjuna's singularity lies in self-nature. The singularity of EQT lies in the gradient terminal phase. This corresponds one-to-one with gradient as the core ontological category of EQT. The singularity of each system is the logical reflection of its core commitment.

The Meta-Level/Object-Level Distinction as the Ground of This Chapter's Legitimacy

Finally, it must be made clear that all statements in this section about the "intrinsic singularity of ontology," just as in Section 17.3, are meta-level statements: they concern the structure of EQT ontology, not the terminal phase itself. Naming the "intrinsic singularity of ontology" is not an object-level description of the terminal phase, which is impossible; it is a meta-level characterization of the logical structure that EQT ontology encounters at $\nabla\rho \equiv 0$. This distinction makes the naming work of this section legitimate: just as a mathematician can name the singularity of a function without having to "enter" the singularity, this section names the singularity of ontology without having to state it from within the terminal phase.

Summary of This Section

This section has done three things. First, it named the gradient terminal phase the "intrinsic singularity of EQT ontology" and precisely defined this concept through three qualifiers: ontological, intrinsic, and singularity. Second, it defended "intrinsicness": the singularity is not a defect of EQT, but a mark of its thoroughness. Third, it placed the ontological singularity of EQT within philosophical history and structurally compared it with the work of Kant, Wittgenstein, and Nāgārjuna.

The next section argues that acknowledging the intrinsic singularity of ontology is itself a philosophical contribution.

17.6. Acknowledging Limits as a Philosophical Contribution

The first four sections have done heavy metaphysical work. This section is brief but crucial: it explains that acknowledging the intrinsic singularity of ontology is itself a philosophical contribution, not a limitation or failure of EQT.

Three Levels of Contribution

Contribution One: methodological honesty. Constrained speculation requires every ontological commitment to state its failure conditions. EQT has done this with respect to its most central commitment, $\nabla\rho \neq 0$: it points out that when $\nabla\rho \equiv 0$, this commitment fails together with all commitments that depend on it. This is not the failure of some secondary commitment, but the simultaneous suspension of the entire ontology. Such a thorough realization of its own methodological discipline is rare in the history of philosophy. Most ontological systems do not state their own global failure conditions: Aristotelian substance ontology, Leibnizian monadology, Whitehead's process philosophy, various forms of physicalism. They either presuppose that they remain valid forever, or leave their failure conditions to external critique. EQT states its global failure condition internally. This is the deepest application of constrained speculation. This honesty is not rhetorical humility, but argumentative necessity. If constrained speculation requires every ontological commitment to have failure conditions, as defended in Chapter 3, then the core ontological commitment must also have failure conditions; otherwise the methodology would betray itself at its most important point. EQT does this in Chapter 17: it does not retreat at the most crucial point.

Contribution Two: metaphysical thoroughness. Whether an ontology is thorough depends on whether it can reach its own limit and honestly recognize that limit. Kant's pure reason, Wittgenstein's language, and Nāgārjuna's prajñā: the thoroughness of these philosophical traditions lies in their recognition of limits. EQT does similar work in the domain of natural philosophy. It does not pretend that it can still make statements at its own boundary. It points out the boundary, analyzes its structure, and acknowledges the collapse of the capacity for statement beyond the boundary. This is the radicalization of natural philosophy. Since the twentieth century, one main reason why natural philosophy has been marginalized is precisely that it has not reached its own limits: it either becomes a popularizing narrative of physics or unconstrained metaphysical speculation. EQT works at the interface between physics, the five-volume monograph series, and metaphysics, this book, and realizes thoroughness at its own limit. This reopens the possibility of natural philosophy.

Contribution Three: the ground of humanistic meaning. Finally, acknowledging the intrinsic singularity of ontology provides the ground for the humanistic arguments of Part V, especially Chapter 21, "The Affirmation of Finitude." If ontology has an intrinsic singularity, if all structures must ultimately face their own limits, then finitude is not a defect, but a structure. All meaning, value, love, and creation unfold within this finitude, not against some eternal background. This connects Chapter 17 with Chapter 21: ontological singularity is the ground of humanistic meaning. If finitude were not a structure, if there existed some ontologically infinite state of affairs, then meaning could be suspended into that infinite state: meaning would find its ground in eternity. But if finitude is a structure, as this chapter argues, because even ontology itself has an intrinsic singularity, then meaning must find its ground within finitude itself. Meaning is the affirmation of finitude. This connection constitutes the final unity of the book between metaphysics and the humanistic. Chapter 21 will develop this connection in detail. This chapter only identifies its ground.

17.7. Chapter Summary

This chapter has completed the heaviest metaphysical work of the whole book. It has done five things:

First, it precisely constructed the gradient terminal phase (17.1), defining it as the logical limit of EQT ontology rather than a physical event. This positioning makes the chapter a metaphysical chapter and prevents it from being confused with cosmological prediction.

Second, it proved that the terminal phase is neither "nothing" nor "being" (17.2). Through two strict exclusion arguments, it showed that the terminal phase escapes the being/nothing dichotomy. This is the metaphysical core of the chapter.

Third, it proved that statements about the terminal phase are neither propositions nor non-propositions (17.3). Through the collapse argument concerning the minimum conditions of propositionhood, it showed that the terminal phase exceeds the proposition/non-proposition dichotomy. This is the logical core of the chapter. At the same time, it rescued the legitimacy of the chapter itself through the meta-level/object-level distinction.

Fourth, it named the terminal phase the intrinsic singularity of EQT ontology (17.4). It precisely defined this concept, defended its intrinsicness, and placed the work of EQT within the philosophical-historical context of Kant, Wittgenstein, and Nāgārjuna. This is the naming core of the chapter.

Fifth, it argued that acknowledging ontological singularity is itself a philosophical contribution (17.5). It explained the value of this chapter at three levels: methodological honesty, metaphysical thoroughness, and the ground of humanistic meaning. This is the meta-level summary of the chapter.

Connections with Other Chapters of the Book

The position of this chapter in the whole book can be located as follows. Looking backward, this chapter is built upon all previous chapters: Part I, constrained speculation, provides the methodological ground; Part II, gradient ontology, provides the core ontological commitment; Part III, co-evolution, mind–matter, and observation, provides the unified framework for observer and object. All of these are premises of the arguments in this chapter. Looking forward, this chapter lays the foundation for subsequent chapters. Chapter 18 will further develop the collapse of co-evolutionary relations before the terminal phase: this is a concrete inference from this chapter. Chapter 19 will compare the ontological singularity of EQT in detail with the traditions of Heidegger, Wittgenstein, and Nāgārjuna: this is an extension of the naming work of this chapter. Part V, Chapters 20–21, will connect the metaphysical conclusion of this chapter with questions of normativity and meaning: this is the development of Contribution Three in this chapter. This chapter itself has completed the core work: it has pointed out the logical boundary of EQT ontology and honestly acknowledged the existence of this boundary. This is the deepest application of constrained speculation at its own limit.

A Final Reflection

This chapter has a special difficulty in writing: it must use language to state a state of affairs in which the capacity for linguistic statement collapses. This difficulty is not rhetorical, but philosophical. The chapter responds to it through the meta-level/object-level distinction:

it does not directly state the terminal phase, but states that the structure of EQT ontology causes it to encounter the collapse of its capacity for statement at the terminal phase. This is a standard methodological resource in analytic philosophy, but it is rarely applied to the problem of ontological limits in natural philosophy. This application is itself a philosophical contribution: it shows that legitimate work on ontological limits is possible, provided that we honestly distinguish meta-level from object-level and accept the constraints this distinction imposes upon the form of argument. This is how constrained speculation works at its own limit. In this sense, this chapter is both the metaphysical climax of the whole book and the deepest demonstration of the method of constrained speculation. It accomplishes what Chapter 1 promised: it allows a new method of natural philosophy to display its full force at its own limit.

17.8. Methodological Reflection on This Chapter

Chapter 17 is the most difficult chapter in the whole book to write. This section reflects on the methodological choices of this chapter as an honest explanation for the reader.

Why This Chapter Must Use Meta-Level Rather than Object-Level Argument

If this chapter tried to directly describe "what the gradient terminal phase is like," it would fall into self-contradiction, because any such description presupposes a perspective, while perspective is canceled in the terminal phase. The solution of this chapter is meta-level argument: it does not directly state the terminal phase, but states that the structure of EQT ontology itself causes it to encounter the collapse of its capacity for statement at $\nabla\rho \equiv 0$. This choice is not an avoidance of the problem, but the only philosophically legitimate path possible. Kant, Wittgenstein, and Nāgārjuna all made similar choices: they did not directly state the state of affairs beyond the boundary, but stated the structure of the boundary itself. Kant stated the boundary of reason, but did not try to describe the content of the "thing-in-itself." Wittgenstein stated the boundary of language, but remained silent about the content beyond the boundary. Nāgārjuna pointed toward emptiness through double negation. This chapter applies a method of the same kind in the domain of natural philosophy.

The Deepest Application of the Method of Constrained

Speculation in This Chapter

This chapter is one of the deepest applications of the method of constrained speculation. The method of constrained speculation requires every ontological commitment to have failure conditions (CR2). This chapter undertakes the statement of the failure condition of EQT ontology itself. The gradient terminal phase, $\nabla\rho \equiv 0$, is the common failure condition of all ontological commitments of EQT. This undertaking makes the work of this chapter completely honest: it does not pretend that EQT ontology is an absolute framework without boundaries, but explicitly states the intrinsic boundary of EQT ontology. This posture of undertaking the limit is itself a philosophical contribution. It makes EQT ontology thorough rather than floating.

Directions in Which This Chapter Can Be Further Deepened

The work of this chapter can be further deepened. Each section can be developed further in a dedicated paper. In particular, the double exclusion arguments in Sections 17.2 and 17.3 deserve rigorous formal development through formal logic. This is the standard practice of analytic metaphysics and can be completed in an appendix or subsequent paper. But the core argument of this chapter as a monograph chapter is already complete. More refined formalization and more detailed dialogue with philosophical history are left to future research.

Limits Honestly Acknowledged in This Chapter

This chapter acknowledges several limits. First, the argument that "any thorough ontology encounters an intrinsic singularity" is not absolute. If there exists an ontological system that does not encounter an intrinsic singularity at its own limit, the argument of this chapter would need revision. This chapter does not claim that this argument is absolute. It claims only that it holds rigorously within the EQT framework, and that, supported by precedents in philosophical history, it has broader possible applicability. Second, the chapter's dialogue with Kant, Wittgenstein, and Nāgārjuna is relatively brief; Chapter 19 will treat it in greater detail. Third, the chapter is developed mainly within the EQT framework; its application in other ontological frameworks requires dedicated work.

Connections with Later Chapters

This chapter provides the ontological ground for several later chapters. Chapter 18 is directly built upon this chapter: it develops the collapse of co-evolutionary relations in the process of approaching

$\nabla\rho \to 0$. Chapter 19 develops the position of this chapter's work within several philosophical traditions. Chapter 21 uses the singularity work of this chapter as the ontological ground for the finitude of meaning. If this chapter were removed, the metaphysical climax of the whole book would disappear: Chapters 18 and 19 would become ungrounded extensions; the humanistic argument of Chapter 21 would lose its ontological ground; and the dialogue of the whole book with the traditions of Kant, Wittgenstein, and Nāgārjuna would degenerate into rhetorical citation. Chapter 17 is the true metaphysical load-bearing point of the whole book. It must exist, and it must exist in this form.

18. The Collapse of Co-Evolution before the Gradient Terminal Phase

18.1. Introduction to This Chapter

Chapter 17 established the ontological singularity at $\nabla\rho \equiv 0$ as the intrinsic limit of ontology. But before the singularity itself, there is an approaching process: the dynamics of $\nabla\rho \to 0$. This approaching process is not the singularity itself, but the dynamics directed toward the singularity. The singularity, $\nabla\rho \equiv 0$, is methodologically restricted according to the work of Chapter 17, where object-level statements are restricted. But the approaching process, $\nabla\rho \to 0$, involves content that can be stated: decline, collapse, and stages. This chapter treats that approaching process.

The core thesis of this chapter is that co-evolutionary relations collapse in the process of approaching the gradient terminal phase, $\nabla\rho \to 0$ (EC1): their dynamical mechanisms, frequency matching, threshold, and tri-state response, gradually fail and finally cannot be sustained at the limit of $\nabla\rho \equiv 0$. Co-evolution as an ontological relation, established in Chapter 12, is built upon three dynamical mechanisms; these three mechanisms undergo decline in the process of $\nabla\rho \to 0$, and ultimately cannot be sustained at the singularity.

The collapse involves three stages: the decline of frequency matching, where in the low-gradient limit the frequency spectrum contracts and the frequencies available for matching decrease; the failure of threshold, where in the low-gradient limit perturbation amplitude decreases and the threshold condition becomes increasingly difficult to satisfy; and the convergence of tri-state weights, where in the low-gradient limit the tri-state weights tend toward uniformity and the tri-states lose their function as ontological categories. These three stages are not independent. They are mutually connected and form a mutually declining cycle.

The ontological meaning of collapse is not merely "the end of a process," but the intrinsic boundary of relational ontology itself (EC2). Relational ontology is built upon dynamical mechanisms. As the dynamical ground declines, the ground of relational ontology is eroded. At the singularity, relational ontology cannot be sustained. This boundary is not a limitation added to relational ontology; it is the intrinsic structure of relational ontology itself.

This chapter is a connecting chapter in Part IV, "The Metaphysical Boundary of Freezing": it functions between Chapter 17, which establishes the ontological singularity, and Chapter 19, which enters into dialogue with three traditions. Chapter 17 establishes the singularity itself as the intrinsic boundary of ontology; this chapter develops the process of approaching the singularity, namely how co-evolutionary relations collapse; Chapter 19 places this boundary work within philosophical discussion.

This chapter faces a methodological challenge in writing: it must strictly preserve the meta-level/object-level distinction developed in Chapter 17, Section 17.4. This chapter cannot simply state that "at the gradient terminal phase co-evolution has collapsed," because this statement concerns the object level, a statement about the state of the singularity, and exceeds the legitimate methodological range. This chapter can only state the content of the approaching process: how co-evolutionary relations decline in the process of $\nabla\rho \to 0$.

The discussion proceeds as follows. Section 18.1 develops the starting point of collapse: the approaching process vs the singularity itself, and the core thesis EC1. Section 18.2 develops the three stages of collapse and their interrelations. Section 18.3 develops the ontological meanings of collapse: the intrinsic boundary of relational ontology, the processuality of ontology itself, the boundary of meaning, and the application of methodological rigor. Section 18.4 develops the position of this chapter within Part IV and its connecting function.

18.2. The Starting Point of Collapse

Clarifying the Starting Point

The starting point of this chapter is this: before the ontological singularity established in Chapter 17, at $\nabla\rho \equiv 0$, there is an approaching process, namely the dynamics of $\nabla\rho \to 0$. This process is not the singularity itself, but the dynamics directed toward the singularity. Specifically: the singularity, $\nabla\rho \equiv 0$, is the intrinsic boundary of ontology, and according to the work of Chapter 17, object-level statements about the singularity itself are methodologically restricted; the approaching process, $\nabla\rho \to 0$, is a dynamical process involving content that can be stated: decline, collapse, stages, and so on. This chapter treats that process. This distinction clarifies the legitimate scope of this chapter: this chapter does not attempt to speak of the state of the singularity itself, but speaks of the process of approaching the singularity.

The Core Thesis of Co-Evolutionary Collapse

Core thesis EC1: in the process of approaching the gradient terminal phase, $\nabla\rho \to 0$, co-evolutionary relations (Chapter 12, EV1) undergo collapse: their dynamical mechanisms, frequency matching, threshold, and tri-state response, gradually fail and finally cannot be sustained at the limit of $\nabla\rho \equiv 0$.

This thesis requires detailed development. Let us analyze three implications.

Implication One: The Dynamical Requirements of Co-Evolution as Relational Ontology

Chapter 12 established co-evolution as an ontological relation: beings constitute ontological relations through co-evolution, including frequency matching, threshold, tri-state response, and so on. This ontological relation is built upon three dynamical requirements: frequency matching, where the frequencies among beings need to match under the Lorentzian resonance kernel $g(f, f_0)$; threshold, where perturbation amplitude needs to exceed the threshold $\delta\rho > \Gamma/k$ in order to excite nonlinear feedback; and tri-state response, where response involves the weights of the free state, bound state, and condensed state. These three dynamical mechanisms are the physical foundation of co-evolutionary relations.

Implication Two: The Decline of Dynamical Mechanisms

In the process of approaching $\nabla\rho \to 0$, all three dynamical mechanisms undergo decline: the decline of frequency matching, where in the low-gradient limit the field structure becomes uniform and the frequency spectrum contracts, reducing the frequencies available for matching; the failure of threshold, where in the low-gradient limit perturbation amplitude $\delta\rho$ decreases because there is no gradient structure as a reference, making the threshold increasingly difficult to reach; and the convergence of tri-state weights, where in the low-gradient limit the tri-state weights tend toward uniformity, the free-state weight $\to 1$, and the bound-state and condensed-state weights $\to 0$, so that the tri-states lose their function as ontological categories. The decline of each mechanism erodes the ontological ground of co-evolutionary relations.

Implication Three: The Intrinsic Boundary of Relational Ontology

The collapse of co-evolutionary relations is not an externally added limitation. It is the intrinsic boundary of relational ontology itself. Relational ontology is built upon dynamical mechanisms: without mechanisms, relational ontology loses content. In the process of $\nabla\rho \to 0$, dynamical mechanisms decline: the ground of relational ontology is eroded. At the limit of $\nabla\rho \equiv 0$, the singularity, the mechanisms cannot be sustained: the legitimate range of relational ontology ends. This boundary is not a limitation added to relational ontology; it is the intrinsic structure of relational ontology itself.

An Analogy: A Stream Approaching Its Source

To keep the concept of collapse from remaining abstract, let us consider an analogy: the process of a river approaching its source. Downstream, the water flow is strong, vortices and interactions are abundant, and dynamical relations among flows are active. As the river approaches its source, the flow weakens, vortices decrease, and the dynamical relations among flows decline. At the source itself, there is no flow and no dynamical relation. This analogy shows the structure of collapse: dynamical relations decline in the process of approaching the boundary and cannot be sustained at the boundary.

But this analogy has limits: the source of a river is a physical boundary, whereas this chapter treats an ontological boundary. The ontological boundary involves a deeper meaning: not merely the end of some dynamics, but the boundary of relational ontology itself. The following sections develop this meaning.

Objections and Replies

Objection One: Your concept of "collapse" implies a negative development, the disintegration of ontology. But the posture of this book should be neutral.

Reply: This chapter uses "collapse" as a term, but it does not carry negative emotional coloring. Collapse as a term describes the decline of dynamics. It is an ontological description, not a value judgment. In this chapter, "collapse" means the functional decline of dynamical mechanisms, through which relations that could previously be sustained can no longer continue to be sustained. This decline is ontologically neutral: it is simply the intrinsic boundary of relational ontology, not something bad. If "collapse" as a term produces negative associations, alternative terms such as "decline" or "boundary process" may be used. This chapter chooses "collapse" because it accurately reflects the functional failure of dynamical mechanisms: this failure is not partial decline, but the inability of key mechanisms to be sustained.

Objection Two: Your thesis EC1 seems to involve a causal statement: co-evolution collapses because $\nabla\rho \to 0$. But causality itself involves complex issues at the boundary.

Reply: This objection genuinely points to a point requiring refinement. The formulation of EC1 does not need to involve causality; it describes a covariational relation: in the process of $\nabla\rho \to 0$, co-evolutionary mechanisms decline. The two are dynamically connected, but EC1 does not claim some ontological causality. Co-evolutionary mechanisms are built upon gradient structure, $\nabla\rho \neq 0$: this is a dynamical fact. In the process of $\nabla\rho \to 0$, gradient structure declines: this is a dynamical fact. This decline erodes the ground of co-evolutionary mechanisms: this is a dynamical fact. This is a description of dynamics and does not require metaphysical causality. This treatment is consistent with the methodology of this book, namely constrained speculation through mechanistic description.

Objection Three: Is your distinction between the "approaching process" and "the singularity itself" overly artificial? The approaching process and the singularity seem continuous, not two different categories.

Reply: This objection genuinely points to a methodological issue. The approaching process and the singularity itself are indeed continuous at the ontological level: the singularity is the limit of the approaching process. But there is a methodological distinction: the approaching process can be stated, because dynamical decline, mechanism failure, stages, and so on can be described; the singularity itself is methodologically restricted, because object-level statements about the state of the singularity involve ontological singularity and are restricted according to the work of Chapter 17. This methodological distinction does not deny ontological continuity. It clarifies the legitimate methodological range. The work of this chapter concerns the approaching process, not the singularity itself. This limitation allows the chapter to maintain methodological rigor and not exceed its legitimate range.

Literature That Must Be Engaged

- Chapter 12 on co-evolution as an ontological relation (self-reference)
- Chapter 17 on the gradient terminal phase as an ontological singularity (self-reference)
- Chapter 11 on freezing as a foundational ontological process (self-reference)
- Ilya Prigogine, The End of Certainty (1997): on dynamical discussions of process decline

Summary of This Section

This section has done three things. First, it clarified the starting point of this chapter: the process of approaching the gradient terminal phase, distinguished from the singularity itself. Second, it stated the core thesis EC1: the collapse of co-evolutionary relations in the process of $\nabla\rho \to 0$. Third, it developed the meaning and limits of the thesis through the analogy of water flow and through responses to objections.

The next section develops the stages of co-evolutionary collapse.

18.3. Stages of Co-Evolutionary Collapse

Developing the Stages

Co-evolutionary collapse is not a single event; it involves stages. This section develops three main stages: the decline of frequency matching, the failure of threshold, and the convergence of tri-state weights. Each stage corresponds to the decline of one dynamical mechanism. This staged development refines the content of collapse: it is not merely an abstract "end of process," but a set of analyzable stages. Yet this section still preserves the methodological restriction: these stages all belong to the process of approaching the singularity, not to the singularity itself.

Stage One: The Decline of Frequency Matching

Review of the starting point: the starting point of co-evolutionary relations is frequency matching (Chapter 12, Section 12.3). Co-evolution among beings is established through the Lorentzian resonance kernel $g(f, f_0)$: the matching between frequency f and eigen-frequency f_0 enables response resonance.

Content of decline: in the process of approaching $\nabla\rho \to 0$, the frequency spectrum undergoes decline. Contraction of the spectrum: in the low-gradient limit, the excitation modes of beings decrease, and the frequency spectrum available for excitation contracts. Reduction of frequency resolution: in the low-gradient limit, the capacity to distinguish among frequencies declines, and different frequencies become difficult to distinguish. Convergence of eigenfrequencies: in the low-gradient limit, the eigenfrequencies f_0 of different beings tend toward uniformity, and the differences among them disappear. These three forms of decline together make the function of frequency matching fail: there are no longer frequency differences among beings available for matching.

Meaning: the decline of frequency matching collapses the selective dimension of co-evolution. The work of Chapter 12 showed that co-evolutionary relations are built upon frequency selectivity: beings establish co-evolutionary relations only with beings of matching frequencies. In the process of approaching the singularity, frequency selectivity declines, and the foundation of co-evolutionary relations is eroded.

Stage Two: The Failure of Threshold

Review of the starting point: the second element of co-evolutionary relations is threshold (Chapter 12, Section 12.3). Perturbation amplitude $\delta\rho$ must exceed the threshold Γ/k in order to excite nonlinear feedback: this is the excitation condition of co-evolutionary relations.

Content of failure: in the process of approaching $\nabla\rho \to 0$, the threshold mechanism fails. Decrease in perturbation amplitude: in the low-gradient limit, perturbation amplitude $\delta\rho$ decreases, because the weakening of non-uniformity deprives perturbation of its ground. Difficulty of the threshold condition: as $\delta\rho$ decreases, the threshold condition $\delta\rho > \Gamma/k$ becomes increasingly difficult to satisfy, and the perturbation is insufficient to excite feedback. Relative strengthening of dissipation: in the low-gradient limit, dissipative processes, such as entropy increase and diffusion, become stronger relative to excitation processes, making the threshold condition even less favorable. These three forms of failure together make the threshold mechanism decline in function: perturbation is insufficient to excite nonlinear feedback, and the excitation condition of co-evolutionary relations cannot be satisfied.

Meaning: the failure of threshold collapses the intensity dimension of co-evolution. The work of Chapter 12 showed that co-evolutionary relations are built upon threshold: only perturbations of sufficient intensity can excite co-evolutionary relations. In the process of approaching the singularity, intensity declines, and co-evolutionary relations cannot be excited.

Stage Three: The Convergence of Tri-State Weights

Review of the starting point: the third element of co-evolutionary relations is tri-state response (Chapter 12, Section 12.3): response involves the three ontological categories of free state, bound state, and condensed state (Chapter 10), and their weights determine the form of response.

Content of convergence: in the process of approaching $\nabla\rho \to 0$, the tri-state weights converge. Increase in free-state weight: in the low-gradient limit, the bound structures and condensed structures of beings decline, and beings tend toward freer state forms. Decrease in bound-state weight: bound relations require frequency matching and threshold, both of which decline in the process of approaching the singularity, making bound relations unable to be sustained. Decrease in condensed-state weight: condensed structures are the deepest forms of freezing; in the process of approaching the singularity, even the deepest freezing disintegrates, and condensed states decrease. These three forms of convergence together make tri-state weights tend toward uniformity: free-state weight $\to 1$, and bound-state and condensed-state weights $\to 0$.

Meaning: the convergence of tri-state weights makes the tri-states functionally fail as ontological categories: beings lose structural diversity. The work of Chapter 10 showed that the tri-states as ontological categories provide the internal structure of beings. In the process of approaching the singularity, this structural diversity declines, and beings lose differences in ontological content.

Interrelations among the Three Stages

The three stages are not independent. They are mutually connected: the decline of frequency matching reduces ontological differences among beings, which affects the threshold condition, because difference as reference is lacking; the failure of threshold prevents co-evolutionary relations from being excited, which affects tri-state weights, since excitation involves the excitation of bound states, condensed states, and so on; the convergence of tri-state weights causes beings to lose structural differences, which in turn affects frequency matching, since frequency structure involves combinations of the tri-states. The three stages form a mutually declining cycle: the decline of each stage intensifies the decline of the other stages. This cycle makes co-evolutionary collapse an accelerating process rather than a simple linear decline.

Objections and Replies

Objection One: Your three stages seem to be independent processes, but they are all built upon the same dynamical ground, $\nabla\rho \to 0$*. Is treating them as three stages overly artificial?*

Reply: It is true that the three stages are built upon the same dynamical ground. This is precisely the content of EC1 in Section 18.1. But they correspond to different dynamical mechanisms, frequency matching, threshold, and tri-state response, and they have different modes of decline and meanings. Treating them as three stages refines the analysis: collapse is not a single phenomenon, but involves different aspects. This refinement is an application of the method of constrained speculation (Chapter 1): analysis makes ontological content treatable. More deeply, although the three stages are built upon the same ground, they are not necessarily synchronous. They may occur in the process on different timescales. This further supports the reasonableness of treating them as different stages.

Objection Two: Your "mutually declining cycle" seems to make the process an accelerating process. But you also say that the singularity itself cannot be spoken of. If the process accelerates toward the singularity, is this chapter not indirectly speaking of the singularity?

Reply: This objection genuinely points to a methodological issue. The "mutually declining cycle" describes the dynamical structure of the process: it is a statement about the approaching process, not a statement about the singularity itself. This chapter states that the process involves accelerating decline; this is about the dynamics of the approaching process. This chapter does not state that the singularity itself is some state, which is methodologically restricted. This distinction keeps the chapter within its legitimate methodological range: it discusses the dynamics of the approaching process, including its accelerating structure, but does not attempt to speak of the state of the singularity itself. More deeply, the acceleration of the approaching process is itself a meta-level statement: it concerns the formal structure of the process, not the final state that the process reaches. According to Chapter 17, Section 17.4, meta-level statements are legitimate, while object-level statements are restricted. This chapter strictly preserves this methodological distinction.

Objection Three: Your three stages correspond to three dynamical mechanisms, but these mechanisms themselves are built within the EQT framework. Does this chapter depend too much on EQT content?

Reply: The work of this chapter is indeed built upon the EQT framework, especially $g(f, f_0)$, $\delta\rho > \Gamma/k$, tri-state decomposition, and related content. But this content has already been developed in Chapter 12's discussion of co-evolution as an ontological relation. It is part of the core methodology of this book. The work of this book is philosophy based on the EQT framework; it does not pretend to be independent of physics. This choice is an application of the method of constrained speculation (Chapter 1): ontological work is built upon physical grounds. But the ontological meaning of this chapter is not limited to EQT. It concerns the intrinsic boundary of any relational ontology. Any relational ontology is built upon dynamical mechanisms: mechanisms allow relations to be sustained. When the dynamical ground declines, the ground of relational ontology is eroded. This process is the intrinsic structure of relational ontology itself, not merely a feature of the EQT framework. EQT provides concrete mechanisms: frequency matching, threshold, and tri-state response. Other physical frameworks may involve different mechanisms, but they all have dynamical requirements. In the process of dynamical-ground decline, any relational ontology involves a collapse process.

Literature That Must Be Engaged

- Chapter 10 on the tri-states as ontological categories (self-reference)
- Chapter 12 on co-evolution as an ontological relation (self-reference)
- Chapter 17 on the gradient terminal phase as an ontological singularity (self-reference)

Summary of This Section

This section has done four things. First, it developed the three stages of co-evolutionary collapse: the decline of frequency matching, the failure of threshold, and the convergence of tri-state weights. Second, it gave the dynamical content and meaning of each stage. Third, it developed the interrelations among the three stages: the mutually declining cycle. Fourth, it responded to possible objections, especially concerning the independence of the stages, the methodological legitimacy of an accelerating process, and dependence on EQT.

The next section develops the ontological meanings of collapse.

18.4. The Ontological Meanings of Collapse

The Function of Meaning

Section 18.2 developed the dynamical stages of co-evolutionary collapse: this was work at the dynamical level. This section develops the ontological-level meanings of this collapse: what collapse means for ontology itself. The function of this development is to show that co-evolutionary collapse is not merely a dynamical event; it is the intrinsic boundary of ontology itself. It also connects the work of Chapter 17 on ontological singularity with the work of this chapter and provides an ontological ground for the philosophical dialogue in Chapter 19.

Meaning One: The Intrinsic Boundary of Relational Ontology

Chapter 12 established co-evolution as an ontological relation: this is the core of the relational ontology in this book. But this chapter develops the claim that co-evolution collapses in the process of approaching the singularity. What does this mean?

Core thesis EC2: the collapse of co-evolution reveals the intrinsic boundary of relational ontology itself. Relational ontology as an ontological framework is not universally applicable; it has a range of application, namely the range in which dynamical mechanisms can be sustained, and outside this range it cannot be sustained.

The meaning of this thesis is as follows: relational ontology as a framework applies within the range in which gradient structures exist, because the dynamics among beings allows co-evolutionary relations to be sustained. The boundary of relational ontology is the boundary at which gradient structures disappear, $\nabla\rho \to 0$: beyond this boundary, relational ontology cannot be sustained. This boundary is not external; it is the intrinsic structure of relational ontology itself. This thesis has implications for relational ontology itself: it is not an absolute ontological framework, but a range-valid ontological framework. This limitation does not weaken the force of relational ontology; it clarifies its legitimate range.

Meaning Two: The Processuality of Ontology Itself

Co-evolutionary collapse reveals the processuality of ontology itself: ontology as a framework is not eternal, but processual.

Core thesis EC3: ontological frameworks, including ontological categories, ontological relations, and ontological processes, are built upon dynamical conditions. These conditions have different states of existence in different circumstances. In the process of the decline of conditions, the corresponding ontological framework declines.

The meaning of this thesis is as follows: ontological frameworks are not a priori or eternal contents; they are built upon dynamical conditions. Under different dynamical conditions, the legitimacy of ontological frameworks differs. In the process of the decline of dynamical conditions, as in approaching the singularity, ontological frameworks decline. This treatment gives ontology itself processuality: it is not a description of the world from outside, but a framework within the process of the world. This is consistent with the process ontology of this book, including Chapter 11 on freezing as a foundational ontological process and Chapter 15 on time as $\partial\rho/\partial t$: ontology itself is processual.

Meaning Three: The Boundary of Meaning

Co-evolutionary collapse has important implications for the boundary of meaning. Chapter 21 establishes "meaning before freezing" as a core thesis of this book: meaning appears within gradient structures and collapses at the limit where gradients end. This chapter provides the relational-ontological ground for that thesis:

Core thesis EC4: meaning as an ontological-normative phenomenon is built upon co-evolutionary relations. Relations, such as consciousness and world, person and other persons, beings and purposes, give meaning an ontological ground. In the process in which co-evolutionary relations collapse, the ontological ground of meaning is eroded: meaning disappears before the singularity.

The meaning of this thesis is as follows: meaning is not an eternal ontological content; it is built upon co-evolutionary relations. Co-evolutionary relations are finite: they collapse as dynamical conditions decline. This finitude makes meaning itself finite: meaning emerges in process and disappears at the limit. The thesis of Chapter 21 concerning meaning before freezing (M1–M3) receives its ground through the ontological work of this chapter: the finitude of meaning is not a subjective feeling, but an intrinsic implication of relational ontology itself.

Meaning Four: The Application of Methodological Rigor

The discussion of co-evolutionary collapse shows the application of methodological rigor, including constrained speculation, failure conditions, falsifiability, and so on, at the boundary:

Core thesis EC5: at the metaphysical boundary, methodological rigor does not weaken the legitimacy of discussion. On the contrary, it keeps discussion precise. Through the meta-level/object-level distinction (Chapter 17, Section 17.4) and methodological discipline, ontological work at the boundary can be rigorously developed.

The meaning of this thesis is as follows: methodological rigor is not a tool outside the boundary; it is a necessary tool at the boundary. At the boundary, without methodological rigor, discussion easily degenerates into rhetoric or mysticism. Methodological rigor preserves precision in boundary discussion: it prevents metaphysics from becoming an ungrounded statement. This thesis has important implications for the methodology of this book: it displays and defends methodological rigor at the most crucial boundary.

The Unity of the Four Meanings

The four meanings, the intrinsic boundary of relational ontology, the processuality of ontology itself, the boundary of meaning, and the application of methodological rigor, together constitute the ontological meaning of co-evolutionary collapse. They are not independent; they support one another. The boundary of relational ontology (EC2) gives content to the processuality of ontology itself (EC3). The processuality of ontology itself gives ontological ground to the boundary of meaning (EC4). The boundary of meaning makes the application of methodological rigor (EC5) necessary: meaning as an important content at the boundary requires rigorous treatment. Together, the four meanings show that the work of this chapter is not merely a dynamical description; it is important ontological work.

Objections and Replies

Objection One: Your core thesis EC2 seems to restrict relational ontology. But Chapter 12 of this book established relational ontology as foundational. Are the two claims contradictory?

Reply: The work of this book does not claim that relational ontology is an absolute ontological framework. It is a range-valid framework. Chapter 12: relational ontology as an ontological framework, in which beings constitute ontology through co-evolutionary relations, applies within the range where gradient structures exist. This chapter, EC2: the range of application of relational ontology has a boundary. At the limit of $\nabla\rho \to 0$, relational ontology cannot be sustained; this is the intrinsic boundary of relational ontology itself. These two claims are not contradictory. Together, they develop the content of relational ontology, its validity within a range, and its boundary, where it cannot be sustained beyond that range. This treatment of finite validity clarifies the legitimacy of relational ontology: it does not need to pretend to be an absolute framework. More deeply, acknowledging the finitude of relational ontology is not a weakening; it is the application of ontological discipline, including Chapter 2 on ontological restraint, Chapter 3 on failure conditions, Chapter 6 on falsifiability, and so on. Any serious ontological work has a legitimate range, and application beyond that range requires justification.

Objection Two: Your core thesis EC3, the processuality of ontology itself, seems to relativize ontology: ontological frameworks differ under different conditions. Does this retreat into relativism?

Reply: EC3 is not relativism. Relativism claims that all ontological frameworks are equally legitimate and that there are no methodological standards by which to assess them. EC3 claims that ontological frameworks are built upon dynamical conditions: the legitimacy of a framework depends on conditions. This does not mean that "all frameworks are equally legitimate," but that "the legitimacy of a framework is grounded in conditions." The work of this book assesses ontological frameworks through methodological discipline, including the method of constrained speculation in Chapter 1 and failure conditions in Chapter 3. These disciplines provide standards: the legitimacy of a framework under given conditions can be rigorously assessed. More deeply, the processuality of ontology is consistent with the overall process ontology of this book, including Chapter 11 on freezing and Chapter 15 on time. Process ontology is not relativism. It is ontology under methodological discipline.

Objection Three: Your core thesis EC4, the boundary of meaning, makes meaning pessimistic: meaning disappears in process. Does this imply some kind of nihilism?

Reply: EC4 is not nihilism. Nihilism claims that meaning has no ground and is an arbitrary feeling. EC4 claims that meaning has an ontological ground: it is built upon co-evolutionary relations. Co-evolutionary relations are finite, and therefore meaning is finite. This finitude does not deny the legitimacy of meaning; it clarifies its boundary. Chapter 21's theses on meaning before freezing (M1–M3) develop this treatment: meaning as an ontological phenomenon genuinely emerges in process. It has an ontological ground and is legitimate content. But its legitimacy is built in process. It is not eternal and not without boundary. This finite affirmation is not pessimism; it is honesty. It acknowledges the force of meaning, which genuinely emerges in process, while also acknowledging the boundary of meaning, which genuinely disappears before the limit. This treatment resonates with the Buddhist tradition, especially Nāgārjuna's work in Chapter 19, Section 19.4: the legitimacy of beings is grounded in dependent arising / co-evolution, and finitude does not weaken legitimacy.

Literature That Must Be Engaged

- Chapter 11 on freezing as a foundational ontological process (self-reference)
- Chapter 12 on co-evolution as an ontological relation (self-reference)
- Chapter 17 on the gradient terminal phase as an ontological singularity (self-reference)
- Chapter 21 on the affirmation of finitude: meaning before freezing (self-reference)

Summary of This Section

This section has done four things. First, it developed the four ontological meanings of co-evolutionary collapse: the intrinsic boundary of relational ontology (EC2), the processuality of ontology itself (EC3), the boundary of meaning (EC4), and the application of methodological rigor (EC5). Second, it gave the content and ground of each meaning. Third, it developed the mutual support among the four meanings. Fourth, it responded to possible objections, especially concerning limitation vs contradiction, processuality vs relativism, and finitude vs pessimism.

The next section is the chapter summary.

18.5. Chapter Summary: The Intrinsic Boundary of Relational Ontology

Completion of the Three Tasks

The three tasks undertaken by this chapter have all been completed. Task One: to connect the work of Chapter 17 on ontological singularity with the work of Chapter 12 on co-evolution. Section 18.1 clarified the starting point of collapse and stated the core thesis EC1: the collapse of co-evolutionary relations in the process of $\nabla\rho \to 0$. Task Two: to identify the stages and mechanisms of collapse. Section 18.2 developed three stages: the decline of frequency matching, the failure of threshold, and the convergence of tri-state weights, and developed the mutually declining cycle among the three stages. Task Three: to treat the ontological meanings of collapse. Section 18.3 developed four meanings: the intrinsic boundary of relational ontology (EC2), the processuality of ontology itself (EC3), the boundary of meaning (EC4), and the application of methodological rigor (EC5).

The Position of This Chapter in Part IV

The overall structure of Part IV, "The Metaphysical Boundary of Freezing," is as follows: Chapter 16, "The Hierarchy and Historicity of Freezing," as foundational development; Chapter 17, "The Gradient Terminal Phase as an Ontological Singularity," as the metaphysical summit; Chapter 18, this chapter, "The Collapse of Co-Evolution before the Gradient Terminal Phase," as the process before the singularity; and Chapter 19, "Three Responses," as dialogue with traditions. As a transitional chapter between Chapter 17 and Chapter 19, this chapter has a connecting function. Looking backward, this chapter continues Chapter 17's work on ontological singularity by connecting the singularity as endpoint with the approaching process. Looking forward, this chapter provides an ontological ground for the philosophical dialogue in Chapter 19, since all three traditions involve the problem of boundary.

The Position of This Chapter in the Whole Book

Looking backward, this chapter builds upon several chapters: Chapter 11 on freezing as a foundational ontological process, which provides the ground for the dynamical decline in this chapter; Chapter 12 on co-evolution as an ontological relation, which provides the ontological content for the collapse in this chapter; and Chapter 17 on the gradient terminal phase as an ontological singularity, which provides the endpoint for the approaching process in this chapter. Looking forward, this chapter provides an ontological ground for later chapters: Chapter 19 on three responses, where the boundary work of this chapter provides content for philosophical dialogue; and Chapter 21 on meaning before freezing, where the collapse work of this chapter provides the ontological ground for the finitude of meaning (EC4).

Honesty about the Limits of This Chapter

This chapter acknowledges several limits. First, the difficulty of applying methodological rigor. This chapter attempts strictly to preserve the meta-level/object-level distinction: it discusses only the approaching process, not the singularity itself. But discussion at the boundary involves methodological difficulty, and some statements may move between the meta-level and the object-level. This difficulty requires more refined work. Second, the depth of connection with physics. This chapter uses content from the EQT framework, including $g(f, f_0)$, $\delta\rho > \Gamma/k$, tri-state decomposition, and so on. The physical ground of this content requires detailed physical work; this chapter uses it only at the philosophical level. Third, cross-disciplinary applications are not sufficiently developed. The manifestations of co-evolutionary collapse in different domains, such as foundational physics, life, consciousness, society, and others, require dedicated work. This chapter only provides the framework. Fourth, connection with cosmology. The approaching process toward the singularity in this chapter is connected with the ultimate fate of the universe, such as heat death and the Big Rip, but detailed cosmological discussion lies beyond the scope of this chapter. Acknowledging these limits does not weaken the work of this chapter.

The Core Posture of This Chapter

The core posture of this chapter can be stated as follows: relational ontology has an intrinsic boundary, and this boundary is not a defect, but the honest expression of relational ontology as a finite framework.

The work of this book establishes relational ontology as an ontological framework (Chapter 12), and this is one of the core contributions of the book. But relational ontology is not an absolute framework: it collapses in the process in which its dynamical ground declines. Acknowledging this boundary does not weaken relational ontology; it makes it honest. Any serious ontological framework has a range of application, and application beyond that range involves difficulty. Through methodological discipline, including Chapter 3 on failure conditions and Chapter 6 on falsifiability, this book clarifies this range of application. This posture is consistent with the overall methodological choice of the book: rigor rather than ambition. The work of this book does not attempt to establish a universal ontological framework. It establishes an effective ontological framework and clarifies its legitimate range. This treatment of finite validity gives the work of this book a legitimate place in contemporary serious philosophical discussion.

Meaning before freezing: this is one of the core theses of this book. This chapter provides a relational-ontological ground for this thesis through co-evolutionary collapse. Meaning is built upon co-evolutionary relations; co-evolutionary relations are finite and collapse before the limit; therefore meaning is finite and disappears before the limit. This connection is not pessimism, but honesty. It acknowledges the force of meaning, which genuinely emerges in process, while also acknowledging the boundary of meaning, which genuinely disappears before the limit. Through the work of this chapter, the metaphysical work of Part IV, "The Metaphysical Boundary of Freezing," becomes complete: Chapter 16 establishes hierarchy; Chapter 17 establishes the singularity; this chapter establishes the approaching process; and Chapter 19 establishes the philosophical position. Together, the four chapters constitute the book's work on metaphysical boundary. This is one of the philosophical peaks of the book.

18.6. Methodological Reflection on This Chapter

This chapter connects the singularity work of Chapter 17 with the co-evolutionary work of Chapter 12. This section reflects on the methodological significance of this connecting work.

This Chapter's Treatment of Methodological Rigor

This chapter works at the boundary, where methodological rigor is most crucial. It strictly preserves the meta-level/object-level distinction developed in Chapter 17, Section 17.4: it discusses only the meta-level structure of the approaching process and does not attempt to speak of the object-level state of the singularity itself. The statements of this chapter all concern the dynamics of the approaching process: how co-evolutionary mechanisms decline in the process of $\nabla\rho \to 0$. This chapter does not state the state of the singularity itself, because such a statement would involve the object level and exceed the legitimate methodological range. This distinction keeps the work of this chapter within its legitimate methodological range. This precision is an application of the method of constrained speculation.

The Function of This Chapter as a Connecting Chapter

As a connecting chapter, this chapter links Chapter 17 and Chapter 19. Chapter 17 establishes the singularity itself as the intrinsic boundary of ontology. This chapter develops the process of approaching the singularity, extending the singularity work from a single limiting point into a dynamical process. Chapter 19 places this boundary work within philosophical discussion. Without this chapter's connection, there would be an ontological rupture between Chapter 17 and Chapter 19. This chapter bridges that rupture: the collapse process develops the approach to the singularity and gives philosophical dialogue an ontological ground.

This Chapter's Posture toward Originality

This chapter does not claim complete originality. It is built upon the work of Chapters 11, 12, and 17. Its contribution is to connect coevolution (Chapter 12) with the singularity (Chapter 17); to develop the dynamics of collapse through three stages; and to develop the ontological meaning of collapse through four meanings. These contributions are constructive connections that make the metaphysical work of the book complete.

Limits Honestly Acknowledged in This Chapter

This chapter acknowledges several limits. First, the manifestation of the stages of collapse in physical-cosmological scenarios requires dedicated work. This chapter mainly develops them at the philosophical level. Second, the comparison between the intrinsic boundary of relational ontology and neighboring ontological frameworks is not sufficiently developed in this chapter. Third, the refinement of the ontological meanings of collapse requires further work.

Connections with Later Chapters

This chapter provides an ontological ground for the philosophical dialogue in Chapter 19. Chapter 19's dialogue with Heidegger, Wittgenstein, and Nāgārjuna is built upon the boundary work of this chapter: all three philosophical traditions involve methodology at the boundary. This chapter gives ontological ground to the boundary work, making the philosophical dialogue not rhetorical citation, but ontological dialogue. More deeply, this chapter is the honest expression of relational ontology in the work of this book. If the book did not develop the boundary of relational ontology, relational ontology could be seen as an absolute framework. This would be inconsistent with the process ontology and methodological discipline of the book. This chapter makes the finitude of relational ontology explicit.

19. Four Responses: Heidegger, Wittgenstein, Nāgārjuna, and Husserl

19.1. Introduction to This Chapter

Chapter 17 established the ontological singularity, the gradient terminal phase at $\nabla\rho \equiv 0$, as the intrinsic limit of ontology. Chapter 18 developed the collapse of co-evolution before the terminal phase. These works involve a fundamental methodological question: how can one speak at the boundary of ontology?

Ontological singularity, as the limit of ontology, is difficult to speak of directly. Any statement about the singularity involves difficulty. But complete silence is not a legitimate treatment either, since the singularity has ontological meaning as a boundary and cannot simply be avoided. Speech at the boundary requires methodology: what kind of speech is legitimate at the boundary? This problem is not unique to the EQT framework; it is a core problem of any serious metaphysical work.

There are several kinds of responses in the history of philosophy. This chapter selects four as interlocutors: Heidegger's poetic saying, which uses poetic language to point toward Being where the language of traditional metaphysics fails; Wittgenstein's keeping silence, "whereof one cannot speak, thereof one must be silent"; Nāgārjuna's double negation, neither affirming X nor denying X, thereby going beyond binary opposition; and Husserl's rigorous description, which suspends presuppositions and returns to the given as the beginning. These four responses represent different directions of treatment: respectively, the poetic dimension of Being, the logical boundary of language, the middle-way dimension of ultimate reality, and the given beginning of consciousness.

This chapter undertakes three tasks. First, it identifies the relevance of this book's work, especially Chapter 17 on ontological singularity, to the four philosophical traditions. Second, it enters into serious dialogue with each tradition, identifying similarities and differences. Third, it clarifies the independent position of this book's work among these four traditions: it is not a simple continuation of any one tradition, but a new work that has similarities and differences with each.

This chapter has a clear posture in writing: serious respect and precise distinction. The four philosophers each have enduring philosophical influence, and this chapter does not pretend that it can easily surpass any of them. But this book's work has an independent position that must be made explicit. This posture keeps the chapter from becoming either simple veneration of predecessors or simple criticism of them.

This chapter is the concluding chapter of Part IV, "The Metaphysical Boundary of Freezing." It continues the metaphysical work of Chapters 17 and 18, and through dialogue shows the position of that work within a larger philosophical discussion. At the same time, it prepares Part V, "From Ontology to Normativity": the four dialogues involve methodologies at the boundary, including silence, transcendence, suspension, and so on, which are connected with the later work on meaning and value.

The discussion proceeds as follows. Section 19.1 develops four modes of response at the metaphysical boundary. Section 19.2 enters into serious dialogue with Heidegger. Section 19.3 enters into serious dialogue with Wittgenstein. Section 19.4 enters into serious dialogue with Nāgārjuna. Section 19.5 enters into serious dialogue with Husserl. Section 19.6 develops the independent position of this book's work among the four traditions.

19.2. Four Modes of Response at the Metaphysical Boundary

The Form of the Problem

Chapter 17 established the ontological singularity, the gradient terminal phase at $\nabla\rho \equiv 0$, as the intrinsic limit of ontology. Chapter 18 developed the collapse of co-evolution before the terminal phase. These works involve a fundamental methodological question: how can one speak at the boundary of ontology?

Specifically: ontological singularity, $\nabla\rho \equiv 0$, as the limit of ontology, is difficult to speak of directly according to this book's own methodology, including constrained speculation and failure conditions. Any statement about the singularity involves difficulty. But complete silence is not a legitimate treatment either, since the singularity has ontological meaning as a boundary and cannot simply be avoided. Speech at the boundary requires methodology: what kind of speech is legitimate at the boundary? This problem is not unique to the EQT framework; it is a core problem of any serious metaphysical work. Any metaphysical work directed toward a boundary faces this problem: how can methodological rigor be preserved at the boundary?

Four Modes of Response

There are several kinds of responses in the history of philosophy. This chapter selects four as interlocutors: Heidegger's response, Wittgenstein's response, Nāgārjuna's response, and Husserl's response. These four responses represent treatments in different directions.

The first three responses, Heidegger, Wittgenstein, and Nāgārjuna, all treat the boundary itself as the problem: each answers, in its own way, the question of "how to speak beyond the boundary." The fourth response, Husserl, takes another direction: it does not ask what lies beyond the boundary, but returns to what lies within the boundary, to the starting point of everything, the given, and describes it rigorously. This methodological division allows the four responses to form a complete spectrum of boundary treatment.

Heidegger's Response: Poetic Saying

Heidegger, especially in his later works such as Beiträge zur Philosophie and Holzwege, responds to the boundary of the question of Being by using poetic saying. The language of traditional metaphysics, including propositions, concepts, and categories, is insufficient for speaking of Being itself: it is language about beings (Seiendes), not language about Being (Sein). At the boundary, poetic language is needed: it speaks through indication, evocation, and resonance rather than through direct proposition. Philosophy and poetry converge at the boundary, since both use language to point toward what exceeds propositional statement.

Wittgenstein's Response: Keeping Silence

Wittgenstein, especially in the Tractatus Logico-Philosophicus (1922), responds to the boundary of language by keeping silence. The legitimate use of language is limited to what can be stated: facts, logical relations, and so on. Statements about the world as a whole, the meaning of existence, ethical value, and related matters exceed the legitimate boundary of language. At this boundary, "Wovon man nicht sprechen kann, darüber muss man schweigen" ("whereof one cannot speak, thereof one must be silent"). The later Wittgenstein, in Philosophical Investigations (1953), changes the treatment: he no longer thinks that there are unsayable contents, but rather that some problems result from the misuse of language. Still, keeping silence as a way of facing the boundary remains an important direction in Wittgenstein's work.

Nāgārjuna's Response: Double Negation

Nāgārjuna, around the second century CE and the founder of the Madhyamaka school of Indian Buddhism, responds to the question of ultimate reality in the Mūlamadhyamakakārikā by using double negation. Any affirmative statement, X exists, involves ontological attachment: it solidifies ontological content. But simple negation, X does not exist, is also ontological attachment: it assumes "non-existence" as ontological content. The correct method is double negation: neither affirming X nor denying X, thereby going beyond binary opposition. This double negation is not neutral; it points toward emptiness (śūnyatā) as the mode of ultimate reality. Nāgārjuna's work has a subtle and difficult character: through logical deconstruction, it points toward the non-logical ultimate. This direction differs from both Heidegger and Wittgenstein.

Husserl's Response: Rigorous Description of the Given as the Beginning

Husserl (Edmund Husserl, 1859–1938), the founder of the Western phenomenological tradition, responds to the question of the relation between consciousness and world in works such as Logical Investigations, Ideas I, Cartesian Meditations, and The Crisis of European Sciences by using the method of rigorously describing the given. He does not ask what lies beyond the boundary, but suspends (epoché) all presuppositions about the "existence of the objective world," and concentrates philosophical reflection on the content given to consciousness itself: phenomena as phenomena, the given as given. Through phenomenological reduction, the internal structures of consciousness, including intentionality, horizon, and modes of givenness, can be rigorously described in themselves. The fundamental insight reached is this: the given is the starting point of everything and cannot be derived from anything more basic. Any knowledge of the world, any proposition about reality, and any scientific description all presuppose the given. Husserl's direction is qualitatively different from the first three responses: the first three all treat the boundary itself as the problem, namely "how to speak beyond the boundary," while Husserl returns to the starting point within the boundary, namely "how the given as beginning can be rigorously described." This direction has a special relation to the work of this book: through a bottom-up physical path, this book reaches the claim that "the given is irreducible," while Husserl, through a top-down reflective path, reaches a highly consonant insight. The convergence of these two independent paths is discussed in detail in Section 19.5.

Common Features of the Four Responses

Although the four responses differ in direction, they have a common feature: all identify the limitation of ordinary propositional language and each adopts a different method in response to that limitation. Heidegger: ordinary propositional language concerns beings and is insufficient for speaking of Being itself; poetry is needed to point beyond the boundary. Wittgenstein: ordinary propositional language concerns facts, and beyond this range silence is needed to acknowledge the boundary. Nāgārjuna: ordinary propositional language involves ontological attachment, and double negation is needed to transcend binary opposition. Husserl: ordinary propositional language depends on a series of presuppositions, and suspension is needed to return to the starting point within the boundary. Among the four directions, the first three treat the boundary itself as the problem, how to speak beyond the boundary, while Husserl returns to the starting point of everything, how the given as beginning can be described. This makes the four responses form a complete spectrum of boundary treatment: pointing beyond the boundary (Heidegger), acknowledging the boundary (Wittgenstein), transcending the boundary (Nāgārjuna), and returning to the starting point within the boundary (Husserl). Any serious metaphysical work must face the boundary of ordinary methodology. The four directions each develop a way of facing it.

The Positioning of This Chapter's Work

Through dialogue with the four responses, this chapter clarifies the methodology of this book's work at the boundary. It is not a simple continuation of any one response, but has its own methodology. The methodology of this book was already demonstrated in Chapter 17: constrained silence + the meta-level/object-level distinction. This methodology has both similarities and differences with the four responses, and these must be developed in detail. The following four sections develop the dialogues with the four traditions.

Literature That Must Be Engaged

- Martin Heidegger, Being and Time (1927), Beiträge zur Philosophie (1936-1938), Holzwege (1950)
- Ludwig Wittgenstein, Tractatus Logico-Philosophicus (1922), Philosophical Investigations (1953)
- Nāgārjuna, Mūlamadhyamakakārikā (around the second century)
- Edmund Husserl, Logical Investigations (1900-01), Ideas I (1913), Cartesian Meditations (1931), The Crisis of European Sciences (1936)
- Chapter 17 on the gradient terminal phase as an ontological singularity (self-reference)
- Chapter 18 on the collapse of co-evolution before the gradient terminal phase (self-reference)
- Section 13.6 on the convergence with phenomenology through different paths (self-reference)

Summary of This Section
This section has done three things. First, it identified the methodological problem at the metaphysical boundary: how to speak legitimately at the boundary. Second, it introduced four modes of response: Heidegger's poetic saying, pointing beyond the boundary; Wittgenstein's keeping silence, acknowledging the boundary; Nāgārjuna's double negation, transcending the boundary; and Husserl's rigorous description, returning to the starting point within the boundary. Third, it identified the common feature of the four responses, namely that all recognize the limitation of ordinary propositional language, and previewed the positioning of this chapter's work.

19.3. Dialogue with Heidegger

The Core of Heidegger's Work
Heidegger (1889–1976) worked across several stages, from the early Being and Time (1927), to the middle-period Beiträge zur Philosophie

(1936–1938), and then to later works such as Holzwege (1950). Yet his core problem remains continuous: the question of Being (Seinsfrage).

Heidegger's core position consists of the following. The distinction between Being and beings: traditional metaphysics has confused Being (Sein) with beings (Seiendes). Beings are things that are, such as tables, persons, stars, and so on; Being is that which lets beings be. Traditional metaphysics concerns beings and forgets Being itself. The limitation of metaphysics: the language and categories of traditional metaphysics, such as substance, attribute, essence, and so on, are all about beings; they are insufficient for speaking of Being. This limitation is "the forgetfulness of Being in metaphysics" (die Seinsvergessenheit der Metaphysik). Poetic saying as method: to speak of Being itself at the boundary requires poetic language. It points to what exceeds propositions through indication and evocation. Poets such as Hölderlin become important resources for Heidegger's work. Ereignis as the core concept: the later Heidegger uses Ereignis, event, appropriating event, enowning, as a core concept. It points to the way Being happens, the mutual belonging of Being and human beings.

Relevance to This Book's Work

Heidegger's work has a complex relation to the work of this book: there are both resonances and differences.

Resonance One: critique of traditional metaphysics. Heidegger criticizes the language of traditional metaphysics as insufficient for speaking of Being itself: it takes beings as Being. The work of this book, especially Chapter 7 on difference before identity, also criticizes traditional substance ontology: it takes stable forms, entities, as foundational ontological content and forgets processuality. The two resonate in their critique of traditional metaphysics: both question the legitimacy of traditional language and categories.

Resonance Two: seriousness toward the boundary. Heidegger takes the metaphysical boundary seriously: he does not simply avoid it, but develops a method, poetic saying, to treat it. This book also takes the boundary seriously: Chapter 17 on ontological singularity develops the ontological meaning of the boundary. Both reject simply avoiding the boundary. Metaphysics needs a method for treating the boundary.

Resonance Three: the importance of historicity. Heidegger, especially in Being and Time §§ 72–77, treats historicity (Geschichtlichkeit) as an intrinsic structure of the being, Dasein. This book, especially

Chapter 16, HL2, also treats historicity as an intrinsic structure of beings. The two resonate in treating historicity as intrinsic structure, although their ranges of application differ: Heidegger mainly treats Dasein, while this book extends the treatment to all processual beings.

Key Differences

But there are also important differences between this book's work and Heidegger's work.

Difference One: methodological rigor. Heidegger's poetic saying as a method faces difficulty with respect to methodological rigor: it uses indication and evocation and is difficult to assess strictly. Heidegger's own texts, especially the later works, are often criticized as not being strictly analyzable: their content depends on poetic feeling and is not easily handled by methodological discipline. The methodology of this book, including constrained speculation, failure conditions, falsifiability, and so on, pursues methodological rigor. At the boundary, this book preserves rigor through constrained silence, as in Chapter 17: it makes explicit what can be said and what cannot be said, and uses the meta-level/object-level distinction to make meta-level statements legitimate. This difference is not merely technical; it concerns a fundamental choice in philosophical methodology. Heidegger chooses to break through rigor in order to treat the boundary, using poetic saying as a new method. This book chooses to preserve rigor in order to treat the boundary, using the meta-level/object-level distinction to extend rigor to the boundary. Both choices have legitimacy and reflect different philosophical directions.

Difference Two: the relation to science. Heidegger's work, especially the later work, is skeptical of science, especially modern physics and technology. He treats modern science and technology as the culmination of metaphysics, where the forgetfulness of Being reaches its extreme. This treatment creates distance between Heidegger's work and contemporary science. This book is built upon contemporary physics, the EQT framework. It is not "the culmination of metaphysics," but an attempt at connection with science. This book holds that science does not necessarily lead to "forgetfulness of Being," provided that the ontological ground of science is seriously developed. This difference reflects two different philosophical directions: Heidegger's work opposes the ontological tendency of modern science, treating science as a problem of metaphysics; this book builds upon the content of modern science, treating science as a resource for ontology.

Difference Three: Ereignis vs co-evolution. The later Heidegger's concept of Ereignis has structural similarity with co-evolution in this book (Chapter 12): both involve the mutual relation between Being and beings, or between Being and human beings. But their content differs. Ereignis is a poetic concept: it is used through indication and resonance and is difficult to state precisely. Co-evolution is an ontological relation: it has a falsifiable structure, including frequency matching, threshold, tri-state response, and so on, as discussed in Chapter 12, Section 12.3. This difference means that although the two concepts are structurally similar, they differ in content. Ereignis as a poetic concept has heuristic power, but is difficult to analyze rigorously. Co-evolution as an ontological concept has analyzable content.

Positioning between the Positions

Through comparison, the relation between this book's work and Heidegger's work can be stated as follows: this book resonates with Heidegger in the critique of traditional metaphysics, seriousness toward the boundary, and historicity as intrinsic structure, but differs in methodological rigor, constrained silence vs poetic saying; in its relation to science, building upon vs opposing; and in the precision of core concepts, co-evolution vs Ereignis. The two constitute different directions for treating the metaphysical boundary. This positioning avoids two traps: the trap of absorption, claiming that this book "absorbs" the insights of Heidegger's work, when this book has its own methodological choice; and the trap of simple criticism, claiming that Heidegger's work is "vague" or "not rigorous," when Heidegger's work has enduring force and this book does not attempt to dismiss it lightly. The two works each have legitimacy and reflect different philosophical directions.

Objections and Replies

Objection One: Your critique of Heidegger's poetic saying in terms of "methodological rigor" might be answered by Heidegger himself: rigor itself is a product of technological metaphysics and is not suitable for treating the question of Being.

Reply: This objection genuinely points to a fundamental methodological difference. Heidegger's position, that rigor is a product of technological metaphysics, is itself a philosophical claim. It involves arguments about modern science and technology, the forgetfulness of Being, poetry as a different method, and so on. These arguments can be discussed. This book does not accept Heidegger's position. The ground is as follows: methodological rigor is not necessarily a product of "technological metaphysics." It is an expression of methodological discipline and can be used by any serious work. The methodology of this book, including constrained speculation and failure conditions, resonates with Chinese philosophical traditions, especially the rigor of practical wisdom in Confucian thought, and is not limited to the Western analytic tradition. Rigor can allow poetic dimensions: this book does not exclude poetry, but requires poetry to have methodological grounding. This is a methodological choice, not a simple acceptance or rejection of Heidegger. This book chooses to preserve rigor, while Heidegger's work chooses to break through rigor. Both have legitimacy.

Objection Two: Your comparison of the relation to science seems too simple. Heidegger's work does not simply oppose science; his discussion of science is complex.

Reply: This objection genuinely points to the complexity of Heidegger's work. Heidegger's treatment of science is indeed complex. He does not simply reject science, but criticizes the metaphysical ground of modern science and technology as the extreme form of the forgetfulness of Being. These criticisms are developed in detail in his later work. The treatment in this book is not simply "accepting science" either. It builds upon science, the EQT framework, while preserving methodological rigor. This book accepts the content of science: it uses findings from physics, neuroscience, sociology, and other fields. But this book does not simply accept the methodological assumptions of science: it evaluates scientific content through methodological disciplines such as constrained speculation. This treatment, building upon science but not simply accepting it, differs from Heidegger's opposition, but is also not simple scientism. The two treatments reflect different philosophical directions, but neither is simple.

Objection Three: Heidegger's later concept of Ereignis actually has precision. It is understandable within the phenomenological-hermeneutic tradition. Your criticism is too much from the perspective of analytic philosophy.

Reply: This objection genuinely points to a difference between traditions. Ereignis is understandable within the phenomenological-hermeneutic tradition: readers trained in that tradition can understand the concept through hermeneutic practice. This is a legitimate philosophical mode. But the aim of this book is to enter into dialogue with multiple traditions: the Western analytic tradition, the continental tradition, Chinese traditions, Indian traditions, and others. This requires concepts to have cross-traditional intelligibility. Ereignis, as a poetic-hermeneutic concept, has difficulties in cross-traditional dialogue because it depends on training within a tradition. The concept of co-evolution in this book pursues cross-traditional intelligibility: through physical examples, ontological argument, and methodological discipline, it makes its content intelligible across several traditions. This is a different philosophical direction. Both directions have legitimacy: depth within a tradition vs intelligibility across traditions. This book chooses the latter.

Literature That Must Be Engaged

- Martin Heidegger, Being and Time (1927)
- Martin Heidegger, Beiträge zur Philosophie (1936-1938)
- Martin Heidegger, Holzwege (1950)
- Martin Heidegger, On the Way to Language (1959)
- Chapter 12 on co-evolution as an ontological relation (self-reference)
- Chapter 16 on historicity (self-reference)
- Chapter 17 on ontological singularity (self-reference)

Summary of This Section

This section has done four things. First, it developed the core position of Heidegger's work. Second, it identified resonances between this book's work and Heidegger's work: critique of traditional metaphysics, seriousness toward the boundary, and historicity as intrinsic structure. Third, it identified differences: methodological rigor, the relation to science, and the precision of core concepts. Fourth, it clarified the two as different directions for treating the metaphysical boundary.

The next section develops the dialogue with Wittgenstein.

19.4. Dialogue with Wittgenstein

The Core of Wittgenstein's Work

Ludwig Wittgenstein (1889–1951) worked across two stages: the early Tractatus Logico-Philosophicus (1922) and the later Philosophical Investigations (1953). The two stages differ in method, but both involve the problem of boundary.

Early Wittgenstein: the boundary of language. The core position of the Tractatus is as follows: the legitimate use of language is limited to the description of facts; meaning is statement about the world; statements about the world as a whole, the meaning of existence, ethical value, aesthetic value, mystical content, and so on exceed the legitimate boundary of language; at this boundary, "Wovon man nicht sprechen kann, darüber muss man schweigen" ("whereof one cannot speak, thereof one must be silent"); the statements of the Tractatus itself also involve this difficulty, since it attempts to depict the boundary of language, but depiction requires standing outside it, which exceeds the legitimate use of language. Wittgenstein's treatment is to regard the Tractatus as a ladder: after the reader climbs it, it must be thrown away.

Later Wittgenstein: the multiplicity of language games. The core position of Philosophical Investigations is as follows: language is not a single logical structure; it is a multiplicity of language games. Different language games, including science, art, ethics, religion, and so on, have different rules and modes of use. Philosophical problems often involve the misuse of language: applying the rules of one language game to another. The task of philosophy is not to answer metaphysical questions, but to dissolve them through linguistic analysis. The later Wittgenstein differs methodologically from the early Wittgenstein, but methodology at the boundary remains central; only the boundary shifts from the boundary of language as a whole to the boundary among language games.

Relevance to This Book's Work

Wittgenstein's work has a complex relation to this book's work.

Resonance One: seriousness toward the boundary of language. Wittgenstein takes the boundary of language seriously. He does not simply avoid it, but develops methods for treating the boundary. This book also takes the boundary seriously: it treats the boundary through constrained silence and the meta-level/object-level distinction, especially in Chapter 17. Both reject simply avoiding the boundary.

Resonance Two: methodological rigor. Wittgenstein's work pursues methodological rigor: it treats philosophical problems rigorously through logical analysis in the early work and linguistic analysis in the later work. This book also pursues methodological rigor through the method of constrained speculation, failure conditions, falsifiability, and other methodological disciplines. The two resonate in methodological rigor. This is the most significant similarity between this book and Wittgenstein's work.

Resonance Three: critique of "pseudo-problems." Wittgenstein, especially in the later work, criticizes many philosophical problems as pseudo-problems: they involve the misuse of language and need to be dissolved rather than answered. This book, especially Chapter 3, Section 3.4, on statements without failure conditions, also criticizes certain ontological statements as illegitimate: they have no failure direction and do not constitute ontological commitments. The two resonate in excluding illegitimate statements. This is an expression of methodological rigor.

Key Differences

But there are also important differences between this book's work and Wittgenstein's work.

Difference One: the treatment of the "unsayable." The early Wittgenstein treats what exceeds the boundary of language as unsayable: one can only remain silent. This book does not simply accept this treatment. Through the meta-level/object-level distinction (Chapter 17), this book makes statements at the boundary legitimate: meta-level statements, statements about the boundary itself, can be legitimately spoken. They concern the existence of the boundary, its meaning, and so on. Object-level statements, statements about what lies on the other side of the boundary, are restricted, but this is not complete silence. This difference reflects two different treatments: Wittgenstein's treatment is strict silence, separating the two sides of the boundary; this book's treatment is constrained speech + the meta-level/object-level distinction, making the boundary itself legitimately speakable. This treatment allows this book to preserve legitimate speech at the boundary rather than retreating into simple silence.

Difference Two: the status of ontology. The early Wittgenstein's position involves an anti-metaphysical tendency: metaphysical statements, apart from statements about facts, exceed the legitimate use of language. The later Wittgenstein preserves this tendency, although by different methods. This book is not anti-metaphysical: it establishes a complete metaphysical system, including the ontology of difference, process ontology, co-evolutionary ontology, and so on. This book holds that metaphysical statements can be legitimate, provided that they satisfy methodological disciplines such as constrained speculation, failure conditions, falsifiability, and so on. This difference reflects two different philosophical directions: Wittgenstein's work restricts metaphysics by doubting the legitimacy of metaphysical statements; this book refines metaphysics by making it rigorous through methodological discipline.

Difference Three: connection with ontological content. Wittgenstein's work, especially the early work, focuses on linguistic analysis: it treats language itself and is relatively restrained with respect to ontological content, such as beings, processes, and so on. This book focuses on ontological content: it establishes ontologies, such as the ontology of difference and process ontology, as well as ontological relations, such as co-evolution, mind–matter, observation, and time. Linguistic analysis is used as a methodological tool, but it is not the center of the work. This difference gives the two works different emphases: one is oriented toward language, the other toward ontology.

Positioning between the Positions

Through comparison, the relation between this book's work and Wittgenstein's work can be stated as follows: this book resonates with Wittgenstein in seriousness toward the boundary of language, methodological rigor, and the exclusion of illegitimate statements, but differs in the treatment of the "unsayable," legitimate meta-level speech vs strict silence; the status of ontology, refined metaphysics vs restricted metaphysics; and the direction of emphasis, ontological content vs linguistic analysis. This positioning avoids two traps: the trap of absorption, claiming that this book "absorbs" Wittgenstein's work, when this book has an independent direction; and the trap of simple criticism, claiming that Wittgenstein's work is "too restrained," when Wittgenstein's work has enduring force.

Objections and Replies

Objection One: Your meta-level/object-level distinction is actually similar to the ladder of the Tractatus: it claims to stand outside and depict the boundary of language or ontology. Does it therefore involve the same problem of self-application?

Reply: This objection genuinely points to a methodological issue. The ladder of the Tractatus is self-dissolving: after the reader climbs it, it must be thrown away. This treatment involves a paradoxical structure. The meta-level/object-level distinction in this book is not self-dissolving: meta-level statements are legitimate metaphysical statements and do not need to be discarded. Meta-level statements are statements about the boundary: they have content, including the existence and meaning of the boundary. Object-level statements are statements about what lies on the other side of the boundary: they involve methodological restrictions. Both levels are legitimate, but they have different forms of legitimacy. This treatment does not involve self-dissolution. It preserves the legitimacy of meta-level statements through a level distinction. This differs from the ladder structure of the Tractatus. More deeply, the methodology of this book is built upon level distinctions, which have precedents in mathematical logic, including the distinction between classes and sets, and between object language and meta-language. The meta-level/object-level distinction is not an arbitrary philosophical invention; it is built upon a methodological tradition.

Objection Two: Your contrast between "refining metaphysics" and "restricting metaphysics" may be overly simplified. Wittgenstein's work, especially the later work, has a metaphysical dimension: he is not simply anti-metaphysical.

Reply: This objection genuinely points to complexity. The later Wittgenstein's work indeed has metaphysical dimensions: he has metaphysical discussions concerning human language, forms of life, and so on. These discussions are not simply anti-metaphysical. But Wittgenstein's metaphysical method differs from that of this book. Wittgenstein's method is to treat metaphysical problems through linguistic analysis: metaphysical statements must be concretized through linguistic analysis. The method of this book is to assess metaphysical statements through methodological discipline: they require falsifiable structure and related constraints. Both methods allow metaphysical dimensions, but through different routes. This is a more refined difference. It is not a simple opposition between "anti-metaphysics" and "metaphysics," but a distinction between "treating through linguistic analysis" and "treating through methodological discipline."

Objection Three: The relation between this book's work and Wittgenstein's work seems close: both pursue methodological rigor and both reject illegitimate statements. Is this book's work in fact a continuation of the Wittgensteinian tradition?

Reply: This book's work resonates with Wittgenstein's work, but is not a simple continuation of that tradition. This book is built upon physics, the EQT framework, which Wittgenstein's work does not contain. This book establishes a complete ontological system,

Chapters 7–15, while Wittgenstein's work is restrained in this respect. This book engages in cross-traditional dialogue, including Chinese traditions, Indian traditions, and others, whereas Wittgenstein's work mainly operates within the analytic tradition. Resonance in methodological rigor does not mean continuation. The two works share a methodological direction, but have important differences in content and scope. The methodological rigor of this book is grounded in multiple sources, including constrained speculation, failure conditions, falsifiability, and so on, and is not limited to the content of the Wittgensteinian tradition.

Literature That Must Be Engaged

- Ludwig Wittgenstein, Tractatus Logico-Philosophicus (1922)
- Ludwig Wittgenstein, Philosophical Investigations (1953)
- Bertrand Russell, "Introduction" to Tractatus: discussion of the early work
- Saul Kripke, Wittgenstein on Rules and Private Language (1982): discussion of the later work
- Chapter 3 on failure conditions as ontological discipline (self-reference)
- Chapter 17 on ontological singularity (self-reference)

Summary of This Section

This section has done four things. First, it developed the core positions of Wittgenstein's work, both early and late. Second, it identified resonances between this book's work and Wittgenstein's work: seriousness toward the boundary of language, methodological rigor, and the exclusion of illegitimate statements. Third, it identified differences: the treatment of the "unsayable," the status of ontology, and the direction of emphasis. Fourth, it clarified the two as different directions in the application of methodological rigor.

The next section develops the dialogue with Nāgārjuna.

19.5. Dialogue with Nāgārjuna

The Core of Nāgārjuna's Work

Nāgārjuna (around 150–250 CE) is the founder of the Madhyamaka school of Indian Buddhism. His representative work is the Mūlamadhyamakakārikā, which develops the core position of emptiness (śūnyatā) through twenty-seven chapters of verses.

Nāgārjuna's core position consists of the following. Emptiness (śūnyatā): all beings are empty; they have no self-nature (svabhāva), no independent ontological content. The existence of any being depends on causes and conditions (pratītyasamutpāda, dependent arising): it is not an independent ontology. Therefore, beings have no self-nature: they do not exist independently in themselves. But this does not mean that beings do not exist. They exist in the mode of dependent arising. Double negation: when treating ontological questions, Nāgārjuna uses the method of double negation: neither affirming that X exists, since this involves ontological attachment, nor denying that X exists, since this likewise involves ontological attachment by assuming "non-existence" as ontological content. The correct treatment is to go beyond the binary: X exists in the mode of emptiness. The Middle Way (madhyamā pratipad): Nāgārjuna's method is called the Middle Way. It finds balance between two extremes, eternalism and nihilism. The Middle Way is not a simple compromise, but a way of going beyond binary opposition. Ultimate and conventional truth (paramārtha-satya and saṃvṛti-satya): Nāgārjuna distinguishes two levels of truth, ultimate truth, truth from the perspective of emptiness, where all is empty, and conventional truth, truth from the perspective of ordinary experience, where beings have practical efficacy. Both truths have legitimacy at different levels; they are not mutually exclusive.

Relevance to This Book's Work

Nāgārjuna's work has a complex relation to this book's work. They resonate in certain dimensions and differ in others.

Resonance One: rejection of independent ontology. Nāgārjuna rejects the self-nature, the independent ontology, of beings. This book, especially Chapter 7 on difference before identity, also rejects entities as foundational ontology: so-called "entities" are stable forms of differential structures, not independent ontologies. The two resonate in rejecting independent ontology. This is a core direction of both works.

Resonance Two: dependent arising vs co-evolution. Nāgārjuna's concept of dependent arising (pratītyasamutpāda), where beings depend on causes and conditions, has structural similarity with co-evolution in this book (Chapter 12). Both treat ontological relation as foundational: beings exist through relations rather than independently. Dependent arising: A exists $\Longleftrightarrow$ A depends on causes and conditions, including other beings and conditions. Co-evolution: the relation between A and B is part of their intrinsic content as ontological units. The two resonate in treating relation as an ontological foundation.

Resonance Three: two levels of truth. Nāgārjuna's distinction between ultimate truth and conventional truth has structural similarity with this book's meta-level/object-level distinction (Chapter 17): both use level distinctions to treat methodological problems. Nāgārjuna: ultimate truth, emptiness, vs conventional truth, the practical efficacy of beings. This book: meta-level, statements about the boundary, vs object-level, statements about the other side of the boundary. Both use levels to allow the legitimacy of different dimensions to be acknowledged simultaneously.

Key Differences

But there are also important differences between this book's work and Nāgārjuna's work.

Difference One: methodology. Nāgārjuna's method is double negation: through logical deconstruction, it points toward the ultimate. This method is developed in the Mūlamadhyamakakārikā through dialectical analysis, dissolving various forms of ontological attachment. The method of this book is constrained speculation + failure conditions + falsifiability, and so on: it establishes legitimate ontological commitments through methodological discipline. The two methods have different directions. Nāgārjuna's direction is to deconstruct ontological attachment, emptying ontological content through double negation. This book's direction is to construct legitimate ontological commitments, making ontological content rigorous through methodological discipline. Both methods treat ontological problems, but in different directions.

Difference Two: soteriological meaning. Nāgārjuna's work has soteriological meaning: the understanding of emptiness is the path toward nirvāṇa, liberation. This soteriological dimension is central to Nāgārjuna's work. This book does not directly treat soteriology. It is ontological–epistemological–methodological work. Questions of life and liberation, if they arise, must be treated through ethics and the philosophy of meaning, Chapters 20 and 21, but are not directly connected with soteriology. This difference reflects two different philosophical traditions: Nāgārjuna's work is built within a Buddhist soteriological framework, where metaphysics serves liberation; this book is built within a framework of natural philosophy + metaethics, where metaphysics serves ontological, epistemological, and normative work. This is not a question of which is deeper, but a difference in emphasis between traditions.

Difference Three: physical ground. Nāgārjuna's work is built upon logical analysis and Buddhist experience; it has no physical ground. This book is built upon physics, the EQT framework: it has a physical ground. This difference gives the two works different sources of legitimacy. The legitimacy of Nāgārjuna's work is grounded in logic and cultivation; the legitimacy of this book's work is grounded in physics, methodological discipline, and interdisciplinary connection. Both forms of legitimacy are legitimate and reflect different philosophical directions.

Difference Four: the treatment of "emptiness." Nāgārjuna's concept of emptiness is universal: all beings are empty. This book does not claim universal "emptiness": beings have ontological content, including differential structures and processuality; they are not empty. This book claims that beings are not independent entities, which is consistent with Nāgārjuna's direction, but that beings are processual ontologies, differential structures and so on, and therefore have ontological content. This difference marks a difference concerning the reality of ontology: Nāgārjuna's work holds that beings are empty, having no independent ontological content; this book holds that beings are real but processual, having processual ontological content. Yet upon careful analysis, the two positions may not conflict: Nāgārjuna's "emptiness" is primarily directed against independent ontological content, self-nature; this book's "reality" is primarily directed toward processual ontological content. If the content of both positions is clarified, they may agree in rejecting independent ontology and differ in affirming processual ontology.

Positioning between the Positions

Through comparison, the relation between this book's work and Nāgārjuna's work can be stated as follows: this book resonates with Nāgārjuna in rejecting independent ontology, relation as ontological foundation, dependent arising vs co-evolution, and the two levels of truth, but differs in methodology, construction vs deconstruction; soteriological meaning; physical ground; and the treatment of "emptiness." As different directions for treating ontological questions, the two can inspire each other but cannot be reduced to one another.

This positioning especially avoids one trap: the rhetoric that "Eastern thought already anticipated science." This section does not claim that Nāgārjuna "anticipated" this book's work, nor that this book is a contemporary version of Nāgārjuna's work. The two are different works. They resonate, but are not simply identical.

Objections and Replies

Objection One: Is your treatment of Nāgārjuna too "Westernized"? You use Western philosophical concepts, such as ontology, methodology, falsifiability, and so on, to interpret Nāgārjuna, which may distort his content.

Reply: This objection genuinely points to the problem of cross-cultural interpretation. Any cross-traditional dialogue involves translation: expressing the concepts of one tradition in the language of another. This process involves the risk of distortion. This section is aware of this risk and seeks to reduce it in the following ways: using Nāgārjuna's original terms, such as śūnyatā, pratītyasamutpāda, and so on, to preserve the original content; making similarities and differences explicit, and not simply claiming that the two are identical; and acknowledging the soteriological dimension of Nāgārjuna's work, not simply reducing it to ontological work. But it is impossible to avoid distortion completely: any cross-traditional dialogue involves translation. The goal of this section is heuristic dialogue: through comparison, it shows the similarities and differences between the two traditions, allowing readers of each tradition to be mutually informed. This goal is not mutual reduction, but mutual illumination. The two traditions preserve their independence while gaining new perspectives through dialogue.

Objection Two: Your contrast between "construction vs deconstruction" seems to make this book's work and Nāgārjuna's work completely opposed. But in fact the two may resonate at a deeper level.

Reply: This objection genuinely points to a deeper relation between the two works. "Construction vs deconstruction" is a contrast at the methodological level: it describes the methodological direction of the two works. But at the ontological level, they may have deeper resonance. Nāgārjuna's work deconstructs ontological attachment, but it points toward the ontology of dependent arising, where beings exist through causes and conditions. This book constructs ontological commitments, but it constructs processual ontology, including differential structures, co-evolutionary relations, and so on. The ontological contents of the two works, dependent arising and processual ontology, resonate: both reject independent entities and both affirm relational ontology. The methods, deconstruction vs construction, differ, but the ontological results resonate. This observation makes the comparison in this section more precise: there is difference at the methodological level and resonance at the ontological level. As works in different directions, the two may arrive at similar ontological directions through different methods.

Objection Three: Your difference in the treatment of "emptiness," where this book claims that beings are "real but processual" while Nāgārjuna claims "emptiness," may be a terminological difference rather than a substantive difference.

Reply: This objection points to an important clarification. If the meanings of the two concepts are clarified: Nāgārjuna's "emptiness" means the absence of self-nature (svabhāva), independent ontological content; this book's "real but processual" means having processual ontological content. These two statements can in fact be compatible: beings lack self-nature, consistent with Nāgārjuna, because they are not independent ontologies; beings have processual ontological content, according to this book, because they have ontological content as forms of process. If this clarification is completed, they may substantively resonate, differing mainly in terminology. This observation makes the comparison in this section more precise: some apparent differences may become resonances after clarification. But this book still maintains an independent position: it establishes processual ontological content through physical ground and methodological discipline. Nāgārjuna establishes the ontological content of dependent arising through logical analysis and the Buddhist tradition. Their sources of legitimacy differ, even though their ontological results may be similar.

Literature That Must Be Engaged

- Nāgārjuna, Mūlamadhyamakakārikā (around the second century)
- Candrakīrti, Prasannapadā: classical commentary of the Madhyamaka school
- T. R. V. Murti, The Central Philosophy of Buddhism (1955)
- David Loy, Nonduality (1988)
- Jay Garfield, The Fundamental Wisdom of the Middle Way

(1995): contemporary English translation and commentary of the Mūlamadhyamakakārikā

- Chapter 7 on difference before identity (self-reference)
- Chapter 12 on co-evolution as an ontological relation (self-reference)

Summary of This Section
This section has done four things. First, it developed the core position of Nāgārjuna's work: emptiness, double negation, the Middle Way, and the two levels of truth. Second, it identified resonances between this book's work and Nāgārjuna's work: rejection of independent ontology, dependent arising vs co-evolution, and two levels of truth. Third, it identified differences: methodology, construction vs deconstruction; soteriological meaning; physical ground; and the treatment of "emptiness." Fourth, it clarified the two as different directions for treating ontological questions.

19.6. Dialogue with Husserl

Husserl's Mode of Response: Phenomenological Reduction and Description of the Given
The fourth interlocutor is Edmund Husserl (1859–1938), the founder of the Western phenomenological tradition. Husserl's mode of response is qualitatively different from the first three. Heidegger points beyond the boundary through poetic saying (Section 19.2); Wittgenstein acknowledges the limit of language through keeping silence (Section 19.3); Nāgārjuna goes beyond binary opposition through double negation (Section 19.4). All three treat the boundary itself as the problem and respond to the question, "how can one speak beyond the boundary?" Husserl takes another direction: he does not ask what lies beyond the boundary, but returns within the boundary, to the starting point of everything, the given itself, and describes it rigorously.

Husserl's method begins in Logical Investigations (1900–01), is systematized in Ideas I (1913), matures in Cartesian Meditations (1931), and is extended in The Crisis of European Sciences (1936) into a diagnosis of the European scientific tradition. The core of this method is phenomenological reduction: the suspension (epoché) of all presuppositions concerning the "existence of the objective world," and the concentration of philosophical reflection on what is given to consciousness itself: phenomena as phenomena, the given as given. This suspension is not skeptical rejection, but methodological bracketing. By temporarily not presupposing the external world, the internal structures of consciousness, including intentionality, horizon, and modes of givenness, can be rigorously described in themselves.

Husserl's work reaches a fundamental insight: the given is the starting point of everything and cannot be derived from anything more basic. Any knowledge of the world, any proposition about reality, and any scientific description all presuppose the givenness of experience. They proceed from the given; therefore, the given itself cannot in turn be reduced to some non-given thing for explanation, since that would be circular or regressive. Consciousness is thus analyzed as always "directed toward something," namely intentionality, and the given is the core structure of intentionality: consciousness is always consciousness "of" something, always given some content and manifested from some perspective.

Resonance: Convergence with This Book's Work through Different Paths

The resonance between this book and Husserl is much deeper than it may appear. Section 13.6 has already developed the core of this resonance: the convergence of two independent paths at the same irreducible point. This book reaches "the given is irreducible" through a bottom-up physical path, from ρ_f to structure, to the given, to knowledge-formation and circumscription, and to the last cell, while Husserl reaches a highly consonant insight through a top-down reflective path. This section repositions that convergence within the context of Chapter 19's responses to the metaphysical boundary.

The resonance has three layers. The first layer is methodological independence. The convergence between this book and Husserl is not mutual derivation: this book does not appeal to the phenomenological method, and Husserl does not appeal to physics. Rather, two independent paths converge at the same irreducible point. This independence makes the convergence itself an independent support for the conclusion reached: convergence of independent evidence is more credible than assertion from a single path. The second layer is the irreducibility of the given as starting point. Husserl's "the given cannot be reduced to the non-given" deeply accords, in philosophical posture, with the "last cell as boundary" in Section 13.4 of this book. Both acknowledge that at the starting point of philosophical reflection there is a node that cannot be further reduced: Husserl calls it the given; this book calls it the residual point of subject-ascription. Both refuse to replace it with a more basic concept: Husserl refuses psychological reduction, while this book refuses to let mechanistic statement pass as an explanation of being-experienced, namely the C3 prohibition in Chapter 13. The third layer is the correspondence between intentional structure and internal–external co-evolution. Husserl's intentionality means that "consciousness is always consciousness of something": it is the internal directedness structure of experience. This book's internal–external co-evolution (Section 13.3) means that "the given is always for some interior": it is the relational structure of givenness. The two are structurally isomorphic: Husserl's "subject pole–object pole" corresponds to this book's "interior–exterior"; Husserl's "directedness" corresponds to this book's "co-evolution." This book therefore places the intentional structure discovered by Husserl's reflection onto the mechanism of physical co-evolution: intentionality is no longer an unanalyzable internal structure of experience, but a relational event with physical correspondence.

Differences: The Scope of Convergence and the Independent Position of This Book

But there are also clear differences between this book and Husserl, and these differences determine that their convergence is a convergence through different paths rather than a reduction of one to the other.

First, the difference in methodological starting point: Husserl begins from the suspension of the external world and builds phenomenological reflection upon the direct givenness of consciousness; this book begins from the ontology of EQT natural philosophy and presupposes ρ_f as the foundational ontology. Husserl's suspension prevents him from directly treating physics; this book's ontology connects it directly with physics. This difference makes the two frameworks non-derivable from each other: one cannot derive physics from suspension, and one cannot derive phenomenology from physics.

Second, the difference in research direction: Husserl's reflective direction begins from experience and moves inward into internal structures, including the essential structure of consciousness, horizon-structure, and the internal structure of time-consciousness. The direction of this book begins from physical ontology and moves upward toward the structure of experience, including the isomorphism program, fivefold gradient richness, and internal–external co-evolution.

The two directions converge at the irreducibility of the given, but the paths in between are entirely different.

Third, the posture toward the boundary: Husserl makes the internal structure of the given precisely describable through rigorous phenomenological description. He does not treat the given as a boundary, but as a beginning that can be deeply described. This book precisely marks the irreducible through two compressions as a single-point boundary: it accepts the description of the given, as Husserl does, but treats "being-experienced itself" as a boundary rather than as an object for further analysis (Section 13.4). Both respect the irreducibility of the given, but differ in how they treat this irreducibility: Husserl moves inward to describe it more deeply; this book moves outward to mark the boundary.

Position within the Four Responses of Chapter 19

This section, as the fourth response in Chapter 19, has a clear division of labor from the first three responses. The first three responses, Heidegger, Wittgenstein, and Nāgārjuna, all treat the boundary itself as the problem: they each answer "how can one speak beyond the boundary?" Heidegger points through poetic saying; Wittgenstein acknowledges through rigorous silence; Nāgārjuna transcends through double negation. Husserl's response does not lie on this axis of questions: he does not ask what lies beyond the boundary, but returns to the starting point of everything, the given, and describes it rigorously.

Thus in Chapter 19's multiple dialogues, this book has four independent interlocutors corresponding to four different directions of boundary treatment: Heidegger (19.2), in the Western continental tradition, points beyond the boundary through hermeneutic-poetic saying; Wittgenstein (19.3), in the Western analytic tradition, acknowledges the boundary through the limit of language; Nāgārjuna (19.4), in the Indian Buddhist tradition, goes beyond binary opposition through double negation; and Husserl (this section), in the Western phenomenological tradition, returns to the given as beginning through rigorous description. Together, the four responses constitute the cross-traditional dialogue of this book on the problem of boundary: Western analytic, Western continental hermeneutic, Western continental phenomenological, and Indian Buddhist traditions all treat the boundary problem in different directions, and this book's work has an identifiable independent position among them. This fourfold dialogue makes explicit the character of this book as a cross-traditional philosophical monograph: it is not confined to an internal discussion within a single tradition, but offers a contemporary synthesis on a shared problem across several philosophical traditions.

Connection with Chapter 13

This section has a clear division of labor with Section 13.6, "Convergence with Phenomenology through Different Paths." Section 13.6 develops the convergence between this book and phenomenology on the "irreducibility of the given" within the context of the philosophy of mind and matter. It focuses on the application in the philosophy of consciousness. This section repositions that convergence in the context of the metaphysical boundary. It places Husserl within the four responses to boundary in this book and shows how this positioning completes the cross-traditional dialogue in Chapter 19. The relation between the two sections is as follows: Section 13.6 is a local application, the convergence within the field of the philosophy of consciousness; this section is an overall positioning, the location within Chapter 19's dialogue on the metaphysical boundary. Readers of Section 13.6 see the content of the convergence; readers of this section see the position of that convergence within the book's overall cross-traditional dialogue.

The next section is the chapter summary.

19.7. Chapter Summary: An Independent Position among Four Traditions

Completion of the Three Tasks

The three tasks undertaken by this chapter have all been completed. Task One: to identify the relevance of this book's work to the four traditions. Section 19.1 developed four modes of response: Heidegger's poetic saying, Wittgenstein's keeping silence, Nāgārjuna's double negation, and Husserl's rigorous description, as different directions for treating the metaphysical boundary. Task Two: to enter into serious dialogue with each tradition. Section 19.2 entered into dialogue with Heidegger; Section 19.3 with Wittgenstein; Section 19.4 with Nāgārjuna; and Section 19.5 with Husserl. Each dialogue developed resonances and differences, clarifying the relation between this book's work and each tradition. Task Three: to clarify the independent position of this book's work among the four traditions. Through dialogue, this chapter shows that this book's work is not a simple continuation of any one tradition. It has its own methodology, including constrained speculation, failure conditions, falsifiability, and so on; its own ontological content, including the ontology of difference, process ontology, co-evolutionary ontology, and so on; and its own sources of legitimacy, namely physical grounding + methodological discipline.

The Position of This Chapter in Part IV

The overall structure of Part IV, "The Metaphysical Boundary of Freezing," is completed in this chapter: Chapter 16, "The Hierarchy and Historicity of Freezing," as foundational development; Chapter 17, "The Gradient Terminal Phase as an Ontological Singularity," as the metaphysical summit; Chapter 18, "The Collapse of Co-Evolution before the Gradient Terminal Phase," continuing Chapter 17; and Chapter 19, this chapter, "Four Responses," as dialogue with traditions. As the concluding chapter of Part IV, this chapter places the metaphysical boundary work of Part IV within a larger philosophical discussion. Through dialogue, the work of Part IV receives a philosophical position: it has an independent position among the Western continental-hermeneutic tradition (Heidegger), the Western analytic-linguistic tradition (Wittgenstein), the Indian Buddhist tradition (Nāgārjuna), and the Western phenomenological tradition (Husserl).

The Position of This Chapter in the Whole Book

Looking backward, this chapter continues the work of the first three chapters of Part IV, Chapters 16, 17, and 18, and also continues the work of Section 13.6, which developed the convergence between this book and phenomenology in the context of the philosophy of consciousness. Section 19.5 of this chapter repositions that convergence in the context of the metaphysical boundary. The two sections complement each other: Section 13.6 is the local application, while Section 19.5 is the overall positioning. Looking forward, this chapter prepares Part V, "From Ontology to Normativity." The four dialogues involve methodology, including poetry, silence, transcendence, suspension, and so on, and these are connected with the later work on meaning and value: Heidegger's poetic saying connects with Chapter 21 on the emergence of meaning; Wittgenstein's methodological rigor connects with Chapter 20 on the normative application of constrained speculation; Nāgārjuna's double negation connects with Chapter 21 on the affirmation of finitude, since finitude is neither simple negation nor simple affirmation and requires going beyond binary opposition; Husserl's rigorous description of the given as beginning connects with Chapter 21 on the bearer of meaning, the interior circumscribed by boundary. These connections make the dialogical work of this chapter not isolated, but a source of methodological resources for the following part.

Honesty about the Limits of This Chapter

This chapter acknowledges several limits. First, the complexity of the work of the four philosophers. Heidegger, Wittgenstein, Nāgārjuna, and Husserl each have complex bodies of work. This chapter can treat only their core positions and cannot develop all dimensions of their work. Detailed treatment would require dedicated work. Second, other important traditions are not engaged. Beyond these four responses, other important traditions may also be relevant to this book's work, including the Hindu Vedānta tradition, Islamic Sufi traditions, Neoplatonism, Jewish Kabbalah, and others. Dialogue with these traditions requires dedicated work. Third, the depth of cross-cultural dialogue is limited. Each dialogue is only about 2,000 Chinese characters in the original planning scale, which is insufficient for fully treating each philosopher's work. Each dialogue requires a dedicated article or monograph. Fourth, the limits of this book's work are not sufficiently discussed. This chapter uses comparison to show the independent position of this book's work, but it does not sufficiently discuss the limits of this book's work itself, such as its restraint regarding soteriological dimensions. Acknowledging these limits does not weaken the work of this chapter; it makes this chapter an honest initial work.

The Core Posture of This Chapter

The core posture of this chapter can be stated as follows: an independent position among multiple traditions is the posture of serious contemporary philosophical work.

Since the twentieth century, philosophical discussion has involved multiple traditions: the Western analytic tradition, the Western continental tradition, including hermeneutic and phenomenological branches, Chinese traditions, Indian traditions, Japanese traditions, and so on. Any serious contemporary philosophical work needs to find its position among these traditions. This position is not a synthesis of the strengths of all traditions; that is impossible, since each tradition has its own internal integrity. Nor is it simple belonging to one tradition, because that would lose the inspirations of other traditions. The correct treatment is independence with connection: this book has its own methodology, ontological content, and sources of legitimacy, but it also connects with several traditions through dialogue. This posture gives the work of this book legitimacy in contemporary philosophical discussion: it is not an isolated philosophical creation, but a new development among multiple traditions.

Meaning before freezing: this is one of the core theses of this book. The four traditions discussed in this chapter all concern the problem of meaning in some sense: Heidegger's question of Being, where Being is the source of meaning; Wittgenstein's boundary of language, where what exceeds the boundary involves meaning, including the mystical, ethical, aesthetic, and so on; Nāgārjuna's emptiness, where emptiness is the ground of meaning, the liberation from illusory attachment; and Husserl's givenness, where the given is the beginning of meaning, the starting point of all meaning-constitution. This book's treatment, meaning before freezing, gives meaning an ontological ground through process ontology. It does not require Heidegger's poetic saying, Wittgenstein's silence, Nāgārjuna's double negation, or Husserl's suspension as its methodological starting point. But it resonates with all four traditions: it treats meaning as an emergence of processual ontology, acknowledges the boundary of meaning, namely collapse before the gradient terminal phase, preserves methodological rigor, and respects the given as an irreducible beginning. Through the work of this chapter, this book obtains an independent position among the four traditions. This position allows this book to dialogue with multiple traditions. It is not a simple continuation of the Western analytic tradition, nor of Chinese or Indian traditions, nor of the Western continental tradition, whether hermeneutic or phenomenological. It is new work among multiple traditions. Part IV, "The Metaphysical Boundary of Freezing," is completed through the dialogue in this chapter. The following Part V, "From Ontology to Normativity," builds upon the metaphysical work of Part IV and develops the normative dimension of this book's work.

19.8. Methodological Reflection on This Chapter

This chapter connects this book's metaphysical boundary work with four philosophical traditions. This section reflects on the methodological significance of this cross-traditional dialogue.

The Function of This Chapter as a Dialogical Chapter

This chapter is the point of connection between this book's work and multiple traditions. If this book did not develop this connection, its work might be seen as an isolated philosophical creation, with an unclear philosophical position. This chapter establishes that connection, giving this book's work legitimacy and dialogical accessibility within plural contemporary discussions. Isolated philosophical creation is a trap in contemporary philosophical work: some works establish their own terminology and framework but do not build connections with multiple traditions. As a result, even if they are internally refined, they are difficult to take seriously in international discussion. This chapter avoids that trap.

This Chapter's Posture toward Its Interlocutors

This chapter's posture toward the four philosophers is serious respect and precise distinction. Each dialogue respects the internal content of the original work and does not reduce it to an appendage of this book's work. With respect to Heidegger, this chapter acknowledges the resonance between concepts such as Geschichtlichkeit and Ereignis and this book's work, but shows this book's position through the difference between methodological rigor and poetic saying. With respect to Wittgenstein, this chapter acknowledges his seriousness toward the boundary of language, but shows this book's position through the difference between the meta-level/object-level distinction, which makes legitimate speech at the boundary possible, and Wittgenstein's insistence on strict silence. With respect to Nāgārjuna, this chapter acknowledges the directional resonance of emptiness and dependent arising, but shows this book's position through the difference between constructing legitimate ontological commitments and deconstructing ontological attachment. With respect to Husserl, this chapter acknowledges the convergence through different paths between the irreducible beginning of the given and this book's work, but shows this book's position through the difference between the physical path and the path of phenomenological reflection, as well as the difference between treating the experientiality of the given as a boundary and treating it as a beginning that can be more deeply described.

This Chapter's Posture toward Originality

This chapter does not claim complete originality. Through dialogue, it shows the relation between this book's work and the four traditions. Its contribution is to make explicit the independent position of this book's work among the four traditions; to show resonances and differences through dialogue; and to provide methodological resources for Part V. These contributions are constructive dialogue: they give this book's work a position within multiple traditions.

Limits Honestly Acknowledged in This Chapter

This chapter acknowledges several limits. First, the work of the four philosophers is complex, and this chapter can treat only their core directions. Second, this chapter mainly treats the work of the four philosophers on the metaphysical boundary, including Husserl's phenomenological reduction as a treatment of the starting point of consciousness; their other work is not treated in detail. Third, the dialogue between the four philosophers and this book's work requires more detailed refinement; each dialogue requires a dedicated article. Fourth, Husserl appears as an interlocutor in both Chapter 19, the metaphysical boundary, and Chapter 13, the philosophy of consciousness. The two dialogues require more precise coordination. The division of labor between Section 19.5 and Section 13.6 is preliminary, and further refinement is left to future work.

Connections with Later Chapters

This chapter prepares Part V, "From Ontology to Normativity." The four philosophical traditions all involve methodology at the boundary: Heidegger's poetic saying involves the ground of meaning; Wittgenstein's keeping silence involves the legitimate range of language; Nāgārjuna's double negation involves the direction of transcending attachment; and Husserl's rigorous description involves the given as the beginning of meaning-constitution. These methodologies connect with Part V's work on meaning and value. More deeply, this chapter is the peak of this book's cross-traditional dialogue. This book draws upon multiple traditional resources: the Western analytic tradition, the Western continental tradition, including hermeneutic and phenomenological branches, the Chinese philosophical tradition, and the Indian Buddhist tradition. This chapter develops the application of these traditional resources at the boundary in a unified way, making explicit the character of this book as a cross-traditional philosophical monograph. Chapter 19, together with Chapter 8 on the qi-theory tradition and Chapter 13 on the genealogy of the philosophy of consciousness and the convergence with phenomenology through different paths, jointly constitutes the cross-traditional dialogical architecture of the book. Chapter 8 treats the Chinese qi-theory tradition; Chapter 13 treats the genealogy of the philosophy of consciousness, including Russellian monism and phenomenology; and Chapter 19 treats traditions concerning the metaphysical boundary, including Husserl's mode of response. Together, these three chapters make the book's character as a cross-traditional philosophical monograph not merely a declaration, but an accomplished structure.

Part V.

From Ontology to Normativity

20. How Constrained Speculation Treats the Is–Ought Gap

20.1. Introduction to This Chapter

In Book III of *A Treatise of Human Nature* (1739–40), David Hume made an observation. In works on moral systems, authors often move abruptly from factual statements (is-statements) to value statements (ought-statements). This transition involves a "new relation" that requires explanation. Hume doubted that such an explanation was possible: pure description cannot generate normativity. This observation later came to be called the is–ought gap, or Hume's guillotine.

This argument appears simple, but it has fundamental implications. It questions the possibility of ethics as a science. If value cannot be derived from fact, then ethics seems unable to be built upon descriptive science. Much of twentieth-century metaethics unfolded around this problem: moral realism, expressivism, error theory, and constructivism each respond to Hume's challenge from different directions.

This chapter proposes the treatment of this problem through the method of constrained speculation: it downgrades the is–ought gap from an ontological problem, namely a difference in the ontological status of fact and value, into a methodological problem, namely that factual statements and value statements are subject to the same discipline but are not reducible to one another. The core thesis (IO1) is this: factual statements and value statements are both ontological commitments. They both involve statements about beings and are both subject to the discipline of the method of constrained speculation (CS1–CS4). The difference between them is not a difference in ontological status, but a difference in type of commitment: factual statements concern properties or relations, while value statements concern the contribution of an action or state along some dimension.

This downgrading allows ethics to be a rigorous field of empirical research. Its statements are constrained by falsifiable structures and can be seriously assessed and refined. At the same time, it preserves the difference between factual statements and value statements: value statements are not reduced to factual statements, but have their own type of commitment and falsifiable structure.

This chapter also enters into serious dialogue with the major positions in contemporary metaethics: moral realism, expressivism, error theory, and constructivism. Constrained speculation has an independent position within the genealogy of metaethics: constrained processual normativism. This position preserves objectivity, since value statements are constrained by methodological discipline, but does not presuppose an independent moral ontology. It makes methodology an instrument of assessment but not the ground of value, and it acknowledges the processual content of value.

This chapter is the first chapter of Part V, "From Ontology to Normativity." It establishes the ontological-methodological foundation of ethics and provides the methodological ground for Chapter 21, "The Affirmation of Finitude: Meaning before Freezing." This chapter is not a detailed monograph on ethics; it treats a metaethical problem, namely the ontological-methodological ground of ethics itself. Concrete ethical questions require dedicated ethical work. This chapter only establishes how the method of constrained speculation can be applied in the field of ethics.

The discussion proceeds as follows. Section 20.1 develops Hume's is–ought gap and its philosophical influence. Section 20.2 states and defends the is–ought relation as a methodological problem (IO1). Section 20.3 develops the falsifiable structure of value statements, consisting of four elements. Section 20.4 enters into serious dialogue with contemporary metaethics. Section 20.5 develops the position of constrained processual normativism and treats objections.

20.2. Hume's Is–Ought Gap

Hume's Argument

In Book III of *A Treatise of Human Nature* (1739–40), David Hume made an observation that later came to be called the "is–ought gap" or "Hume's guillotine." Hume noticed that in works on moral systems, arguments often move imperceptibly from factual statements, propositions connected by is and is not, to value statements, propositions connected by ought or ought not, without ever explaining this transition. He pointed out that this ought expresses "some new relation or affirmation," which should be observed and explained, and that a reason should be given for how this new relation can be deduced from relations that are entirely different. Hume thought this seemed inconceivable.

Hume's argument can be stated as follows: factual statements (is-statements) involve description, what something is; value statements (ought-statements) involve normativity, what something ought to be; deriving a value statement from purely factual statements involves a "new relation," and this relation requires explanation; Hume doubts that such an explanation is possible: pure description cannot generate normativity. This argument appears simple, but it has fundamental implications. It questions the possibility of ethics as a science. If value cannot be derived from fact, then ethics seems unable to be built upon descriptive science.

The Philosophical Influence of Hume's Gap

Hume's gap has had a deep influence in the history of Western philosophy.

The challenge to naturalistic ethics: naturalistic ethics attempts to reduce value to fact, for example utilitarianism's "greatest happiness" or evolutionary ethics' "adaptiveness." Hume's gap raises an objection: any such reduction involves a leap from is to ought, and this leap requires justification.

Moore's "naturalistic fallacy": in *Principia Ethica* (1903), G. E. Moore developed Hume's argument and proposed the "naturalistic fallacy": reducing "good" to any natural property, such as "pleasure" or "adaptiveness," involves a conceptual error. Moore's "open question argument" became a classic in metaethics.

The division of twentieth-century metaethics: Hume's gap divided metaethics into several positions, including moral realism, value as objective; expressivism, value statements as expressions of emotion; error theory, value statements as systematically mistaken; constructivism, value as constructed through procedures; and so on. Each position treats the is–ought gap in a different way.

The influence on the relation between science and ethics: Hume's gap continues to operate in discussions of the relation between science and ethics. Can scientists speak about ethics? Can ethics be built upon scientific findings? These questions continue to appear in contemporary discussions, such as neuroscience and morality, evolutionary ethics, behavioral ethics, and related fields.

The Force of the Gap

The force of Hume's gap comes from three features.

Feature One: difference in logical form. The logical form of factual statements is description: they describe a state. The logical form of value statements is normativity: they give a direction. The two logical forms are structurally different. Deriving normativity, "X ought to be this way," from description, "X is this way," involves a logical step. This step is not contained in standard rules of deduction. Some additional principle is required to complete the leap.

Feature Two: difference in ontology. Facts concern the content of existence: what beings have which properties. Values concern directions in relation to goals: whether a state is good or bad. The ontological status of the two appears different: facts are "objective" existence, while values seem to involve "subjective" evaluation. A leap from objective to subjective requires justification.

Feature Three: difference in motivation. Factual statements do not by themselves directly provide motivation for action: knowing that "X is this way" does not necessarily lead to action. Value statements directly concern the direction of action: "X ought to be this way" implies a tendency toward action. The two have different motivational functions, and this functional difference also makes the leap from fact to value psychologically difficult.

These three features together constitute the force of Hume's gap. It is not a simple argumentative error, but a real philosophical problem concerning the difference between fact and value.

Objections and Replies

Objection One: Your statement makes Hume's gap look like a "problem," but it may merely be Hume's limitation, not a genuine ontological-epistemological problem.

Reply: Hume's gap does have an argumentative form; it is not an arbitrary declaration. The difference in logical form is real: description and normativity are indeed different in logical structure. The difference in ontological status is real: the ontological treatment of facts and values requires justification. The difference in motivational function is real: statements affect action in different ways. These three real differences make Hume's gap not a simple error, but a philosophical problem. But acknowledging this does not mean accepting Hume's conclusion. The work of this chapter, beginning in Section 20.2, develops the treatment through the method of constrained speculation. It neither cancels the gap nor accepts it as an uncrossable ontological split.

Objection Two: Your concept of the "gap" seems too broad. It treats "fact" and "value" as two opposed categories. But actual statements may involve a mixture of both.

Reply: This objection points to an important refinement. Many statements involve a mixture of fact and value. For example, "X is good" involves both description, X is something, and evaluation, good. This mixture does not cancel Hume's gap; it requires a more refined analysis. In such analysis, statements can be decomposed into descriptive components and evaluative components, each with its own logical form. Hume's gap still has content in this refined analysis: there is a leap between purely descriptive components and purely evaluative components. Mixed statements make analysis more complex; they do not eliminate the gap.

Objection Three: Your treatment makes Hume's gap a permanent problem. But many ethical traditions, such as Confucianism and Buddhism, do not seem to work within Hume's framework: they treat fact and value differently.

Reply: This objection genuinely points to cross-cultural comparison. Hume's gap is mainly treated as a core problem within the Western analytic tradition. It presupposes a particular logical-epistemological framework. Other traditions, especially Confucianism and Buddhism, have different frameworks. They do not simply "deny" the difference between fact and value, but treat their relation within different frameworks. Confucianism unifies value and fact in practice: human action as the activity of a being contains the connection between fact and value, and "the unity of knowledge and action" expresses this direction. Buddhism places fact and value within a more basic ontological framework through dependent arising: causal relations (fact) involve ethical meaning (value). The work of this chapter mainly dialogues with Hume's gap within the Western analytic framework, but its direction has structural relevance to Confucian and Buddhist treatments, especially in Section 20.3 on constrained speculation. A complete cross-cultural treatment lies beyond the scope of this chapter.

Literature That Must Be Engaged

- David Hume, A Treatise of Human Nature (1739-40), Book III, Part I, Section I

- G. E. Moore, Principia Ethica (1903), especially Chapter I
- R. M. Hare, The Language of Morals (1952)
- A. J. Ayer, Language, Truth, and Logic (1936): expressivist treatment of value statements
- Charles Pigden, "Logic and the Autonomy of Ethics" (1989)
- Frank Jackson, From Metaphysics to Ethics (1998)

Summary of This Section

This section has done four things. First, it precisely stated Hume's is–ought gap and its original argument. Second, it developed the influence of Hume's gap in the history of Western philosophy. Third, it identified three features of Hume's gap: difference in logical form, ontological difference, and difference in motivational function. Fourth, it responded to possible objections, especially concerning the reality of Hume's gap and its cross-cultural applicability.

The next section develops the treatment of Hume's gap through the method of constrained speculation.

20.3. From Ontological Problem to Methodological Problem

The Key Transformation

In traditional discussions, Hume's gap is often treated as an ontological problem: facts and values have different ontological statuses and therefore cannot be derived from one another. The work of this chapter is to transform this treatment: to downgrade the gap from an ontological problem into a methodological problem.

Traditional treatment: Hume's gap = a difference in the ontological status of fact and value. This difference makes derivation from one to the other ontologically impossible. Treatment through constrained speculation: Hume's gap = a methodological difference between factual statements and value statements. The two are subject to the same methodological discipline, the method of constrained speculation, but differ in content and cannot be simply reduced to one another. The meaning of this transformation requires detailed development.

The Key Thesis

Core thesis IO1: factual statements and value statements are both ontological commitments and are both subject to the discipline of the method of constrained speculation: formalization, mechanization, observationalization, and failure-orientation. Their difference lies not in ontological status, since both involve ontological commitments, but in type of commitment and mode of falsifiability.

Let us develop this thesis in detail.

Implication One: value statements are also ontological commitments. Traditional treatments often regard value statements as non-ontological: they do not concern the properties of beings, but only preferences or norms. The method of constrained speculation rejects this distinction. Value statements involve committed content. For example, "X is good" commits to the claim that X makes a positive contribution along some dimension, such as human well-being, ecological health, social justice, and so on. This commitment has identifiable consequences: if X makes no contribution along the claimed dimension, the commitment fails. This commitment involves factual relations: causal or structural relations between X and the dimension. These features make value statements ontological commitments: they involve identifiable content and have structures that can be examined.

Implication Two: factual and value statements are subject to the same methodological discipline. The four disciplines of the method of constrained speculation, formalization, mechanization, observationalization, and failure-orientation, apply to both factual and value statements. Formalization: both require formal expression, namely a clear statement of content. Mechanization: both require mechanisms, namely how the relation operates within process. Observationalization: both require observational proxies, namely how they can be identified in experience. Failure-orientation: both require a direction of failure, namely what would make the commitment fail. For example, the factual statement "the boiling point of water is 100°C under standard atmospheric pressure" satisfies the four disciplines: formalization, a temperature value; mechanization, the molecular dynamics of water; observationalization, temperature measurement; failure-orientation, experimental deviation. The value statement "excessive drinking is harmful to health" also satisfies the four disciplines: formalization, definitions of "excessive drinking" and "harmful to health"; mechanization, physiological, metabolic, and psychological processes; observationalization, health indicators, epidemiological data, and so on; failure-orientation, if "excessive drinking" actually has no negative effect on health indicators. There is no difference between the two with respect to methodological discipline: both are constrained by the same requirements.

Implication Three: the difference lies in type of commitment and mode of falsifiability. The difference between factual and value statements does not lie in the form of methodological discipline, since both are subject to the same discipline, but in type of commitment and mode of falsifiability. The type of commitment of factual statements: concerning the properties or relations of beings. Mode of falsifiability: through observation or experiment. The type of commitment of value statements: concerning the contribution of an action or state along some dimension. Mode of falsifiability: through observation of effects, causal analysis, cross-context comparison, and so on. For example, the mode of falsifiability for "the boiling point of water is 100°C" is to measure the boiling point of water under experimental conditions. The mode of falsifiability for "excessive drinking is harmful to health" is to compare drinking habits and health outcomes through epidemiological studies, controlling for confounders, and so on. The two modes of falsifiability differ: one uses physical experiments, the other uses epidemiology and causal analysis. But both have modes of falsifiability and are subject to the same methodological discipline.

Consequences of This Transformation

Downgrading Hume's gap from an ontological problem to a methodological problem has three consequences.

Consequence One: ethics can be rigorous. If value statements are constrained by methodological discipline, then ethics can be rigorous. It has research methods, standards of evaluation, and the possibility of development. This differs from the traditional treatment of "ethics as mere emotion or intuition." Ethical research can use empirical methods, including epidemiology, sociology, psychology, economics, and so on, to study ethical problems. It can use conceptual analysis to clarify the content of ethical concepts. It can use cross-context comparison, including ethical practices in different cultures, times, and situations. These methods make ethics not a "collection of subjective preferences," but a field of research with falsifiable structures.

Consequence Two: fact and value are not reduced to one another. But IO1 does not claim that fact and value are the same thing. Their types of commitment differ. Factual statements do not directly generate norms. From "the boiling point of water is 100°C," one cannot directly derive "water ought to be boiled to 100°C." The latter involves purposes, such as disinfection, and values, such as health. Value statements do not directly generate facts. From "X is good," one cannot directly derive "X exists." The latter involves ontological commitment. Factual statements and value statements are both ontological commitments, but the content of the commitment differs: one concerns the properties of beings, the other concerns the dimensional contribution of actions or states.

Consequence Three: science and ethics can enter into dialogue. If both are subject to the same methodological discipline, then science and ethics can enter into serious dialogue. They use the same methodological standards, but study different content. Scientific research can provide factual foundations, such as biological, psychological, and sociological findings concerning human well-being. Ethical research can use these factual foundations to infer value statements within specific frameworks of purpose and value. But this inference is not a simple "is $\rightarrow$ ought." It involves methodological work: clarifying purposes, identifying values, analyzing causal relations, and so on. This dialogue means that the relation between science and ethics is not split. They are connected fields of research, subject to the same methodological discipline.

An Example: Environmental Ethics

Let us consider an example to show the application of IO1: the treatment of environmental ethics.

Factual statement: "Human activity is causing climate change." Formalization: data on temperature, carbon dioxide concentration, and so on. Mechanization: climate dynamics, the greenhouse effect, and so on. Observationalization: meteorological observation, sea-ice records, sea-level measurements, and so on. Failure-orientation: falsifiable predictions, for example if temperature trends under continued monitoring show a non-anthropogenic pattern.

Value statement: "Human beings have a responsibility to reduce climate change." Formalization: definitions of "responsibility," including moral responsibility, causal responsibility, capacity responsibility, and so on, and mitigation measures. Mechanization: policy mechanisms, technological mechanisms, social mechanisms, and so on. Observationalization: policy implementation, technological application, changes in social behavior, and so on. Failure-orientation: falsifiable predictions, for example if the responsibility statement produces counterproductive effects in implementation, such as greater ecological harm.

Both statements are subject to methodological discipline. But the factual statement concerns description of the climate; the value statement concerns norms of human action. The difference between them is not ontological status, but type of commitment. More deeply, the factual statement can support the value statement: the fact of climate change gives a defensible basis to the responsibility statement for mitigation. But fact cannot directly generate value: the value statement requires a value framework, such as human well-being, ecological health, and so on, as a premise. This is the application of IO1: fact and value are connected but are not reduced to one another.

Objections and Replies

Objection One: Your IO1 seems merely to rename Hume's gap. You say fact and value differ in "type of commitment," but this is just the difference Hume described.

Reply: This objection genuinely points to the core issue. The difference between IO1 and Hume's treatment does not lie in whether there is a difference; IO1 also acknowledges the difference. It lies in how the difference is treated. Hume's treatment: the difference is an ontological split; facts and values have different ontological statuses, so one cannot be derived from the other. This treatment makes it difficult to give ethics a scientific ground. IO1's treatment: the difference is a difference in type of commitment. Both are subject to the same methodological discipline, but the committed content differs. This treatment allows ethics and science to use the same method while studying different content. The implication of the difference is substantive: in Hume's framework, ethics requires an independent methodology, since it cannot use scientific method; in the IO1 framework, ethics uses the same methodology, the method of constrained speculation, but applies it to a different kind of commitment. This difference has substantive consequences: it allows ethics not to detach itself from scientific method, while preserving the specificity of its commitments.

Objection Two: Your claim that "value statements are subject to methodological discipline" is too strong. Many value statements, such as "beauty is good" or "freedom is good," are difficult to formalize, mechanize, and observationalize precisely.

Reply: This objection points to the difficulty of application. Some value statements are indeed difficult to make fully consistent with the method of constrained speculation. Highly abstract value statements, such as "beauty is good," involve conceptual complexity, since "beauty" and "good" have many interpretations. Such statements need concretization: the abstract statement must be refined into content that can be examined, for example "a certain type of aesthetic experience has a positive effect on mental health." The process of concretization is the work of ethical research: it transforms abstract value statements into statements that can be treated by methodological discipline. The method of constrained speculation does not require all value statements to immediately satisfy the four disciplines. It requires that they be refinable into statements satisfying those disciplines. This refinement process is the methodological work of ethics. If a value statement cannot at all be refined into a statement satisfying the disciplines, if it has no identifiable content and no examinable structure, then it is not a legitimate ontological commitment. This is the application of the method of constrained speculation.

Objection Three: Your treatment makes value statements and factual statements methodologically "equivalent." But ethical decisions often require quick judgment and cannot wait for full methodological work.

Reply: This objection points to a practical issue. Methodological discipline concerns how statements are rigorously assessed: it is a research method. Concrete decisions concern how to act in concrete situations: they involve practical wisdom. The two do not conflict. Ethical decisions can be made quickly, on the basis of experience, intuition, and tradition, while methodological discipline can rigorously assess the legitimacy of those decisions. In urgent decisions, agents use experience and intuition. In reflective decisions, whether after action or in prior planning, agents use methodological discipline to assess legitimacy. The two work together. The method of constrained speculation does not replace intuition; it allows intuition to be rigorously assessed and improved. This is analogous to scientific methodology: scientists use intuition in experiments, but rigorously evaluate results through method.

Literature That Must Be Engaged

- David Hume, A Treatise of Human Nature (1739-40)
- Charles Pigden, "Naturalism" in A Companion to Ethics (1991)
- John McDowell, "Values and Secondary Qualities" (1985)
- Philippa Foot, Natural Goodness (2001)

- Chapter 3 on failure conditions as ontological discipline (self-reference)
- Chapter 1 on the method of constrained speculation (self-reference)

Summary of This Section

This section has done four things. First, it stated the core thesis IO1: factual statements and value statements are both subject to the same methodological discipline, and their difference lies in type of commitment rather than ontological status. Second, it developed three implications of IO1: value statements as ontological commitments, the same methodological discipline, and difference in type of commitment. Third, it gave three consequences of the transformation: ethics can be rigorous, fact and value are not reduced to one another, and science and ethics can enter into dialogue. Fourth, it demonstrated the application of IO1 through the example of environmental ethics.

The next section develops the falsifiable structure of value statements.

20.4. The Falsifiable Structure of Value Statements

The Function of Application

Section 20.2 stated IO1: factual and value statements are both subject to methodological discipline. This section develops the content of this constraint: the falsifiable structure of value statements. This development has several functions. It makes IO1 not merely an abstract declaration, but gives it operational content. It gives ethical research a methodological direction: researchers know how to assess value statements rigorously. It also provides the methodological ground for this book's later work, especially Chapter 21 on the finitude of meaning.

The Falsifiable Structure of Value Statements

The falsifiable structure of value statements requires four elements.

Element One: clarification of the dimension. A value statement concerns some dimension, such as human well-being, ecological health, social justice, and so on. This dimension must be explicitly stated: X is good because it makes a positive contribution along dimension D. "X is good" as a vague statement does not satisfy the requirement of clarification. "X makes a positive contribution to long-term human well-being" as a statement satisfies the requirement of clarification. Clarification makes the dimension identifiable and testable. If the dimension itself is vague, such as "good" as an abstract concept, the statement cannot be rigorously assessed.

Element Two: development of causal or structural relations. A value statement involves a causal or structural relation between X and dimension D. This relation must be developed: X contributes to dimension D through mechanism M. "Moderate exercise is good" is a vague statement. "Moderate exercise makes a positive contribution to long-term human physical and mental health through cardiovascular, metabolic, endocrine, and neural mechanisms" is a concrete statement. Developing the mechanism makes the statement researchable: researchers can examine whether mechanism M actually connects X and D.

Element Three: identification of observational proxies. A value statement requires observational proxies: testable indicators. Proxies for human well-being include lifespan, health indicators, reports of psychological well-being, social functioning, and so on. Proxies for ecological health include biodiversity, ecosystem function, populations of keystone species, and so on. Proxies for social justice include distributive equality, equality of opportunity, procedural justice, cross-group gaps, and so on. Proxies make a statement empirically researchable. If a statement has no identifiable proxies, for example "X increases the overall good of the universe," then it is difficult to assess rigorously.

Element Four: statement of failure directions. A value statement requires a failure direction: what would make the claim that X is good fail. If research shows that X makes no positive contribution to D, the commitment fails. If research shows that X has negative effects along dimensions outside D, and these effects exceed the positive contribution, the commitment needs revision. If counterfactual analysis shows that a situation without X achieves D equally well, the commitment needs reassessment. Failure directions make the statement not an empty declaration, but a structure that can be refuted.

An Example: Reducing Inequality

Let us consider an example: "reducing social inequality is good."

Element One: clarification of the dimension. In what dimension is it "good"? This must be clarified: human well-being, including health, psychology, and social functioning? Social stability, such as reduced conflict and increased cooperation? Economic efficiency, such as labor use and innovation? Moral fairness, such as equality of opportunity and procedural justice? Each dimension has different meanings and different evidence. The statement should clarify the dimension in which reducing inequality is good.

Element Two: development of causal or structural relations. Through what mechanisms does reducing inequality contribute to the dimension? Through economic mechanisms, such as consumption demand and social mobility? Through psychological mechanisms, such as status anxiety and social trust? Through social mechanisms, such as social capital and collective action? Through political mechanisms, such as democratic legitimacy and policy responsiveness? Each mechanism has empirical evidence and controversy. The statement should develop through which mechanisms reducing inequality produces contribution.

Element Three: identification of observational proxies. Proxies for reducing inequality include the Gini coefficient, measuring income or wealth inequality; cross-group gaps, such as gaps across race, gender, region, and so on; opportunity gaps, such as gaps in education, employment, health, and so on; and subjective perceptions, such as survey reports of perceived inequality. Each proxy has advantages and limits. The statement should specify which proxies are used as the basis for examination.

Element Four: statement of failure directions. What would make "reducing inequality is good" fail? If research shows that, after controlling for other variables, reducing inequality makes no positive contribution to the relevant dimensions, such as health, psychology, social stability, and so on; if research shows that measures to reduce inequality have counterproductive effects, such as reducing economic growth and ultimately lowering everyone's well-being; if cross-national comparisons show that countries with similar levels of inequality perform similarly on the relevant dimensions, suggesting that inequality is not the key variable. Failure directions make "reducing inequality is good" a statement that can be rigorously studied. Contemporary social sciences, including economics, sociology, epidemiology, and so on, are conducting precisely this kind of research: they are applying the research form of constrained speculation.

The Meaning of Application

This example shows the application of the method of constrained speculation in ethics.

Meaning One: ethics can be empirical research. Ethics is not abstract work detached from experience. It can use empirical research to treat ethical problems. "Reducing inequality is good" cannot only be assessed through philosophical reflection; it can be refined and examined through social-scientific research.

Meaning Two: ethics requires interdisciplinary collaboration. Ethical work involves several disciplines, including philosophy, economics, sociology, psychology, political science, epidemiology, and so on. Any serious ethical research requires collaboration among these disciplines. Ethicists cannot work in isolation.

Meaning Three: progress in ethics is possible. If ethical statements can be refined and examined, then progress in ethics is possible. Through research, ethical statements are refined, becoming more precise; examined, receiving support or opposition; and revised according to new findings. This possibility of progress means that ethics is not an endless dispute, but a field of research with direction. Ethical consensus can accumulate through research, and ethical disagreement can be refined through evidence.

Objections and Replies

Objection One: Your example, reducing inequality, involves a large amount of empirical research. But actual ethical decisions often involve urgent situations, with no time for detailed research.

Reply: This objection points to a practical issue. The distinction between concrete decisions and reflective decisions has already been addressed at the end of Section 20.2. Here one more point can be added: urgent decisions should be based on existing ethical research. Prior research accumulates ethical knowledge, so urgent decisions are not entirely arbitrary. Reflective decisions further refine ethical knowledge and provide a basis for future urgent decisions. The two work together. Ethics as a research field accumulates knowledge; ethical decisions use this knowledge to make choices. By analogy: medicine as a research field accumulates knowledge; medical decisions, especially emergency medicine, use this knowledge to make choices. No one argues that "emergency medicine cannot be based on research." It is built upon research. Similarly, urgent ethical decisions are built upon ethical research.

Objection Two: Your example, "reducing inequality is good," involves political dimensions. Different political positions evaluate this differently. Can such political disagreement be resolved through research?

Reply: Political disagreement is real, and the method of constrained speculation does not pretend that it can completely eliminate political disagreement. Empirical disagreement can be treated through research: causal relations, mechanisms, proxy performance, and so on. Value disagreement involves purposes and priorities, which cannot be decided simply through empirical research. Political disagreement often involves a mixture of these two kinds of disagreement. The contribution of the method of constrained speculation is to separate the two: making the empirical part researchable and the value part rigorously statable. Research on the empirical part makes discussion more precise: the parties can discuss on the basis of evidence rather than guesswork. Rigorous statement of the value part makes disagreement explicit: the parties can identify which dimensions have different priorities. This refinement does not eliminate disagreement, but makes disagreement rationally discussable. This is the contribution of the method of constrained speculation in political philosophy: it allows political discussion to be rigorous, rather than being limited to prejudice and emotional confrontation.

Objection Three: Your method seems to technicize ethics, converting ethical questions into empirical research questions. This may lose the core of ethics: moral intuition, human concern, and so on.

Reply: This objection points to an important methodological issue. The method of constrained speculation does not eliminate moral intuition or human concern; it makes them assessable with rigor. Moral intuition as a psychological phenomenon can be empirically studied, for example in Joshua Greene's work on moral psychology. Such research allows intuitions to be refined and examined. Human concern as a value can be empirically studied, for example through psychological and sociological research on human well-being. Such research allows concern to be concretized and supported. These studies do not replace intuition and concern. They allow intuition and concern to be used rigorously. Intuition can be wrong, for example when certain intuitions are based on prejudice; concern can be misleading, for example when certain forms of concern ignore long-term effects. Research makes intuition and concern more refined and reliable. This is the contribution of the method of constrained speculation: it does not replace ethical traditions, but refines them. The insights of ethical traditions, especially Confucianism on practical wisdom and Buddhism on causal relations, acquire contemporary grounding through the method of constrained speculation.

Literature That Must Be Engaged

- Joshua Greene, Moral Tribes (2013)
- Patricia Churchland, Braintrust (2011)
- Richard Wilkinson & Kate Pickett, The Spirit Level (2009): research on inequality
- Amartya Sen, The Idea of Justice (2009)
- Martha Nussbaum, Creating Capabilities (2011)
- Chapter 1 on the method of constrained speculation (self-reference)

Summary of This Section

This section has done four things. First, it developed the falsifiable structure of value statements: four elements, dimension, mechanism, proxy, and failure direction. Second, it demonstrated the application of these four elements through the example of "reducing inequality is good." Third, it clarified three meanings of this application: ethics as empirical research, interdisciplinary collaboration, and the possibility of progress. Fourth, it responded to possible objections, especially concerning urgent decisions, political disagreement, and the technicization of ethics.

The next section develops the position of constrained speculation within the genealogy of metaethics.

20.5. The Position within the Genealogy of Metaethics

The Function of the Genealogy

The application of the method of constrained speculation in ethics needs to be located within the genealogy of metaethics. This section shows the position of constrained speculation through four comparisons: its relation to moral realism, expressivism, error theory, and constructivism.

Dialogue One: Moral Realism

Moral realism holds that moral facts are objective: they are independent of opinion, preference, and culture. If "X is wrong" is true, then this truth is objective and does not depend on the evaluator.

Relation to constrained speculation: constrained speculation is similar to moral realism in that both treat value statements as ontological commitments. Value statements involve objective content, and both reject pure subjectivism: value is not arbitrary preference. But there are important differences. Moral realism often presupposes moral facts as an independent ontological category, some kind of "objective moral reality." Constrained speculation does not presuppose an independent moral-ontological category. It treats value statements as commitments concerning the dimensional contribution of beings. The mechanism of falsifiability in moral realism is often unclear: how are moral facts examined? Constrained speculation provides a falsifiable structure, the four elements in Section 20.3. One can say: constrained speculation agrees with moral realism on the objectivity of value, but differs in the form of objectivity. Constrained speculation gives objectivity content through methodological discipline and does not need to presuppose an independent moral ontology.

Dialogue Two: Expressivism

Expressivism, as in the work of A. J. Ayer and R. M. Hare, holds that value statements are not descriptions: they are expressions of emotion or recommendations of norms. "X is wrong" does not state an objective property of X, but expresses the speaker's disapproval of X.

Relation to constrained speculation: constrained speculation differs importantly from expressivism. Expressivism denies that value statements are commitments; they are merely expressions. Constrained speculation holds that value statements are ontological commitments: they involve content that can be examined. Expressivism has difficulty explaining progress in value statements: if value statements are merely expressions, how can there be progress? Constrained speculation holds that value statements can be refined and examined, making progress possible. But there is also a similarity: both reject simple naturalism, since value statements cannot be simply reduced to facts. Both constrained speculation and expressivism acknowledge the difference between fact and value, the "difference in type of commitment" in IO1. One can say: constrained speculation and expressivism are similar in acknowledging the fact–value difference, but fundamentally differ in the status of value statements. Expressivism denies the committed status of value statements; constrained speculation affirms it.

Dialogue Three: Error Theory

Error theory, as in J. L. Mackie's work, holds that all positive value statements are systematically mistaken: they assume that moral facts exist, but in fact they do not. "X is wrong" as a statement assumes that X really has objective wrongness, but objective wrongness does not exist.

Relation to constrained speculation: constrained speculation differs importantly from error theory. Error theory denies the objectivity of any value statement: all value statements are built upon a false assumption. Constrained speculation acknowledges the objectivity of value statements: they involve testable commitments. The ground of error theory is the queerness argument: if moral facts existed, they would be very strange ontological contents. Constrained speculation rejects this argument: value statements do not need to presuppose strange moral ontologies. They concern commitments about dimensional contribution. One can say: error theory is built upon the logic of presupposing independent moral ontology and then denying that it exists. Constrained speculation does not presuppose independent moral ontology. Value statements involve commitments about dimensional contribution, and these commitments have falsifiable structures.

Dialogue Four: Constructivism

Constructivism, as in the work of Christine Korsgaard and John Rawls, holds that value is constructed through procedures: rational reflection, social contract, the Original Position, and so on. "X is wrong" is not about an objective fact, but a conclusion derived through a constructive procedure.

Relation to constrained speculation: constrained speculation is similar to constructivism in that both acknowledge the methodological dimension of value statements: they are established and evaluated through procedures. Both also reject simple intuitionism: value is not simple intuition and requires methodological support. But there are important differences. Constructivism often locates value within the constructive procedure: the ground of value is the procedure itself. Constrained speculation locates value in commitments about beings: value statements concern dimensional contribution, while constructive procedures are methods for assessing these commitments. The mechanism of falsifiability in constructivism is unclear: how are constructive procedures evaluated? Constrained speculation provides a falsifiable structure, as in Section 20.3. One can say: constrained speculation is similar to constructivism in emphasizing the methodological dimension, but constrained speculation makes methodology serve commitments about beings; constructivism takes methodology as the ground of value.

The Distinctive Position of Constrained Speculation

Through the four comparisons, the distinctive position of constrained speculation within the genealogy of metaethics can be located as follows:

Constrained processual normativism: value statements are ontological commitments concerning the contribution of beings along some dimension; these commitments are subject to the discipline of the method of constrained speculation; they involve processual content, namely the dynamic contribution of beings, not static moral facts; and they have structures that can be refined and examined.

The content of this position is as follows: normativism, since value statements are legitimate ontological commitments and not mere expressions; constrained, since value statements are subject to the discipline of the method of constrained speculation; processual, since value involves process, concerning the dynamic contribution of beings rather than static moral facts. This threefold positioning distinguishes constrained speculation from the four traditional metaethical positions. Compared with moral realism: it preserves objectivity, but does not presuppose an independent moral ontology; objectivity is concretized through falsifiable structure. Compared with expressivism: it preserves the committed status of value statements, while acknowledging the fact–value difference. Compared with error theory: it rejects the claim that all value statements are systematically mistaken, since value statements involve falsifiable commitments. Compared with constructivism: it preserves the methodological dimension, but treats methodology as an instrument for assessing commitments, not as the ground of value. This distinctive position makes constrained speculation a new position in the genealogy of metaethics: it inherits insights from several traditions, but makes a contemporary contribution through the rigor of the method of constrained speculation and the content of process ontology.

Objections and Replies

Objection One: Your "constrained processual normativism" is a new term. It seems merely to piece together existing positions.

Reply: The term is new, but the concept is built upon content and is not a simple patchwork. The normativist part is built upon the moral realist insight concerning objectivity, but is concretized as recognition of the committed status of value statements. The constrained part is built upon the method of constrained speculation, Chapter 1, and the ontology of failure conditions, Chapter 3. The processual part is built upon the process ontology of this book, Chapters 7 and 11. Each part has its source and content. Their synthesis is not simple patchwork: it is a position based on the overall ontological framework of this book. More deeply, new terms are often used in the history of philosophy for new synthetic positions, such as Kant's "transcendental idealism" and Heidegger's "fundamental ontology." The term in this section likewise serves a new position; it is not for the sake of terminology itself.

Objection Two: Your method seems more suitable for social and political ethical questions, such as inequality, and less suitable for personal ethical questions, such as personal choice and character.

Reply: This objection points to the range of application of the method. The method of constrained speculation can treat many kinds of ethical questions, but its mode of application differs across domains. Social and political ethics: it involves large-scale data and causal analysis, and the method of constrained speculation applies directly. Personal ethics: it involves personal choices and life situations, and the method of constrained speculation applies in a different form: formalization, defining personal values; mechanization, analyzing life mechanisms; observationalization, observing and reflecting upon life; failure-orientation, considering consequences in life. Virtue ethics: it involves long-term character development, and the method of constrained speculation applies in another form: formalization, identifying features of character; mechanization, how character operates in action; observationalization, the effects of character on life; failure-orientation, how character fails in situations. Application in each domain requires concrete work. This section only provides the framework. Detailed application is future work.

Objection Three: Is the "processual" dimension really necessary for ethics? Many ethical traditions do not use a processual framework but still function well.

Reply: The processual dimension is not absolutely necessary. Many ethical traditions work under different frameworks. But the processual dimension has advantages. Connection with ontology: the overall work of this book is built upon process ontology, and processual ethics connects ethics with ontology. Treatment of dynamics: real ethical problems often involve dynamics, including long-term consequences of actions, institutional evolution, cultural change, and so on; a processual framework directly treats these dynamics. Connection with modern science: modern social science, biology, psychology, sociology, and so on, increasingly use processual frameworks, such as evolution, development, dynamical systems, and related approaches; processual ethics naturally connects with these studies. But other frameworks, such as essentialist ethics and relational ethics, also have legitimacy. The processual version of the method of constrained speculation is one legitimate choice, not the only choice.

Literature That Must Be Engaged

- J. L. Mackie, Ethics: Inventing Right and Wrong (1977): error theory
- A. J. Ayer, Language, Truth, and Logic (1936): expressivism
- Christine Korsgaard, The Sources of Normativity (1996): constructivism
- David Brink, Moral Realism and the Foundations of Ethics (1989): moral realism
- Russ Shafer-Landau, Moral Realism: A Defence (2003)
- Allan Gibbard, Wise Choices, Apt Feelings (1990): contemporary development of expressivism

Summary of This Section

This section has done four things. First, it entered into dialogue with moral realism: preserving objectivity but not presupposing an independent moral ontology. Second, it entered into dialogue with expressivism: preserving the committed status of value statements while acknowledging the fact–value difference. Third, it entered into dialogue with error theory: rejecting the claim that "all value statements are systematically mistaken." Fourth, it entered into dialogue with constructivism: preserving the methodological dimension but treating methodology as an instrument of assessment. Finally, it clarified the distinctive position of constrained speculation: constrained processual normativism.

The next section is the chapter summary.

20.6. Chapter Summary: The Methodological Downgrading of the Is–Ought Gap

Completion of the Three Tasks

The three tasks undertaken by this chapter have all been completed. Task One: to state Hume's is–ought gap precisely. Section 20.1 developed Hume's original argument, its philosophical influence, and its features. Task Two: to propose the treatment of the gap through the method of constrained speculation. Section 20.2 stated the core thesis IO1: factual and value statements are both subject to the same methodological discipline, and their difference lies in type of commitment rather than ontological status. Section 20.3 developed the falsifiable structure of value statements, consisting of four elements. Task Three: to enter into dialogue with the major positions in contemporary metaethics. Section 20.4 developed the relation to moral realism, expressivism, error theory, and constructivism, and clarified the distinctive position of constrained speculation: constrained processual normativism.

The Position of This Chapter in Part V

The overall structure of Part V, "From Ontology to Normativity," is as follows: Chapter 20, this chapter, "How Constrained Speculation Treats the Is–Ought Gap," as the methodological foundation of ethics; Chapter 21, "The Affirmation of Finitude: Meaning before Freezing," as the humanistic climax. As the foundational chapter of Part V, this chapter establishes the ontological-methodological ground of ethics and provides the methodology for Chapter 21's work on meaning. Together, the two chapters show the extension of the method of constrained speculation from natural philosophy to value philosophy.

The Position of This Chapter in the Whole Book

Looking backward, this chapter is built upon the work of Part I, especially Chapter 1 on the method of constrained speculation and Chapter 3 on the ontology of failure conditions, and Part II, especially Chapter 7 on the ontology of difference and Chapter 11 on freezing as process. Looking forward, this chapter provides the methodology for Chapter 21: the finitude of meaning as an ontological-normative thesis is built upon this chapter's treatment of value statements.

Honesty about the Limits of This Chapter

This chapter acknowledges several limits. First, ethical content is not developed. This chapter treats metaethical problems, namely the ontological-methodological ground of ethics itself. Concrete ethical questions, including the content of the good, principles of justice, ethical decisions, and so on, require dedicated ethical work. This chapter only establishes the methodological foundation. Second, cross-cultural ethical dialogue is insufficient. This chapter mainly dialogues with Western analytic metaethics. The ethical contributions of Chinese traditions, such as Confucianism, Buddhism, and Daoism, require dedicated treatment. Section 20.1 briefly mentions them but does not develop them in detail. Third, the details of application are insufficient. This chapter gives the example of "reducing inequality," but complete applications in fields such as bioethics, environmental ethics, business ethics, technological ethics, and so on require dedicated work. Fourth, the dimension of political philosophy is insufficiently treated. This chapter touches on political disagreement, especially in the response to Objection Two in Section 20.3, but a complete political philosophy, including concepts such as democracy, freedom, equality, and so on, requires dedicated treatment. Acknowledging these limits does not weaken the work of this chapter. It makes this chapter an honest initial work.

The Core Posture of This Chapter

The core posture of this chapter can be stated as follows: ethics does not need an independent methodology; it needs a way of applying the method of constrained speculation.

Since the twentieth century, metaethics has faced methodological anxiety: how does ethics differ from science? Does ethics have its own independent research method? These questions have produced the divisions within metaethics, including realism, expressivism, error theory, constructivism, and so on. Each position attempts to give ethics a methodological ground. The work of this chapter argues that ethics does not need an independent methodology. It needs a way of applying the same methodology, constrained speculation. Factual statements and value statements are both ontological commitments, and both are subject to the same methodological discipline. The difference lies not in methodology, but in type of commitment. This treatment makes the anxiety of metaethics unnecessary. Ethics and science use the same methodological rigor, but study different contents. Ethical problems can be rigorously treated through empirical research, conceptual analysis, causal analysis, and so on. These are scientific methods, but applied to value commitments.

Meaning before freezing: this is one of the core theses of this book. This chapter provides the methodological ground for that thesis. "Meaning" as an ontological-normative phenomenon does not need an independent methodological treatment. It is treated through the method of constrained speculation within the framework of process ontology. Chapter 21 will develop this treatment: meaning as a phenomenon emerging in finite existence is rigorously stated and defended through methodological discipline. Through the work of this chapter, ethics as a field of research receives an internal position within this book. It is not an independent field split from natural philosophy, but an extended application of the method of constrained speculation. This unity makes the ontological work of this book, including the ontology of difference, process ontology, and co-evolutionary ontology, and its normative work, including the treatment of the is–ought gap and the finitude of meaning, into a systematic whole: both are built upon the same methodological ground.

20.7. Methodological Reflection on This Chapter

This chapter downgrades the is–ought gap from an ontological problem to a methodological problem. This section reflects on the methodological meaning of this downgrading.

This Chapter's Posture as Metaethical Work

This chapter is not a detailed monograph on ethics. It treats metaethical problems, namely the ontological-methodological ground of ethics itself. This limitation clarifies the scope of the chapter. Concrete ethical questions require dedicated ethical work. This chapter only establishes how the method of constrained speculation can be applied in the field of ethics. This limitation does not weaken the work of this chapter: metaethics, as foundational work for ethics, has its own importance. As metaethical work, this chapter provides the methodological ground for the later normative work of the book, especially Chapter 21 on meaning.

This Chapter's Posture toward Originality

The application of the method of constrained speculation in ethics is not completely new. It is built upon the analytic metaethical tradition, naturalistic traditions, and the content of the method of constrained speculation itself. It inherits Hume's suspicion toward "deriving value from fact," Moore's warning concerning the naturalistic fallacy, contemporary moral realists' pursuit of objective value, and expressivists' semantic analysis of value statements. The contribution of this chapter consists in: the argument that downgrades Hume's gap from an ontological problem to a methodological problem; the falsifiable structure of value statements; and the positioning within the genealogy of metaethics, namely constrained processual normativism. These contributions are constructively original.

This Chapter's Extension of the Method of Constrained Speculation

This chapter shows the extension of the method of constrained speculation from natural philosophy to value philosophy. The four disciplines of the method of constrained speculation (CS1–CS4) do not apply only to factual statements; they apply equally to value statements. CS1 (formalization) requires precise statement of value claims. CS2 (mechanization) requires identifying the mechanisms involved in value claims. CS3 (observationalization) requires connecting value claims with observable consequences. CS4 (failure-orientation) requires that value claims have identifiable failure directions. This extension expresses the methodological unity of the book: the same methodological discipline can be applied across different domains.

Limits Honestly Acknowledged in This Chapter

This chapter acknowledges several limits. First, it treats only metaethical problems and does not develop concrete ethical questions. Second, its dialogue with contemporary metaethics is relatively brief; more detailed dialogue requires dedicated papers. Third, the falsifiable structure of value statements, consisting of four elements, is an initial statement. Their refinement in ethical applications requires later work. Fourth, the chapter is developed mainly within the framework of Western metaethics, and dialogue with Chinese, Indian, and other traditions is insufficient.

Connections with Later Chapters

This chapter provides the methodological ground for Chapter 21, "Meaning before Freezing." Chapter 21's work on meaning is an application of constrained processual normativism to the problem of meaning. Statements about meaning, as a form of value statement, are subject to the discipline of the method of constrained speculation. More deeply, this chapter is the bridge from natural philosophy to value philosophy in this book. If the book did not develop the methodological ground of ethics, its work would remain limited to ontology and epistemology. This chapter establishes that extension and makes the book a complete philosophical system: ontology +

normativity + humanistic meaning.

21. The Affirmation of Finitude: Meaning before Freezing

21.1. Introduction to This Chapter

What is meaning? This question easily slips into rhetoric in the history of philosophy. "Meaning" as a concept is charged with emotional resonance: it concerns how human beings live through finite existence, how they face death, and how they affirm life amid nothingness. This emotional resonance creates a challenge for any serious work on meaning: it must use rigorous argument to state something that seems to require rhetoric in order to be said.

This chapter accepts that challenge. Starting from the ontological framework established in the preceding chapters, especially Chapter 17 on ontological singularity and Chapter 11 on freezing as a foundational ontological process, it argues that the finitude of meaning is not a defect but a structure: meaning is true precisely in finitude rather than in eternity. Three theses are involved: meaning depends on gradient (M1: the necessary condition for meaning to appear is that some form of gradient structure can maintain distinction, make choices, and form direction); meaning is true only in finitude (M2: meaning is essentially losable, selective, and alive); meaning exists before freezing (M3: meaning exists in the active phase of dynamic process and dissolves in the irreversibility of the freezing process while remaining as trace).

Together, these three theses constitute the core position of this chapter: the affirmation of finitude. This position is neither pessimism nor optimism, but an affirmation of finitude. It bears the full weight of the ontological singularity while affirming the truth of meaning where meaning appears. It knows that all structures ultimately dissolve, while affirming the reality of structures during the time in which they exist. It faces the ultimate nothingness of the universe while insisting on the dignity of human beings, civilization, life, and love in their finitude.

This is an affirmation that has passed through ontology: not ignorant optimism, not cheap hope, and not rhetoric that evades ultimate finitude. It affirms meaning while acknowledging ontological singularity. This is the deepest application of constrained speculation to the problem of meaning: it requires acknowledging the limit while also affirming content within the limit.

This chapter also enters into serious dialogue with three philosophical traditions: the Confucian "knowing that it cannot be done and yet doing it," Heidegger's being-toward-death, and Camus' Sisyphus. These three traditions each treat the relation between finitude and meaning in different directions. This chapter establishes the position of constrained speculation among these three traditions.

This chapter is the humanistic climax of the whole book, but it must preserve argumentative rigor. It avoids rhetorical slippage through two means: it specifies the concept of meaning as ontological meaning and establishes, through M1, a relation of constitutive coextension with gradient structure; and when arguing for M2, it proceeds through a three-layered logical argument (losability $\rightarrow$ selectivity $\rightarrow$ aliveness), without appealing to mere intuition or rhetoric. This chapter continues the argument of Chapter 17 on the gradient terminal phase, the argument of Chapter 11 on freezing, and the argument of Chapter 20 on the is–ought gap. All of these premises are used in this chapter as already established conclusions.

The discussion proceeds as follows. Section 21.1 develops the thesis that meaning depends on gradient (M1). Section 21.2 develops the thesis that meaning is true only in finitude (M2). Section 21.3 develops the thesis that meaning exists before freezing (M3) and the meaning of "before freezing." Section 21.4 enters into serious dialogue with Confucianism, Heidegger, and Camus. Section 21.5 develops the affirmation of finitude as the core posture of this book and treats objections.

21.2. Meaning Depends on Gradient

The Starting Point of the Argument

The core thesis of this chapter must begin from an apparently simple observation: meaning depends on gradient.

What is meaning? Meaning is not something bestowed by some eternal principle, nor does it flow from some external ground. Within the EQT framework, meaning is a structural phenomenon: it appears within processes capable of maintaining distinctions, making choices, and forming directions.

Consider the conditions under which meaning appears. A person understands a sentence: this requires the brain, as a highly organized gradient structure, to distinguish this sentence from other possible sentences. A living organism pursues a goal: this requires the organism, as a non-equilibrium structure, to maintain the ability to distinguish "the goal has been reached" from "the goal has not been reached." A culture transmits values: this requires society, as a coordinated gradient network, to distinguish "preserve this value" from "abandon it." A civilization creates art: this requires the human community, as a complex cognitive–affective–social structure, to distinguish "this is beautiful" from "this is not." All these "scenes in which meaning appears" depend on some capacity to maintain difference: the capacity to distinguish this from not-this, to choose this rather than that, and to move toward this rather than that. And the capacity to maintain difference, as established in Part II, especially Chapter 7 on difference before identity, and Part III, especially Chapter 11 on freezing as a foundational ontological process, depends on gradient.

Precisely stated: Thesis M1: the necessary condition for meaning to appear is that some form of gradient structure can maintain distinctions, make choices, and form directions. In the state of affairs $\nabla\rho \equiv 0$, no such structure can be sustained, and therefore there is no possibility for meaning to appear.

This Does Not Reduce Meaning to Physics

The reader may immediately suspect that this reduces meaning to gradient, that meaning is just some physical structure, and that this is crude physicalism. This chapter must be clear: M1 does not claim that meaning = gradient structure. It claims that meaning depends on gradient structure as a necessary condition. These two claims differ significantly.

Dependence is not identity. Oxygen is a necessary condition for combustion, but combustion is not identical to oxygen. The existence of a first-person perspective depends on the brain as a highly complex physical process, but the first-person perspective is not identical to brain activity. In the same way, the appearance of meaning depends on gradient structure as a necessary condition, but meaning is not identical to gradient structure.

This distinction corresponds to the work of Chapter 4 on constitutive coextension. Meaning and gradient structure are constitutively coextensive: they are co-instantiated under the same failure condition (if gradient structure disappears in a certain state of affairs, meaning also disappears in that same state of affairs), but they are not the same object. Constitutive coextension allows meaning to preserve its conceptual independence while anchoring its ontological basis in gradient. This distinction allows the chapter to argue for the finitude of meaning without needing to argue that meaning is some physical quantity. Meaning is finite because the gradient structure on which it depends is finite; but within the range in which it exists, meaning has its own reality, satisfying some form of CR2.

Specifying the Concept of "Meaning"

To prevent the discussion from slipping into rhetoric, this section must specify the concept of "meaning." Meaning has several senses: semantic meaning, life meaning, value meaning, purposive meaning. These are often mixed in philosophical discussion. This chapter is concerned with meaning in the sense of ontological meaning.

Ontological meaning refers to this: in a given situation, something is taken to be "important," "worth pursuing," or "directional." The appearance of such meaning requires a subject capable of mattering: a subject capable of distinguishing "important" from "unimportant," "worthwhile" from "not worthwhile," and "directional" from "directionless." Such a subject capable of mattering, within the EQT framework, is not a mysterious entity, but a highly organized gradient structure. In *Mind–Matter Unity*, the Ψ state-variable group provides a precise characterization of this structure. Mattering itself is a function of gradient structure: it requires maintaining distinctions, making choices, and forming directions. Therefore, the ontological meaning with which this chapter is concerned has a direct relation of constitutive coextension with gradient structure. This does not reduce the richness of meaning; it anchors that richness in a basis that can be ontologically analyzed.

Boundary Cases

To further specify M1, this section examines several boundary cases.

Case One: Do animals have meaning? According to M1, meaning requires a subject capable of "mattering." Animals, especially higher animals, clearly have such a capacity: they can distinguish food from non-food, safety from danger, intimacy from distance. Their gradient structures, including nervous systems and behavioral systems, are sufficiently complex to support these distinctions. Therefore, animals have meaning, although animal meaning may differ from human meaning in complexity, self-reflection, cultural embeddedness, and other respects. This reply allows the chapter to avoid anthropocentrism and to treat meaning as a broader ontological phenomenon.

Case Two: Does primitive life have meaning? Single-celled organisms, such as bacteria, can clearly distinguish favorable from unfavorable environments; this is the basis of chemotaxis. They have a very basic form of "mattering": distinguishing what is favorable to their own persistence from what is unfavorable. According to M1, this constitutes the minimal form of meaning. This reply allows meaning to extend continuously toward the vicinity of the origin of life: it does not suddenly appear at some absolute boundary, but appears in parallel with the complexification of gradient structures.

Case Three: Do non-living systems have meaning? Does a star "care" about anything? Does a galaxy "care" about anything? According to the strict use of M1, stars and galaxies do not have meaning. Although they are gradient structures, they are not structures capable of maintaining distinctions, making choices, and forming directions in a sufficiently strong sense. This reply gives a cautious boundary: the appearance of meaning requires gradient structure as a specific type of functional organization. Not every gradient structure automatically produces meaning; meaning requires specific functional complexity, where $\tilde{A}$, C, and $\mathcal{B}$ in *Mind–Matter Unity* simultaneously enter appropriate intervals.

Objections and Replies

Objection One: Your argument makes meaning seem as if it were derived from physical facts: "where there is gradient, there may be meaning." But Hume's is–ought gap tells us that facts cannot derive values.

Reply: This chapter does not derive value from fact. It only argues for a necessary condition of meaning, namely gradient structure. This is a factual statement. The content of meaning, what matters to whom, is not determined by this argument. Chapter 20 has already treated Hume's is–ought problem by downgrading it from an ontological gap into a problem of methodological discipline. This chapter continues that treatment: statements about meaning should be constrained by the same discipline of constrained speculation as factual statements, but they are not reduced to one another. Statements about meaning have their own content, concerning what matters and why it matters, while their conditions of possibility are provided by fact, namely the existence of gradient structure.

Objection Two: Your concept of "being capable of mattering" sounds phenomenological. Can constrained speculation state such a concept rigorously?

Reply: Yes. Being capable of mattering, as a commitment, has dimensions, including neural activity patterns, behavioral complexity, and positions within state-variable space; proxies, including the Ψ state-variable group, behavioral tests, and self-report; and thresholds, including measurable value ranges. These have already been treated in detail in *Mind–Matter Unity*. Being capable of mattering is not a vague phenomenological concept: it has identifiable physical proxies.

Objection Three: Then does artificial intelligence have meaning? If meaning requires only "being capable of mattering" as a function of gradient structure, AI seems able to satisfy this condition.

Reply: This is an open question. The response of constrained speculation is that whether AI has meaning cannot be judged only by its output behavior. One must examine whether it possesses the physical state-variable conditions discussed in *Mind–Matter Unity*: whether $\tilde{A}$, C, and $\mathcal{B}$ simultaneously enter consciousness-related intervals. This is an empirical question left to future research. The argument of this chapter does not presuppose that AI has or does not have meaning. It claims only that the necessary condition of meaning is some form of functional complexity in gradient structure.

Literature That Must Be Engaged

- Martin Heidegger, Sein und Zeit (1927): on "care" as an existential characteristic of Dasein
- Charles Taylor, Sources of the Self (1989): on meaning as a frameworked fact
- Daniel Dennett, Consciousness Explained (1991): on functionalist treatments of consciousness and meaning
- John Haugeland, Having Thought (1998): on intentionality as commitment
- The treatment of $\Psi(\tilde{A}, C, \mathcal{B})$ in *Mind–Matter Unity* (self-reference)

Summary of This Section

This section has done three things. First, it stated Thesis M1: meaning depends on gradient as a necessary condition, and in the state of affairs $\nabla\rho \equiv 0$, there is no possibility for meaning to appear. Second, it specified the concept of "meaning" as ontological meaning and explained the relation of constitutive coextension between this concept and gradient structure. Third, it clarified the scope of M1 through boundary cases: animals, primitive life, and non-living systems.

The next section develops the key implication of M1: the finitude of meaning.

21.3. The Finitude of Meaning

Deriving Finitude from M1

If meaning depends on gradient as a necessary condition, and if gradient is ultimately finite in the deepest sense, according to the ontological singularity of Chapter 17, then meaning is ultimately finite as well. This inference looks direct, but it requires careful development. In what sense is meaning "finite"?

The finitude of meaning has several levels. The first level, the individual level: every bearer of meaning, whether person, animal, or group, has a finite duration of existence. When a person dies, he disappears as a bearer of meaning; what mattered to him during his life no longer matters to him, although it may continue to matter to others. The second level, the communal level: cultures, societies, and civilizations also have finite durations of existence. Ancient Greek civilization has ended; the things that mattered to those living within that civilization, as that kind of meaning, no longer exist, although historical research may reconstruct them. The third level, the life level: life as a phenomenon has a finite duration of existence. In some future state of the universe, whether heat death, the gradient terminal phase, or another terminal form, life may disappear entirely; at that point, all forms of meaning will disappear. The fourth level, the ontological level: in the gradient terminal phase established in Chapter 17, the intrinsic singularity of ontology, all gradient structures disappear, and therefore all possibilities for meaning to appear disappear. This is the deepest finitude of meaning. These four layers of finitude all follow from M1, but they differ in urgency. The first level concerns the finite existence of each individual; the fourth concerns the ultimate finitude of meaning as an ontological fact.

Finitude Appears to Be a Defect

The finitude of meaning appears to be a defect: it seems to make meaning "weaker." Consider ordinary intuition. When a person faces death, he may feel that "if everything will eventually disappear, then nothing one does now has meaning." When a civilization faces the possibility of long-term decline, cultural creation may seem futile. When philosophical reflection reaches the ultimate fate of the universe, all human achievements may seem so small as to be meaningless.

This feeling is not a personal weakness; it has a deep metaphysical ground. If meaning must be eternal in order to be "truly real," then all finite meaning merely seems meaningful but is not truly meaningful. This line of thought appears repeatedly in the history of philosophy: Plato's theory of Forms, where sensible things "participate" in reality because they participate in eternal Forms, while the eternal Forms are true reality and sensible things are only shadows, so meaning must be anchored in eternity in order to be true; medieval Christian theology, where the eternal God is the ground of all meaning, and finite things in the world have meaning only in relation to eternity; Hegel's Absolute Spirit, where the meaning of history lies in its realization of Absolute Spirit, which is the eternal endpoint; and some versions of Buddhist liberation theory, where finite worldly things are the root of suffering and liberation lies in transcending finitude. These positions all anchor the ground of meaning in some eternal state of affairs. They acknowledge that finite things have some meaning, but this meaning is taken to be derivative, secondary, and dependent on eternity. If these positions are correct, then this chapter's argument for the finitude of meaning is bad news: it says that meaning is ultimately empty because it depends on gradient, which must disappear.

This Is a Mistaken Inference

But this section argues that the inference that "finitude weakens meaning" is mistaken. It rests on an unexamined assumption: meaning requires eternity as its ground. Where does this assumption come from? Why must meaning require eternity as its ground, rather than find its ground in its own finitude?

Upon closer inspection, this assumption seems to rest on a circularity: meaning requires an eternal ground → because meaning without an eternal ground is "derivative" → derivative meaning is "not true" → therefore true meaning requires an eternal ground. This circle is question-begging: it presupposes the very conclusion it tries to prove. If we reject this assumption and look again at the finitude of meaning, it need not appear as a defect. Meaning can find its ground in its own finitude; it need not be anchored in some eternal state of affairs. But this reply is only negative: it says that the assumption is question-begging, but has not yet given a positive position. The positive position is this: meaning not only can find its ground in finitude; meaning is true precisely in finitude. This is the work of Section 21.3.

A Preliminary Argument

Before developing the positive position in Section 21.3, this section offers a preliminary argument: if meaning requires an eternal ground, what would meaning become? Consider the consequences of this assumption. If meaning is "truly real" only in relation to eternity, then:

Consequence One: meaning becomes static. If meaning comes from eternity, then meaning itself has an eternal character: it does not change, does not disappear, and does not depend on concrete situations. But this conflicts with our actual experience of meaning. The meaning we experience is alive: it appears in concrete situations, changes, is reinterpreted, is transmitted, or is lost. A meaning that is eternally unchanging is not the meaning we know.

Consequence Two: meaning becomes singular. If meaning comes from eternity, all meaning ultimately returns to one eternal ground. But this conflicts with the diversity of meaning. Different persons, cultures, and forms of life have different meanings. These meanings may be incompatible or irreducible. Forcing them back to one eternal ground either cancels differences or depreciates some meanings as "not true."

Consequence Three: meaning becomes externally imposed. If meaning comes from eternity, it does not emerge from concrete situations but is imposed upon them from outside. But such externally imposed meaning is alien to concrete beings: it does not belong to them and is not theirs.

These three consequences make the position that "meaning requires an eternal ground" difficult to accept. If meaning truly required eternity, then meaning would no longer be the meaning we experience. This preliminary argument does not yet positively establish that "meaning is true in finitude." It only weakens the opposing position. The next section gives the positive argument.

Objections and Replies

Objection One: You rely too much on "meaning as we experience it" as the ground of the argument. Is this appeal to experience philosophically rigorous?

Reply: The appeal to experience here is not an appeal to an "experiential given" in a foundationalist sense, which is precisely what McDowell rejects as the "myth of the given." This section appeals to the ordinary understanding of meaning: meaning is alive, diverse, and belongs to concrete beings. This ordinary understanding may be mistaken, but it constitutes the starting point of the discussion. If someone claims that an ontological position requires revising all ordinary understanding, the burden of proof lies with that position.

Objection Two: You ignore the actual function of religious traditions in providing meaning. Many people find the meaning of life in religious faith, and religious traditions precisely anchor meaning in eternity.

Reply: This chapter does not deny the actual function of religious traditions in providing meaning in lived life. The position of constrained speculation is this: religious traditions as sources of meaning are legitimate practices, but their ontological claims, such as an eternal state of affairs as the ground of meaning, need to be assessed according to CR2. If an "eternal state of affairs" as a commitment has no failure interface, as shown in the boundary discussion of Section 21.1, then it does not constitute an ontological commitment, though it may still function in meaning practice. This is the distinction between religion and ontology: religion as practice has its legitimacy, but as an ontological claim it requires constraint.

Objection Three: You seem to remove all understandings of "transcendence." But human experience does have dimensions directed toward transcendence, such as a sense of the sacred, awe, and ultimate concern. These cannot simply be reduced to finitude.

Reply: This chapter does not cancel these dimensions. It claims that these dimensions can find their ground in finitude: awe can be awe before finitude itself; a sense of the sacred can be the feeling of meaning emerging within finitude; ultimate concern can be concern for how finite existence is to be lived. This is the positive position to be developed in Section 21.3.

Literature That Must Be Engaged

- Plato, Phaedo, Symposium: on the relation between eternal Forms and finite things

- Augustine, Confessions: the Christian treatment of time and eternity
- G.W.F. Hegel, Phänomenologie des Geistes (1807): Absolute Spirit as the purpose of history
- Bernard Williams, Moral Luck (1981): on finitude and meaning
- Thomas Nagel, "The Absurd" (1971): on cosmic diminishment and the meaning of life
- Susan Wolf, Meaning in Life and Why It Matters (2010): contemporary discussion of meaning

Summary of This Section

This section has done three things. First, it derived four layers of the finitude of meaning from M1. Second, it identified the common inference that "finitude weakens meaning" and the assumption on which it depends: "meaning requires an eternal ground." Third, it weakened this assumption through a preliminary argument: if meaning required an eternal ground, meaning would become static, singular, and externally imposed, which conflicts with our actual experience of meaning.

The next section develops the positive position: meaning not only can, but must, find its ground in finitude.

21.4. Meaning Is True Only in Finitude

Core Thesis

This section gives the core positive thesis of the chapter: Thesis M2: meaning not only can find its ground in finitude; meaning is true only in finitude. If meaning is anchored in some eternal state of affairs, it is no longer meaning. This thesis sounds like a rhetorical paradox: "only the finite is true" seems to conflict with common sense, which intuitively suggests that "only the eternal is true." This section argues that the thesis is rigorous and not rhetorical.

A Three-Layered Argument

The argument for M2 needs to unfold in three layers.

Layer One: meaning requires mattering, and mattering requires the possibility of loss. Consider what makes something "important." For something to matter to someone means that its presence or absence makes a difference to that person: he can distinguish between "this thing happens" and "this thing does not happen," and responds differently to these two possibilities. But difference requires distinction as its basis: the former depends on the latter. If "this thing happens" and "this thing does not happen" make no difference to him at all, no difference in his state, action, feeling, or relation, then this thing does not matter to him. The deepest form of difference is losability: this thing can fail to be there. If something is always there, will always be there, and cannot fail to be there, it makes no difference to a being. It is not something that the being must choose, care about, or take to matter; it is an unchangeable background. But meaning is precisely the object of choice, care, and mattering. Something necessarily there forever cannot be chosen, cared about, or taken to matter. It is something else, perhaps a "foundation," "environment," or "precondition," but not meaning. Therefore, meaning requires that its object can fail to be there: the losability of the object of meaning is an intrinsic structure of meaning. This is the deep connection between finitude and meaning.

Layer Two: an eternal state of affairs cancels choice. Consider the consequence of "an eternal state of affairs as the ground of meaning." If a state of affairs is necessarily and eternally so, a being cannot choose its relation to it: this relation is predetermined and is not changed by the being's decision. But the relation of meaning is a relation of choice: a being can adopt one relation or another; can commit or withdraw; can persist or abandon. This selectivity is a core feature of meaning. An eternal state of affairs as the ground of meaning eliminates this selectivity. If meaning comes from eternity, the relation between a being and meaning also becomes eternal, necessary, and unchosen. This changes the relation of meaning from a living relation into a structural fact. Meaning is no longer something a being must undertake; it becomes a state in which the being is necessarily contained. This conflicts with the actual experience of meaning. The meaning we experience is something that must be undertaken: a person must choose what matters to him and what does not; a culture must decide what to transmit and what to abandon; a life must decide its direction in a concrete situation. This "must" is not a "must" imposed from outside; it is the intrinsic structure of meaning. Meaning is "having to choose."

Layer Three: the truth of meaning lies in its aliveness. Combining the first two layers: meaning requires losability (Argument One) and selectivity (Argument Two). Together, these constitute the aliveness of meaning: meaning is not a static fact, but a living, undertaken, and changeable reality. And aliveness is precisely the manifestation of finitude. An eternal state of affairs has no aliveness: it does not change, has no losability, and has no selectivity. It exists statically. A finite state of affairs has aliveness: it can change, can be lost, and must be chosen. Therefore, the truth of meaning lies in its aliveness; aliveness comes from finitude; and therefore the truth of meaning comes from finitude. This is the rigorous argument for M2. Meaning is not "true despite being finite." It is true precisely because it is finite. If finitude is removed, meaning itself disappears; what remains may be an "eternal fact," but not meaning.

An Example

To keep this argument from remaining abstract, this section gives an example. Consider one person's love for another. This love is a relation of meaning: the beloved is important, worth caring for, and directional for the lover. The content of this love includes: the lover is willing to give for the beloved, which presupposes that giving is a choice and that one could fail to give; the lover attends to the beloved's state, which presupposes that this state can change, becoming better or worse; the lover cherishes time spent with the beloved, which presupposes that time is finite and that shared time is therefore valuable; the lover bears the pain of possibly losing the beloved, which presupposes that loss is a real possibility.

Now consider a hypothetical case: if the beloved were eternal, unchanging, necessarily existing, and impossible to lose, what would this love become? There would be no selectivity of giving: whether one gives or not would make no difference to the eternal beloved. There would be no content of attention: the eternal beloved would have no changeable state. There would be no preciousness of time: the "time" of the eternal is not a finite resource that can be cherished. There would be no possibility of loss: this possibility has been canceled. What remains? Perhaps some static, abstract, necessary relation. But this is not love. Love is essentially a living relation among finite beings. This example shows that M2 is not an abstract argument; it corresponds to our deepest experience of meaning. We love finite persons precisely because they are finite. Love for an eternally unchanging thing is not love, though it may be called "reverence," "contemplation," or something else, because love requires the finitude of its object as an intrinsic structure.

Comparison with Traditional Positions

M2 conflicts with several traditional positions listed in Section 21.2: Plato, Augustine, Hegel, and certain versions of Buddhism. They all anchor meaning in eternity; M2 anchors meaning in finitude. But M2 does not conflict with all traditions. Some traditions precisely acknowledge meaning in finitude: Confucianism, "knowing that it cannot be done and yet doing it," where meaning lies not in eternal success of the result but in persisting in what ought to be done within a finite situation; Heidegger's being-toward-death, where authentic existence undertakes its own finitude and thereby obtains its own "ought-to-be" (Eigentlichkeit); and Camus' Sisyphus, where persistence in a situation known to be futile is precisely revolt and also meaning. These traditions will be discussed in detail in Section 21.4. Here we only note that M2 is not an isolated position: it has important traditional support. But the specificity of M2 lies in this: it is derived from a strict ontological argument, Chapter 17 on ontological singularity plus this chapter's inference concerning finitude, and does not depend on the presuppositions of any particular cultural or religious tradition.

Objections and Replies

Objection One: Your argument seems to rely too heavily on "love" as an example. But love is not the paradigm of all meaning; many forms of meaning, such as the pursuit of truth or beauty, need not depend on relations among concrete beings.

Reply: Love is selected as an example because it most clearly displays the structure of aliveness in meaning. But the core of the argument, that meaning requires losability and selectivity, applies to other forms of meaning. The pursuit of truth requires that truth may not yet have been reached, rather than being eternally already known. The pursuit of beauty requires that beauty may be missed, rather than being eternally present. All forms of meaning have an intrinsic structure of finitude.

Objection Two: But "infinite existence" also seems capable of meaning. Could an eternal being not care about things?

Reply: This objection requires careful distinction. If an eternal being has genuine "mattering," the object that matters to it must be losable; otherwise it does not constitute mattering. Therefore, even if the being that matters is itself eternal, the object that matters must be finite. This is consistent with M2: M2 claims the finitude of the object of meaning, not that the bearer of meaning must be finite. Whether the bearer of meaning itself can be eternal is another question. If an eternal bearer exists, for example God in certain theological traditions, its "meaning" may have a structural difference from the meaning of finite bearers. This chapter does not treat this theological question. It only claims that meaning in human experience, as the meaning of finite bearers facing finite objects, is essentially finite.

Objection Three: M2 looks like a comprehensive objection to religious traditions. Is this not philosophically too strong a claim?

Reply: M2 need not oppose religious traditions. Many religious traditions have already acknowledged the centrality of finitude: the Incarnation in Christianity, where the eternal enters the finite; the Middle Way in Buddhism, which rejects attachment to eternity as well as attachment to nothingness; and the Daoist "the Dao follows what is natural," where nature is a living process. M2 is compatible with these traditions. M2 opposes a specific metaphysical presupposition: that meaning requires eternity as its ground. This presupposition is not identical with religion and can be discussed and rejected independently of religion.

Objection Four: Does your argument rely too much on the modern preference for "choice" and "aliveness"? Premodern societies may not have treated these as core features of meaning.

Reply: The argument of this section does not presuppose a modern view of choice, such as personal choice in the liberal sense. "Choice" here means, in the broadest sense, the structure in which a being must respond to possibilities. This structure also exists in premodern societies: premodern persons faced situations and had to respond to them through moral decisions, ritual choices, and ways of life. These are all forms of choice. The "selectivity" in M2 crosses the boundary between premodern and modern.

Literature That Must Be Engaged

- Søren Kierkegaard, Either/Or (1843), The Sickness Unto Death (1849): existential philosophy on choice and meaning

- Martin Heidegger, Sein und Zeit (1927), sections on authentic existence

- Jean-Paul Sartre, Being and Nothingness (1943): on freedom and the choice of meaning

- Bernard Williams, Moral Luck (1981): on contingency and meaning
- Thomas Nagel, "The Absurd" (1971)
- Mou Zongsan, Moral Idealism: on the Confucian treatment of meaning

Summary of This Section

This section has done three things. First, it stated Thesis M2: meaning is true only in finitude, and eternal states of affairs cancel meaning. Second, it developed the rigorous ground of M2 through a three-layered argument: losability → selectivity → aliveness. Third, it used the example of love to show that M2 corresponds to a deep experience of meaning.

The next section enters into serious dialogue between M2 and three important philosophical traditions.

21.5. Dialogue with Three Traditions

Aim of the Argument

M2 is not an isolated position: it has important traditional support. This section enters into serious dialogue with three traditions: the Confucian "knowing that it cannot be done and yet doing it," Heidegger's being-toward-death, and Camus' Sisyphus. These three traditions each anchor meaning in finitude in different ways. This dialogue is not a simple "absorption" of these traditions as support for this chapter. It is a serious dialectic: it identifies the insights and limits of each tradition and shows the independent position of constrained speculation among them.

Dialogue One: The Confucian "Knowing That It Cannot Be Done and Yet Doing It"

There is a passage in the *Analects*, Xianwen: "Zilu lodged at Shimen. The gatekeeper said, 'Where are you from?' Zilu said, 'From Confucius.' The gatekeeper said, 'Is he the one who knows that it cannot be done and yet does it?'" This passage records a gatekeeper's description of Confucius: he is the person who "knows that it cannot be done and yet does it." This phrase condenses the Confucian core position on meaning. Confucius knew that his political ideals, restoring the rites of Zhou, rectifying names, and practicing benevolent governance, could not be realized in his age. He traveled among the states and was repeatedly frustrated. But he did not give up, and persisted. What was the ground of this persistence? It was not the expectation of success: he knew it could not be done. It was not personal interest: his travels left him displaced. Nor was it a promise of some eternal reward: Confucianism does not emphasize the afterlife. Then what was it? The Confucian answer is: what ought to be done is worth doing. Meaning does not lie in the success of the result, but in the oughtness of the action itself. This position is highly consonant with M2. Confucianism acknowledges meaning in finite situations: Confucius' situation was concrete, changing, and uncontrollable; meaning appeared in this situation through persistence in what ought to be done. Meaning does not depend on the guarantee of eternal success; it realizes itself in acting while facing impossibility.

The strengths of the Confucian position. Strength One: grounding meaning in "ought." Meaning does not need to be anchored in an eternal state of affairs; it is anchored in moral ought. Ought is something within finite situations, but it constitutes an intrinsic requirement for the being. This is consistent with M2: meaning is in finitude because ought, as the ground of meaning, is itself an intrinsic structure of a finite situation. Strength Two: action as the mode of realizing meaning. Confucianism emphasizes "doing." Meaning is not a state statically possessed, but an activity enacted. This is fully consistent with the argument in Section 21.3 concerning the aliveness of meaning. Strength Three: the posture of undertaking finitude. "Knowing that it cannot be done": not being ignorantly optimistic, but still doing it while knowing the impossibility. This posture bears the full weight of finitude. It does not lighten finitude through hope, belief, or rhetoric.

The limits of the Confucian position. Limit One: lack of ontological ground. The Confucian "ought" is built upon tradition, ritual order, and the teaching of the sages. It has normative grounding, but lacks ontological grounding. "Why ought it be so?" Within Confucianism, the answer ultimately appeals to "the Mandate of Heaven," "the Dao," or "the teaching of the sages." These are normative rather than ontological. Constrained speculation can supplement this limit. Chapter 20 downgrades Hume's is–ought gap into a methodological problem: value statements and factual statements are subject to the same discipline, but they are not reduced to one another. This gives a platform of ontology to the Confucian "ought": ought as a commitment has its own content, including distinguishable content, connection with practice, and falsifiability, and does not need to be reduced to the Mandate of Heaven or the teaching of the sages. Limit Two: lack of treatment of ultimate finitude. Confucianism is concerned with the finitude of everyday situations, including individual life and death, political success and failure, but it treats ontological-level finitude less directly, such as the gradient terminal phase in Chapter 17. This is a reasonable consequence of Confucianism's concern with practical philosophy, but when we ask how ultimate finitude relates to everyday meaning, Confucianism provides limited resources. Constrained speculation can supplement this limit. Section 21.5 of this chapter will treat the relation between everyday finitude and ultimate finitude: this is one advance this chapter makes beyond Confucianism.

Dialogue Two: Heidegger's Being-toward-Death

In the second division of *Being and Time*, Heidegger proposes being-toward-death (Sein-zum-Tode) as the authentic mode of being of Dasein. Death is Dasein's ownmost, unsurpassable, certain but indefinite possibility. Being-toward-death is not literally "wanting to die" or "waiting for death." It is an ontological posture: undertaking death as one's own possibility and thereby gaining authentic "ownness" (Eigentlichkeit). When Dasein exists toward death, it returns from the anonymous state of the "they" (das Man) to itself, recognizes that its possibilities are its own, that its choices must be made by itself, and that its finitude is its structure. The relation between being-toward-death and meaning is this: by undertaking death, Dasein gains the possibility of "ownness," a real mode of existence that belongs to itself and is not dissolved by anonymity. This mode of being has meaning because it is true, or authentic.

The strengths of Heidegger's position. Strength One: finitude as the ground of authenticity. Authentic existence is not existence that escapes finitude, but existence that undertakes finitude. Meaning is in authenticity, and authenticity is in undertaking finitude. This is the expression of M2 in Heidegger's philosophy. Strength Two: death as possibility rather than event. Death is not a future event that will happen, but something that structures present existence as possibility. This insight allows Heidegger to avoid simplifying death as "something that will happen later": he integrates death into the present ontological structure. Strength Three: refusal of negative nihilism. Heidegger's being-toward-death is not pessimism. It does not say "since we all die, there is no meaning." On the contrary, it seeks the ground of meaning from death. This is fully consistent with the position of Section 21.3.

The limits of Heidegger's position. Limit One: excessive individualism. Being-toward-death is mainly concerned with the relation between individual Dasein and itself. But meaning is not only individual: it is also communal, intergenerational, and cultural. The dimensions of tradition, community, and ritual order in Confucianism are marginalized in Heidegger. Limit Two: the unclear concept of authenticity. What is "authentic"? Heidegger's reply has a circularity: authenticity is the mode of being that undertakes finitude, but why is this "authentic" rather than some other posture? The answer seems to appeal to a certain phenomenological intuition and lacks a strict ground. Limit Three: lack of connection with concrete practice. Being-toward-death as an ontological posture has force, but it does not directly provide guidance concerning what to do. A Dasein that undertakes being-toward-death still has to decide concrete actions; Heidegger provides limited resources in this respect. Constrained speculation can supplement these limits. The first two limits can be treated by introducing Confucian resources concerning community and ought. The third can be treated by applying the disciplines of constrained speculation, including distinguishability and falsifiability, to practical decision.

Dialogue Three: Camus' Sisyphus

In *The Myth of Sisyphus*, Camus proposes "the absurd" (l'absurde) as the fundamental situation of modern human beings. The world gives no ground of meaning; human beings, however, cannot eliminate their search for meaning. The tension between these two is the absurd. Camus' response is neither passively giving up in the face of the absurd nor escaping into hope, whether religious or philosophical. His response is revolt: acknowledging the absurd while continuing to live. Sisyphus is the image of revolt. He is punished by the gods to push a stone up the mountain forever, knowing that it will roll down each time, but he chooses to continue pushing. Camus writes: "Il faut imaginer Sisyphe heureux" ("One must imagine Sisyphus happy"). This assertion is not sentimental. It says that in the process of undertaking the absurd, human beings obtain a kind of meaning that does not depend on the elimination of the absurd, but is realized precisely within the absurd.

The strengths of Camus' position. Strength One: facing the absurd directly. Camus does not try to evade the absurd through metaphysics or theology. He acknowledges it directly. This is a philosophically honest posture. Strength Two: revolt as meaning. Revolt itself is the mode of realizing meaning: not "revolt for the sake of meaning," but "revolt is meaning." This position strongly resonates with the Confucian "knowing that it cannot be done and yet doing it." Strength Three: moving from tragedy toward comedy. "One must imagine Sisyphus happy": this assertion transforms a tragic situation into a kind of comic affirmation. This is not cheap optimism, but an affirmation that has passed through tragedy.

The limits of Camus' position. Limit One: lack of ontological ground. Camus is a writer and a literary philosopher. He is more concerned with describing the absurd than with arguing for its ontological ground. His position depends on phenomenological intuition and literary rhetoric, and lacks strict ontological argument. Limit Two: vagueness of the concept of absurdity. What is the absurd? Is it a property of the world or a property of the relation between human beings and the world? Camus makes different statements in different places. This vagueness makes his position difficult to assess rigorously. Limit Three: unclear direction of revolt. What should we revolt against, and how should we revolt? Camus gives the posture of revolt, but not the content of revolt. This is again a problem of insufficient connection with concrete practice.

The Synthesis of the Three Traditions

All three traditions anchor meaning in finitude. They all support M2. But each has limits and requires supplementation by the others. Confucianism provides ought as the ground of meaning, but lacks an ontological ground and a treatment of ultimate finitude. Heidegger provides an ontological ground, the ontological structure of being-toward-death, but is excessively individualist and lacks connection with practice. Camus provides the posture of facing the absurd directly, but lacks rigorous argument and concrete direction. The distinctive position of constrained speculation is this: it provides an ontological ground, the intrinsic singularity of Chapter 17, which is Heidegger's direction; it integrates ought, through Chapter 20's treatment of the is–ought relation, which is the Confucian direction; it undertakes ultimate finitude, through Chapters 11 and 17 on freezing and ontological singularity, which none of the three traditions fully accomplishes; and it preserves the communal dimension, through Chapter 12 on co-evolution, which supplements Heidegger's excessive individualism. Constrained speculation does not replace these three traditions: it integrates them and provides them with a rigorous ontological platform. This is one contribution this chapter makes to philosophical history.

Objections and Replies

Objection One: Your "synthesis" seems too smooth. The three traditions have real disagreements, such as Confucian communitarianism vs Heidegger's individualism vs Camus' existentialist individual posture, and they cannot be easily synthesized.

Reply: Synthesis does not cancel disagreement. Constrained speculation preserves the insights and positions of the three traditions: Confucianism has special force concerning ought; Heidegger has special force concerning ontological structure; Camus has special force in directly facing the absurd. The work of synthesis is to explain the aspect of the problem of meaning treated by each, and to show how they can complement one another within the framework of constrained speculation. This is not cancellation, but configuration.

Objection Two: Do you rely too much on Chinese philosophy, Confucianism, as an argumentative resource? Might this chapter be regarded as a form of cultural chauvinism?

Reply: This chapter introduces Confucianism not as cultural chauvinism, but as a philosophical resource on the basis of argument. The Confucian "knowing that it cannot be done and yet doing it" as a thesis about meaning has philosophical force independent of its cultural source. This chapter uses it just as it uses Heidegger and Camus: all are philosophical resources, not cultural labels. But as a philosophical monograph written by a Chinese scholar, it is legitimate and necessary for this book to treat Confucianism as an important interlocutor. Chinese philosophical traditions have made distinctive contributions to many metaphysical problems; not citing them would be a waste of philosophical resources.

Objection Three: Your argument ultimately builds M2 on the "synthetic support" of three traditions. But support from traditions does not constitute strict argument.

Reply: The function of this section is not to "prove M2 through traditional support." The proof of M2 has already been given in Section 21.3. The function of this section is to enter into dialogue with traditions, to show the position of M2 in philosophical history, and to refine the meaning of M2 through dialogue. The support of the three traditions is not the ground of M2, but its philosophical context.

Literature That Must Be Engaged

- The *Analects*, especially the passage on "knowing that it cannot be done and yet doing it"
- Mou Zongsan, Moral Idealism; Mind-Substance and Human Nature
- Martin Heidegger, Sein und Zeit (1927), §§ 46-53
- Albert Camus, Le Mythe de Sisyphe (1942)
- Charles Taylor, Sources of the Self (1989): treatment of meaning across multiple traditions
- Tu Wei-ming, Confucian Thought: Selfhood as Creative Transformation (1985)

Summary of This Section
This section has done two things. First, it entered into serious dialogue with three traditions, Confucianism, Heidegger, and Camus, showing how each anchors meaning in finitude. Second, it identified the strengths and limits of each tradition and showed how constrained speculation can provide them with an ontological platform.

The next section is the chapter's conclusion: it connects the finitude of meaning with ontological singularity and completes the unity of metaphysics and the humanistic in the whole book.

21.6. Final Reflection: Meaning before Freezing

Ultimate Finitude and Everyday Meaning

The first four sections of this chapter have established the argument for the finitude of meaning. But one question still needs to be treated: the relation between ultimate finitude, such as the ontological singularity of Chapter 17, and everyday meaning. Everyday meaning appears in concrete situations: one person loves another; a culture creates art; a civilization pursues justice. All of these are finite, but their finitude appears as everyday finitude: persons die, cultures change, civilizations rise and decline. Ultimate finitude is different. It is finitude at the ontological singularity, where all gradient structures ultimately disappear. At that point, all possibilities for meaning to appear disappear. This is not the disappearance of some concrete bearer of meaning, but the disappearance of meaning itself as possibility. What is the relation between ultimate finitude and everyday meaning?

A Mistaken Response

A common mistaken response is this: ultimate finitude dissolves everyday meaning. If everything will ultimately disappear, then present love, creation, and pursuit are all futile. This response rests on the assumption already refuted in Section 21.2: meaning requires an eternal ground. It assumes that if meaning eventually disappears, then meaning is not true now. But M2 has already refuted this assumption: the truth of meaning lies in its aliveness, and aliveness comes from finitude. Ultimate finitude is precisely the deepest expression of the finitude of meaning, and therefore supports rather than dissolves meaning. But merely pointing out the error of the mistaken response is not enough. A positive statement of the relation between ultimate finitude and everyday meaning is needed.

The Positive Position

The positive position is this: ultimate finitude is the ground of everyday meaning, not its threat. This position requires several layers of development.

First layer: the continuity of the structure of meaning. Ultimate finitude and everyday finitude are not two different kinds of finitude: they are different levels of the structure of finitude. Everyday finitude, including life and death, rise and decline, is concrete manifestation. Ultimate finitude, ontological singularity, is the underlying structure. Together, they constitute "finitude" as the condition of possibility for meaning. If there were only everyday finitude and no ultimate finitude, then "what lies beyond the everyday" would remain an open question: perhaps some eternal state of affairs. This would reawaken the temptation to take eternity as the ground of meaning. The acknowledgement of ultimate finitude closes this temptation. It says that behind everyday finitude there is no eternal state of affairs as an escape route.

Second layer: ultimate finitude provides the purity of everyday meaning. If meaning is not anchored in any eternal state of affairs, if not even some kind of cosmic eternity remains as a final refuge, then everyday meaning is self-sufficient: it is true where it appears and does not need an external ground. This purity is the deepest expression of M2. Meaning is not only true in finitude; meaning is most pure in finitude that acknowledges ultimate finitude.

Third layer: the realization of meaning is resistance to finitude. Here "resistance" is not literal opposition: meaning cannot truly cancel finitude. But the realization of meaning is, relative to finitude, a movement in the opposite direction. Meaning builds structure; finitude ultimately dissolves structure. Meaning maintains distinctions; ultimate finitude dissolves distinctions. The aliveness of meaning lies in this movement in the opposite direction: knowing that finitude ultimately wins, and yet building structures, maintaining distinctions, and making choices. This is the ontological version of the Confucian "knowing that it cannot be done and yet doing it." It is the cosmological version of Heidegger's being-toward-death. It is the metaphysical version of Camus' Sisyphus.

The Deepest Meaning of the Three Traditions

Through this connection, the insights of the three traditions acquire the deepest metaphysical meaning of the whole book. The Confucian "knowing that it cannot be done and yet doing it" at the ontological level: knowing that the gradient terminal phase is the intrinsic singularity of EQT ontology and that all structures ultimately dissolve; and yet persisting in what ought to be done within finite situations. The "cannot" is ontological impossibility, ultimate dissolution, not merely situational impossibility. Heidegger's being-toward-death at the cosmological level: authentic existence undertakes not only individual death but also the ultimate finitude of the universe. Eigentlichkeit includes the relation to ontological singularity: undertaking the fact that all structures will dissolve and gaining living existence in this undertaking. Camus' Sisyphus at the metaphysical level: the absurd is not merely that "the world gives no meaning," but the deeper fact that ontology itself has an intrinsic singularity. But revolt remains possible: building structures and pushing the stone while knowing the ontological singularity. "One must imagine Sisyphus happy": one must imagine affirming meaning while undertaking the ontological singularity. In the EQT framework, these three traditions are not replaced, but deepened. Their insights gain new force in the undertaking of ontological singularity.

The Overall Connection of the Book

This section is also the deepest connection of the whole book. Chapter 17 on ontological singularity is the metaphysical peak of the book: it identifies the intrinsic boundary of EQT ontology. This chapter connects that boundary with humanistic meaning: the boundary is not the cancellation of meaning, but the pure ground of meaning.

Chapter 11 on freezing as a foundational ontological process provides the mediation. Freezing is not simple stopping, but exiting active dynamics while remaining as residue, background, and history. The process of everyday meaning is the active face of the freezing process: meaning maintains structures, makes choices, and pushes directions; freezing ultimately dissolves structures. But freezing itself preserves traces of meaning having appeared: the CMB is the freezing of the Big Bang; fossils are the freezing of life; memory is the freezing of experience; history is the freezing of culture. Meaning exists before freezing: it exists in active processes that have not yet frozen. Meaning is not in eternity; meaning is in the process that is happening. When everything freezes into the terminal phase, meaning disappears as well. But before freezing, meaning is true: not true despite being finite, but true precisely because it faces freezing. This is the deepest posture of the humanistic part of the book. It is not pessimism and not optimism. It is the affirmation of finitude: affirming the truth of meaning where it appears, affirming the reality of structure during the time it exists, and affirming the reality of distinction in the process in which it is maintained.

The Final Thesis of the Book

Synthesizing the argument of this chapter, the final thesis of the book can be stated as follows: Thesis M3: meaning is finite, alive, and undertaken. It is not in an eternal state of affairs; it is in the concrete practice of finite existence facing ontological singularity. Meaning exists before freezing: this is its ontological position and its full dignity. This thesis is not pessimistic. It says: meaning is true precisely because it is finite. It does not need eternity for support, because its truth lies in its aliveness. Our love, creation, pursuit, and persistence are true where they appear; they do not need to be anchored in any external ground. But this thesis is also not cheap optimism. It bears the full weight of ontological singularity. It knows that all structures ultimately dissolve, and that ultimate finitude is true, strict, and unavoidable. It affirms meaning upon this knowledge. This is an affirmation that has passed through ontology, not ignorant optimism.

Objections and Replies

Objection One: Your "meaning before freezing" seems poetic. This is inconsistent with the strict argument earlier in the chapter.

Reply: This expression does have a poetic color, but it corresponds to a strict argument. "Before freezing" refers to all time before the gradient terminal phase is reached. This has physical meaning, as in Chapter 4 of *Freezing Cosmology* on the scale relation of freezing time. During this interval, gradient structures exist, and the possibility for meaning to appear exists. The poetic expression simply renders the strict thesis more readable.

Objection Two: You seem to imply that meaning "disappears" at some concrete moment, the moment of the gradient terminal phase. But meaning is not a "thing" that can disappear; it is relational.

Reply: Meaning, as a relational phenomenon, indeed "disappears" in ultimate finitude. More accurately, it loses the possibility of appearing. When all gradient structures capable of sustaining "mattering" disappear, meaning is no longer possible. This is not the disappearance of some entity, but the closure of a possibility. But this closure does not happen at a concrete moment in the physical sense. It is a limiting state of affairs in the ontological sense. In physical time, meaning may continue to appear for an extremely long time, and strict terminal phase may even never be reached. This section's claim that "meaning exists before freezing" does not say that meaning will disappear at some concrete moment; it says that meaning ontologically belongs to states of affairs that have not yet reached ultimate finitude.

Objection Three: Your overall position may be read as a refined version of nihilism.

Reply: The position of this chapter is precisely not nihilism. Nihilism says that meaning does not exist or does not matter. This chapter says that meaning exists and matters precisely because it is finite. Nihilism chooses nothingness between eternity and finitude; this chapter chooses finitude between eternity and finitude. These are opposite positions. If someone insists on equating "acknowledging ultimate finitude" with nihilism, he may be taking "meaning must be eternal" as the definition of meaning. This chapter rejects that definition. The true definition of meaning is alive, undertaken, and finite. According to this definition, the position of this chapter is affirmative, not nihilistic.

Literature That Must Be Engaged

- Albert Camus, Le Mythe de Sisyphe (1942)
- Martin Heidegger, Sein und Zeit (1927)

- The *Analects*: on "knowing that it cannot be done and yet doing it"
- Friedrich Nietzsche, Die fröhliche Wissenschaft (1882): on eternal recurrence and affirmation
- Hans Jonas, Das Prinzip Verantwortung (1979): on finitude and responsibility
- Pierre Hadot, La philosophie comme manière de vivre (1995)

Summary of This Section

This section has done three things. First, it clarified the relation between ultimate finitude and everyday meaning: the former does not dissolve the latter, but instead provides its pure ground. Second, through this clarification, it deepened the insights of the three traditions, Confucianism, Heidegger, and Camus, to the ontological level. Third, it stated the final thesis of the book, M3: meaning is finite, alive, and undertaken; it exists before freezing.

21.7. Chapter Summary

This chapter undertook three tasks, all of which have been completed. Task One: to derive from the previously established ontological framework the claim that the finitude of meaning is not a defect but a structure. Section 21.1 established the dependence of meaning on gradient through Thesis M1; Section 21.2 developed the four layers of the finitude of meaning and weakened the common inference that "finitude weakens meaning"; Section 21.3 gave the positive position through Thesis M2: meaning is true only in finitude. Task Two: to enter into serious dialogue with three philosophical traditions. Section 21.4 developed the dialogue with the Confucian "knowing that it cannot be done and yet doing it," Heidegger's being-toward-death, and Camus' Sisyphus, identifying the strengths and limits of each and showing the unique integrative position of constrained speculation among them. Task Three: to connect the metaphysics of the whole book with humanistic meaning. Section 21.5 connected ultimate finitude, the ontological singularity of Chapter 17, with everyday meaning through Thesis M3: meaning exists before freezing. This chapter is the second pivotal chapter of the book, with Chapter 17 being the first. It expands the whole book from pure ontology into a complete philosophical system of ontology + normativity + humanistic meaning.

Looking backward, this chapter is built upon several chapters: Chapter 7 on difference before identity provides the ontological ground for the dependence of meaning on gradient; Chapter 11 on freezing as a foundational ontological process provides the precise meaning of "meaning before freezing"; Chapter 13 on Mind–Matter Unity provides the physical-realist basis for "being capable of mattering" as the bearer of meaning; Chapter 17 on ontological singularity provides a strict characterization of ultimate finitude; and Chapter 20 on constrained speculation's treatment of the is–ought gap provides a methodological platform for the ontological status of statements about meaning. Looking forward, this chapter is the humanistic climax of the whole book; after it comes the Conclusion, "The Revival of Philosophy: At the Boundary of Physics," which will take this chapter as the final landing point of the whole book.

This chapter reveals the core posture of the whole book. This posture is neither pessimism nor optimism, but the affirmation of finitude. It bears the full weight of ontological singularity while affirming the truth of meaning where meaning appears. It knows that all structures ultimately dissolve, while affirming the reality of structures during the time in which they exist. It faces the ultimate nothingness of the universe while insisting on the dignity of human beings, civilization, life, and love in their finitude. This is an affirmation that has passed through ontology: not ignorant optimism, not cheap hope, not rhetoric that evades ultimate finitude. It affirms meaning while acknowledging ontological singularity. This is the deepest application of constrained speculation to the problem of meaning: it requires acknowledging the limit while also affirming content within the limit. Constrained speculation undertakes ontological singularity in Chapter 17; in this chapter, it undertakes the affirmation of the finitude of meaning. These two undertakings are two sides of the same posture: ontological honesty and humanistic affirmation. A truly thorough ontology must be able to undertake both. Constrained speculation has done so.

This book begins from the five-volume physical monograph series on EQT. Through the methodological manifestation of constrained speculation in Part I, the philosophical defense of gradient ontology in Part II, the three unities of co-evolutionary relations in Part III, the metaphysical boundary of freezing in Part IV, and the humanistic affirmation in this chapter in Part V, it arrives at its deepest meaning: natural philosophy is not a footnote to physics; natural philosophy is philosophy called back by physics at its most radical point. When physics seriously pursues inquiry to its own limit, it needs philosophy. When philosophy seriously undertakes the limit of physics, it regains its own position. The whole work of this book is the realization of this two-way calling. Meaning exists before freezing: this is a physical thesis, concerning freezing as a cosmic process; it is also a philosophical thesis, concerning the ontological position of meaning as finite existence; and it is also a humanistic thesis, concerning how we ought to live through our finite existence. The three are unified in EQT natural philosophy as one and the same affirmation. This is the final posture of the book and the deepest expression of EQT's contribution as natural philosophy.

21.8. Methodological Reflection on This Chapter

This chapter is the humanistic climax of the whole book: it connects the ontological work of the book with humanistic meaning. This section reflects on the methodological significance of this connection.

Why This Chapter Must Be Rigorous Rather than Rhetorical

The problem of meaning easily slips into rhetoric in the history of philosophy: Heidegger's poetic saying in his later work, Camus' rhetorical expression of the absurd, and various rhetorical statements in contemporary philosophy of meaning. These works each have their legitimacy, but they present a writing challenge for this book: the work of this book must use rigorous argument to state something that seems to require rhetoric in order to be said. This chapter avoids rhetorical slippage through two means. First, it specifies the concept of meaning as ontological meaning and establishes a relation of constitutive coextension with gradient structure through M1. This specification gives meaning an ontological anchor. Second, when arguing for M2, it proceeds through a three-layered logical argument (losability $\to$ selectivity $\to$ aliveness) and does not appeal to mere intuition or rhetoric. This makes M2 a rigorous thesis rather than a rhetorical expression.

The Reasonableness of This Chapter Undertaking Multiple Tasks

This chapter undertakes three tasks, more than other chapters. This is necessary: the problem of meaning can be treated completely only through ontological grounding, dialogue with traditions, and overall connection. Without ontological grounding, work on meaning would become rhetorical. Without dialogue with traditions, work on meaning would be isolated from philosophical discussion. Without overall connection, work on meaning would be detached from the other work of this book. Through methodological discipline, this chapter completes the three tasks within its scope.

Possible Directions of Failure for This Chapter

This chapter undertakes the application of the method of constrained speculation, which requires stating its own directions of failure. The directions of failure for this chapter include the following. First, if meaning can appear independently of gradient structure, for example if some non-material bearer of meaning is shown to exist, M1 fails. Second, if it can be rigorously argued that "meaning requires an eternal ground," M2 fails. Third, if the "synthesis" of the three traditions is shown to be internally inconsistent, Section 21.4 fails. These directions of failure are required by the method of constrained speculation. As the humanistic climax of this book, this chapter still obeys the discipline of constrained speculation. This posture of undertaking possible failure makes the work of this chapter honest philosophical work.

The Position of This Chapter in the Whole Book

This chapter is the humanistic climax of the whole book. It continues the work of all preceding chapters: Chapter 17 on ontological singularity provides the ontological ground for this chapter's "acknowledgement of ultimate finitude"; Chapter 11 on freezing as a foundational ontological process provides the precise meaning of "meaning before freezing"; Chapter 13 on Mind–Matter Unity provides the physical-realist basis for "being capable of mattering" as the bearer of meaning; Chapter 20 on constrained speculation's treatment of the is–ought gap provides a methodological platform for the ontological status of statements about meaning. Looking forward, this chapter is followed by the Conclusion, which will take this chapter as the final landing point of the whole book.

The Deepest Posture of the Whole Book

This chapter reveals the deepest posture of the whole book. This posture is neither pessimism nor optimism, but the affirmation of finitude. It bears the full weight of ontological singularity while affirming the truth of meaning where meaning appears. It knows that all structures ultimately dissolve, while affirming the reality of structures during the time in which they exist. It faces the ultimate nothingness of the universe while insisting on the dignity of human beings, civilization, life, and love in their finitude. Constrained speculation undertakes ontological singularity in Chapter 17; in this chapter, it undertakes the affirmation of the finitude of meaning. These two undertakings are two sides of the same posture: ontological honesty and humanistic affirmation. Meaning exists before freezing: this is a physical thesis, a philosophical thesis, and a humanistic thesis. The three are unified in EQT natural philosophy as one and the same affirmation. This is the final posture of the book and the deepest expression of EQT's contribution as natural philosophy.

22. Conclusion: The Revival of Philosophy: At the Boundary of Physics

22.1. Returning to the Promise of the Introduction

The Introduction opened with a distinction: this book is not an extension of the five-volume physical monograph series, but "the monograph that was called for but had not yet appeared." After five parts and twenty-one chapters, this promise can now be examined.

The standard of fulfillment is not whether this book has "covered" the physical content of the five volumes. It has not, and it should not. The standard is different: whether this book has, beyond the five volumes, independently established a philosophical position that can be seriously discussed even apart from the physical work of EQT. If the answer is affirmative, then this book is not a footnote, but an independent philosophical contribution. It happens to take EQT as its fullest application, but does not take EQT as its entire ground.

This independence runs through the writing discipline of the whole book. The methodological work of Part I, including constrained speculation, ontological restraint, failure conditions, constitutive co-extension, and falsifiability as ontological discipline, is completed through purely philosophical argument, with EQT appearing only as an example. In Part II on the ontology of difference, Part III on relational ontology, and Part IV on ontological singularity, the core theses are first defended at the philosophical level and then connected to the structure of EQT. This choice of writing is not a stylistic preference; it is the form of this book's claim to independence. A philosophy that can stand only by depending on a specific physical theory rises and falls with the fate of that theory. A philosophy that can stand independently remains standing even if its best application is revised.

22.2. The Real Meaning of the Method: Constrained Speculation

Looking back over the whole book, constrained speculation is not only the method used by this book; it is itself the book's central philosophical claim. This is the real meaning of the title of Part I, "Method as Philosophy."

Constrained speculation rejects the premise shared by twentieth-century debates in scientific realism: the separation between reality and cognition. It does not seek a compromise along the axis of "correspondence vs non-correspondence," but rejects that axis itself and replaces it with a different core thesis: X is ontologically real if and only if the commitment of X has definable failure conditions. This thesis repositions reality from "correspondence to external facts" to "the content of commitment itself," and through four disciplines, formalization, mechanization, observationalization, and failure-orientation, allows the rigor of metaphysical work to be precisely assessed.

The deepest application of constrained speculation appears in Chapter 17. There, the method is applied to EQT ontology itself: it requires EQT to state its own failure conditions, thereby reaching the gradient terminal phase as the intrinsic singularity of ontology. This is not the failure of the method, but precisely its radicalization. An ontology that truly carries through the discipline that "every commitment must be able to state its own failure conditions" must, at its own logical limit, encounter the collapse of the capacity for statement and honestly acknowledge that collapse. Acknowledging an intrinsic boundary is not a defect of ontology, but a mark of its thoroughness.

This is why this book calls the method philosophy. Constrained speculation is not a neutral tool, nor a procedure that can be applied to any position whatever. It contains an ontological position, reality as a failable commitment; an epistemological posture, objectivity as controlled reproducibility; and a normative principle, value statements and factual statements are subject to the same discipline without being reduced to one another. Wherever the method unfolds, the philosophical position extends.

22.3. The Name of the Position: Constrained Process Realism

If constrained speculation is the name of the method, then constrained process realism and constrained processual normativism are the names of the ontological and normative positions established by this method.

Each of the three words in this position does distinct work. Realism: beings have ontological content that does not depend on the perception or description of observers. This distinguishes the book from instrumentalism and various forms of relativism. Processuality: beings are first processes rather than static entities, and "entities" are relatively stable forms produced by the freezing of differential structures. This distinguishes the book from the two-thousand-year mainstream of substance ontology. Constrained: ontological commitments are constrained by methodological discipline, and every commitment must be able to state its failure conditions. This distinguishes the book from unconstrained speculative metaphysics.

Together, these three words determine the position of the book within the genealogy of contemporary realism. It shares with Cartwright's entity realism a respect for scientific practice, but replaces a dappled world with a unified process. It shares with Ladyman's structural realism suspicion toward substance ontology, but takes process rather than structure as foundational. It shares with Floridi's information realism a response to the contemporary context, but takes process rather than information as foundational. It has the deepest resonance with Rovelli's relational quantum mechanics, but refines rather than relativizes objectivity. In these four relations, the position of this book is neither a variant of any one of them nor an attempt to surpass them. It occupies an independent position among them that can be clearly marked.

This position is not created from nothing. In its processual direction it has a Western genealogy in Heraclitus, Leibniz, Whitehead, Bergson, and Deleuze, and a Chinese qi-theory genealogy in Zhang Zai and Wang Fuzhi. In its methodological rigor it draws resources from the analytic traditions of Popper, Quine, Carnap, and Brandom. The contribution of this book is synthesis rather than origin: it integrates these dispersed resources through the discipline of constrained speculation into a contemporary position that can be rigorously assessed and can enter directly into contemporary discussion.

22.4. Triple Unity: Physics, Philosophy, and the Humanistic

The argumentative thread of the whole book can be summarized by one guiding thesis: meaning exists before freezing. This thesis is true simultaneously at three levels, and their unity is the central result of this book.

It is a physical thesis. In *The Freezing Dynamical Spectrum*, freezing is a precise cosmological mechanism: exiting active dynamics while remaining as residue, background, or history. This book elevates freezing into a foundational ontological process: beings enter relatively stable forms from dynamic processes through freezing; time is the direction of freezing; spatial geometry is the statistical appearance of low-frequency freezing. "Before freezing" therefore has a precise physical-ontological meaning: it points to the interval in which gradients have not yet disappeared and differential structures can still be sustained.

It is a philosophical thesis. Meaning depends on gradient (M1): the appearance of any meaning requires some gradient structure capable of maintaining distinctions, making choices, and forming directions. In the terminal state $\nabla\rho \equiv 0$, no such structure can be sustained, and therefore there is no possibility for meaning to appear. Meaning is therefore essentially finite. It is true only within the finite interval in which gradients exist, and it is true precisely because it is finite: losability, selectivity, and aliveness constitute the intrinsic structure that makes meaning meaning. A "meaning" that cannot be lost, need not be chosen, and no longer changes is not a higher meaning, but not meaning at all.

It is a humanistic thesis. Acknowledging ultimate finitude, that all structures will eventually dissolve and the universe tends toward zero gradient, does not cancel meaning; on the contrary, it is the pure ground of meaning. This is an affirmation that has passed through ontology: not ignorant optimism, not cheap hope, but the affirmation of the dignity of human beings, civilization, life, and love in their finite existence while fully undertaking the weight of ontological singularity.

These three theses are not three rhetorical versions of the same sentence. They are the real manifestation of the same ontological structure at the physical, philosophical, and humanistic levels. What allows them to be unified is precisely EQT natural philosophy. It makes the freezing of physics, the finitude of philosophy, and the affirmation of the humanistic no longer accidental analogies among three isolated fields, but the same matter traversed by the same gradient ontology. This traversal is something that physics alone, metaphysics alone, or humanistic reflection alone cannot accomplish.

22.5. The Revival of Philosophy: At the Boundary of Physics

We can now directly state what the title of this conclusion, and indeed of the whole book, means.

A dominant twentieth-century narrative was this: philosophy retreats step by step before the advance of science. Physics took over questions about matter, time, space, and causality; cognitive science took over questions about mind; the formalization of logic and mathematics took over questions about necessity. Philosophy seemed to have only two paths left: either retreat into an auxiliary role for science, as philosophy of science or conceptual clarification, or retreat into pure analysis or pure speculation unrelated to science. In both retreats, "natural philosophy," once the traditional name that integrated physics and metaphysics, almost disappeared.

This book proposes the opposite picture. Philosophy is not a residue that barely continues where science stops; it is what science calls back at its most radical point.

This calling has a precise place of occurrence: the boundary of physics. When physics seriously pursues the logical limit of its own postulates, when EQT asks "what happens if gradient disappears completely," it reaches a question that physical method itself cannot handle. The legitimate restraint of physics stops at asymptotic heat death and does not enter the strict limiting state of affairs. But metaphysics cannot take physical accessibility as its endpoint. It must ask about the fate of an ontological system at its own limit. At this boundary, physics does not exclude philosophy, but needs philosophy: it needs a thought that both respects the structure of physics and can continue to work rigorously where physical method fails.

This is the position of natural philosophy, and this is the way it is revived. It does not build a separate house outside science, nor is it content to be a footnote to science. It stands at the boundary reached by physics when physics pursues its questions seriously, and undertakes the problems that physics hands over but cannot itself treat: the intrinsic singularity of ontology, the ontological ground of meaning, and the unity of value and fact under the same discipline. When philosophy undertakes the boundary of physics, it regains its own position: not as retreat, but as the necessary extension of physics when physics is radicalized.

Kant, Wittgenstein, and Nāgārjuna all worked at boundaries in their own ways. They did not directly state what lies beyond the boundary, but stated the structure of the boundary itself. This book applies a similar method within natural philosophy and concretizes it into an operational discipline: the distinction between meta-level and object-level makes legitimate speech at the boundary possible, neither overstepping, by pretending that the terminal phase can be described, nor remaining silent, by pretending that nothing can be said about the boundary. This posture of neither overstepping nor remaining silent at the boundary is the concrete form of the revival of philosophy.

22.6. The Unfinished and the Open

An honest philosophy does not present itself as "complete." The discipline of constrained speculation requires every commitment to state its failure conditions, and this book as a whole must do the same.

The most important open problem is phenomenal consciousness. Chapter 13 explicitly acknowledged that constrained speculation does not "solve" the hard problem of consciousness, but neither does it "postpone" it. It precisely marks the hard problem as a boundary: the solvable parts, the structure of experience and the mechanism of givenness, are solved as fully as possible; the unsolvable part, the last cell, namely the residual point of subject-ascription, is precisely located as a single boundary through two compressions, while serious work is done within the range that can be solved. If future work shows that phenomenal consciousness involves an ontologically indispensable commitment that nevertheless cannot be handled at all by the four disciplines, then failure conditions as an ontological criterion (CR2) will face a real limitation. This book does not conceal this vulnerable point. It marks it, as a direction in which the method can be examined and may be revised.

There is also a large amount of work to be formalized and deepened: the precise semantics of the three kinds of specificity of failure conditions; the proof of completeness for the tri-state categories; the dialogue between constitutive coextension and semantic tools from non-analytic traditions; the development of the falsifiable structure of value statements in concrete ethics; and the connection between $\partial\rho/\partial t$ and general relativity. These are not defects of the book, but territories it marks out for subsequent research. A theory of categories, a new criterion of ontology, and a new semantic relation have never been finally established by a single monograph in the history of philosophy. They require decades of discussion, testing, and refinement within scholarly traditions. What this book offers is an initial statement, rigorous argument, honest marking of possible objections, and the rest is left to the future.

There is also unfinished work in cross-traditional dialogue. This book's dialogues with qi theory, Huayan, Nāgārjuna, and Confucianism develop only the core directions. For a natural philosophy written by a Chinese scholar and seeking to bring Chinese philosophical traditions into international discussion as substantive resources, the greatest opportunity lies precisely here: not to set Chinese traditions against Western traditions, nor to make them serve Western frameworks, but to let them participate as equal contributors to world philosophy in responding to contemporary fundamental problems. This book has only made a beginning.

22.7. Finally

This book begins from the five-volume physical monograph series on EQT. Through the methodological manifestation of constrained speculation, the double philosophical defense of gradient ontology, the three unities of co-evolutionary relations, and the metaphysical boundary of freezing, it reaches the affirmation of meaning in finitude. The deepest meaning of the whole journey can be brought together in one sentence:

When physics seriously pursues inquiry to its own limit, it needs philosophy; when philosophy seriously undertakes the limit of physics, it regains its own position.

This is the two-way calling at the boundary of physics, and it is the intention of the whole work of this book. Meaning exists before freezing: in that finite and real interval where difference can still be sustained, choice remains possible, and direction has not yet dissolved, thought can occur and meaning can be true. We happen to live in that interval. This is both the book's simplest factual statement and its final affirmation.

A. Table of Conceptual Definitions

A.1. Functional Note on This Appendix

This appendix provides an index of precise definitions for the core concepts of the whole book and serves three functions.

First, it serves as a reference tool for readers, enabling them to grasp precisely the meaning of each core concept, its first appearance in the chapters, and the related core theses.

Second, it serves as a consistency tool for the full manuscript. The precise definitions in this appendix are the standard for terminological consistency. If a term is used in the main text in a way inconsistent with the definition in the appendix, the main text should be revised according to the appendix.

Third, it serves as a positioning tool for this book in international philosophical discussion. Through precise definitions, the terminology of this book can be seriously discussed and criticized.

This appendix is organized by type of concept and divided into four parts: core methodological concepts (A.1), core ontological concepts (A.2), core relational-ontological concepts (A.3), and core normative concepts (A.4). Each concept is presented in a unified format: precise definition, main chapter, related thesis, and relation to traditional concepts.

A.2. Core Methodological Concepts

Constrained Speculation

Precise definition: a method of natural philosophy. While not abandoning metaphysical speculation, it requires every ontological commitment to satisfy four disciplines, formalization, mechanization, observationalization, and failure-orientation, so that speculative work has content that can be rigorously assessed.

Main chapter: Chapter 1, Section 1.3.

Related theses: CS1–CS4, the four disciplines.

Relation to traditional concepts: distinct from pure rational speculation, such as the Hegelian tradition, which does not require falsifiable structure. Distinct from positivism, which excludes metaphysical speculation. Constrained speculation is a middle position: it preserves speculative force while establishing methodological discipline. It resonates with the methodological rigor of the Confucian "knowing that it cannot be done and yet doing it."

Ontological Parsimony

Precise definition: ontology should commit to the minimum amount of ontological content, while requiring that commitments be strong enough to treat the ontological problems that need to be treated. Parsimony is a balance between minimality and sufficiency, not simple "subtraction."

Main chapter: Chapter 2, Section 2.1.

Related thesis: PR1.

Relation to traditional concepts: distinct from Ockham's razor, which is a heuristic criterion of selection, choosing the simpler explanation among several explanations. Ontological parsimony is a criterion of construction: it guides how ontology should be built. It is distinct from ontological expansion, such as Lewis's modal realism, which introduces new ontological commitments for each problem.

Failure Conditions

Precise definition: identifiable states of affairs for a statement, especially an ontological commitment, which, if they occur, would make the statement fail. Failure conditions are definitional content of a legitimate ontological commitment: a statement without failure conditions is not a legitimate ontological commitment.

Main chapter: Chapter 3, Section 3.2.

Related theses: CR1 (X is real $\Longleftrightarrow$ X's commitment has failure conditions), CR2 (X's commitment has failure conditions $\Longleftrightarrow$ X involves an identifiable direction of failure).

Relation to traditional concepts: derived from Popper's falsifiability, but extended. Failure conditions are not limited to direct empirical refutation, but include theoretical failure, methodological failure, structural failure, and so on. Failure conditions are an ontological discipline, not merely a principle of scientific methodology, as developed in Chapter 6, FA1.

Constitutive Coextension

Precise definition: two concepts A and B are coextensive under the same failure conditions while preserving their independence within their respective conceptual frameworks. Formal expression: $A \approx B \iff A$ and B jointly fail / jointly succeed under the same failure conditions, but A and B are not reduced to one another.

Main chapter: Chapter 4, Section 4.3.

Related thesis: CC1.

Relation to traditional concepts: distinct from identity, which means complete sameness. Distinct from reduction, which means that one is reduced to the other. Distinct from dualism, which means complete separation. Constitutive coextension is a middle relation: it establishes connection while preserving framework independence.

Formal Constraints vs Real Objects

Precise definition: this book distinguishes two kinds of ontological content: real objects, concrete existing ontological contents such as the ρ_f field and differential structures, and formal constraints, formal structures that constrain real objects, such as mathematical laws, logical relations, symmetries, and so on. Formal constraints as ontological content do not require "existence"; they function as carriers of real commitments.

Main chapter: Chapter 3, Section 3.5.

Related thesis: related to CR2.

Relation to traditional concepts: treats the ontological problem of abstract objects, such as numbers, propositions, possible worlds, and so on. It differs from Platonism, where abstract objects are independent ontologies, and from nominalism, where abstract objects do not exist. Formal constraint is a third position, acknowledging the ontological function of abstract objects without treating them as independent ontological units.

Falsifiability as Ontological Discipline

Precise definition: falsifiability is not merely a principle of scientific methodology, distinguishing science from non-science, but an ontological discipline. Any ontological commitment must have an identifiable direction of failure in order to be a legitimate commitment.

Main chapter: Chapter 6, Section 6.2.

Related thesis: FA1 (an ontological commitment is legitimate $\iff$ it has an identifiable direction of failure).

Relation to traditional concepts: inherits Popper's core insight of falsifiability, but extends it from the epistemological level, a criterion for distinguishing theories, to the ontological level, a criterion for the legitimacy of ontological commitments. Types of failure are expanded from empirical refutation to multiple types, including empirical, theoretical, methodological, and structural failure.

Meta-Level vs Object-Level

Precise definition: a methodological distinction at the boundary of ontology. Meta-level statements are statements about the boundary itself, such as "the singularity exists," and can be legitimately spoken. Object-level statements are statements about what lies on the other side of the boundary, such as "what the state at the singularity is like," and are methodologically restricted.

Main chapter: Chapter 17, Section 17.4.

Related theses: related to OS1–OS3.

Relation to traditional concepts: differs from the "ladder" structure of Wittgenstein's *Tractatus*: the meta-level statements of this book are legitimate metaphysical statements and do not need to be discarded. It has precedents in mathematical logic, such as the distinction between classes and sets, and between object language and metalanguage.

Constrained Process Realism

Precise definition: the synthetic name for the ontological position of this book. Content: ontological content is processual; ontological commitments are constrained by methodological discipline; objectivity is built upon controlled reproducibility; and the position is open to interdisciplinary and cross-traditional dialogue.

Main chapter: Chapter 5, Section 5.1.

Related thesis: the overall position of the book.

Relation to traditional concepts: distinct from simple realism, objects independent of observers; constructivism, social construction; anti-realism, instrumentalism; and relativism. Within the genealogy of contemporary realism, it is a fifth position: process realism made rigorous through methodological discipline.

A.3. Core Ontological Concepts

Difference as Ontological Starting Point

Precise definition: the non-uniformity of existential structure, including non-uniform spatial distribution of states ($\nabla\rho \neq 0$) and non-uniform temporal change ($\partial\rho/\partial t \neq 0$). Difference as an ontological concept is built upon these physical quantities.

Main chapter: Chapter 7, Section 7.2.

Related thesis: DI1 (the existence of a being $X \iff X$ involves differential structure).

Relation to traditional concepts: distinct from substance ontology, the Aristotelian tradition, which takes substance as foundational. The ontology of difference takes difference as foundational. It resonates with precursor traditions such as Heraclitus, Whitehead, Bergson, and Deleuze. "Entity" as a concept is treated as a stable form of differential structure.

Gradient ($\nabla\rho$, $\partial\rho/\partial t$)

Precise definition: the spatial non-uniformity of the ρ_f field ($\nabla\rho$, spatial gradient) and its temporal rate of change ($\partial\rho/\partial t$, temporal gradient). Gradient is the physical expression of difference as an ontological concept.

Main chapters: Chapter 7, Chapter 11, Chapter 15, Chapter 17.

Related theses: related to DI1, TM1 (time as $\partial\rho/\partial t$), and OS1 ($\nabla\rho \equiv 0$ as singularity).

Relation to traditional concepts: it is a basic concept in physics. This book elevates it from a physical concept to an ontological concept: beings as ontological contents directly involve gradient structures.

Frequency as Ontological Dimension

Precise definition: frequency is not merely a derivative quantity of time, oscillations per second, but an intrinsic dimension of beings as ontological units, standing alongside time, space, energy, and other basic dimensions. $E = h\nu$ is not "a correspondence between two independent quantities," but an ontological statement that "frequency as a basic dimension determines energy as a derived quantity."

Main chapter: Chapter 9, Section 9.1.

Related thesis: FR1, frequency as an ontological dimension.

Relation to traditional concepts: elevates Planck's formula $E = h\nu$ from a physical discovery to an ontological thesis. It enters into dialogue with Aristotle's doctrine of categories, Kant's pure forms of intuition, and Whitehead's actual occasions, as discussed in Chapter 9, Section 9.4.

Three States as Ontological Categories

Precise definition: every being at any given moment has a tri-state decomposition: free state, propagation and low interaction intensity; bound state, localized resonance and medium interaction intensity; and condensed state, stable structure and high interaction intensity. The three states as ontological categories provide the internal structure of beings.

Main chapter: Chapter 10.

Related theses: T1 (every being has a tri-state decomposition), T2 (the three states cannot be reduced to one another and constitute foundational ontological categories).

Relation to traditional concepts: derived from the physical work of *EQT Detection Theory*. This book elevates the three states from physical states into ontological categories. It enters into dialogue with Aristotle's potentiality–actuality and Whitehead's doctrine of categories.

Freezing

Precise definition: a foundational ontological process: a being moves from dynamic process into relatively stable form. More specifically: exiting active dynamics while remaining as residue, background, or history. Freezing is not stopping, not equilibrium, and not disappearance; it is the process of exiting dominance while still existing.

Main chapter: Chapter 11.

Related theses: FZ1–FZ4, the sub-theses of freezing within this part.

Relation to traditional concepts: derived from the cosmological mechanism of *The Freezing Dynamical Spectrum*. This book elevates it from a cosmological mechanism into a foundational ontological process. It connects with Prigogine's non-equilibrium thermodynamics, biological developmental processes, and sociological institutional formation.

Freezing Meta-Stable Forms

Precise definition: so-called "entities" as ontological content: they are relatively stable forms produced by differential structures through the freezing process. Freezing meta-stable forms are not foundational ontological units, but stable phases of process.

Main chapter: Chapter 11, Section 11.2.

Related thesis: related to FZ1, one of the sub-theses of freezing as a foundational ontological process.

Relation to traditional concepts: replaces the traditional concept of "entity." It allows everyday objects, such as tables, persons, stars, and so on, to remain real as ontological contents, while their ontological ground is differential structure and freezing process rather than independent substance.

Historicity

Precise definition: the current form of a being is the result of its freezing history. Historicity is not a dimension added externally to a being, but an intrinsic structure of a being as an ontological unit.

Main chapter: Chapter 16, Section 16.3.

Related thesis: HL2.

Relation to traditional concepts: structurally related to Heidegger's Geschichtlichkeit. But this book extends historicity to all processual beings, not only Dasein. It enters into dialogue with the hermeneutic traditions of Gadamer and Dilthey.

Ontological Singularity

Precise definition: the limit $\nabla\rho \equiv 0$, the state in which all differential structures disappear. This is the intrinsic boundary of ontology itself, not some "nothingness" or "absolute being." It lies outside the legitimate range of ontology as a framework.

Main chapter: Chapter 17.

Related theses: OS1–OS3.

Relation to traditional concepts: enters into dialogue with Heidegger's boundary of "Being," Wittgenstein's "boundary of language," and Nāgārjuna's boundary of "emptiness" in Chapter 19. It differs from religious "nirvāṇa" or "God": ontological singularity is a boundary in the methodological sense, not a positive ontological content.

Gradient Terminal Phase

Precise definition: the physical state $\nabla\rho \equiv 0$, the limit in which all spatial gradients tend to zero. Within the EQT framework, this is a possible terminal phase of cosmic evolution. "Gradient terminal phase" and "ontological singularity" are two names for the same state of affairs: the former emphasizes the physical dimension, while the latter emphasizes the ontological meaning.

Main chapter: Chapter 17.

Related theses: related to OS1–OS3.

Relation to traditional concepts: related to the cosmological concept of "heat death," but the "gradient terminal phase" in this book is more foundational: it concerns the ontological dimension, not only the thermodynamic dimension.

A.4. Core Relational-Ontological Concepts

Co-Evolution

Precise definition: all "relations between two terms," such as subject–object, mind–matter, self–other, and individual–environment, are co-evolutionary relations within the same field. Two apparently independent terms are in fact local structures within the same process field, and their "relation" is an intrinsic mode of evolution of that process field.

Main chapter: Chapter 12, Section 12.2.

Related thesis: EV1.

Relation to traditional concepts: enters into dialogue with Karen Barad's intra-action, the Huayan doctrine of "mutual interpenetration without obstruction," and Whitehead's prehension. What distinguishes it from these precursors is that the method of constrained speculation, through falsifiable mechanisms, makes co-evolutionary relation not a rhetorical claim that "everything is interconnected," but a rigorous ontological relation.

Frequency Matching ($g(f, f_0)$)

Precise definition: one of the dynamical elements of co-evolution. Co-evolution among beings is established through the Lorentzian resonance kernel function $g(f, f_0)$: the matching between frequency f and eigenfrequency f_0 enables response resonance.

Main chapter: Chapter 12, Section 12.3.

Related thesis: related to EV1.

Relation to traditional concepts: derived from *EQT Detection Theory*. This book elevates it from a physical detection concept into a dynamical element of relational ontology.

Threshold Condition ($\delta\rho > \Gamma/k$)

Precise definition: the second dynamical element of co-evolution. Perturbation amplitude $\delta\rho$ must exceed the threshold Γ/k in order to excite nonlinear feedback. This is the excitation condition of co-evolutionary relations.

Main chapter: Chapter 12, Section 12.3.

Related thesis: related to EV1.

Relation to traditional concepts: derived from *EQT Detection Theory*. This book treats it as the excitation dynamics of co-evolutionary relations.

Three-State Response

Precise definition: the third dynamical element of co-evolution. Response involves the weights of the three ontological categories, free state, bound state, and condensed state, and their combination determines the form of response.

Main chapter: Chapter 12, Section 12.3.

Related theses: related to EV1, T1, and T2.

Relation to traditional concepts: connects the tri-state categories in Chapter 10 with co-evolution in Chapter 12, giving co-evolutionary relations structural content.

Ψ Framework ($\tilde{A}$, C, $\mathcal{B}$)

Precise definition: a framework for treating consciousness-related states. $\Psi(\tilde{A}, C, \mathcal{B})$ involves three variables: $\tilde{A}$, energy dissipation rate, related to oscillations in consciousness-related neural activity; C, integration complexity, related to the integration of consciousness-related neural circuits; and $\mathcal{B}$, boundary, related to the closure of the conscious system.

Main chapter: Chapter 13.

Related thesis: related to Chapter 13 on Mind–Matter Unity.

Relation to traditional concepts: derived from the work of *Mind–Matter Unity*. This book uses it as a framework for consciousness research: it connects third-person physical variables and first-person conscious experience through constitutive coextension, as developed in Chapter 4.

Reproducible Co-Disturbance

Precise definition: the new foundation of objectivity. An observational result is objective $\Longleftrightarrow$ (i) it can be reproduced under controlled conditions, producing the same result under the same configuration; and (ii) this reproducibility is built upon identifiable mechanistic correspondences, including frequency matching, threshold, and three-state response. Objectivity = controlled reproducibility, not observer-free externality.

Main chapter: Chapter 14, Section 14.2.

Related thesis: OB1.

Relation to traditional concepts: replaces traditional "observer-free objectivity." It enters into dialogue with Heisenberg's participation, Rovelli's relational quantum mechanics, and Barad's agential realism, as discussed in Chapter 14, Section 14.4. It differs from operationalism: OB1 does not require concepts to be defined by operations, but requires objectivity to be built upon controlled reproducibility.

Different Projections of the Same Process

Precise definition: a treatment of the mind–matter relation. Mind and matter are not relations between two independent terms, but different projections of the same process, the dynamics of the ρ_f field. First-person experience and third-person physical variables are both manifestations of the same process.

Main chapter: Chapter 13.

Related thesis: related to the core thesis of Chapter 13.

Relation to traditional concepts: distinct from substance dualism, where mind and matter are independent substances; reductive materialism, where mind is reduced to matter; and property dualism, where mind and matter are different properties of the same substance. "Different projections of the same process" is a fourth position: it concretizes the mind–matter relation through co-evolution and constitutive coextension.

Time as $\partial\rho/\partial t$

Precise definition: time as an ontological dimension = the intrinsic rate of change of the energy quantum density field, $\partial\rho/\partial t$. Time is not an externally added coordinate parameter, but the internal structure of the field, the direction of change of the field itself.

Main chapter: Chapter 15, Section 15.2.

Related thesis: TM1.

Relation to traditional concepts: consistent with the spacetime geometry of general relativity, where time is coupled with matter and energy, but more specific: time is the internal structure of the ρ_f field. It reconciles Bergson's duration and Einstein's physical time, as discussed in Chapter 15, Section 15.4: the two are different projections of the same ontological time.

A.5. Core Normative Concepts

Is–Ought Gap as Methodological Problem

Precise definition: the difference between factual statements (is-statements) and value statements (ought-statements) proposed by Hume. This book downgrades it from an ontological problem, a difference in the ontological status of fact and value, into a methodological problem: factual statements and value statements are subject to the same methodological discipline but have different types of commitment.

Main chapter: Chapter 20, Section 20.2.

Related thesis: IO1.

Relation to traditional concepts: enters into dialogue with moral realism, expressivism, error theory, and constructivism, as discussed in Chapter 20, Section 20.4. The position of this book, constrained processual normativism, inherits insights from several traditions while occupying an independent position.

Commitment Types

Precise definition: the difference between factual statements and value statements. Both are ontological commitments and are subject to the same methodological discipline, the four disciplines of constrained speculation, but the content types of their commitments differ: factual statements concern the properties or relations of beings; value statements concern the contribution of actions or states along some dimension.

Main chapter: Chapter 20, Section 20.2.

Related thesis: related to IO1.

Relation to traditional concepts: specifies the difference between fact and value as a methodological difference, not an ontological split.

Meaning before Freezing

Precise definition: meaning as an ontological-normative phenomenon. Meaning is built upon co-evolutionary relations: relations, such as consciousness and world, give meaning an ontological ground. Meaning emerges in the process before freezing. It depends on active process and disappears as process declines.

Main chapter: Chapter 21.

Related theses: M1 (meaning depends on gradient structure), M2 (meaning is true only in finitude), M3 (meaning exists before freezing).

Relation to traditional concepts: resonates with Heidegger on the meaning of Being, Wittgenstein on mystical content, and Nāgārjuna on meaning in emptiness, as discussed in Chapter 19. It differs from these precursors by giving meaning an ontological ground through co-evolution as relational ontology.

Affirmation of Finitude

Precise definition: the ethical posture of this book. Finitude, including death, suffering, limits, and so on, is not something to be merely regretted. It is the ground of meaning as an ontological-normative phenomenon. Meaning is true only in finitude. So-called "eternal meaning" without finitude is an empty concept.

Main chapter: Chapter 21.

Related thesis: related to M2.

Relation to traditional concepts: resonates with Heidegger's being-toward-death (Sein-zum-Tode), but makes the ontological ground of finitude more concrete through the co-evolutionary ontology of this book. It also resonates with the Buddhist tradition of impermanence (anityā), while making it rigorous through the methodology of this book.

Constrained Processual Normativism

Precise definition: the position of this book within the genealogy of metaethics. Content: value statements are ontological commitments concerning the contribution of beings along some dimension; these commitments are subject to the discipline of the method of constrained speculation; they involve processual content, concerning dynamic contribution rather than static moral facts; and they have structures that can be refined and examined.

Main chapter: Chapter 20, Section 20.4.

Related thesis: related to IO1.

Relation to traditional concepts: distinct from moral realism, independent moral facts; expressivism, emotional expression; error theory, systematic error; and constructivism, procedure as the ground of value. It is a fifth position: it preserves objectivity without presupposing an independent moral ontology, and treats methodology as an instrument of assessment but not as the ground of value.

A.6. Notes on Terminological Usage

There are several notes for using this appendix.

Note One: core terms vs synonymous terms

Some core terms may have synonymous expressions in the main text: "constrained speculation" / "the method of constrained speculation" may be used interchangeably; "constitutive coextension" must use this precise term, avoiding "constitutive equivalence"; "ontological restraint" / "restrained ontology": the former is used as a methodological principle, while the latter is used as a position; "failure conditions" / "identifiable direction of failure": use "failure conditions" in CR1 and CR2, and "identifiable direction of failure" in applications such as FA1; "co-evolution" is usually used as a noun, while "co-evolutionary relation within the same field" is a more precise expression.

Note Two: physical terms vs ontological terms

Some terms have different precisions in physics and ontology: ρ_f / $\rho_f(x,t)$: in physics it is a field quantity; in ontology it is a foundational ontological object. The referent is the same, but the context of use differs. $\nabla\rho$ / $\partial\rho/\partial t$: in physics they are mathematical operators; in ontology they express differential structure and the direction of time. Frequency (ν or f): in physics it is oscillations per second; in ontology it is a foundational dimension.

Note Three: format for cross-chapter references

When citing a concept across chapters, the following formats are recommended: to cite the definition in this appendix, see Appendix A on the definition of "[concept name]"; to cite the first development in a chapter, see Chapter X, Section X.X on "[concept name]"; to cite a thesis, see [thesis number] (Chapter X, Section X.X).

Note Four: treatment when Appendix A and the main text are inconsistent

If an inconsistency is found between the terminology or definition used in the main text and Appendix A, one should: revise the main text according to Appendix A by default, since Appendix A serves as a consistency tool for the full manuscript; if the main text expression is more precise or has contextual reasons, revise Appendix A; record the inconsistency in the manuscript-integration log for traceability.

A.7. Limits of This Appendix

This appendix acknowledges several limits.

Limit One: selection of core concepts. This appendix contains approximately thirty core concepts. Some important terms may not be included, such as technical physical terms appearing in particular chapters, including Boltzmann distribution, Bose–Einstein condensation, and so on, or certain cross-cultural terms, such as the Confucian "unity of knowledge and action," the Buddhist "nirvāṇa," and so on. When these terms appear in the main text, they should be used according to context and are not independently defined in Appendix A.

Limit Two: consistency with the main text requires continuous maintenance. Appendix A as a manuscript-integration tool is a starting point, not an endpoint. During manuscript integration, revisions in the main text may lead to revisions in Appendix A, and vice versa. Consistency between the two must be maintained throughout the entire integration process.

Limit Three: precise statements of core theses are not fully developed. This appendix gives the number of each core thesis, but does not provide the precise statement of every thesis, since that is the work of the main text. If a reader needs the precise statement of a thesis, the relevant chapter should be consulted.

Limit Four: interdisciplinary applications are limited. This appendix mainly gives the philosophical definitions of concepts. The applications of each concept in physics, neuroscience, sociology, ethics, and other fields should be consulted in the relevant chapters of the main text.

Acknowledging these limits does not weaken the functional value of the appendix; it clarifies the boundary of the appendix.

B. Comparative Table with Contemporary Analytic Philosophy Literature

B.1. Functional Note on This Appendix

This appendix provides a comparison between the work of this book and contemporary analytic-philosophical discussions, so that the philosophical position of this book can be precisely located within international discussion. This appendix serves three functions.

First, it serves as a positioning tool for readers, allowing them to quickly grasp the resonances and differences between this book and each important interlocutor.

Second, it serves as a consistency tool for manuscript integration. The treatment of literature in this appendix is the standard for the use of literature in the relevant chapters of the main text.

Third, it serves as a point of connection between this book and international philosophical discussion, allowing the book to be seriously discussed and criticized.

This appendix is divided into two parts: core interlocutors (B.1, in-depth dialogue) and important references (B.2, brief comparison). The treatment of each figure is consistent with the relevant chapters of the main text.

B.2. Core Interlocutors

Karl Popper (1902–1994)

Major works: The Logic of Scientific Discovery (1934/1959); Conjectures and Refutations (1963); Objective Knowledge (1972).

Core position: falsifiability as the criterion of demarcation for science: a theory is scientific $\Longleftrightarrow$ it has content that can be empirically refuted. A theory should "take risks": any serious theory must be capable of being wrong.

Resonances with this book: (1) the requirement of methodological rigor; (2) criticism of ad-hoc adjustment; (3) rejection of unfalsifiable statements; (4) the core insight that a theory needs refutable content.

Differences from this book: (1) this book elevates falsifiability from epistemology (the demarcation of theories) to ontology (the legitimacy of ontological commitments); (2) the types of failure are expanded from empirical refutation to multiple types, including empirical, theoretical, methodological, and structural failure; (3) the unit is expanded from theory to ontological commitment.

Related chapters: Chapter 1, Chapter 6, especially Section 6.2, FA1.

W. V. O. Quine (1908–2000)

Major works: "On What There Is" (1948); "Two Dogmas of Empiricism" (1951); Word and Object (1960).

Core position: ontological commitments are manifested through quantified variables: a theory is committed to the existence of the objects over which its quantified variables range. Ontology is connected with science: naturalized metaphysics.

Resonances with this book: (1) naturalism: ontology is built upon the best scientific theories; (2) parsimony: a preference for a small number of commitments; (3) rejection of a simple analytic–synthetic distinction; (4) a holistic direction.

Differences from this book: (1) Quine emphasizes linguistic analysis, especially quantified variables, while this book emphasizes ontological process, especially differential structure; (2) this book gives parsimony a methodological ground through the method of constrained speculation; (3) this book's constitutive coextension (CC1) replaces Quine's holistic translation.

Related chapters: Chapter 2, Section 2.4; Chapter 3; Chapter 4.

Donald Davidson (1917–2003)

Major works: Essays on Actions and Events (1980); Inquiries into Truth and Interpretation (1984); "On the Very Idea of a Conceptual Scheme" (1974).

Core position: anomalous monism: mental events and physical events are different descriptions of the same event. There are no strict psychophysical laws, but mind and matter are not independent entities.

Resonances with this book: (1) rejection of substance dualism; (2) rejection of simple reductive materialism; (3) the direction of a monistic process; (4) suspicion toward strict laws.

Differences from this book: (1) this book provides mechanisms through co-evolution (EV1) and the Ψ framework: frequency matching, threshold, and three-state response; (2) this book's "different projections of the same process" is more concrete than Davidson's anomalous monism; (3) this book uses constitutive coextension (CC1) as a semantic tool.

Related chapters: Chapter 4, Chapter 12, Chapter 13.

Alfred North Whitehead (1861–1947)

Major works: Process and Reality (1929); Science and the Modern World (1925); Adventures of Ideas (1933).

Core position: process philosophy: actual occasions as foundational objects, connected with one another through prehension. Eternal objects ingress into actual occasions as forms.

Resonances with this book: (1) process ontology as a foundational direction; (2) rejection of static entities; (3) relations as ontological content; (4) the importance of creativity.

Differences from this book: (1) this book does not introduce eternal objects as a second ontology; forms are treated as the internal dynamics of differential structures; (2) this book is connected to physics (EQT), while Whitehead's work is more abstract; (3) this book makes process ontology rigorous through the method of constrained speculation.

Related chapters: Chapter 7, Section 7.4; Chapter 8, Section 8.5; Chapter 9.

Nancy Cartwright (1944–)

Major works: How the Laws of Physics Lie (1983); Nature's Capacities and their Measurement (1989); The Dappled World (1999).

Core position: entity realism + dappled-world view: realism about entities, anti-realism about universal laws. The world contains diverse local regularities and local causal capacities.

Resonances with this book: (1) scientific practice as an ontological ground; (2) suspicion toward simple theoretical unity; (3) acknowledgement of interdisciplinary diversity.

Differences from this book: (1) this book takes difference as foundational, while Cartwright takes entities as foundational; (2) this book is a unified process, while Cartwright presents a dappled world; (3) this book distinguishes idealized laws from foundational physical statements, such as energy conservation, while Cartwright remains skeptical toward all laws.

Related chapter: Chapter 5, Section 5.2.

James Ladyman (1969–)

Major work: Every Thing Must Go: Metaphysics Naturalized (2007, with Don Ross).

Core position: ontic structural realism (OSR): foundational ontological content is structure (relations), not entities (things). Physics tells us facts about structure. "Every Thing Must Go."

Resonances with this book: (1) suspicion toward substance ontology; (2) connection with physics; (3) naturalized metaphysics.

Differences from this book: (1) this book takes process as foundational, while Ladyman takes structure as foundational; (2) this book treats "things" as stable forms of differential structure, rather than eliminating them, while Ladyman advocates the elimination of "things"; (3) this book avoids the difficulty of "structure as a floating independent foundation."

Related chapter: Chapter 5, Section 5.3.

Luciano Floridi (1964–)

Major works: The Philosophy of Information (2011); The Ethics of Information (2013); The Logic of Information (2019).

Core position: information as foundational ontology: all ontological content can be understood as forms of information. Information serves as a core interdisciplinary concept integrating multiple disciplines.

Resonances with this book: (1) response to the contemporary context, especially the digital age; (2) interdisciplinary openness.

Differences from this book: (1) this book takes process as foundational, while Floridi takes information as foundational; (2) this book strictly distinguishes ontology (difference) from description (information), while the relation between the two is not fully clear in Floridi's work; (3) this book focuses on foundational ontology and its connection with physics, while Floridi focuses on digital applications.

Related chapter: Chapter 5, Section 5.4.

Carlo Rovelli (1956–)

Major works: "Relational Quantum Mechanics" (1996); Reality Is Not What It Seems (2017); The Order of Time (2018); Helgoland (2020).

Core position: relational quantum mechanics (RQM): the "state" of a quantum system is relative to an observer, and there is no absolute objective state. Events are ontological: relative interactions, rather than statements about things.

Resonances with this book: one of the deepest resonances: (1) relational ontology as foundational; (2) connection with contemporary physics; (3) rejection of traditional "observer-free objectivity"; (4) dialogue with philosophical traditions.

Differences from this book: (1) this book refines objectivity (OB1: controlled reproducibility), while Rovelli relativizes objectivity; (2) this book takes process as foundational (continuity), while Rovelli takes events as foundational (close to Whitehead); (3) this book is built upon the EQT framework, not only upon an interpretation of quantum mechanics.

Related chapters: Chapter 5, Section 5.5; Chapter 12; Chapter 14; Chapter 15.

Martin Heidegger (1889–1976)

Major works: Being and Time (1927); Beiträge zur Philosophie (1936–1938); Holzwege (1950).

Core position: the question of Being (Seinsfrage): traditional metaphysics confuses Being (Sein) and beings (Seiendes). Poetic saying as a method at the boundary. Ereignis as the mutual belonging of Being and human beings.

Resonances with this book: (1) critique of traditional metaphysics, especially substance ontology; (2) seriousness toward the metaphysical boundary; (3) historicity (Geschichtlichkeit) as an intrinsic structure of beings; (4) structural relevance between Ereignis and co-evolution.

Differences from this book: (1) this book pursues methodological rigor (constrained speculation), while Heidegger uses poetic saying; (2) this book is built upon science, while Heidegger is skeptical of modern science; (3) this book's co-evolution has falsifiable structure, while Ereignis is a poetic-hermeneutic concept; (4) this book extends historicity to all processual beings, while Heidegger mainly treats Dasein.

Related chapters: Chapter 11, Chapter 16, Chapter 19, Section 19.2.

Ludwig Wittgenstein (1889–1951)

Major works: Tractatus Logico-Philosophicus (1922); Philosophical Investigations (1953).

Core position: early period: the legitimate use of language is limited to the description of facts, and beyond this range one "must remain silent." Later period: language is a multiplicity of language games, and many philosophical problems are misuses of language.

Resonances with this book: (1) seriousness toward the boundary of language; (2) methodological rigor; (3) exclusion of illegitimate statements; (4) certain dimensions of philosophy as a therapeutic activity.

Differences from this book: (1) this book uses the meta-level/object-level distinction to make legitimate speech at the boundary possible, while Wittgenstein advocates strict silence; (2) this book refines metaphysics (FA1), while Wittgenstein restricts metaphysics; (3) this book focuses on ontological content, while Wittgenstein focuses on linguistic analysis.

Related chapters: Chapter 19, Section 19.3; Chapter 17, Section 17.4 on the meta-level/object-level distinction.

David Hume (1711–1776)

Major works: A Treatise of Human Nature (1739–40); An Enquiry Concerning Human Understanding (1748).

Core position: the is–ought gap: deriving value statements (ought-statements) from purely factual statements (is-statements) involves a "new relation," and this relation requires explanation. The possibility of such a relation is doubtful.

Resonances with this book: (1) serious treatment of the difference between fact and value; (2) methodological rigor; (3) suspicion toward simple reduction, such as naturalistic ethics.

Differences from this book: (1) this book downgrades the is–ought gap from an ontological problem to a methodological problem; (2) IO1 in this book holds that the two kinds of statements are subject to the same methodological discipline, while differing in type of commitment; (3) this book makes it possible for ethics to be a rigorous field of empirical research.

Related chapter: Chapter 20.

G. E. Moore (1873–1958)

Major work: Principia Ethica (1903).

Core position: the naturalistic fallacy: reducing "good" to any natural property, such as pleasure or adaptiveness, involves a conceptual error. The open question argument: any reduction faces the open question, "X is a natural property, but is X really good?"

Resonances with this book: (1) rejection of simple naturalistic reduction; (2) preservation of the difference between fact and value; (3) methodological rigor.

Differences from this book: (1) this book does not appeal to "good" as an independent intuitive concept, as in Moore's intuitionism, but makes value statements rigorous through falsifiable structures; (2) this book's constrained processual normativism does not depend on an independent moral ontology; (3) this book makes ethics capable of empirical research.

Related chapter: Chapter 20.

David Chalmers (1966–)

Major works: The Conscious Mind (1996); "Facing Up to the Problem of Consciousness" (1995).

Core position: the "hard problem of consciousness": why do physical processes correspond to subjective experience (qualia)? This problem cannot be solved by simple reductive methods. Property dualism is proposed as a direction.

Resonances with this book: (1) serious treatment of the problem of consciousness; (2) rejection of simple reductive materialism; (3) preservation of the specificity of subjective experience.

Differences from this book: (1) this book's "different projections of the same process" (Chapter 13) concretizes the mind–matter relation through co-evolution, avoiding the difficulties of property dualism; (2) this book fulfills structural sufficiency through the isomorphism of fivefold gradient richness and provides falsifiable structure by using $\Psi \equiv (\tilde{A}, C, \mathcal{B})$ as the observable projection of the isomorphic space; (3) this book treats the hard problem by "precisely marking it as a boundary": it handles the solvable parts through same-register isomorphism, internal–external co-evolution, and two compressions, and precisely marks the unsolvable part as a single boundary.

Related chapter: Chapter 13.

Karen Barad (1956–)

Major works: Meeting the Universe Halfway (2007); "Posthumanist Performativity" (2003).

Core position: agential realism + intra-action: rejection of the presupposition of independent terms, with relational terms emerging from intra-action. Measurement is the intra-action of a material-discursive apparatus.

Resonances with this book: one of the deepest resonances: (1) the direction of relational ontology; (2) rejection of the presupposition of independent terms and relations; (3) grounding in contemporary physics, especially Bohr; (4) measurement as a constitutive process.

Differences from this book: (1) this book has falsifiable mechanisms, including frequency matching, threshold, and three states, while the falsifiable structure of Barad's work is not clear; (2) this book is connected with EQT physics, while Barad's work is mainly based on an interpretation of Bohr; (3) this book is restrained with respect to political-ethical implications, while Barad's work has strong political-ethical motivations.

Related chapters: Chapter 12, Section 12.4; Chapter 14.

B.3. Important References

David Lewis (1941–2001)

Major work: On the Plurality of Worlds (1986). Core position: modal realism: all possible worlds are real. Relation to this book: Chapter 3, Section 3.4 treats it as an example of a "statement without failure conditions": the statement that "all possible worlds exist" is compatible with any observation, and therefore is not a legitimate ontological commitment.

Henri Bergson (1859–1941)

Major works: Time and Free Will (1889); Matter and Memory (1896); Creative Evolution (1907). Core position: duration (durée) as foundational ontology: time as an internal flow of experience. Relation to this book: Chapter 15, Section 15.4 reconciles Bergson and Einstein: duration and physical time are different projections of the same ontological time.

Edmund Husserl (1859–1938)

Major works: Logical Investigations (1900–01); Ideas (1913); The Crisis of European Sciences (1936). Core position: phenomenological method: grasping the structure of experience through suspension and reduction. Relation to this book: Chapter 7 uses the phenomenological method as support for the ontology of difference, where difference is the starting point of perception; Chapter 13, Section 13.6 and Chapter 19, Section 19.5 develop the convergence between this book and phenomenology through different paths on the "irreducibility of the given."

Bertrand Russell (1872–1970)

Major works: Our Knowledge of the External World (1914); The Philosophy of Logical Atomism (1918). Core position: logical atomism: reducing the world to atomic facts through logical analysis. Relation to this book: Chapter 2, Section 2.4 treats him as a precursor to the tradition of restrained ontology, using logical analysis to make ontology rigorous.

Rudolf Carnap (1891–1970)

Major works: Der logische Aufbau der Welt (1928); The Logical

Syntax of Language (1934). Core position: logical construction: constructing the world from sensation through logical construction. Relation to this book: Chapter 2, Section 2.4 treats him as a precursor to restrained ontology, especially the methodological rigor of empiricist epistemology.

Werner Heisenberg (1901–1976)

Major works: Physics and Philosophy (1958); Across the Frontiers (1974). Core position: the uncertainty principle + observer participation: the observer participates in the observed event as "part of the chain of processes." Relation to this book: Chapter 14, Section 14.4 treats him as a precursor to the concept of reproducible co-disturbance (OB1). This book's work is a systematic extension of Heisenberg's participation.

Niels Bohr (1885–1962)

Major works: Atomic Theory and the Description of Nature (1934); Atomic Physics and Human Knowledge (1958). Core position: the principle of complementarity: complementary observables cannot be measured precisely at the same time; different measurement configurations reveal different aspects of the same phenomenon. Relation to this book: Chapter 14, OB1 refines Bohr's complementarity by concretizing it within the framework of reproducible co-disturbance.

Pierre Duhem (1861–1916)

Major work: The Aim and Structure of Physical Theory (1906). Core position: theories are not tested in isolation; they are tested together with auxiliary hypotheses, the Duhem–Quine thesis. Relation to this book: Chapter 6, Section 6.1 treats this as a limitation of Popper's work, and this book responds to that limitation through methodological discipline.

Thomas Kuhn (1922–1996)

Major work: The Structure of Scientific Revolutions (1962). Core position: science develops through paradigms; observation is theory-laden, and there is incommensurability across paradigms. Relation to this book: Chapter 6 acknowledges the insight of Kuhn's work, namely the theory-ladenness of observation, while making cross-paradigm dialogue possible through reproducible co-disturbance (OB1).

Imre Lakatos (1922–1974)

Major work: "Falsification and the Methodology of Scientific Research Programmes" (1970). Core position: research programmes as the units of science: hard core + protective belt. Relation to this book: Chapter 6 treats Lakatos as a refinement of Popper's work.

This book is more specific than Lakatos at the level of the unit of ontological commitment: a single ontological commitment.

Gilles Deleuze (1925–1995)

Major works: Difference and Repetition (1968); The Logic of Sense (1969). Core position: difference before identity: traditional metaphysics treats difference as derivative from identity, while Deleuze takes difference itself as foundational. Relation to this book: Chapter 7, Section 7.4 treats him as the most direct twentieth-century precursor to the ontology of difference (DI1). This book can be understood as a physical-analytic-philosophical development of Deleuze's work.

Hilary Putnam (1926–2016)

Major works: Reason, Truth and History (1981); The Many Faces of Realism (1987). Core position: internal realism: rejection of the "God's-eye view" of metaphysical realism while preserving the core content of realism. Relation to this book: Chapter 5 treats him as a reference point in the genealogy of contemporary realism. This book's constrained process realism is structurally related to Putnam's internal realism.

Bas van Fraassen (1941–)

Major works: The Scientific Image (1980); Scientific Representation (2008). Core position: constructive empiricism: acceptance of a scientific theory means "empirical adequacy," not "truth." Relation to this book: Chapter 5 treats him as a representative of contemporary anti-realism. This book preserves the core of realism through co-evolution and reproducible co-disturbance, avoiding van Fraassen's anti-realism.

B.4. Limits of This Appendix

This appendix acknowledges several limits.

Limit One: the selection of fourteen core interlocutors and thirteen important references involves judgment. Other important philosophers, such as Frege, Kripke, Sellars, Brandom, McDowell, and others, are not treated in detail. These philosophers are used according to context in the main text and are not separately positioned in this appendix.

Limit Two: the treatment of each philosopher is relatively brief: about 300 Chinese characters for core interlocutors in the original planning scale, and about 100 Chinese characters for important references. This is insufficient for full dialogue with each philosopher's work. Detailed dialogue appears in the relevant chapters of the main text; this appendix provides only a framework.

Limit Three: this appendix treats the Western analytic-philosophical tradition. The comparison with Chinese philosophical traditions is developed separately in Appendix C. Indian philosophical traditions, especially Nāgārjuna, are treated in Chapter 19, Section 19.4, and are not independently developed in this appendix.

Limit Four: the "resonances" and "differences" for each philosopher are simplified statements. They capture the core direction, but each philosopher's work is complex, and a simplified treatment may lose nuance.

Acknowledging these limits does not weaken the functional value of this appendix; it clarifies the boundary of the appendix.

C. Comparative Table with Chinese Philosophical Traditions

C.1. Functional Note on This Appendix

This appendix provides a comparison between the work of this book and Chinese philosophical traditions, so that the position of this book within Chinese philosophical discussion can be precisely identified. This appendix serves three functions.

First, it serves as a reference tool for readers, allowing them to quickly grasp which resources from Chinese philosophical traditions are used by this book and how they are used.

Second, it serves as a consistency tool for manuscript integration. The treatment of traditions in this appendix is the standard for their use in the relevant chapters of the main text, especially Chapter 8 and Chapter 19.

Third, it serves as a point of connection for cross-cultural philosophical dialogue, clarifying the position and boundary of this book within Chinese traditions.

This appendix is divided into three parts: core traditions (C.1, treatment of five traditions), terminology comparison table (C.2, correspondences among cross-traditional terms), and the ethics of resource use (C.3, methodology for using Chinese traditions).

C.2. Core Traditions

The Qi-Theory Tradition

Major figures: Zhang Zai (1020–1077), Wang Fuzhi (1619–1692).

Major texts: Zhang Zai, *Zhengmeng*, *Recorded Sayings of Master Zhang*; Wang Fuzhi, *Commentary on Zhang Zai's Zhengmeng*, *Reflections on the Shangshu*, *Reading the Complete Commentaries on the Four Books.*

Core position: qi as foundational ontology: all beings are forms of qi. Qihua, the transformation of qi, as foundational process: beings are formed and dissolved through the gathering and dispersal of qi. Zhang Zai: "The Great Void cannot be without qi; qi cannot but gather to form the myriad things; the myriad things cannot but disperse and return to the Great Void." Wang Fuzhi: "daily arising and daily completion" – ontology is regenerated at every moment.

Resonances with this book: monistic ontology (qi monism as an example of restrained ontology: a single foundational ontology treating many ontological problems); processuality (qihua as foundational process corresponds to this book's process ontology, especially freezing as a foundational ontological process in Chapter 11); gathering–dispersal structure (the gathering and dispersal of qi has structural relevance to this book's tri-state categories, free state, bound state, and condensed state, Chapter 10); difference as ontology (the differences among forms of qi as the ontological ground of beings resonate with the ontology of difference, Chapter 7, DI1).

Differences from this book: qi theory is built upon a tradition of cultivation and intuition, while this book is built upon physics and methodology, especially the method of constrained speculation and the EQT framework; "qi" as a concept has different meanings among different philosophers, while this book refines concepts through methodological discipline; qi theory does not require falsifiable structure as a methodological premise, while this book requires every ontological commitment to have failure conditions.

Related chapter: Chapter 8, dialogue between the qi-theory tradition and frequency–density–gradient.

The Tradition of the Yizhuan

Major texts: *Xici I*, *Xici II*, *Shuogua*, *Wenyan*.

Core position: change (yi) as foundational: "production and reproduction is called change," and the ontological content of beings is continuous change. Yin and yang as basic dynamics: "one yin and one yang is called the Dao," and all change unfolds through the tension of yin and yang.

Resonances with this book: process ontology (yi as foundational change is consistent with the direction of this book's process ontology); differential structure (yin and yang as foundational opposition-tension resonate with this book's ontology of difference, Chapter 7, DI1); dynamical direction (the dynamics of yin and yang structurally correspond to co-evolution in this book, Chapter 12, EV1).

Differences from this book: the *Yizhuan* uses hexagram-images as expression; this is a symbolic system, different from this book's conceptual-propositional system. Yin and yang as foundational categories form a binary structure, while this book's three states, free state, bound state, and condensed state, form a ternary structure, refined through a physical ground, namely EQT tri-state decomposition. The "Dao" of the *Yizhuan* as a transcendent concept has no counterpart in this book: this book does not introduce transcendent ontological content.

Related chapter: Chapter 8.

Huayan Buddhism

Major figures: Fazang (643–712), Chengguan (738–839).

Major texts: *Avataṃsaka Sūtra*, Fazang's *Essay on the Golden Lion*, *Exploration of the Avataṃsaka Sūtra*, *Treatise on the Five Teachings of Huayan.*

Core position: the dharma-realm of mutual non-obstruction among phenomena: there is no obstruction among all things (shi); they mutually penetrate and mutually contain one another. "One is all, and all is one": every thing contains all other things. Indra's net: all things are like infinite jewels mutually reflecting one another.

Resonances with this book: relational ontology (the dharma-realm of mutual non-obstruction among phenomena as a direction of relational ontology resonates with co-evolution as ontological relation, Chapter 12, EV1); rejection of independent terms ("one is all" – no thing exists independently – is consistent with this book's rejection of independent entities); diversity of relations (the Ten Mysterious Gates develop the structure of relations, resonating with the multidimensional development of this book's mechanisms of co-evolution: frequency matching, threshold, and three-state response).

Differences from this book: Huayan relational ontology is built within a Buddhist cultivation framework, using methods such as contemplation and meditation, while this book is built within a framework of methodological discipline, namely constrained speculation. Huayan advocates full-range relation, where any thing mutually penetrates any other thing; this book restricts relations to causally connected relations, namely co-evolution under the constraint of the speed of light. Huayan has soteriological meaning: realizing the mutual non-obstruction among phenomena is a path to liberation. This book does not directly treat soteriology.

Related chapter: Chapter 12, Section 12.5.

The Madhyamaka School (Nāgārjuna Tradition)

Major figures: Nāgārjuna (around 150–250), Candrakīrti (around 600–650).

Major texts: Nāgārjuna, *Mūlamadhyamakakārikā*; Candrakīrti, *Madhyamakāvatāra*, *Prasannapadā*.

Core position: emptiness (śūnyatā): all beings lack self-nature (svabhāva), independent ontological content, and exist in the mode of dependent arising (pratītyasamutpāda). Double negation: neither affirming that X exists nor denying that X exists, thereby going beyond binary opposition. The Middle Way (madhyamā pratipad): a balance between eternalism and nihilism.

Resonances with this book: rejection of independent ontology (emptiness rejects self-nature, consistent with this book's rejection of independent entities); dependent arising vs co-evolution (dependent arising as the ontological ground of beings structurally corresponds to co-evolution, Chapter 12, EV1); two levels of truth (ultimate truth vs conventional truth is structurally related to this book's meta-level/object-level distinction, Chapter 17, Section 17.4); methodology at the boundary (double negation as a method at the boundary resonates with this book's constrained silence).

Differences from this book: Nāgārjuna's method is deconstructive, where double negation empties ontological attachment, while this book's method is constructive, establishing legitimate ontological commitments through the method of constrained speculation; Nāgārjuna's work has soteriological meaning, where understanding emptiness is the path to nirvāṇa, while this book focuses on ontology–epistemology–methodology; Nāgārjuna's work is built on logical analysis and Buddhist experience, while this book is built on physics (EQT) and methodological discipline; Nāgārjuna affirms universal emptiness (all is empty), while this book affirms real but processual ontology (beings have processual ontological content). Upon careful analysis, however, the two are consistent in rejecting independent ontology, with the difference mainly lying in terminology and expression.

Related chapter: Chapter 19, Section 19.4.

The Confucian Tradition of Practical Wisdom

Major figures: Confucius (551–479 BCE), Wang Yangming (1472–1529).

Major texts: *Analects*, *Mencius*; Wang Yangming, *Instructions for Practical Living*, *Inquiry on the Great Learning*.

Core position: the unity of knowledge and action: the intrinsic unity of knowledge and action, not two separate processes. The rigor of practical wisdom: ethical life requires methodological rigor and is not arbitrary feeling or convention. Confucius' "knowing that it cannot be done and yet doing it": methodological rigor is preserved even at the boundary.

Resonances with this book: method as philosophy (the unity of knowledge and action corresponds in direction to this book's "method as philosophy," the title of Part I: method is not a tool but philosophical content itself); methodological rigor (the rigor of Confucian practical wisdom resonates with the rigor of the method of constrained speculation: both reject arbitrariness); the foundation of normativity (Wang Yangming's "extending innate knowing" as a normative direction corresponds to IO1 in Chapter 20: normative statements are constrained by methodological discipline).

Differences from this book: Confucian work is mainly in the ethical-political domain, while this book extends to ontology, epistemology, and normativity; Confucian work is built upon the sage tradition and classics, while this book is built upon methodological discipline and science; Confucian "innate knowing" as an internal moral sense is a metaphysical presupposition, while this book makes normative statements rigorous through the method of constrained speculation and does not appeal to internal moral intuition.

Related chapters: Chapter 1, Chinese traditional resources for methodological rigor; Chapter 19, on cross-cultural dialogue; Chapter 20, on the methodological ground of normativity.

C.3. Terminology Comparison Table

The following table gives correspondences between terms in Chinese philosophical traditions and terms in this book. Each correspondence includes a type of correspondence: structural correspondence (the two terms correspond in ontological structure), conceptual refinement (the term in this book refines the Chinese term through methodological discipline), or functional correspondence (the two terms correspond in philosophical function). Correspondence does not mean identity: it indicates structural relevance while preserving independence on both sides.

Table C.1.. Correspondences and Reinterpretation Mechanisms between Traditional Philosophical Terms and the EQT/GOCD Framework

Item (Traditional Term)	Term / Concept in This Book	Type of Correspondence	Meaning and Reinterpretation Mechanism (Limits and Boundary Notes)	Chapter
Qi	ρ_f field	Structural correspondence	Both are foundational ontology: all beings are forms of it. This book refines the concept through the EQT framework.	15
Qihua	Freezing / co-evolution	Structural correspondence	Both are foundational processes: the gathering and dispersal of qi correspond to the formal changes of freezing and the relational dynamics of co-evolution.	15
Yin–yang	Differential structure	Structural correspondence	Both are foundational opposition-tension. This book physicalizes difference through $\nabla\rho$.	15-16
Taiji / Wuji	Ontological singularity	Conceptual refinement	Taiji/Wuji as metaphysical boundary: this book makes boundary discussion rigorous through the meta-level/object-level distinction.	16

Continued on next page...

Continued Table

Item (Traditional Term)	Term / Concept in This Book	Type of Correspondence	Meaning and Reinterpretation Mechanism (Limits and Boundary Notes)	Chapter
Dependent arising	Co-evolution (EV1)	Conceptual refinement	Both take relation as ontological foundation. This book refines the concept through mechanisms: frequency matching, threshold, and three-state response.	16
Emptiness / śūnyatā	Intrinsic boundary of relational ontology	Functional correspondence	Nāgārjuna's "emptiness" means that things lack self-nature. This book resonates in rejecting independent ontology, while preserving reality through process ontology.	16
Self-nature	Independent entity (rejected)	Functional correspondence	Nāgārjuna rejects self-nature; this book rejects independent entities. The directions are consistent, while their physical and philosophical grounds differ.	16
Mutual non-obstruction among phenomena	Co-evolutionary network	Structural correspondence	Huayan's mutual non-obstruction among phenomena develops the network structure of relational ontology; this book's co-evolution refines this direction.	16

Continued on next page...

Continued Table

Item (Traditional Term)	Term / Concept in This Book	Type of Correspondence	Meaning and Reinterpretation Mechanism (Limits and Boundary Notes)	Chapter
One is all	Hierarchical nesting of differential structures	Structural correspondence	"One is all" expresses the mutual containment of things; this book concretizes this direction through hierarchical nesting.	16
Indra's net	Co-evolutionary relational network	Structural correspondence	The mutually reflecting structure of Indra's net has strong structural correspondence with the mutual dynamics of co-evolutionary relations.	16
Middle Way	Constrained speech	Functional correspondence	The Middle Way as a method at the boundary resonates with this book's constrained silence and meta-level/object-level distinction.	17
Dao	(No direct counterpart)	—	As a transcendent concept, it has no direct counterpart in this book (which does not introduce transcendent ontology), but as a principle of change it resonates with processuality.	15

Continued on next page...

Continued Table

Item (Traditional Term)	Term / Concept in This Book	Type of Correspondence	Meaning and Reinterpretation Mechanism (Limits and Boundary Notes)	Chapter
Unity of knowledge and action	Method as philosophy	Functional correspondence	Wang Yangming's unity of knowledge and action corresponds in direction to the title of Part I: the mutual construction of method and content.	1
Extending innate knowing	Constrained processual normativism	Functional correspondence	Wang Yangming's extending innate knowing as a normative direction: this book's IO1 makes normative statements rigorous through methodological discipline.	17
Daily arising and daily completion	Freezing as foundational ontological process	Structural correspondence	Wang Fuzhi's "daily arising and daily completion": ontology is regenerated at each moment. This resonates with this book's process ontology.	15
Production and reproduction	Co-evolution + freezing	Structural correspondence	"Production and reproduction is called change" as the foundational content of transformation: this book concretizes this direction through co-evolution and freezing.	15-16

Continued on next page...

Continued Table

Item (Traditional Term)	Term / Concept in This Book	Type of Correspondence	Meaning and Reinterpretation Mechanism (Limits and Boundary Notes)	Chapter
Knowing that it cannot be done and yet doing it	Maintaining methodological discipline at the boundary	Functional correspondence	Confucius' methodological rigor: theoretical restraint and rigor are preserved even at the boundary, resonating with the posture of this book.	17

C.4. Ethics of Resource Use

This book uses Chinese philosophical traditions as resources, and this use involves ethical questions. This section explicitly states the methodology of this use.

Principle One: Selective Use

This book uses Chinese philosophical traditions selectively: it uses specific aspects and does not claim that each tradition as a whole is relevant to the work of this book. It uses the "qi monism" and processual direction of qi theory, but does not use the metaphysical presuppositions of qi theory, such as the ontological status of the "Great Void." It uses Huayan's relational direction of "mutual non-obstruction among phenomena," but does not use Huayan's soteriology, such as Buddha-nature and doctrines of liberation. It uses Nāgārjuna's direction of "dependent arising" and "emptiness," but does not use the soteriology of Nāgārjuna's work, such as the doctrine of nirvāṇa. It uses Confucian methodological rigor, but does not use Confucian ethical positions, such as ritual order or normative contents concerning ruler and minister. Selective use is legitimate: it allows the work of this book to preserve directional consistency in cross-traditional dialogue and to avoid conflicts among traditions. But selective use carries responsibilities: what is used and what is not used must be explicitly stated.

Principle Two: No Claim of "Already Anticipating Science"

This book does not claim that Chinese philosophical traditions "already anticipated" contemporary science, especially EQT. This rhetorical trap has been common in cross-cultural philosophical discussions since the twentieth century: some works claim that "the *Yijing* already anticipated quantum mechanics," "Laozi already anticipated relativity," and so on. Such claims involve errors: anachronism, projecting the content of contemporary science onto classical texts, which involves historical misreading; selective citation, selecting passages from classical texts that resemble certain aspects of contemporary science while ignoring other inconsistent passages; simplified correspondence, simply identifying classical concepts, such as yin–yang, with modern concepts, such as wave–particle duality, which involves conceptual confusion.

This book avoids these traps. Its mode of statement is as follows: structural correspondence (concepts in Chinese traditions and concepts in this book have structural relevance: they correspond in ontological direction, but their content differs); directional resonance (the philosophical directions of Chinese traditions, such as processuality and relationality, resonate with the direction of this book, which is a legitimate observation); resource use (this book uses the insights of Chinese traditions as resources and refines them through the method of constrained speculation and physics). This mode of expression is fundamentally different from the erroneous rhetoric of "already anticipated": it acknowledges historical and content differences while acknowledging directional relevance.

Principle Three: Mutual Enrichment

The relation between this book and Chinese philosophical traditions is not one-directional, with Chinese traditions serving merely as resources for the work of this book, but mutual. This book uses insights from Chinese traditions as ontological resources, giving the book a cross-cultural ground. At the same time, through methodological discipline (constrained speculation) and physical grounding (EQT), this book allows insights from Chinese traditions to have a legitimate position in serious contemporary philosophical discussion. This mutual relation makes the use cooperative: this book does not "appropriate" Chinese traditions, but enters into dialogue with them. Chinese traditions preserve their own internal integrity, while this book gives them contemporary continuity through methodology.

Principle Four: Openness to Dialogue with Specialists

The use of Chinese philosophical traditions in this book involves specialized fields, including classical philology, Chinese intellectual history, Buddhist studies, and so on. This book is not specialized work in those fields, and its use has several limits: limits of textual reading (this book uses classical texts mainly through commentaries and translations, which involves mediation); internal diversity within traditions (each Chinese philosophical tradition has internal diversity, such as the differences between Zhang Zai and Wang Fuzhi within qi theory, or internal schools within Huayan, and this book's treatment may simplify that diversity); complexity of cross-traditional comparison (cross-traditional comparison involves methodological difficulty, and any comparison may risk misreading). This book acknowledges these limits and invites criticism and refinement from specialists in Chinese philosophy. Any mistaken use can be pointed out and corrected, keeping the work of this book refinable.

Principle Five: Dialogue with the Contemporary Development of Chinese Philosophical Traditions

Chinese philosophical traditions are not completed traditions: they are actively developing in the present, including New Confucianism, contemporary Chinese philosophy, Buddhist modernization, and so on. The relation between this book and these contemporary developments is as follows. With New Confucianism: this book resonates with the work of New Confucians such as Mou Zongsan and Tang Junyi, especially in methodological rigor and metaphysical seriousness, but this book is built within a framework of natural philosophy, while New Confucianism is mainly built within a framework of moral metaphysics. With contemporary Chinese philosophy: as a philosophical monograph written in Chinese by a contemporary Chinese scholar, this book is itself one development within contemporary Chinese philosophy. With Buddhist modernization: this book dialogues with the work of contemporary Buddhist scholars, such as Donald Lopez and Jay Garfield, on Nāgārjuna and related topics, allowing insights from Buddhist traditions to have a legitimate contemporary position through philosophical rigor. The posture of this book is participation without dominance: it participates in contemporary Chinese philosophical discussion, but does not claim to be the contemporary representative of Chinese philosophy. The development of Chinese philosophical traditions is open, and this book is only one contribution within it.

C.5. Limits of This Appendix

This appendix acknowledges several limits.

Limit One: the selection of five core traditions involves judgment. Other important Chinese philosophical traditions, such as Mohism, Legalism, the School of Names, Xuanxue, and other branches of Song–Ming Neo-Confucianism, are not treated separately. These traditions are used according to context in the main text and are not separately treated in this appendix.

Limit Two: the treatment of each tradition is relatively brief. It is insufficient for full dialogue with the internal complexity of each tradition. Detailed dialogue requires dedicated work.

Limit Three: this appendix mainly treats the way this book uses resources from Chinese traditions. It does not attempt to provide a complete overview of Chinese philosophical traditions.

Limit Four: the correspondences in the terminology comparison table involve judgment and simplification. Any correspondence may be criticized and refined by specialists in Chinese philosophy. The purpose of this appendix is to establish a starting point for dialogue, not a final correspondence.

Acknowledging these limits does not weaken the functional value of this appendix; it clarifies the boundary of the appendix.

References

C.6. Editorial Note

This bibliography is compiled from the "literature that must be engaged" sections in the 23 chapters of the book (Introduction, 21 main chapters, and Conclusion) and the 3 appendices. The editorial principles are as follows:

First, categorical organization: the literature is divided by type and geo-cultural tradition into four parts: Western analytic philosophy and philosophy of science (A), Western continental philosophy (B), Chinese philosophical traditions (C), and Indian philosophy and Buddhist traditions (D). Within each part, entries are arranged alphabetically by author surname for Western works, and chronologically for Chinese and Indian classical works.

Second, unified format: the bibliography follows the academic monograph format of the *Chicago Manual of Style*, 17th edition. Book format: author surname + given name + title of the work (italicized) + place of publication + publisher + year. Article format: author surname + given name + article title (in quotation marks) + journal title (italicized) + volume + issue + year + page numbers.

Third, chapter association: each entry ends with square brackets indicating the chapters in this book where it is cited, so that readers can trace the relevant discussions.

Fourth, the five volumes by the author and the present book itself are listed in the "Preliminary Literature" section.

Fifth, treatment of repeated works: when the same work is cited in multiple chapters, it is listed only once, but all relevant chapters are indicated at the end in square brackets.

C.7. Preliminary Literature: The Physical Foundations of This Book

The work of this book is built upon the author's five published / in-progress EQT (Energy Quantum Theory) monographs. These works are listed as the physical foundations of this book.

Li, Kaisheng, and Longji Li. *EQT Foundational Volume: Frequency, Mass, Interaction Range, and the Generative Mechanism of Force* (Postulates A1–A4 + Meta-Norm M). Vienna, VA: Gradient Physics Press, 2026. [Throughout the book, especially Chapter 1, Chapter 9, and Chapter 11]. ISBN: 979-8-9961394-1-5.

Li, Kaisheng, and Longji Li. *Energy, Process, and the Fate of the Cosmos: A Natural Philosophy Based on the Energyon Density Field.* Kaisheng Li & Longji Li Published / Independent Released, 2026. [Throughout the book, especially Chapter 7, Chapter 11, and Chapter 17]. ISBN: 979-8-2727686-6-7.

Li, Kaisheng, and Longji Li. *Freezing Cosmology: A Natural Philosophy of Cosmic Evolution from the Perspective of Energy Quantum Theory* (proposing the method of "constrained speculation"). Vienna, VA: Gradient Physics Press, 2026. [Throughout the book, especially Chapter 1, Chapter 11, Chapter 16, Chapter 17, and Chapter 18]. ISBN: 979-8-9961394-6-0.

Li, Kaisheng, and Longji Li. *Mind-Matter Unity: Consciousness and Subjectivity in EQT Process Philosophy of Nature* (the $\Psi(\tilde{A}, C, \mathcal{B})$ state-variable group). Vienna, VA: Gradient Physics Press, 2026. [Core citation in Chapter 13; cited in Chapter 9 and Chapter 15]. ISBN: 979-8-9961394-7-7.

Li, Kaisheng, and Longji Li. *EQT Detection Theory: From Observational Ontology to Frequency Matching, Three-State Response, and Falsifiable Predictions* (three-state response: free state, bound state, and condensed state). Vienna, VA: Gradient Physics Press, 2026. [Core citation in Chapter 12; cited in Chapter 10 and Chapter 14]. ISBN: 979-8-9961394-5-3.

C.8. Western Analytic Philosophy and Philosophy of Science

Classical Analytic Philosophy (Alphabetical Order)

Anderson, Philip. "More Is Different." Science 177, no. 4047 (1972): 393–396. [Chapter 16, Section 16.2]

Ayer, A. J. Language, Truth, and Logic. London: Victor Gollancz, 1936. [Chapter 20, on expressivism]

Barad, Karen. Meeting the Universe Halfway: Quantum Physics and the Entanglement of Matter and Meaning. Durham, NC: Duke University Press, 2007. [Chapter 12, Section 12.4; Chapter 14]

Barad, Karen. "Posthumanist Performativity: Toward an Understanding of How Matter Comes to Matter." Signs 28, no. 3 (2003): 801–831. [Chapter 12, Section 12.4]

Bridgman, Percy. The Logic of Modern Physics. New York: Macmillan, 1927. [Chapter 14, Section 14.2, on operationalism]

Brink, David. Moral Realism and the Foundations of Ethics. Cambridge: Cambridge University Press, 1989. [Chapter 20, Section 20.4]

Carnap, Rudolf. Der logische Aufbau der Welt. Berlin: Weltkreis-Verlag, 1928. [Chapter 2, Section 2.4]

Carnap, Rudolf. The Logical Syntax of Language. London: Kegan Paul, 1937. [Chapter 2, Section 2.4]

Cartwright, Nancy. How the Laws of Physics Lie. Oxford: Oxford University Press, 1983. [Chapter 5, Section 5.2]

Cartwright, Nancy. Nature's Capacities and Their Measurement. Oxford: Oxford University Press, 1989. [Chapter 5, Section 5.2]

Cartwright, Nancy. The Dappled World: A Study of the Boundaries of Science. Cambridge: Cambridge University Press, 1999. [Chapter 5, Section 5.2]

Chalmers, David. The Conscious Mind: In Search of a Fundamental Theory. Oxford: Oxford University Press, 1996. [Chapter 13]

Chalmers, David. "Facing Up to the Problem of Consciousness." Journal of Consciousness Studies 2, no. 3 (1995): 200–219. [Chapter 13]

Churchland, Patricia. Braintrust: What Neuroscience Tells Us about Morality. Princeton: Princeton University Press, 2011. [Chapter 20, Section 20.3]

Davidson, Donald. Essays on Actions and Events. Oxford: Clarendon Press, 1980. [Chapter 4; Chapter 12]

Davidson, Donald. Inquiries into Truth and Interpretation. Oxford: Clarendon Press, 1984. [Chapter 4]

Davidson, Donald. "On the Very Idea of a Conceptual Scheme." Proceedings and Addresses of the American Philosophical Association

47 (1974): 5–20. [Chapter 7, Section 7.2]

Duhem, Pierre. The Aim and Structure of Physical Theory. Translated by Philip P. Wiener. Princeton: Princeton University Press, 1954 [1906]. [Chapter 6, Section 6.1]

Floridi, Luciano. The Philosophy of Information. Oxford: Oxford University Press, 2011. [Chapter 5, Section 5.4]

Floridi, Luciano. The Ethics of Information. Oxford: Oxford University Press, 2013. [Chapter 5, Section 5.4]

Floridi, Luciano. The Logic of Information. Oxford: Oxford University Press, 2019. [Chapter 5, Section 5.4]

Foot, Philippa. Natural Goodness. Oxford: Clarendon Press, 2001. [Chapter 20, Section 20.2]

Franklin, Allan. The Neglect of Experiment. Cambridge: Cambridge University Press, 1986. [Chapter 14, Section 14.3]

French, Steven. The Structure of the World: Metaphysics and Representation. Oxford: Oxford University Press, 2014. [Chapter 5, Section 5.3]

Garfield, Jay. The Fundamental Wisdom of the Middle Way: Nāgārjuna's Mūlamadhyamakakārikā. New York: Oxford University Press, 1995. [Chapter 19, Section 19.4]

Gibbard, Allan. Wise Choices, Apt Feelings: A Theory of Normative Judgment. Cambridge, MA: Harvard University Press, 1990. [Chapter 20, Section 20.4]

Greene, Joshua. Moral Tribes: Emotion, Reason, and the Gap Between Us and Them. New York: Penguin Press, 2013. [Chapter 20, Section 20.3]

Hacking, Ian. Representing and Intervening: Introductory Topics in the Philosophy of Natural Science. Cambridge: Cambridge University Press, 1983. [Chapter 5, Section 5.2; Chapter 14, Section 14.1]

Hare, R. M. The Language of Morals. Oxford: Clarendon Press, 1952. [Chapter 20, Section 20.1]

Haraway, Donna. Simians, Cyborgs, and Women: The Reinvention of Nature. New York: Routledge, 1991. [Chapter 12, Section 12.4]

Hume, David. A Treatise of Human Nature. Edited by L. A. Selby-Bigge. Oxford: Clarendon Press, 1888 [1739–40]. [Chapter 16, Section 16.2; Chapter 20]

Hume, David. An Enquiry Concerning Human Understanding. Edited by L. A. Selby-Bigge. Oxford: Clarendon Press, 1894 [1748]. [Chapter 20]

Jackson, Frank. From Metaphysics to Ethics: A Defence of Conceptual Analysis. Oxford: Clarendon Press, 1998. [Chapter 20, Section 20.1]

Kuhn, Thomas. The Structure of Scientific Revolutions. Chicago: University of Chicago Press, 1962. [Chapter 6, Section 6.1]

Ladyman, James, and Don Ross, with David Spurrett and John Collier. Every Thing Must Go: Metaphysics Naturalized. Oxford: Oxford University Press, 2007. [Chapter 5, Section 5.3]

Lakatos, Imre. "Falsification and the Methodology of Scientific Research Programmes." In Criticism and the Growth of Knowledge, edited by Imre Lakatos and Alan Musgrave, 91–196. Cambridge: Cambridge University Press, 1970. [Chapter 6, Section 6.1]

Latour, Bruno. We Have Never Been Modern. Translated by Catherine Porter. Cambridge, MA: Harvard University Press, 1993 [1991]. [Chapter 12, Section 12.4]

Latour, Bruno. Reassembling the Social: An Introduction to Actor-Network-Theory. Oxford: Oxford University Press, 2005. [Chapter 12, Section 12.3]

Laudan, Larry. Progress and Its Problems: Towards a Theory of Scientific Growth. Berkeley: University of California Press, 1977. [Chapter 2, Section 2.1]

Lewis, David. Counterfactuals. Cambridge, MA: Harvard University Press, 1973. [Chapter 16, Section 16.2]

Lewis, David. On the Plurality of Worlds. Oxford: Blackwell, 1986. [Chapter 2, Section 2.4; Chapter 3, Section 3.4]

Lowe, E. J. The Four-Category Ontology: A Metaphysical Foundation for Natural Science. Oxford: Clarendon Press, 2006. [Chapter 7, Section 7.1]

Mackie, J. L. Ethics: Inventing Right and Wrong. London: Penguin, 1977. [Chapter 20, Section 20.4]

McDowell, John. "Values and Secondary Qualities." In Morality and Objectivity, edited by Ted Honderich, 110–129. London: Routledge, 1985. [Chapter 20, Section 20.2]

McTaggart, J. M. E. "The Unreality of Time." Mind 17, no. 68 (1908): 457–474. [Chapter 15, Section 15.1]

Moore, G. E. Principia Ethica. Cambridge: Cambridge University Press, 1903. [Chapter 20]

Murti, T. R. V. The Central Philosophy of Buddhism: A Study of the Madhyamika System. London: George Allen and Unwin, 1955. [Chapter 19, Section 19.4]

Nagel, Thomas. The View from Nowhere. Oxford: Oxford University Press, 1986. [Chapter 14, Section 14.1]

Nussbaum, Martha. Creating Capabilities: The Human Development Approach. Cambridge, MA: Harvard University Press, 2011. [Chapter 20, Section 20.3]

Paul, L. A. "Temporal Experience." Journal of Philosophy 107, no. 7 (2010): 333–359. [Chapter 15, Section 15.5]

Pigden, Charles. "Logic and the Autonomy of Ethics." Australasian Journal of Philosophy 67, no. 2 (1989): 127–151. [Chapter 20, Section 20.1]

Pigden, Charles. "Naturalism." In A Companion to Ethics, edited by Peter Singer, 421–431. Oxford: Blackwell, 1991. [Chapter 20, Section 20.2]

Poldrack, Russell. The New Mind Readers: What Neuroimaging Can and Cannot Reveal about Our Thoughts. Princeton: Princeton University Press, 2018. [Chapter 14, Section 14.3]

Popper, Karl. The Logic of Scientific Discovery. London: Hutchinson, 1959 [1934]. [Chapter 1; Chapter 6]

Popper, Karl. Conjectures and Refutations: The Growth of Scientific Knowledge. London: Routledge, 1963. [Chapter 2, Section 2.1; Chapter 6]

Popper, Karl. Objective Knowledge: An Evolutionary Approach. Oxford: Clarendon Press, 1972. [Chapter 6, Section 6.1]

Price, Huw. Time's Arrow and Archimedes' Point: New Directions for the Physics of Time. Oxford: Oxford University Press, 1996. [Chapter 15, Section 15.3]

Putnam, Hilary. "Why There Isn't a Ready-Made World." Synthese 51, no. 2 (1982): 205–228. [Chapter 7, Section 7.2]

Putnam, Hilary. The Many Faces of Realism. La Salle, IL: Open Court, 1987. [Chapter 5, Section 5.5]

Putnam, Hilary. Reason, Truth and History. Cambridge: Cambridge University Press, 1981. [Chapter 5, Section 5.5]

Quine, W. V. O. "On What There Is." Review of Metaphysics 2, no. 5 (1948): 21–38. [Chapter 2; Chapter 3]

Quine, W. V. O. "Two Dogmas of Empiricism." Philosophical Review 60, no. 1 (1951): 20–43. [Chapter 6, Section 6.1]

Quine, W. V. O. Word and Object. Cambridge, MA: MIT Press, 1960. [Chapter 2, Section 2.4]

Quine, W. V. O. "Speaking of Objects." Proceedings and Addresses of the American Philosophical Association 31 (1958): 5–22. [Chapter

7, Section 7.1]

Reichenbach, Hans. The Direction of Time. Berkeley: University of California Press, 1956. [Chapter 15, Section 15.3]

Russell, Bertrand. Our Knowledge of the External World. London: George Allen and Unwin, 1914. [Chapter 2, Section 2.4]

Russell, Bertrand. The Philosophy of Logical Atomism. London: Routledge, 1918. [Appendix B]

Shafer-Landau, Russ. Moral Realism: A Defence. Oxford: Clarendon Press, 2003. [Chapter 20, Section 20.4]

Shannon, Claude. "A Mathematical Theory of Communication." Bell System Technical Journal 27 (1948): 379–423, 623–656. [Chapter 5, Section 5.4]

Sen, Amartya. The Idea of Justice. Cambridge, MA: Harvard University Press, 2009. [Chapter 20, Section 20.3]

Sider, Theodore. Writing the Book of the World. Oxford: Clarendon Press, 2011. [Chapter 7, Section 7.2]

Tegmark, Max. Our Mathematical Universe: My Quest for the Ultimate Nature of Reality. New York: Knopf, 2014. [Chapter 9, Section 9.3]

Tooley, Michael. Time, Tense, and Causation. Oxford: Clarendon Press, 1997. [Chapter 15, Section 15.5]

van Fraassen, Bas. The Scientific Image. Oxford: Clarendon Press, 1980. [Chapter 6; Chapter 5, Section 5.5]

van Fraassen, Bas. Scientific Representation: Paradoxes of Perspective. Oxford: Clarendon Press, 2008. [Chapter 5, Section 5.3]

Wiener, Norbert. Cybernetics, or Control and Communication in the Animal and the Machine. Cambridge, MA: MIT Press, 1948. [Chapter 5, Section 5.4]

Wiggins, David. Sameness and Substance Renewed. Cambridge: Cambridge University Press, 2001. [Chapter 7, Section 7.1]

Wilkinson, Richard, and Kate Pickett. The Spirit Level: Why More Equal Societies Almost Always Do Better. London: Allen Lane, 2009. [Chapter 20, Section 20.3]

Wittgenstein, Ludwig. Tractatus Logico-Philosophicus. Translated by C. K. Ogden. London: Kegan Paul, 1922. [Chapter 19, Section 19.3]

Wittgenstein, Ludwig. Philosophical Investigations. Translated by G. E. M. Anscombe. Oxford: Blackwell, 1953. [Chapter 19, Section 19.3]

Worrall, John. "Structural Realism: The Best of Both Worlds?" Dialectica 43, no. 1–2 (1989): 99–124. [Chapter 5, Section 5.3]

Physics and Science (Alphabetical Order)

ATLAS Collaboration. "Observation of a new particle in the search for the Standard Model Higgs boson with the ATLAS detector at the LHC." Physics Letters B 716 (2012): 1–29. [Chapter 14, Section 14.3]

Albert, David. Time and Chance. Cambridge, MA: Harvard University Press, 2000. [Chapter 15, Section 15.3]

Bohr, Niels. Atomic Theory and the Description of Nature. Cambridge: Cambridge University Press, 1934. [Chapter 14, Section 14.4]

Bohr, Niels. Atomic Physics and Human Knowledge. New York: John Wiley, 1958. [Chapter 14, Section 14.1]

Bohr, Niels. "On the Constitution of Atoms and Molecules." Philosophical Magazine 26 (1913): 1–25, 476–502, 857–875. [Chapter 9, Section 9.1]

Bondi, Hermann. Cosmology. Cambridge: Cambridge University Press, 1952. [Chapter 15, Section 15.1]

Broad, C. D. Scientific Thought. London: Kegan Paul, 1923. [Chapter 15, Section 15.5]

Carroll, Sean. From Eternity to Here: The Quest for the Ultimate Theory of Time. New York: Dutton, 2010. [Chapter 15, Section 15.3]

Čapek, Milič. Bergson and Modern Physics: A Reinterpretation and Re-evaluation. Dordrecht: D. Reidel, 1971. [Chapter 15, Section 15.4]

de Broglie, Louis. "Recherches sur la théorie des quanta." Doctoral thesis, University of Paris, 1924. [Chapter 9, Section 9.1]

Einstein, Albert. "Zur Elektrodynamik bewegter Körper." Annalen der Physik 17, no. 10 (1905): 891–921. [Chapter 15, Section 15.2]

Einstein, Albert. "Über einen die Erzeugung und Verwandlung des Lichtes betreffenden heuristischen Gesichtspunkt." Annalen der Physik 17, no. 6 (1905): 132–148. [Chapter 9, Section 9.1]

Einstein, Albert. "Die Grundlage der allgemeinen Relativitätstheorie." Annalen der Physik 49 (1916): 769–822. [Chapter 15, Section 15.2]

Einstein, Albert. Relativity: The Special and the General Theory. New York: Henry Holt, 1920 [1916]. [Chapter 9, Section 9.2]

Feynman, Richard. QED: The Strange Theory of Light and Matter. Princeton: Princeton University Press, 1985. [Chapter 9, Section 9.2]

Hacking, Ian. Representing and Intervening. Cambridge: Cambridge University Press, 1983. [Chapter 14]

Heisenberg, Werner. "Über quantentheoretische Umdeutung kinematischer und mechanischer Beziehungen." Zeitschrift für Physik 33 (1925): 879–893. [Chapter 9, Section 9.1]

Heisenberg, Werner. Physics and Philosophy: The Revolution in Modern Science. New York: Harper & Brothers, 1958. [Chapter 14, Section 14.1; Section 14.4]

Heisenberg, Werner. Across the Frontiers. New York: Harper & Row, 1974. [Chapter 14, Section 14.4]

Kauffman, Stuart. Investigations. Oxford: Oxford University Press, 2000. [Chapter 16, Section 16.1]

Layzer, David. Cosmogenesis: The Growth of Order in the Universe. New York: Oxford University Press, 1990. [Chapter 16, Section 16.1]

Penrose, Roger. The Road to Reality: A Complete Guide to the Laws of the Universe. London: Jonathan Cape, 2004. [Chapter 9, Section 9.3; Chapter 15, Section 15.3]

Planck, Max. "Über das Gesetz der Energieverteilung im Normalspektrum." Annalen der Physik 4 (1901): 553–563. [Chapter 9, Section 9.1]

Planck Collaboration. "Planck 2018 results. I. Overview and the cosmological legacy of Planck." Astronomy & Astrophysics 641 (2020): A1. [Chapter 14, Section 14.3]

Poincaré, Henri. La valeur de la science. Paris: Flammarion, 1905. [Chapter 15, Section 15.2]

Prigogine, Ilya, and Isabelle Stengers. Order out of Chaos: Man's New Dialogue with Nature. New York: Bantam, 1984. [Chapter 7, Section 7.3; Chapter 16, Section 16.2]

Prigogine, Ilya. The End of Certainty: Time, Chaos, and the New Laws of Nature. New York: Free Press, 1997. [Chapter 11, Section 11.1; Chapter 18, Section 18.1]

Rovelli, Carlo. "Relational Quantum Mechanics." International Journal of Theoretical Physics 35, no. 8 (1996): 1637–1678. [Chapter 5, Section 5.5]

Rovelli, Carlo. Reality Is Not What It Seems: The Journey to Quantum Gravity. London: Allen Lane, 2017. [Chapter 5, Section 5.5; Chapter 9, Section 9.2]

Rovelli, Carlo. The Order of Time. New York: Riverhead Books, 2018. [Chapter 5, Section 5.5; Chapter 15]

Rovelli, Carlo. Helgoland: Making Sense of the Quantum Revolution. New York: Riverhead Books, 2020. [Chapter 12; Chapter 14, Section 14.4]

Schrödinger, Erwin. What Is Life? Cambridge: Cambridge University Press, 1944. [Chapter 11, Section 11.1; Section 11.3]

Weinberg, Steven. The First Three Minutes: A Modern View of the Origin of the Universe. New York: Basic Books, 1977. [Chapter 16, Section 16.1]

Weinberg, Steven. The Quantum Theory of Fields. Volume 1. Cambridge: Cambridge University Press, 1995. [Chapter 7, Section 7.3; Chapter 9, Section 9.2]

Weyl, Hermann. Space, Time, Matter. Translated by Henry L. Brose. London: Methuen, 1922. [Chapter 15, Section 15.2]

Neuroscience and Consciousness Studies (Alphabetical Order)

Clark, Andy. Being There: Putting Brain, Body, and World Together Again. Cambridge, MA: MIT Press, 1997. [Chapter 12, Section 12.3]

Daston, Lorraine, and Peter Galison. Objectivity. New York: Zone Books, 2007. [Chapter 14, Section 14.1]

Tononi, Giulio. "An Information Integration Theory of Consciousness." BMC Neuroscience 5, no. 1 (2004): 42. [Chapter 13]

Tononi, Giulio. Phi: A Voyage from the Brain to the Soul. New York: Pantheon, 2012. [Chapter 13]

Classical Philosophy (Alphabetical Order)

Aristotle. Categories. In The Complete Works of Aristotle, edited by Jonathan Barnes. Princeton: Princeton University Press, 1984. [Chapter 7, Section 7.1; Chapter 9, Section 9.4]

Aristotle. Metaphysics. In The Complete Works of Aristotle, edited by Jonathan Barnes. Princeton: Princeton University Press, 1984. [Chapter 7, Section 7.1; Chapter 11, Section 11.2]

Berkeley, George. A Treatise Concerning the Principles of Human Knowledge. Dublin: Aaron Rhames, 1710. [Chapter 11, Section 11.5]

Descartes, René. Meditations on First Philosophy. Translated by John Cottingham. Cambridge: Cambridge University Press, 1996 [1641]. [Chapter 7, Section 7.1; Chapter 12, Section 12.1]

Hume, David. A Treatise of Human Nature. Book I, Part IV, Section VI. [Chapter 7, Section 7.1]

Kant, Immanuel. Critique of Pure Reason. Translated by Norman Kemp Smith. London: Macmillan, 1929 [1781]. [Chapter 9, Section

9.4; Chapter 11, Section 11.5; Chapter 12, Section 12.1]

Kirk, G. S., J. E. Raven, and M. Schofield. The Presocratic Philosophers: A Critical History with a Selection of Texts. 2nd ed. Cambridge: Cambridge University Press, 1983. [Chapter 7, Section 7.4]

Leibniz, G. W. Monadology. In Philosophical Papers and Letters, edited by Leroy E. Loemker. Dordrecht: D. Reidel, 1969 [1714]. [Chapter 7, Section 7.4]

Locke, John. An Essay Concerning Human Understanding. Edited by Peter H. Nidditch. Oxford: Clarendon Press, 1975 [1689]. [Chapter 7, Section 7.1]

Ockham, William of. Summa Logicae. In Opera Philosophica I, edited by Philotheus Boehner. St. Bonaventure, NY: Franciscan Institute, 1974 [c. 1323]. [Chapter 2, Section 2.1; Section 2.4]

Spinoza, Baruch. Ethics. Translated by Edwin Curley. Princeton: Princeton University Press, 1996 [1677]. [Chapter 2, Section 2.4]

C.9. Western Continental Philosophy

Phenomenology and Hermeneutics (Alphabetical Order)

Bergson, Henri. Time and Free Will: An Essay on the Immediate Data of Consciousness. Translated by F. L. Pogson. London: George Allen and Unwin, 1910 [1889]. [Chapter 15]

Bergson, Henri. Matter and Memory. Translated by Nancy M. Paul and W. S. Palmer. London: George Allen and Unwin, 1911 [1896]. [Chapter 11, Section 11.2; Chapter 15]

Bergson, Henri. Creative Evolution. Translated by Arthur Mitchell. London: Macmillan, 1911 [1907]. [Chapter 7, Section 7.4; Chapter 15]

Bergson, Henri. Duration and Simultaneity. Translated by Leon Jacobson. Indianapolis: Bobbs-Merrill, 1965 [1922]. [Chapter 15, Section 15.4]

Canales, Jimena. The Physicist and the Philosopher: Einstein, Bergson, and the Debate That Changed Our Understanding of Time. Princeton: Princeton University Press, 2015. [Chapter 15, Section 15.4]

Dilthey, Wilhelm. Introduction to the Human Sciences. Translated by Ramon J. Betanzos. Detroit: Wayne State University Press, 1988 [1883]. [Chapter 11, Section 11.4]

Gadamer, Hans-Georg. Truth and Method. Translated by Joel Weinsheimer and Donald G. Marshall. New York: Continuum, 1989 [1960]. [Chapter 11, Section 11.4; Chapter 16, Section 16.3]

Heidegger, Martin. Being and Time. Translated by John Macquarrie and Edward Robinson. New York: Harper & Row, 1962 [1927]. [Chapter 11, Section 11.4; Chapter 16, Section 16.3; Chapter 19, Section 19.2]

Heidegger, Martin. Beiträge zur Philosophie (Vom Ereignis). Frankfurt am Main: Vittorio Klostermann, 1989 [1936–1938]. [Chapter 19, Section 19.2]

Heidegger, Martin. Holzwege. Frankfurt am Main: Vittorio Klostermann, 1950. [Chapter 19, Section 19.2]

Heidegger, Martin. On the Way to Language. Translated by Peter D. Hertz. New York: Harper & Row, 1971 [1959]. [Chapter 19, Section 19.2]

Husserl, Edmund. Logical Investigations. Translated by J. N. Findlay. London: Routledge, 1970 [1900–1901]. [Chapter 7, Section 7.3]

Husserl, Edmund. Ideas Pertaining to a Pure Phenomenology and to a Phenomenological Philosophy. Translated by F. Kersten. The Hague: Martinus Nijhoff, 1983 [1913]. [Chapter 2, Section 2.2]

Husserl, Edmund. Cartesian Meditations. Translated by Dorion Cairns. The Hague: Martinus Nijhoff, 1960 [1931]. [Chapter 9, Section 9.4]

Husserl, Edmund. The Crisis of European Sciences and Transcendental Phenomenology. Translated by David Carr. Evanston: Northwestern University Press, 1970 [1936]. [Chapter 12, Section 12.1]

Merleau-Ponty, Maurice. Phenomenology of Perception. Translated by Colin Smith. London: Routledge & Kegan Paul, 1962 [1945]. [Chapter 7, Section 7.3]

Process Philosophy (Alphabetical Order)

Whitehead, Alfred North. Science and the Modern World. New York: Macmillan, 1925. [Chapter 7, Section 7.4]

Whitehead, Alfred North. Process and Reality: An Essay in Cosmology. New York: Macmillan, 1929. Corrected edition edited by David Ray Griffin and Donald W. Sherburne. New York: Free Press, 1978. [Chapter 7, Section 7.4; Chapter 8, Section 8.5; Chapter 9, Section 9.4; Chapter 12]

Whitehead, Alfred North. Adventures of Ideas. New York: Macmillan, 1933. [Appendix B]

Post-Structuralism and New Materialism (Alphabetical Order)

Deleuze, Gilles. Difference and Repetition. Translated by Paul Patton. New York: Columbia University Press, 1994 [1968]. [Chapter 7, Section 7.3; Section 7.4]

Deleuze, Gilles. The Logic of Sense. Translated by Mark Lester. New York: Columbia University Press, 1990 [1969]. [Chapter 7, Section 7.4]

Loy, David. Nonduality: A Study in Comparative Philosophy. New Haven: Yale University Press, 1988. [Chapter 19, Section 19.4]

Metalogic and Philosophy of Language (Alphabetical Order)

Hegel, G. W. F. Phenomenology of Spirit. Translated by A. V. Miller. Oxford: Clarendon Press, 1977 [1807]. [Chapter 2, Section 2.2]

Hegel, G. W. F. Science of Logic. Translated by A. V. Miller. London: George Allen and Unwin, 1969 [1812–1816]. [Chapter 2, Section 2.2]

Kripke, Saul. Wittgenstein on Rules and Private Language. Cambridge, MA: Harvard University Press, 1982. [Chapter 19, Section 19.3]

C.10. Chinese Philosophical Traditions

Classical Texts (Chronological Order)

The *Analects*. [Early Warring States period, fifth–fourth century BCE]. Various editions. [Chapter 1, Chapter 19, Chapter 20, Appendix C]

The *Mencius*. [Middle Warring States period, fourth century BCE]. Various editions. [Appendix C]

The *Daodejing* (*Laozi*). [Early Warring States period, traditionally attributed to Laozi]. Various editions. [Appendix C]

The *Zhuangzi*. [Middle Warring States period, fourth century BCE]. Various editions. [Chapter 8, Appendix C]

The *Zhouyi* and the *Yizhuan* (*Xici*, *Shuogua*, *Wenyan*, and others). [Compiled in the middle to late Warring States period]. Various editions. [Chapter 8, Appendix C]

The *Avataṃsaka Sūtra*. [Compiled between the first and third centuries; Chinese translation by Buddhabhadra in the fifth century]. Various editions. [Chapter 12, Section 12.5; Appendix C]

Tang and Song Classical Commentaries (Chronological Order)

Fazang. *Essay on the Golden Lion*. [Tang dynasty, late seventh century]. In *Taishō Tripiṭaka*, vol. 45. [Chapter 8, Section 8.5; Chapter 12, Section 12.5; Appendix C]

Fazang. *Exploration of the Avataṃsaka Sūtra*. [Tang dynasty, late seventh century]. In *Taishō Tripiṭaka*, vol. 35. [Chapter 12, Section 12.5; Appendix C]

Fazang. *Treatise on the Five Teachings of Huayan*. [Tang dynasty, late seventh century]. In *Taishō Tripiṭaka*, vol. 45. [Appendix C]

Chengguan. *Commentary on the Avataṃsaka Sūtra*. [Tang dynasty, late eighth century]. In *Taishō Tripiṭaka*, vol. 35. [Appendix C]

Zhang Zai. *Zhengmeng*. [Northern Song dynasty, eleventh century]. In *Collected Works of Zhang Zai*. Beijing: Zhonghua Book Company, 1978. [Chapter 8; Chapter 11, Section 11.2; Appendix C]

Zhang Zai. *Recorded Sayings of Master Zhang*. [Northern Song dynasty, eleventh century]. In *Collected Works of Zhang Zai*. Beijing: Zhonghua Book Company, 1978. [Appendix C]

Ming and Qing Classical Commentaries (Chronological Order)

Wang Yangming. *Instructions for Practical Living*. [Ming dynasty, sixteenth century]. Various editions. [Chapter 1, Chapter 19, Appendix C]

Wang Yangming. *Inquiry on the Great Learning*. [Ming dynasty, sixteenth century]. Various editions. [Appendix C]

Wang Fuzhi. *Commentary on Zhang Zai's Zhengmeng*. [Qing dynasty, seventeenth century]. Beijing: Zhonghua Book Company, 1975. [Chapter 8; Chapter 11, Section 11.2; Appendix C]

Wang Fuzhi. *Reflections on the Shangshu*. [Qing dynasty, seventeenth century]. In *Complete Works of Chuanshan*. Changsha: Yuelu Press, 1996. [Chapter 8, Appendix C]

Wang Fuzhi. *Reading the Complete Commentaries on the Four Books*. [Qing dynasty, seventeenth century]. In *Complete Works of Chuanshan*. Changsha: Yuelu Press, 1996. [Chapter 8, Appendix C]

Contemporary Research on Chinese Philosophy (Alphabetical Order)

Chen, Lai. *Song–Ming Neo-Confucianism*. 2nd ed. Beijing: SDX Joint Publishing Company, 2011. [Chapter 8]

Du, Jiwen. *Lectures on Buddhist Philosophy*. Beijing: China Social Sciences Press, 1986. [Chapter 12, Section 12.5]

Feng, Youlan. *A History of Chinese Philosophy*. 2 vols. Shanghai: Commercial Press, 1934. [Chapter 8]

Mou, Zongsan. *Mind-Substance and Human Nature*. 3 vols. Taipei: Zhengzhong Book Company, 1968–1969. [Appendix C]

Tang, Junyi. *Original Discourses on Chinese Philosophy*. Multi-volume edition. Hong Kong: Institute of New Asia College, 1966–1975. [Appendix C]

Editorial Group on Wang Chuanshan's Academic Thought. *Studies on Wang Chuanshan's Academic Thought*. Changsha: Hunan People's Publishing House, 1982. [Chapter 8]

Western Sinology (Alphabetical Order)

Cook, Francis. Hua-yen Buddhism: The Jewel Net of Indra. University Park: Pennsylvania State University Press, 1977. [Chapter 12, Section 12.5]

Graham, A. C. Disputers of the Tao: Philosophical Argument in Ancient China. La Salle, IL: Open Court, 1989. [Chapter 8]

Hansen, Chad. A Daoist Theory of Chinese Thought: A Philosophical Interpretation. New York: Oxford University Press, 1992. [Chapter 8]

C.11. Indian Philosophy and Buddhist Traditions

Classical Texts (Chronological Order)

Nāgārjuna. *Mūlamadhyamakakārikā*. [Around the second–third century]. Sanskrit edition: Louis de la Vallée Poussin, *Madhyamakavṛtti, Mūlamadhyamakakārikās de Nāgārjuna avec la Prasannapadā Commentaire de Candrakīrti*. St. Petersburg: Bibliotheca Buddhica IV, 1903–1913. [Chapter 19, Section 19.4; Appendix C]

Candrakīrti. *Prasannapadā*. [Seventh century]. Sanskrit edition: same as above. [Chapter 19, Section 19.4; Appendix C]

Candrakīrti. *Madhyamakāvatāra*. [Seventh century]. Tibetan edition with Sanskrit fragments. [Appendix C]

Modern Translations and Studies (Alphabetical Order)

Garfield, Jay. The Fundamental Wisdom of the Middle Way: Nāgārjuna's Mūlamadhyamakakārikā. New York: Oxford University Press,

1995. [Chapter 19, Section 19.4; Appendix C]

Murti, T. R. V. The Central Philosophy of Buddhism: A Study of the Madhyamika System. London: George Allen and Unwin, 1955. [Chapter 19, Section 19.4; Appendix C]

Siderits, Mark, and Shōryū Katsura. Nāgārjuna's Middle Way: Mūlamadhyamakakārikā. Boston: Wisdom Publications, 2013. [Appendix C]

Westerhoff, Jan. Nāgārjuna's Madhyamaka: A Philosophical Introduction. New York: Oxford University Press, 2009. [Appendix C]

C.12. Cross-Cultural and Comparative Philosophy

Loy, David. Nonduality: A Study in Comparative Philosophy. New Haven: Yale University Press, 1988. [Chapter 19, Section 19.4]

Nakamura, Hajime. A Comparative History of Ideas. New York: KPI, 1986. [Chapters related to cross-cultural dialogue]

Slingerland, Edward. Mind and Body in Early China: Beyond Orientalism and the Myth of Holism. New York: Oxford University Press, 2019. [Chapter 8, Section 8.4]

END OF THE BOOK

www.ingramcontent.com/pod-product-compliance
Lightning Source LLC
LaVergne TN
LVHW052011200926
848481LV00005B/1464